Springer Natural Hazards

The Springer Natural Hazards series seeks to publish a broad portfolio of scientific books, aiming at researchers, students, and everyone interested in Natural Hazard research. The series includes peer-reviewed monographs, edited volumes, textbooks, and conference proceedings. It covers all categories of hazards such as atmospheric/climatological/oceanographic hazards, storms, tsunamis, floods, avalanches, landslides, erosion, earthquakes, volcanoes, and welcomes book proposals on topics like risk assessment, risk management, and mitigation of hazards, and related subjects.

More information about this series at http://www.springer.com/series/10179

Xiaofeng Li
Editor

Hurricane Monitoring With Spaceborne Synthetic Aperture Radar

Springer

Editor
Xiaofeng Li
National Oceanic and Atmospheric Administration
College Park, MD
USA

ISSN 2365-0656 ISSN 2365-0664 (electronic)
Springer Natural Hazards
ISBN 978-981-10-2892-2 ISBN 978-981-10-2893-9 (eBook)
DOI 10.1007/978-981-10-2893-9

Library of Congress Control Number: 2017936444

Printed on acid-free paper

This Springer imprint is published by Springer Nature
The registered company is Springer Nature Singapore Pte Ltd.
The registered company address is: 152 Beach Road, #21-01/04 Gateway East, Singapore 189721, Singapore

Preface

Synthetic aperture radar (SAR) is a high-resolution (<100 m spatial resolution) and large swath (450 km) microwave radar that can image the ocean surface under all weather conditions, day and night. The first spaceborne SAR on board the National Aeronautics and Space Administration (NASA) SEASAT was launched in 1978. Since then, tropical cyclones have been observed on many SAR images. However, utilization of SAR imagery is a relatively new tool for hurricane research and forecasting due to its limited temporal and spatial coverage, lack of operational analysis tools and high cost. All these impediments are in the process of being swept away. National Oceanic and Atmospheric Administration (NOAA) is implementing an operational SAR wind operational processing system. European Space Agency (ESA) launched the first operational SAR satellite, Sentinel-1A, in 2014, and then the Sentinel-1B in 2016. Canadian Space Agency (CSA) will launch the RADARSAT Constellation Mission SAR missions in 2018 time frame. Currently, the Sentinel-1 SAR data are freely available on ESA's internet hub at near real time. Other SAR data are available either through commercial purchase or research proposals from the CSA RADARSAT-2, the Japan Aerospace Exploration Agency (JAXA) ALOS-2, the German DLR TerraSAR-X and TanDEM-X, Italian COSMO-SkyMed, among others.

Unlike conventional polar-orbiting and geostationary weather satellite hurricane images, which are acquired by passive remote sensing instruments operating in the visible and infrared channels and show the spiral cloud patterns, SAR images provide information of the intense air-sea interaction right at the ocean surface with much higher spatial resolution since the active microwave SAR signal penetrates the clouds. Information of sea surface wind, swell wave, boundary layer roll, eye structure, mesovortices, rainbands, and arc clouds can all be retrieved from SAR with some uncertainties. Due to SAR's great imaging capabilities, the numbers of SAR-based research and the scientists have been growing exponentially in the past decade. In addition, weather forecasters are able to look at hurricanes in such fine scale on SAR imagery for the first time. A lot of researches are devoted to developing new capabilities to use SAR data for hurricane research as it reveals fine-structure of micro- and meso-scale features within a hurricane system.

The book is based on many cutting-edge researches that the chapter authors contributed to SAR hurricane research in recent years. The targeted readers include oceanographers, meteorologists, remote sensing scientists, and graduate students in these subject matters.

Washington, D.C., USA
January 2017

Xiaofeng Li

Acknowledgements

Editorial assistance from Dongliang Shen and Hongyuan Zhang and help from Springer Nature editor Lisa Fan are greatly appreciated. The views, opinions, and findings contained in this book are those of the authors and should not be construed as an official NOAA or U.S. government position, policy, or decision.

Contents

Acronyms

AGW Atmospheric Gravity Wave
ALOS Advanced Land Observing Satellite
AMSR Advanced Microwave Scanning Radiometer-EOS
ASAR Advanced SAR
ASF Alaska Satellite Facility
AVHRR Advanced Very High-Resolution Radiometer
BSEE Bureau of Safety and Environmental Enforcement
BT Best Track
Cal/Val Calibration/Validation
CB Composite Bragg
CMA Chinese Meteorological Administration
CP Compact Polarization
CSA Canadian Space Agency
CSTARS The Center for Southeastern Tropical Advanced Remote Sensing
DCE Discrete Curve Evolution
DMRV Double Modified Rankine Vortex
DSE Discrete Skeleton Evolution
DWH Deepwater Horizon
DWT Discrete Wavelet Transform
ECMWF European Center for Medium range Weather Forecasting
EEZ Exclusive Economic Zone
ERC Eyewall Replacement Cycle
ERS European Remote Sensing
ESA European Space Agency
ETC Extratropical Cyclones
ETM+ Enhanced Thematic Mapper Plus
FFT Fast Fourier Transform
F_r Froude number
FY FengYun
GFS Global Forecast System

GMF	Geophysical Model Function
GNSS-R	Global Navigation Satellite Systems Reflectometry
GOES	Geostationary Orbiting Environmental Satellite
GPS	Global Positioning System
HE	Hurricane Eye
HRD	Hurricane Research Division
H_s	Significant Wave Height
H* Wind	Hurricane Wind Analysis System
IEM	Integral Equation Method
IR	Infrared
JMA	Japan Meteorological Agency
JTWC	Joint Typhoon Warning Center
KNMI	The Royal Netherlands Meteorological Institute
LG	Local Gradient
MABL	Marine Atmospheric Boundary Layer
MetOp	European Meteorological Operational satellite Programme
MMSE	Minimum Mean Square Error
MODIS	Moderate Resolution Imaging Spectroradiometer
M-P	Marshall-Palmer
MRGP	Maximum Radiometric Gradient Points
MSS	Mean Square Slope
MTSAT	Multifunctional Transport Satellites
MVISR-1	Multichannel Visible Infrared Scanning Radiometers
NASA	National Aeronautics and Space Administration
NCEP	National Centers for Environmental Prediction
NDBC	National Data Buoy Center
NEXRAD	Next-Generation Radar
NHC	National Hurricane Centers
NOAA	National Oceanic and Atmospheric Administration
NPOESS	National Polar-orbiting Operational Environmental Satellite System
NRCS	Normalized Radar Cross-Section
NSCAT	The NASA Scatterometer
NWP	Numerical Weather Prediction
OEDA	Oil Emulsion Detection Algorithm
OHMRV	One-Half Modified Rankine Vortex
PALSAR	Phased-Array L-band Synthetic Aperture Radar
PBL	Planetary Boundary Layer
PCC	Polarimetric Correlation Coefficient
PDF	Probability Density Function
PPB	Probabilistic Patch Based
PR	Polarization Ratio
PSOA	Particle Swarm Optimization Algorithm
QuikSCAT	Quick Scatterometer
RCM	RADARSAT Constellation Mission
RGB	Red-Green-Blue

RMS	Root Mean Square
RMW	Radius of Maximum Wind
RSMC	Regional Specialized Meteorological Center
RSS	Rain-Induced Surface Scatterings
RTM	Radiative Transfer Model
RVS	Raindrop Volumetric Scattering
SAR	Synthetic aperture radar
SFMR	Stepped-Frequency Microwave Radiometer
SHEW	Symmetric Hurricane Estimates for Wind
SLP	Sea-Level Pressure
SMRV	Single Modified Rankine Vortex
SRA	Scanning Radar Altimeter
SWA	ScanSAR Wide A
SWB	ScanSAR Wide B
SWDA	SAR Wind Direction Algorithm
SWH	Significant Wave Height
TC	Tropical Cyclone
TCBL	Tropical Cyclone Boundary Layer
TCNNA	Textural Classifier Neural Network Algorithm
TD-X	TanDEM-X
TRMM	The Tropical Rainfall Measuring Mission
TS-X	TerraSAR-X
UTC	Coordinated Universal Time
Vis	Visible
VISSR	Visible and Infrared Spin Scan Radiometers
WMLE	Weighted Maximum Likelihood Estimation
WRF	Weather Research and Forecasting
WSS	Wind-Induced Surface Scattering
WW-III	WAVEWATCH III

Chapter 1
Hurricane Precipitation Observed by SAR

D.G. Long and C. Nie

Abstract The SAR-observed backscatter from the ocean's surface is related to the surface wave spectrum, which is in turn related to the near-surface vector wind. This enables retrieval of near-surface winds from SAR images. Rain impacting the surface affects the wind-driven surface wave spectrum and roughens the surface. Rain can be observed in SAR images due to the effects the rain has on the surface and scattering and attenuation of the radar signal by the falling rain. With its high resolution SAR is a useful sensor for studying rain. This Chapter focuses on SAR observation of rain in ocean images. The effect of rain on the SAR backscatter image is modeled. Using a case study of RADARSAT ScanSAR SWA images of Hurricane Katrina, rain effects are analyzed for three different incidence angle ranges using collocated ground-based Doppler weather radar (NEXRAD) rain measurements. The rain-induced backscatter observed by the ScanSAR is consistent with C-band scatterometer-derived wind/rain scattering models when the polarization difference between the sensors are considered. New insights into the temporal behavior of rain effects on the small-scale surface wave spectrum derived from the ScanSAR images are presented.

1.1 Introduction

Synthetic aperture radar (SAR) measurements have been used to study coastal processes, currents, and sea ice with its high spatial resolution and large spatial coverage. Studies confirm that SAR measurements can be used in the retrieval of the near ocean surface winds at ultra high resolution [1]. The normalized radar cross section (σ°) measured by microwave radars over the ocean is mainly from wind-driven gravity-capillary waves due to Bragg scattering. By making multiple near

D.G. Long (✉)
Department of Electrical and Computer Engineering, Brigham Young University,
459 Clyde Building, Provo, UT 84602, USA
e-mail: long@ee.byu.edu

C. Nie
Formerly with Department of Electrical and Computer Engineering,
Brigham Young University, 459 Clyde Building, Provo, UT 84602, USA

X. Li (ed.), *Hurricane Monitoring With Spaceborne Synthetic Aperture Radar*, Springer Natural Hazards, DOI 10.1007/978-981-10-2893-9_1

simultaneous observations of the surface backscatter from different azimuth and/or incidence angles at each point in the observation swath, wind scatterometers such as the European Space Agency (ESA) Earth Remote Sensing (ERS) scatterometer (ESCAT), the ESA Advanced Scatterometer (ASCAT), and the U.S. QuikSCAT employ a geophysical model function to estimate the wind speed and direction over the ocean [2–4]. Since SARs have only one measurement for each geographic location, the wind direction must be inferred from the orientation of the wind-induced streaks visible in most SAR images [1, 5, 6], or obtained from additional information such as numerical wind prediction models [7]. Given the wind direction, the wind speed is retrieved from either the spectral width of the image spectrum in azimuth direction or by inversion of a geophysical model function (GMF) that relates the normalized radar backscatter (denoted σ°) to the wind speed and direction. The GMF is a function of the radar frequency, polarization, incidence angle, and azimuth angle and is used by wind scatterometers as well.

Compared with C-band wind scatterometers, SAR can provide wind estimates at much finer (100–1000 m compared to 25 km) resolution, which is useful for studying micro-scale weather events, including rain. Rain cells are often observed in SAR images over the ocean [8, 9]. Rain-induced backscatter is from two processes: atmospheric attenuation and scattering by falling rain drops. The former is small at C-band; however, rain-induced surface scattering can be significant [10]. Raindrops striking the water and downdraft created by rain cells modify the roughness of the ocean surface; and hence the surface backscatter.

Melsheimer et al. [8] analyzed SAR signatures of rain cells over the ocean using C and X-band SAR data, showing that rain generally reduces the surface backscatter at low incidence angles and enhances the backscatter at high incidence angles. Weinman et al. [11] studied rain over the ocean with dual frequency SAR and derived the differential polarized phase shift. Unfortunately, this technique cannot be used with single frequency SAR systems.

Wind and rain retrieval from radar measurements is well-developed in the scatterometer community. For example, using C-band scatterometer measurements Nie and Long [10] found that rain surface backscatter can dominate the total backscatter from the ocean surface in moderate to heavy rains. While rain can degrade the accuracy of scatterometer wind measurements [10, 12], incorporating rain effects into the GMF permits simultaneous retrieval of both wind and rain at Ku-band [13–15] and at C-band [16].

In this study, we consider the effects of rain on Canadian RADARSAT scanning SAR (ScanSAR) wide A (SWA) mode images and present a case study of rain observation during Hurricane Katrina in 2005. In this mode, the image resolution is fairly coarse (500 m), which precludes wind direction estimation from the SAR image. We thus adopt a wind scatterometer-like approach based on Nie and Long [16] to simultaneously infer wind and rain where wind directions are specified with the aid of a hurricane model [7, 17]. Various rain effects in the SAR images are illustrated and analyzed. The high resolution and rapid storm movement permits us to examine a number of short-time temporal effects of the rain on the surface roughness spectrum.

This analysis requires a wind/rain GMF. Lacking a well-validated GMF model for HH polarization at C-band, we adjust the C-band VV polarization scatterometer GMF (CMOD5) [18] using a polarization ratio correction as described in Nie and Long [7].

1.2 Rain Effects on C-Band SAR Measurements over the Ocean

In the atmosphere, rain-induced volume-scattering increases the power backscattered toward the SAR, while also attenuating the signal to and from the surface. Raindrops striking the water create various splash products including rings, stalks, and crowns from which the signal scatters. The contribution of each of these splash products to the backscattering varies with incidence angle and polarization. Ring waves are found to be the dominant features for VV-polarization. For HH-polarization, the radar backscatter from non-propagating splash products increases with increasing incidence angles while the radar backscatter from ring waves decreases. These splash products are imposed on the wind-generated wave field. Raindrops impinging on the ocean surface also generate turbulence in the upper water layer which attenuate the short gravity wave spectrum [10]. Using multi-frequency SIR-C/X-SAR data and ERS 1/2 SAR (C band, VV-polarization) data, Melsheimer et al. [8] demonstrate that the modification of the sea surface roughness by falling raindrops mainly depends on the wavelength of water waves. The net effect of the raindrops on the ocean surface is a decrease of the amplitude of water waves which have wavelengths above 10 cm and an increase of the amplitude of water waves with a wavelength below 5 cm. For waves with wavelengths between 5 and 10 cm, rain may increase or decrease the amplitude of the Bragg waves, though the critical transition wavelength at which increase turns to decrease is not well defined [8]. The critical wavelength is believed to depend on rain rate, drop size distribution, wind speed, and the temporal evolution of the rain event.

In addition to surface effects induced by raindrops, the sea surface roughness is also affected by the airflow (downdraft) associated with the rain event and the large scale wind flow, as illustrated in Fig. 1.1. When the downdraft reaches the sea surface, it spreads radially outward as a strong local surface wind that increases the sea surface roughness. Note that the gust front is the outer edge of the downdraft. When the mean ocean surface wind is low, the downdraft is often visible on SAR images over the ocean as a nearly circular bright pattern with a sharp edge [9, 19]. When the ocean surface wind is strong, the airflow pattern is distorted; hence the SAR signature shows both bright and dark areas [20].

Using C-band scatterometer (ERS-1/2 VV-polarization) measurements, Nie and Long [10] quantitatively analyzed the rain surface effects on C-band radar signals at incidence angles higher than 40°. Their study demonstrates that rain surface

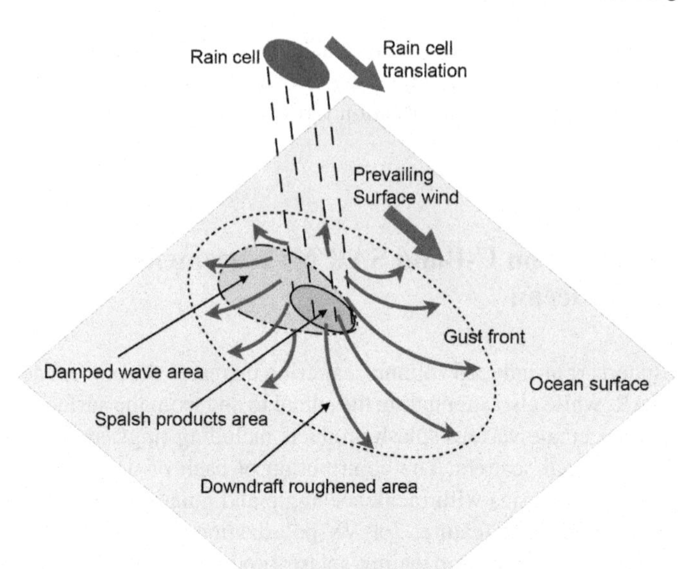

Fig. 1.1 Schematic diagram of the various surface effects caused by a rain cell over the ocean. In the splash area, raindrops striking the water create splash products. The damped wave area is created by rain-generated turbulence in the upper water layer. The *blue arrows* illustrate the airflow of the downdraft, which spreads over and roughens the ocean surface. Note that due to upper atmospheric circulation, the wind cell translates horizontally. In hurricanes, this direction generally coincides with the prevailing surface wind direction

backscatter can dominate the total backscatter in moderate to heavy rains and a simple phenomenological backscatter model can be used to represent rain backscatter with relatively high accuracy. RADARSAT ScanSAR SWA measurements cover wind incidence angle ranges between 20° and 49°, providing a good opportunity to study the effects of rain on C-band HH-polarization SAR measurements at different incidence angles under hurricane conditions. To quantitatively analyze the rain effects on SAR measurements, the wind/rain backscatter model developed in [10] and briefly summarized below is adapted. A SAR response model due to rain atmospheric effects is developed in the following subsections. To estimate SAR wind speed, the recalibration and polarization ratio approach developed by Nie and Long [7] is used. Rain-induced atmospheric attenuation and backscatter are estimated using collocated NEXRAD weather radar data. Finally, rain surface perturbations are estimated and modeled.

1.2.1 Wind/Rain Backscatter Model for SAR

In raining areas, the measured normalized radar cross section by the SAR over the ocean is affected by rain atmospheric effects and various surface effects including splash products, turbulence, and downdraft. As shown in Fig. 1.1, the area affected by downdraft and turbulence is larger than the rain core area. Furthermore, the effect of turbulence varies with the temporal evolution of the rain event at a give location. At the beginning of the rain event, the wave damping effect induced by rain is insignificant because surface turbulence is under development. The dampening grows during the rain event then decays after the rain moves on. Since the turbulence decays slowly due to the molecular viscosity of water and the length scales of the turbulence, the damping effect can exist for some time after a rain event ends [8]. Unfortunately, the lifetime of rain-induced turbulence in water has rarely been studied. As a reference, the lifetime of vortex rings generated by rain drops impinging the water surface is of the order of a minute for a drop diameter of 1 mm [21]. In the analysis of the SAR measurements shown below, the wave damping effect is still observed about five minutes after rain passes and so it is assumed that the lifetime of rain-induced surface turbulence is of this order.

A detailed model of each of the surface effects is beyond the scope of this chapter. Instead, we focus on bulk models for the effects of rain on the Bragg wave field in the rain core area by combining all the surface contributions together into a single rain surface perturbation term, σ_{surf}. σ_{surf} is assumed to be additive with the wind-induced surface backscatter. The rain-modified measured backscatter, σ_m, is represented by a simple additive model [10, 12].

$$\sigma_m = (\sigma_{wind} + \sigma_{surf})\alpha_{atm} + \sigma_{atm} \tag{1.1}$$

where σ_{wind} is the wind-induced surface backscatter, σ_{surf} is the rain-induced surface perturbation backscatter, α_{atm} is the two-way rain-induced atmospheric attenuation, and σ_{atm} is rain-induced atmospheric backscatter.

The σ_{wind} is estimated by projecting H*wind wind speeds (s) and directions (d) through an HH-polarization GMF derived from collocation of H*winds and ScanSAR data [7],

$$\sigma_{wind} = \text{CMOD5}(s, d, \chi, \theta)p(\theta) \tag{1.2}$$

where CMOD5 is the wind-only scatterometer GMF [18], χ is the azimuth angle of SAR measurements, θ is the incidence angle, and $p(\theta)$ is the Thompson et al. [22] polarization ratio model used to convert the VV-pol CMOD5 GMF for use at HH-pol. ScanSAR wind speeds are derived using wind directions from H*wind [7]. Rain-induced atmospheric attenuation and backscatter are estimated using collocated NEXRAD weather radar data.

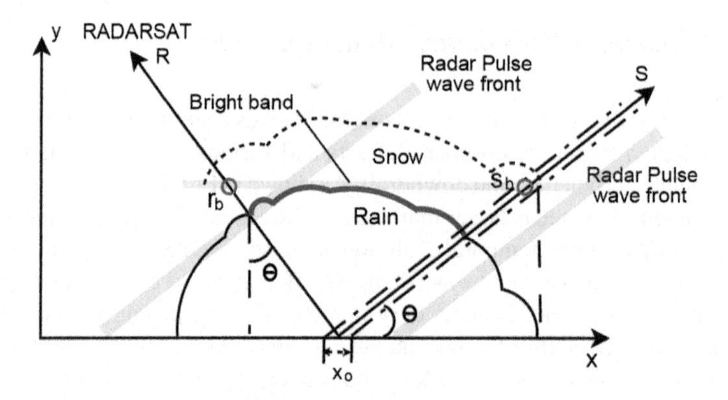

Fig. 1.2 Schematic diagram of the SAR scattering geometry for a rain cell. The *oblique lines* represent the radar pulse under the approximation of plane wave incidence

1.2.2 Evaluation of Atmospheric Attenuation and Backscattering

The SAR measurement geometry is displayed in Fig. 1.2. For simplicity, we use a plane-wave incidence approximation to represent the synthetic aperture radar pulse. We define a new coordinate system $r - s$. r is along the SAR slant range and s is perpendicular to r. For the SAR surface backscatter at x_o, the atmospheric attenuation is contributed by the raindrops along coordinate r from the surface to the bright band altitude and by snow above the bright band. The typical altitude of the bright band is about 5 km.

The attenuation coefficient of rain, K_r, can be estimated using the $k_r - R$ (R is rain rate in mm/h) relationship [23]

$$K_r = aR^b \quad \text{dBkm}^{-1} \tag{1.3}$$

where $a = 0.0018$ dBkm^{-1} and $b = 1.05$ for a 5 cm SAR signal wavelength. R is the rain rate in mm/h. The attenuation coefficient of snow is related to snowfall rate by [23]

$$K_s = 0.0222\frac{R^{1.6}}{\lambda^4} + 0.34\varepsilon_i''\frac{R}{\lambda} \quad \text{dBkm}^{-1} \tag{1.4}$$

where λ is the wavelength, $\varepsilon_i'' \simeq 10^{-3}$ at $-1\,°$C. For $\lambda = 5.6$ cm, $R = 100$ mm/h, $K_s = 0.04$ dBkm^{-1}, while $K_r = 0.227$ dBkm^{-1} under the same conditions. Therefore, the attenuation due to snow is negligibly small and is ignored in the following analysis. The path integrated attenuation (PIA) in dB is the integration of $K_r(r, s)$ through the R axis ($s = 0$), from the bright band altitude, r_b (shown in Fig. 1.2), to the ocean surface, 0,

$$PIA = 2\int_0^{r_b} k_r(r, 0)dr \quad \text{dB} \tag{1.5}$$

where $k_r(r, 0) = aR(r, 0)^b$. Since $r = (x_0 - x)/\sin\theta$ and $k_r(r, 0) = k_r(x, (x_0 - x)/\tan\theta)$, the above equation can be expressed as

$$PIA = 2\frac{1}{\sin\theta}\int_{x_0 - r_b \sin\theta}^{x_0} k_r\left(x, \frac{x_0 - x}{\tan\theta}\right) dx \quad \text{dB} \tag{1.6}$$

The net two way atmospheric attenuation factor α_{atm} is calculated by converting the PIA from dB to normal space,

$$\alpha_{atm} = 10^{-PIA/10} \tag{1.7}$$

In this study the atmospheric backscatter (σ_{atm}) expected for SAR observations is estimated from the rain rate obtained from the NEXRAD measurements using these expressions. For a specific position on coordinate s, the effective reflectivity of the atmospheric rain, $Z_e(0, s)$, is calculated using Eq. (1.13). The volume backscattering coefficient σ_{vc} can be computed from [23]

$$\sigma_{vc}(0, s) = 10^{-10}\frac{\pi^5}{\lambda_\circ^4}|K_w|^2 Z_e(0, s) \quad \text{m}^2/\text{m}^3 \tag{1.8}$$

where $\lambda_\circ = 5.6$ cm is the wavelength of RADARSAT SAR, and $|K_w|^2$ is a function of the wavelength $\lambda_\circ$ and the physical temperature of the material. K_w is assumed to be 0.93 for the water and 0.19 for snow in this paper [24]. The quantity σ_{vc} represents physically the backscattering cross-section (m^2) per unit volume (m^3). According to Fujiyoshi et al. [25], the Z-R relationship for snow is $Z = 427R^{1.09}$. As previously noted, due to its small contribution snow-induced volume backscattering is disregarded in this study.

The volume backscattering cross-section observed by the SAR is attenuated by the two-way attenuation factor, $\alpha_{atm}(0, s)$,

$$\sigma_{vro}(0, s) = \sigma_{vc}(0, s)\alpha_{atm}(0, s) \tag{1.9}$$

where $\alpha_{atm}(0, s)$ is the path integrated two-way attenuation at s on S axis. The total atmospheric rain backscatter as seen by SAR is $\sigma_{vro}(r, s)$ integrated through the radar pulse plane (along the S axis where $r = 0$) from the bright band altitude on the S axis (shown in Fig. 1.2), s_b, to the ocean surface,

$$\sigma_{atm} = \sin\theta \int_0^{s_b} \sigma_{vro}(0, s)ds \quad \text{m}^2/\text{m}^2 \tag{1.10}$$

where θ is the incidence angle. Since $s = (x - x_0)/\cos\theta$ and $\sigma_{vro}(0, s) = \sigma_{vro}(x, (x - x_0)\tan\theta)$, this equation can be transformed to coordinate $x - y$ as

$$\sigma_{atm} = \tan\theta \int_{x_0}^{x_0 + s_b\cos\theta} \sigma_{vro}\left(x, (x - x_0)\tan\theta\right) dx \tag{1.11}$$

After calculating σ_{atm} and α_{atm}, we estimate the surface perturbation backscatter σ_{surf} by

$$\sigma_{surf} = \alpha_{atm}^{-1}(\sigma_m - \sigma_{atm}) - \sigma_{wind} \tag{1.12}$$

where the σ_{surf} can be negative at low incidence angles, corresponding to the loss of the wind-induced backscatter. A positive value is an increase in the net backscatter.

1.3 Data

Hurricane Katrina attained Category 5 status on the morning of August 28 and reached its peak strength at 1:00 p.m. that day. At approximately midnight of August 28, 2007, RADARSAT flew over Katrina, providing an excellent wide swath set of C-band measurements in a hurricane. During the same period, shore-based NEXRAD and air-borne NOAA WP-3D radar also covered Hurricane Katrina from different locations, acquiring 3 dimensional rain. In this section, the data sets used in this study are briefly described. In Fig. 1.3, we show the path of Hurricane Katrina, the outlines of the RADARSAT ScanSAR SWA data, the locations of NEXRAD weather radar stations and the path of the NOAA WP-3D.

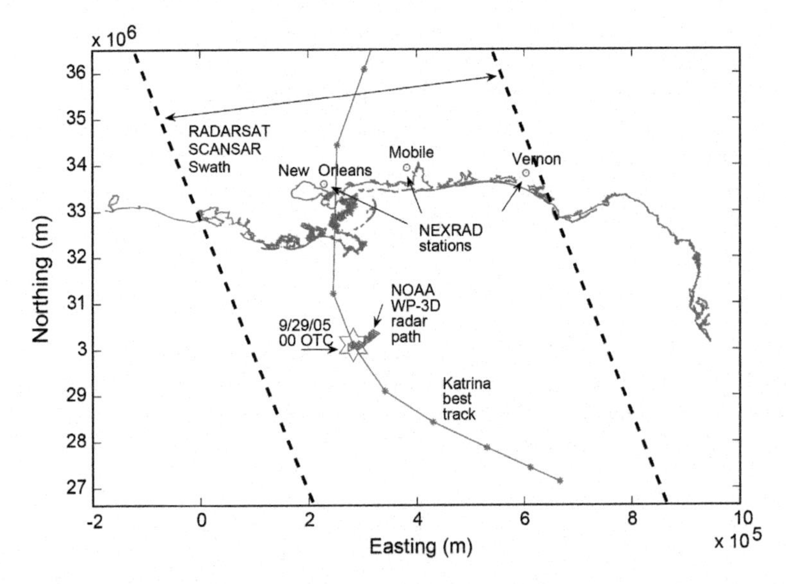

Fig. 1.3 Diagram of the Hurricane Katrina best track as determined by the Hurricane Research Division, the RADARSAT ScanSAR SWA observation swath, and the path of the NOAA WP-3D airplane. Three NEXRAD weather radar stations are plotted as *red circles*. The large star shows the Katrina eye center location at the time of the RADARSAT overpass

1.3.1 RADARSAT ScanSAR SWA Data

The Canadian satellite RADARSAT works at 5.3 GHz in HH polarization. The scanning SAR (ScanSAR) wide A (SWA) mode of RADARSAT provides coverage of a 500 km nominal ground swath at incidence angles between 20° and 49°, with a spatial resolution of 100 m [26].

Two 510 × 510 km calibrated RADARSAT ScanSAR SWA images were acquired over the ocean around New Orleans at 23:49:05 and 23:50:50, on 28 August, 2005, during the period of Hurricane Katrina. At the time of observation, the hurricane was a Category 5 hurricane with a fully developed eye.

The image processed by the Alaska Satellite Facility (ASF) is 510 × 510 km with a pixel spacing of 50 m. The range resolution of the four beams varies from 73.3 to 162.7 m, while the azimuth resolution varies from 93.1 to 117.5 m. The raw ScanSAR SWA data was processed by the ASF into calibrated images. However, the radiometric calibration of ScanSAR SWA images is very difficult due to many limitations including scalloping between the bands, underestimation of σ° [27], and beam overlapping. It is also noted that the calibration at ASF is mainly "tuned" to high latitude areas, which may result in degraded calibration for low latitude areas. The accuracy of the ASF-calibrated SWA images has not been well studied. In Albright [28], the relative radiometric accuracy for SWA is estimated to be about 0.47 dB. The ScanSAR SWA geographic location accuracy is thought to be similar to the overall relative location error of the ScanSAR SWB, about 135 m.

To retrieve vector winds, the parameters needed for wind retrieval are estimated from the SAR image. The incidence angle for each image pixel is calculated from ScanSAR SWA data using a method proposed by Shepherd [29] and the normalized radar cross section σ° is calculated for each pixel [7].

In the two ScanSAR images, rain bands exist next to the eyewall of Katrina and several long rain cell clusters span a wide range of incidence angles, providing a good data source to study rain effects on measurements at various incidence angles.

1.3.2 Hurricane Research Division H*wind Data

To validate the SAR retrieved wind fields and calculate the wind-induced backscatter, coincident H*wind surface wind fields [30] are used in the study. The H*wind Surface Wind Analysis System is an experimental high resolution hurricane research tool developed by the Hurricane Research Division (HRD) at the National Oceanic and Atmospheric Administration (NOAA). The H*wind system assimilates and synthesizes disparate observations into a consistent wind field. The H*wind system uses all available surface weather observations. All data are processed to conform to a common framework for a 10 m height, the same exposure, and the same averaging period using accepted methods from micrometeorology and wind engineering [31]. The analysis provides the maximum sustained 1-min wind speed. Due to the limited coverage of the observations and the smoothing effect of the analysis process, fine scale details of the ocean surface winds are filtered out. The spatial resolution of H*wind estimates is 0.0542° in latitude and longitude, while the time resolution is 3 h. The H*wind-predicted wind fields are trilinearly interpolated in space and time to RADARSAT ScanSAR SWA data times and locations.

1.3.3 NEXRAD Doppler Weather Radar Data

NEXRAD is a collection of ground-based weather radars deployed throughout the U.S. Several NEXRAD stations monitored Hurricane Katrina as it closed in on the coast. NEXRAD observations provide three-dimensional rain rates which we can compare to the SAR-derived rain rates. The NEXRAD radar operates at S-band (2.7–3.0 GHz). During storm events, NEXRAD uses a pre-programmed set of scanning elevations, Volume Coverage Pattern (VCP) 11, to acquire data. The radar successively scans 360° in azimuth angle in 1° increments and from 0.5° to 6.2° in 0.95° increments in elevation angle. Additional circular scans at a 7.5°, 8.7°, 10.0°, 12.0°, 14.0°, 16.7°, and 19.5° elevation angle are performed [32, 33].

In general, rain rates are derived from NEXRAD measurements of reflectivity Z by inversion of the reflectivity to rain rate (Z-R) relationship,

$$Z = aR^b \tag{1.13}$$

where constants a and b are dependent on drop-size distribution. The optimal Z-R constants determined by Jorgensen and Willis [34] in mature hurricanes are $a = 300$ and $b = 1.35$. The NEXRAD Z measurements are estimated at 1 km resolution over the range of 1–460 km from the radar.

To collocate the NEXRAD rain measurements with RADARSAT ScanSAR SWA data, the NEXRAD measurements are converted from Plan Position Indicator (PPI) to Constant Altitude Plan Position Indicator (CAPPI) with 1×1 km resolution in the horizontal and 1 km resolution in the vertical. Interpolation is used to project the measurements from PPI to CAPPI. The ray path is computed using the "four-thirds earth radius model" [35]. The NEXRAD rain rates are then projected to UTM coordinates.

As shown in Fig. 1.3, NEXRAD data from stations at New Orleans (LIX), Mobile (MOB), and Tallahassee (EVX and TLH) are used. In the overlapping area of two radars, we select the rain estimates from the nearest station. To ensure the quality of the rain estimates, we limit the maximum range of NEXRAD radar data to a 200 km radius.

1.4 Results and Analysis

As noted, rain effects vary with incidence angle. In the following we quantitatively analyze the radar backscatter of several rain cells at different incidence angles.

1.4.1 Incidence Angle Between 22° and 23.6°

Figure 1.4 displays the SAR $\sigma°$ of a typical rain cell located near the coast in this dataset. The collocated H*wind speed and vectors are shown in Fig. 1.5. The incidence angles of the SAR measurements are between 22° and 23.6°. At this incidence angle, the dominant rain effect is a dampening of the surface backscatter; hence, the rain cell looks darker than the surrounding rain-free ocean in the SAR image. The H*wind model predicts that the wind speed over the imaged area is essentially constant. Since the LIX NEXRAD station is the closest station to this site, radar data from the LIX station is used to calculate rain rates.

Because the gain spatial response function is not uniform over the NEXRAD footprint, the NEXRAD-observed rain is a weighted spatial average of the rain. To compensate for this, the collocated SAR measurements are averaged over the NEXRAD footprint by weighting with the NEXRAD spatial response function within

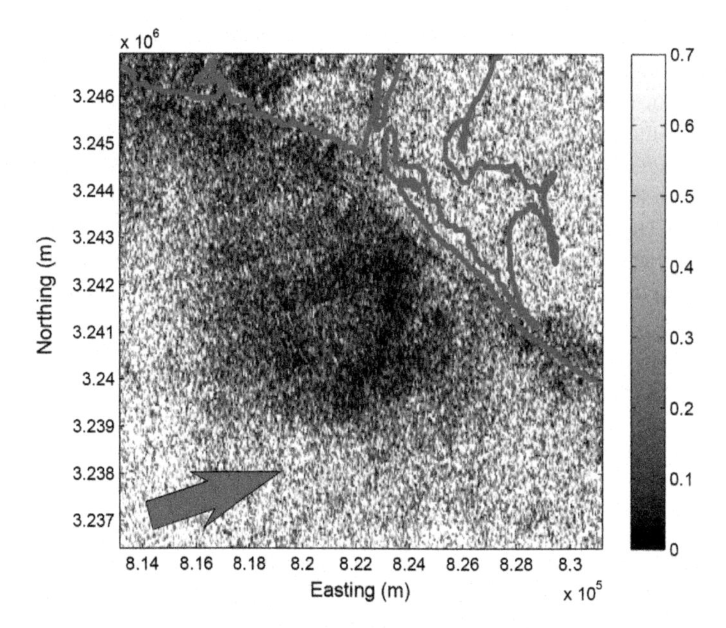

Fig. 1.4 $\sigma°$ of a rain cell located near the sea shore of New Orleans in Hurricane Katrina. The coast line is marked using solid lines and the *red arrow* shows the azimuth direction of RADARSAT ScanSAR observation. The near-surface wind speed is ≈20 m/s

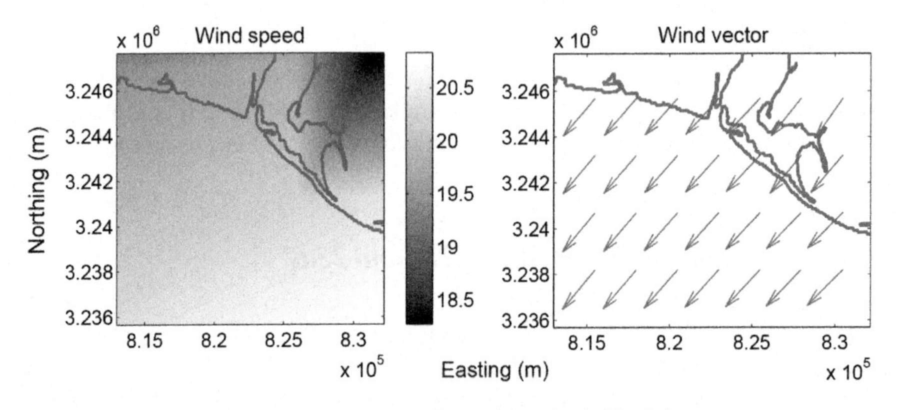

Fig. 1.5 Collocated H*wind winds corresponding to the region in Fig. 1.4

the 3-dB antenna pattern contour. Lacking detailed information for NEXRAD's spatial response function, we use a Gaussian radiation pattern in this study [35]. To minimize the errors introduced by the SAR and NEXRAD data processing, the different map projections, and the spatial and time differences between the two sensors, we assume the rain is uniformly distributed in the vertical direction and use the vertically-averaged rain rate as the surface rain rate. Due to the coarse resolution of the SCANSAR image, we do not attempt to separate atmospheric rain from the surface rain effects.

Figure 1.6a and b displays the atmospheric attenuation and backscatter induced by rain and computed from NEXRAD observations. Compared with the surface σ° at this incidence angle range, the atmospheric backscatter is insignificant, while the atmospheric attenuation is significant in heavy rains. Due to the SAR geometry, the SAR measurements affected by rain atmospheric attenuation and backscattering are not limited to the rain-cell area. Figure 1.7a and b display the collocated σ_{surf} and the NEXRAD surface rain rate, respectively. In Fig. 1.7c and d, the profiles of rain rate and σ_{surf} are plotted along the red solid line in Fig. 1.7a and b. These show that the σ_{surf} generally decreases as rain rate increases. Note that the profile of σ_{surf} is wider than the rain rate profile.

To relate the σ_{surf} with rain rate, we use a power law model [10]. σ_{surf} can be expressed as a polynomial function of rain rate,

$$10log_{10}(\sigma_{surf}(\theta)) \approx f_{sr}(R_{dB}) = \sum_{n=0}^{N} x_{sr}(n)R_{dB}^{n} \tag{1.14}$$

where $R_{dB} = 10log_{10}(R_{surf(ant)})$, and $x_{sr}(n)$ are the corresponding model coefficients. $N = 1$ for the linear model, and $N = 2$ for the quadratic model. Because the estimate of σ_{surf} is relatively noisy, we first make a nonparametric estimate of σ_{surf} as a function of R_{dB} using an Epanechnikov kernel with a 2 mm/h dB bandwidth in rain rate as shown in Fig. 1.8a. Then, we estimate the model coefficients for the

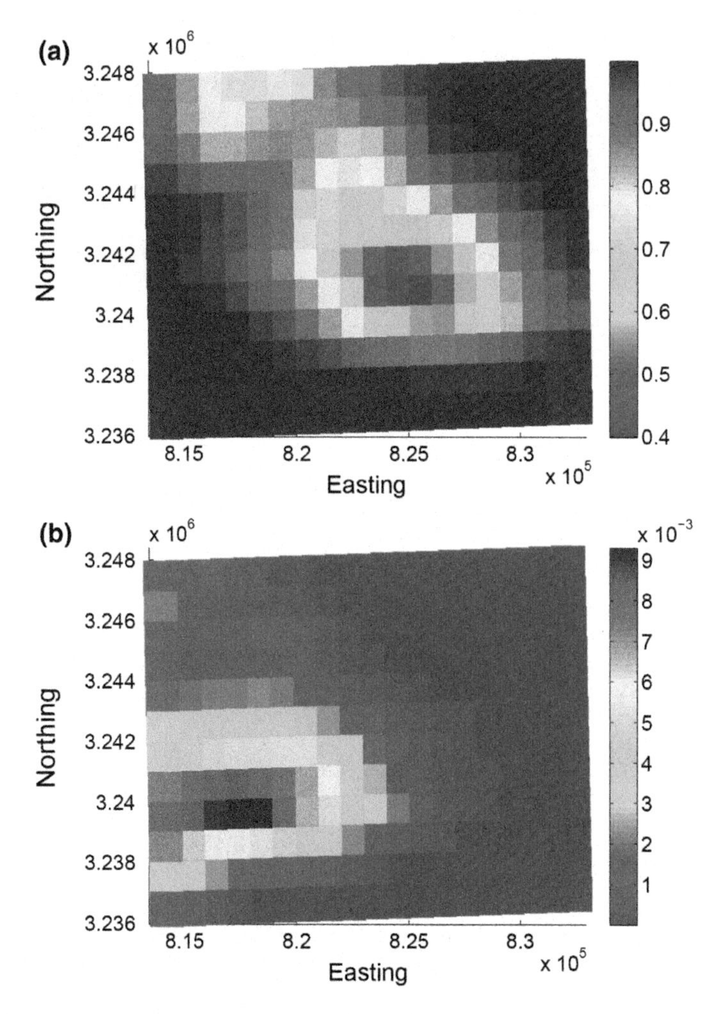

Fig. 1.6 **a** Rain-induced atmospheric attenuation and **b** atmospheric backscatter computed from NEXRAD observations over the region in Fig. 1.4

quadratic model using a linear least-squares fit as shown in Fig. 1.8b. In the following analysis of other rain cells, we use this same method. With the estimated model coefficients it is possible to infer the rain rate from the SAR-derived σ_{surf}.

1.4.2 Incidence Angle Between 28° and 31.7°

Figure 1.9 displays the SAR signature of a rain cell over the ocean about 150 km from the MOB NEXRAD station. Figure 1.10 displays the collocated H*wind speeds and

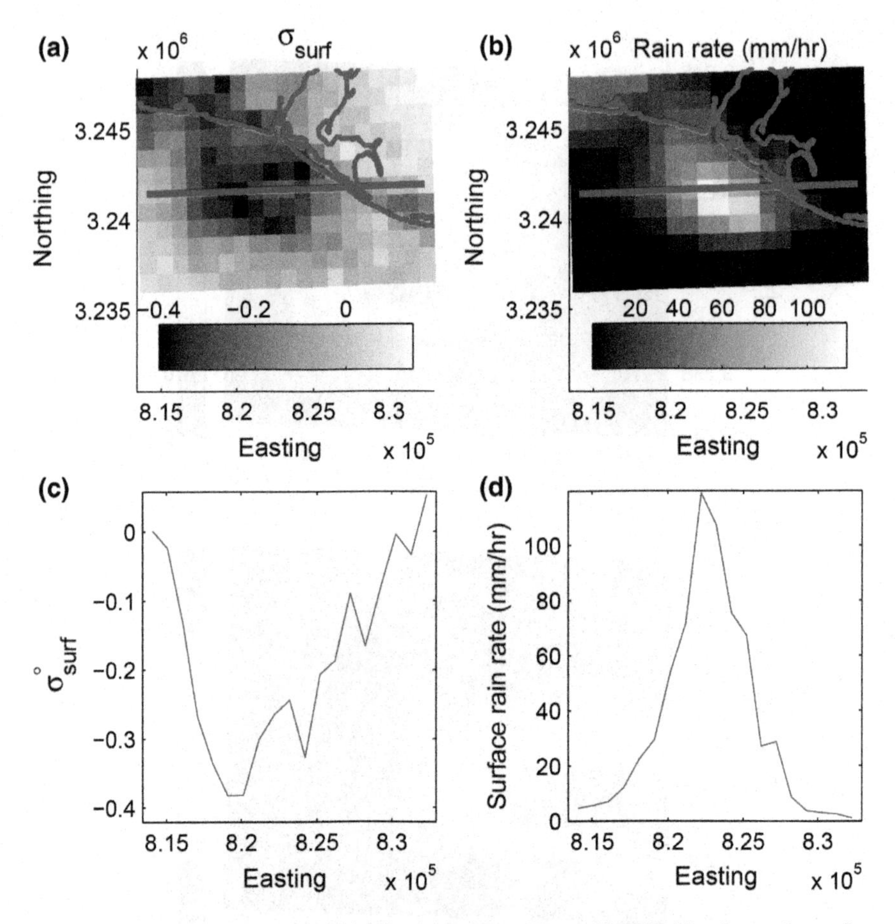

Fig. 1.7 **a** σ°_{surf} surf of the rain cell in Fig. 1.4. **b** The collocated NEXRAD rain rate in mm/h. **c** and **d** the profile of σ° and rain rate along the *solid line* plotted in **a** and **b**

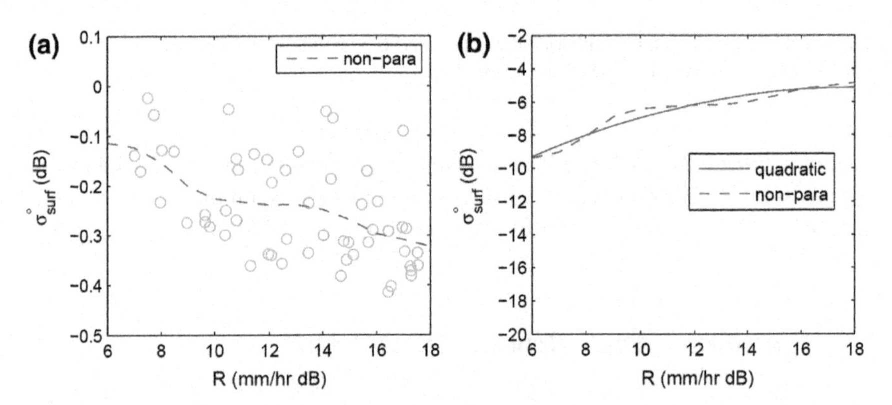

Fig. 1.8 **a** σ°_{surf} versus rain rate nonparametric fit. **b** Quadratic fit to σ_{surf} in log-log space compared to the non-parametric fit

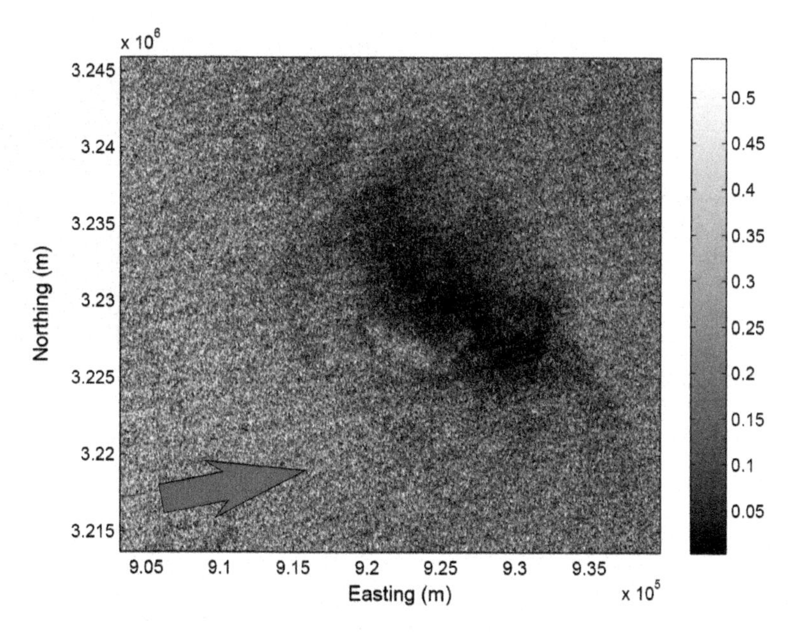

Fig. 1.9 RADARSAT σ° of a rain cell located near the sea shore of New Orleans in Hurricane Katrina. The *red arrow* shows the azimuth direction of RADARSAT ScanSAR observation. The near-surface wind speed is $\approx$22 m/s

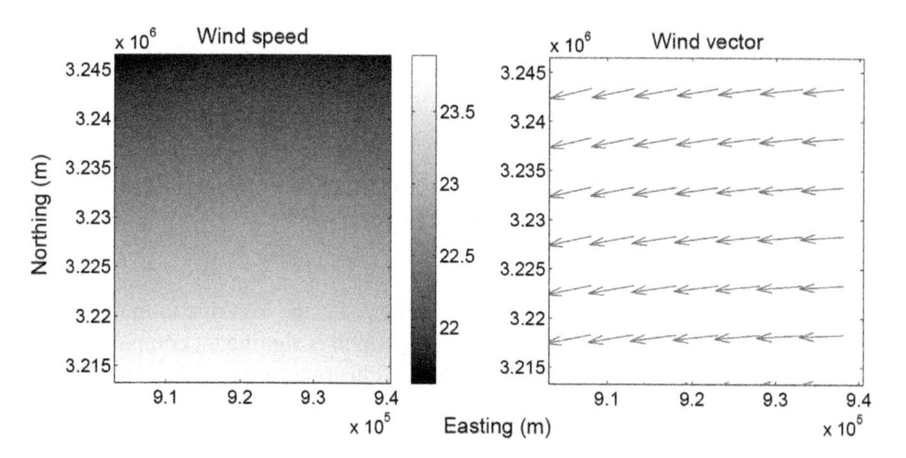

Fig. 1.10 Collocated H*wind winds corresponding to the region in Fig. 1.9

directions. At this SAR incidence angle range, the damping effect of the rain on the surface wave spectrum is dominant. Figure 1.11 analyzes the normalized radar cross-section of this event. The collocated NEXRAD-derived rain rate of the intense rain cell shown in Fig. 1.11b creates the spatially larger SAR signature illustrated in Fig. 1.11a. The rain effect depresses the surface backscatter creating an apparent

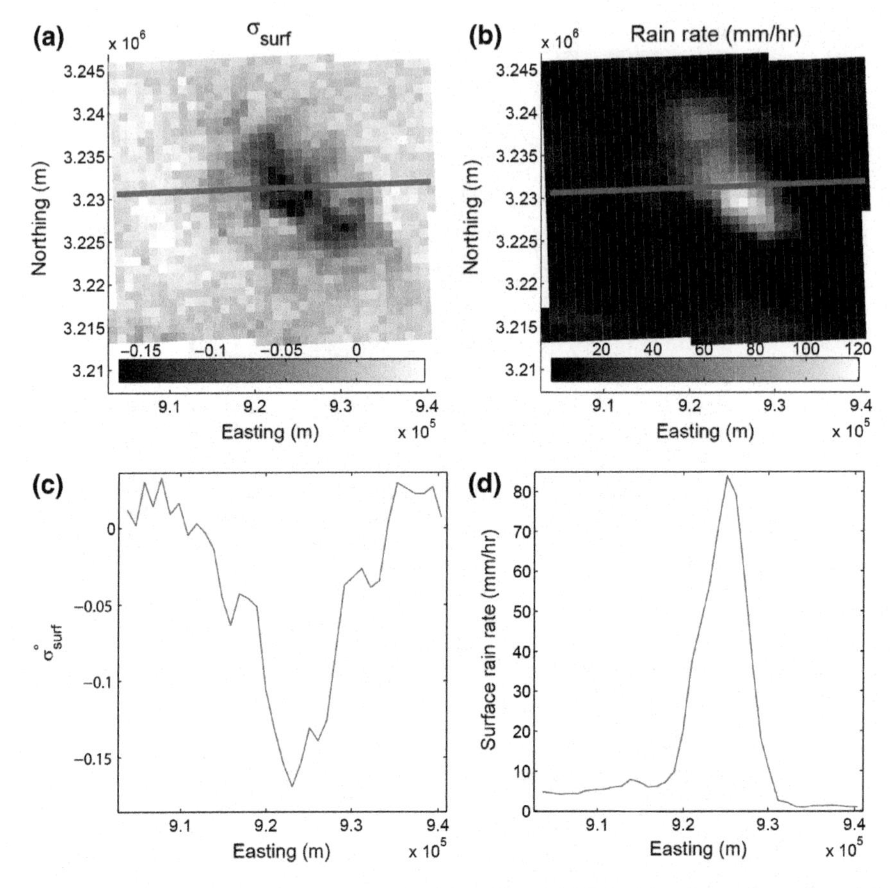

Fig. 1.11 **a** σ°_{surf} of the rain cell in Fig. 1.9. **b** The collocated NEXRAD rain rate in mm/h. **c** and **d** the profile of σ° and rain rate along the *solid line* plotted in **a** and **b**

negative "surface backscatter". As shown in Fig. 1.12, the loss due to the damping effect is as high as -7 dB when $R \approx 63$ mm/h, which is significant compared to the wind-induced surface backscatter. Figure 1.12a illustrates the non-parametric fit to the estimated σ_{surf} derived from the SAR data with respect to R_{dB} while (b) displays the quadratic fit to the non-parametric fit. Due to the relatively large number of collocated data points, the nonparametric fit in Fig. 1.12a is smooth and the quadratic fit agrees well with the nonparametric fit in Fig. 1.12b.

1.4.3 Incidence Angle Between 44° and 45.7°

Figure 1.13 displays the SAR signature of a rain cell over the ocean which is about 70 km from the EVX NEXRAD station. Through comparison between σ_{surf} and rain

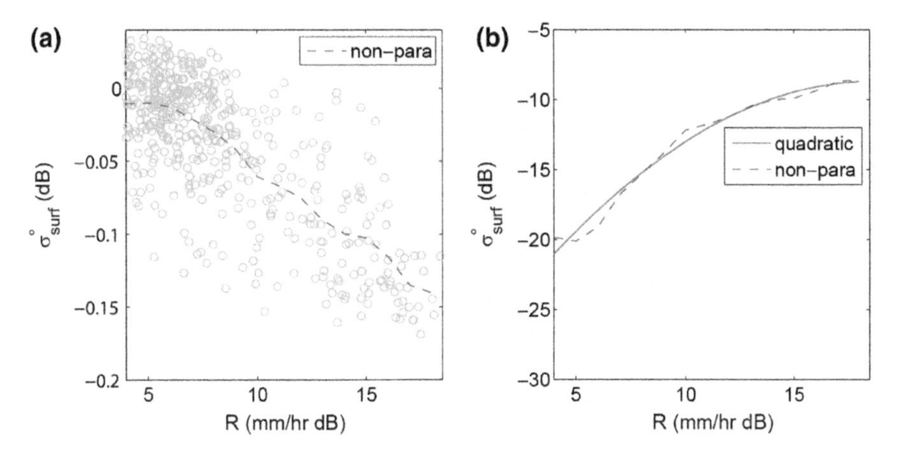

Fig. 1.12 **a** Nonparametric fit to σ_{surf}. **b** Quadratic fit to the non-parametric fit of σ_{surf} in log-log space

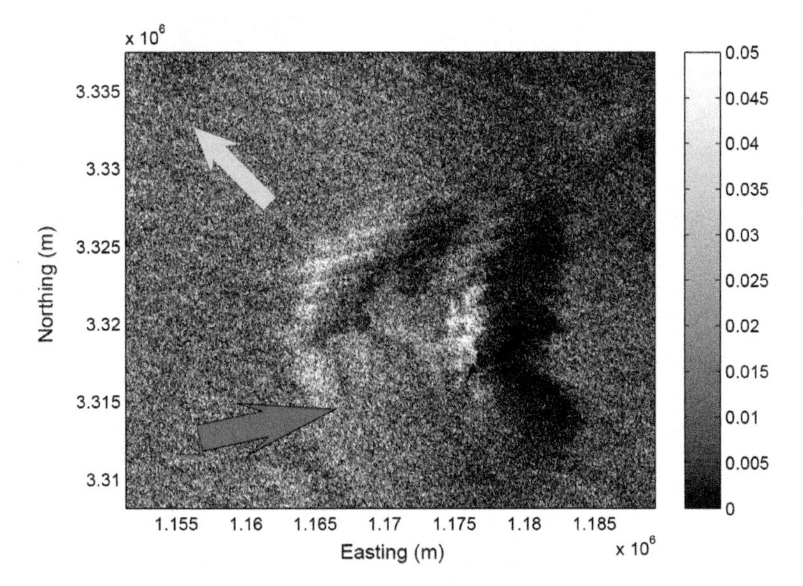

Fig. 1.13 σ° of a rain cell located near the sea shore of New Orleans in Hurricane Katrina. The *red arrow* shows the azimuth direction of RADARSAT ScanSAR observation and the light *blue arrow* shows the wind direction. The near-surface wind speed is $\approx$10 m/s

rate in Fig. 1.15, we find that the enhancing effect of rain is dominant within the rain cells. However, damping areas (which are darker due to reduced σ°) are found next to the rain enhanced areas. The damping areas have shapes similar to the rain cells but are shifted due to the motion of the rain cell. Note that two negative peaks exist in the profile of σ_{surf} along the solid line, as shown in Fig. 1.15. Because the wind direction is pointing in the west-northern direction, as shown in Fig. 1.14, the rain

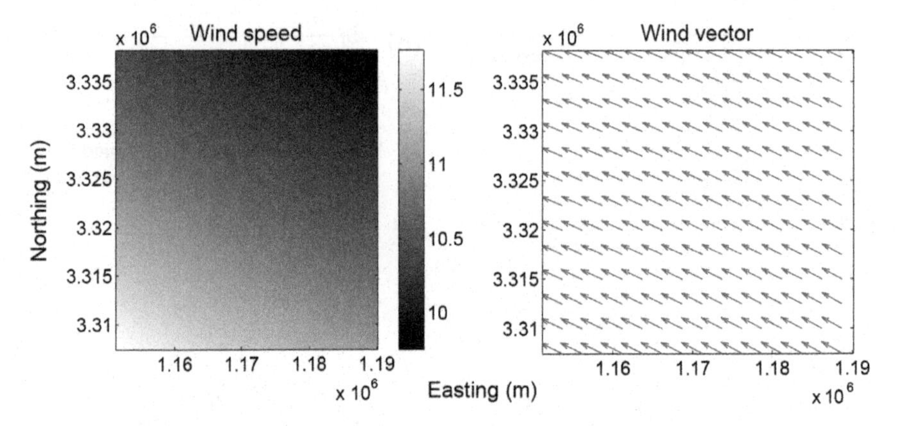

Fig. 1.14 Collocated H*wind winds corresponding to the region in Fig. 1.13

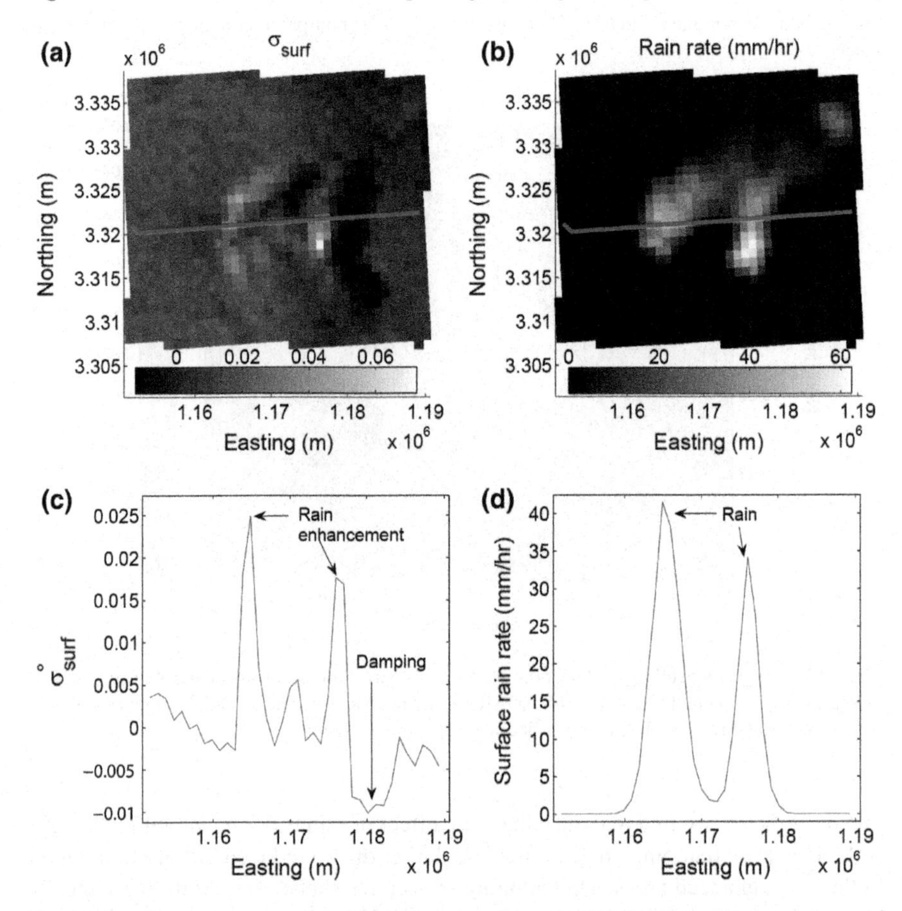

Fig. 1.15 **a** σ°_{surf} of the rain cell in Fig. 1.13. **b** the collocated NEXRAD rain rate in mm/h. **c** and **d** display the profile of σ° and rain rate along the *solid line* plotted in **a** and **b**

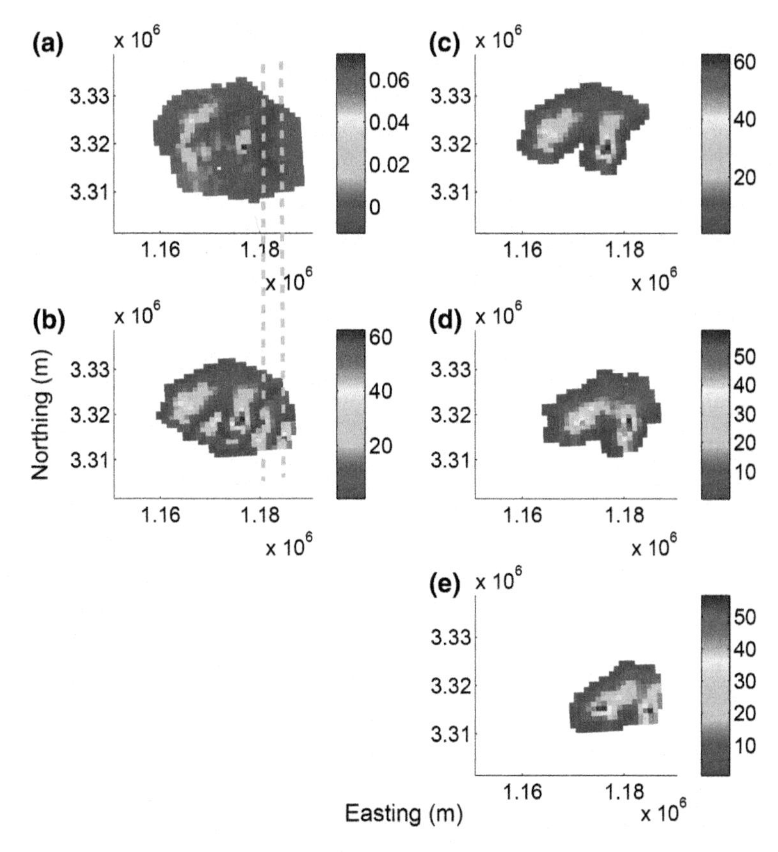

Fig. 1.16 **a** σ_{surf} derived from the RADARSAT image. **b** Overlay of the NEXRAD measurements from **c**–**e**. **c** NEXRAD measurements collocated with the SAR measurement time. **d** NEXRAD measurements about 5 min prior to the SAR observation. **e** NEXRAD measurements about 10 min prior to the SAR observation. The rain cell is moving to the upper left, see Fig. 1.13

cell is moving towards west-north, as shown in Fig. 1.16. The path of the rain cell shown in Fig. 1.16b matches the damping areas shown in Fig. 1.16a. As discussed previously, the damping effect continues after rain events. Hence, the damping area is the result of the rain previously falling in the area. Since the rain cell is moving with the wind, it is leaving a "trail" of damped wave surface, which takes time to "recover".

We note that the lifetime of the rain damping effect has rarely been studied. It is likely that the lifetime depends on many factors such as the type of rain, rain rate, drop size distribution, wind speed, incidence angle, and so on. However, we can infer the lifetime for these particular SAR observation conditions. As shown in Fig. 1.16a and b, the damping area (near Easting 1.18×10^6 m) collocates with the rain measurements acquired 5 and 10 min previously. Based on this, we conclude that the lifetime of the rain damping effect at C-band is approximately between 5 and 10 min

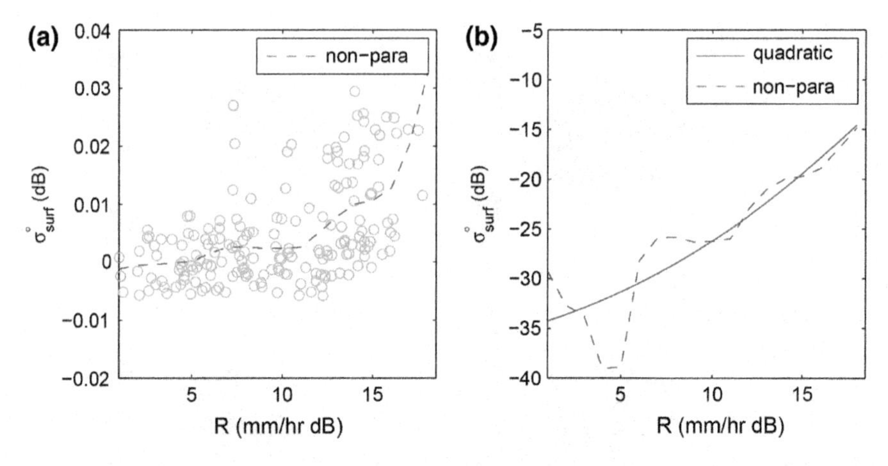

Fig. 1.17 **a** Nonparametric fit to σ_{surf} for Fig. 1.13. **b** Quadratic and linear fits to the non-parametric fits of σ_{surf} in log-log space. Non-parametric fits are also plotted

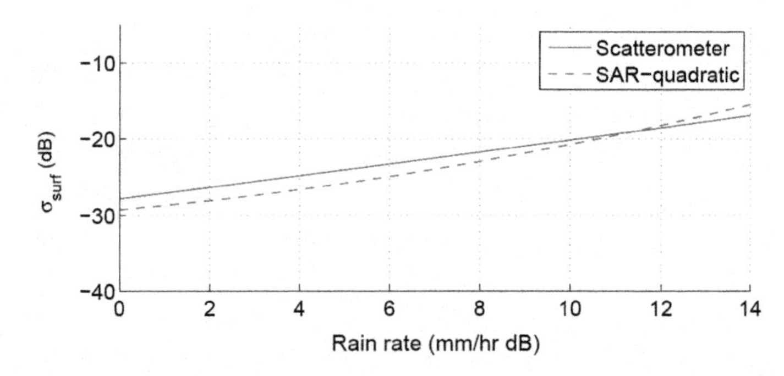

Fig. 1.18 Comparison of SAR-derived and scatterometer-derived surface perturbation, σ_{surf}, versus rain models for VV polarization (see text)

when the wind speed is about 10 m/s, the rain rate is 70 mm/h, and the incidence angle is 45°. This is potentially an important insight into rain/wave interaction.

Figure 1.17a illustrates the non-parametric fit to the estimated σ_{surf} with respect to R_{dB} for this case, while Fig. 1.17b displays the quadratic and linear fits to the non-parametric fit. In Fig. 1.17b, the linear and quadratic model are close, suggesting that σ_{surf} is almost a linear function of surface rain rate in log-log space at this incidence angle. Figure 1.18 compares the scatterometer C-band VV polarization wind backscatter model developed by Nie and Long [10] and the quadratic model derived from the HH polarization SAR measurements for this case. The latter has been adjusted using the Thompson et al. [22] polarization model to VV polarization. The two rain models are close, suggesting that the SAR-derived σ_{surf} versus rain is consistent with the scatterometer derived model when the polarization difference between HH and VV polarizations is considered. Unfortunately, the limited data preclude a systematic comparison of the two models.

Table 1.1 Coefficients of the σ_{surf} model at three incidence angles

Incidence angle (°)	P(0)	P(1)	P(2)
22–23	−14.6081	1.0563	−0.0295
28–31.7	−28.6799	2.1404	−0.0572
44–45.7	−34.79	0.5249	0.0332

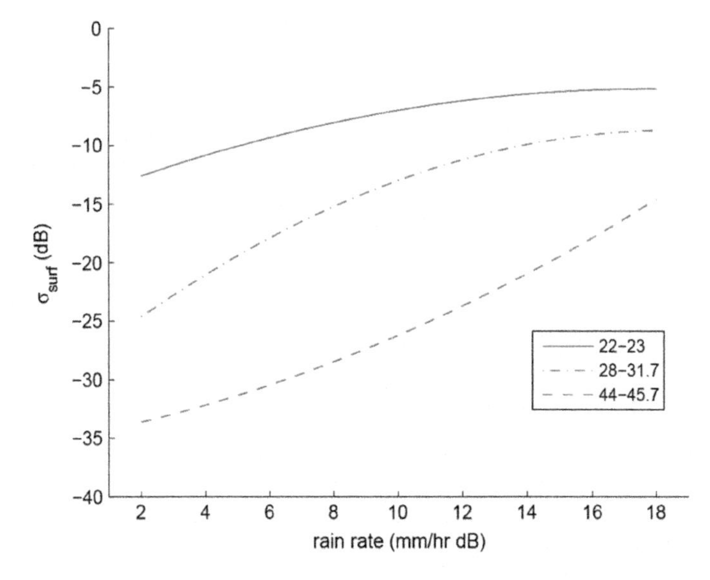

Fig. 1.19 σ_{surf} versus rain rate at different incidence angles. Note that for incidence angle bins 22°–23° and 28°–31° σ_{surf} is negative due to the damping effect. In this case $|\sigma_{surf}|$ in dB is displayed

1.4.4 Rain Model Coefficients

The coefficients of the rain backscatter model for the three incidence angles considered in the previous case studies are listed in Table 1.1. σ_{surf} versus rain rate at the different incidence angles is plotted in Fig. 1.19. The σ_{surf} versus rain model at high incidence angle is close to a linear model in log-log space. Here, we further investigate the relationship between σ_{surf} and incidence angle by plotting the σ_{surf} with respect to incidence angle for a specific surface rain rate in Fig. 1.20. The magnitude of σ_{surf} generally decreases with incidence angles. At heavy rain rates, the decreasing ratio is smaller than at low to moderate rain rates.

At low incidence angles, loss of σ_{surf} occurs due to the damping effect of rain, while rain enhances the backscatter at high incidence angles. As shown in Fig. 1.20, both the loss and enhancement of σ_{surf} can be a significant component of the total backscatter in moderate to heavy rain rates. At extreme rain rates, the wind component of the backscatter may not be significant [16]. Hence, including the rain effects on

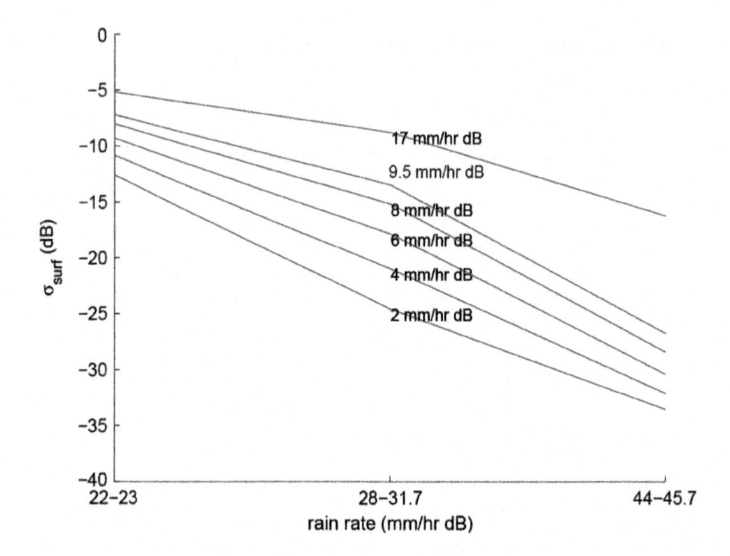

Fig. 1.20 σ_{surf} versus incidence angle for various rain rates at different incidence angles. Note that for incidence angle bins 22°–23° and 28°–31° σ_{surf} is negative due to the rain damping effect. In this case $|\sigma_{surf}|$ in dB is displayed

C-band radar backscatter is very important when attempting SAR wind retrieval in the presence of rain. This is consistent with the wind scatterometer results of Nie and Long [16].

1.5 Conclusion

Rain is clearly visible in C-band RADARSAT ScanSAR SWA images of Hurricane Katrina due to its impact on the radar signal. These include atmospheric effects (attenuation and backscattering) and surface effects. Using a simple wind/rain backscatter model and collocated SAR and NEXRAD data, we quantitatively analyze different rain effects on the ScanSAR measurements for three different incidence angle ranges and estimate the coefficients of a rain GMF. The observed rain signature varies with the incidence angle of the observations. The C-band SAR-derived σ_{surf} is found to be consistent with C-band wind scatterometer-derived models. Rain surface effects on C-band SAR measurements can dominate the surface backscatter in moderate to heavy rains and needs to be considered when retrieving near-surface winds from SAR backscatter data. Based on the pattern rain-induced backscatter damping visible in the imagery, we estimate that the C-band Bragg wave spectrum requires 5–10 min after rain termination to be re-established in moderate winds.

References

1. Wackerman, C.C., C.L. Rufenach, R.A. Shuchman, J.A. Johannessen, and K.L. Davidson. 1996. Wind vector retrieval using ERS-1 synthetic aperture radar imagery. *IEEE Transactions on Geoscience and Remote Sensing* 34 (6): 1343–1352.
2. Attema, E.P.W. 1991. The active microwave instrument on-board the ERS-1 satellite. *Proceedings of the IEEE* 79 (6): 791–799.
3. Figasaldaña, J., J.J. Wilson, E. Attema, R. Gelsthorpe, M. Drinkwater, and A. Stoffelen. 2002. The Advanced scatterometer (ASCAT) on the Meteorological Operational (MetOp) platform: A follow on for European wind scatterometers. *Canadian Journal of Remote Sensing* 28 (3): 404–412.
4. Spencer, M.W., C. Wu, and D.G. Long. 1997. Tradeoffs in the design of a spaceborne scanning pencil beam scatterometer: Application to SeaWinds. *IEEE Transactions on Geoscience and Remote Sensing* 35 (1): 115–126.
5. Vachon, P., and F. Dobson. 1996. Validation of wind vector retrieval from ERS-1 SAR images over the ocean. *The Global Atmosphere and Ocean System* 5 (2): 177–187.
6. Lehner, S., J. Horstmann, W. Koch, and W. Rosenthal. 1998. Mesoscale wind measurements using recalibrated ERS SAR images. *Journal of Geophysical Research: Oceans* 103 (C4): 7847–7856.
7. Nie, C., and D.G. Long. 2008a. RADARSAT ScanSAR wind retrieval and rain effects on ScanSAR measurements under hurricane conditions. In *IEEE International Geoscience and Remote Sensing Symposium*, vol. 2, 493–496. IEEE.
8. Melsheimer, C., W. Alpers, and M. Gade. 2001. By the synthetic aperture radar aboard the ERS satellites and by surface-based weather radars. *Journal of Geophysical Research* 106 (C3): 4665–4677.
9. Atlas, D. 1994a. Footprints of storms on the sea: A view from spaceborne synthetic aperture radar. *Journal of Geophysical Research: Oceans* 99 (C4): 7961–7969.
10. Nie, C., and D.G. Long. 2007. A C-band wind/rain backscatter model. *IEEE Transactions on Geoscience and Remote Sensing* 45 (3): 621–631.
11. Weinman, J., F. Marzano, W. Plant, A. Mugnai, and N. Pierdicca. 2009. Rainfall observation from X-band space-borne synthetic aperture radar. *Natural Hazards and Earth System Sciences* 9 (1): 77–84.
12. Draper, D.W., and D.G. Long. 2004a. Evaluating the effect of rain on SeaWinds scatterometer measurements. *Journal of Geophysical Research: Oceans* 109 (C12).
13. Allen, J.R., and D.G. Long. 2005. An analysis of SeaWinds-based rain retrieval in severe weather events. *IEEE Transactions on Geoscience and Remote Sensing* 43 (12): 2870–2878.
14. Draper, D.W., and D.G. Long. 2004b. Simultaneous wind and rain retrieval using SeaWinds data. *IEEE Transactions on Geoscience and Remote Sensing* 42 (7): 1411–1423.
15. Draper, D.W., and D.G. Long. 2004c. Assessing the quality of SeaWinds rain measurements. *IEEE Transactions on Geoscience and Remote Sensing* 42 (7): 1424–1432.
16. Nie, C., and D.G. Long. 2008b. A C-band scatterometer simultaneous wind/rain retrieval method. *IEEE Transactions on Geoscience and Remote Sensing* 46 (11): 3618–3631.
17. Williams, B.A., and D.G. Long. 2008. Estimation of hurricane winds from SeaWinds at ultra-high resolution. *IEEE Transactions on Geoscience and Remote Sensing* 46 (10): 2924–2935.
18. Hersbach, H., A. Stoffelen, and S. De Haan. 2007. An improved C-band scatterometer ocean geophysical model function: CMOD5. *Journal of Geophysical Research: Oceans* 112 (C3).
19. Atlas, D. 1994b. Origin of storm footprints on the sea seen by synthetic aperture radar. *Science* 266: 1364–1366.
20. Mitnik, L.M. 1992. Mesoscale coherent structures in the surface wind field during cold air outbreaks over the far eastern seas from the satellite side looking radar. *La Mer* 30: 297–314.
21. Hallett, J., and L. Christensen. 1984. Splash and penetration of drops in water. *Journal de Recherches Atmospheriques* 18 (4): 225–242.

22. Thompson, D.R., T.M. Elfouhaily, and B. Chapron. 1998. Polarization ratio for microwave backscattering from the ocean surface at low to moderate incidence angles. In *IEEE International Geoscience and Remote Sensing Symposium*, vol. 3, 1671–1673. IEEE.
23. Battan, L.J. 1973. *Radar Observation of the Atmosphere*. Chicago: The University of Chicago.
24. Ulaby, F.T., R.K. Moore, and A.K. Fung. 1982. Microwave remote sensing active and passive, vol. i.
25. Fujiyoshi, Y., T. Endoh, T. Yamada, K. Tsuboki, Y. Tachibana, and G. Wakahama. 1990. Determination of a Z-R relationship for snowfall using a radar and high sensitivity snow gauges. *Journal of Applied Meteorology* 29 (2): 147–152.
26. Raney, R.K., A.P. Luscombe, E. Langham, and S. Ahmed. 1991. RADARSAT. *IEEE International Geoscience and Remote Sensing Symposium* 79 (6): 839–849.
27. Horstmann, J., W. Koch, S. Lehner, and R. Tonboe. 2000. Wind retrieval over the ocean using synthetic aperture radar with C-band HH polarization. *IEEE Transactions on Geoscience and Remote Sensing* 38 (5): 2122–2131.
28. Albright, W. 2004. Calibration report for RADARSAT ScanSAR Wide A on the ScanSAR processor. *Alaska Satellite Facility*.
29. Shepherd, N. 1998. Extraction of beta nought and sigma nought from RADARSAT CDPF products. *Rep. AS97-5001 Rev*, 2.
30. Powell, M.D., S.H. Houston, L.R. Amat, and N. Morisseau-Leroy. 1998. The HRD real-time hurricane wind analysis system. *Journal of Wind Engineering and Industrial Aerodynamics* 77: 53–64.
31. Powell, M.D., S.H. Houston, and T.A. Reinhold. 1996. Hurricane Andrew's landfall in south Florida. Part I: Standardizing measurements for documentation of surface wind fields. *Weather and Forecasting* 11 (3): 304–328.
32. F. C. for Meteorological Services, and S. Research. 1990. Doppler radar meteorological observations, Part B: Doppler radar theory and meteorology. CM-H11B-1990.
33. F. C. for Meteorological Services, and S. Research. 1991. Doppler radar meteorological observations, Part C: WSR-88D products and algorithms. FCM-H11C-1991.
34. Jorgensen, D.P., and P.T. Willis. 1982. A ZR relationship for hurricanes. *Journal of Applied Meteorology* 21 (3): 356–366.
35. Doviak, R.J., and D.S. Zrnic. 1984. *Doppler Radar and Weather Observations*, vol. I. Dublin: Academic press.

Chapter 2
Tropical Cyclone Multiscale Wind Features from Spaceborne Synthetic Aperture Radar

Jun A. Zhang and Xiaofeng Li

Abstract This study presents multi-scale wind features observed in space-borne synthetic aperture radar (SAR) images in tropical cyclones. Examples of eyewall mesovotices, spiral rainbands, fine-scale-band features, arc clouds, and boundary layer rolls are documented. Although these wind features are strongly tied to tropical cyclone dynamics and intensity based on previous numerical studies, they are not well-observed due to high rainfall and cloudiness that limits remote sensing instrument and severe environment for in-situ observations to survive. Since SAR images view the actual ocean surface responses to the storm-forced winds, they provide clear evidence for the presence of these wind features below clouds and their interaction with the sea surface. Analyses of the characteristics of boundary layer rolls based on SAR images show good agreement with in-situ aircraft observations, suggesting that a SAR image has a great potential to be utilized to study tropical cyclone low-level structure.

2.1 Introduction

Hurricanes account for a significant portion of damage, injury and loss of life from natural hazards, and are the most expensive natural catastrophes in the US [1]. The threat of high winds and storm surge has been known for intense storms. Multiscale wind features such as the "eyewall mesovortices", spiral band features, and boundary layer wind streaks or rolls are believed to largely contribute to the hurricane induced damage during landfalls. These features are also strongly tied to hurricane dynamics according to previous observational and numerical modeling studies.

J.A. Zhang (✉)
National Oceanic and Atmospheric Administration (NOAA)/AOML/Hurricane Research Division, University of Miami/CIMAS, Miami, FL, USA
e-mail: jun.zhang@noaa.gov

X. Li
GST, National Oceanic and Atmospheric Administration (NOAA)/NESDIS, College Park, MD, USA
e-mail: xiaofeng.li@noaa.gov

X. Li (ed.), *Hurricane Monitoring With Spaceborne Synthetic Aperture Radar*, Springer Natural Hazards, DOI 10.1007/978-981-10-2893-9_2

Previous observations have shown that intense transient vorticity features usually exist close to the inside edge of eyewalls of numerous intense hurricanes. Marks and Houze [2] reported the first example of eyewall mesovortices in Hurricane Debby (1982) using airborne Doppler radar data. Mesovortices were also seen in satellite images in Hurricane Isabel (2003) by Kossin and Schubert [3] and analyzed in Doppler radar data in Hurricane Guillermo (1997) by Reasor et al. [4]. Such vortical features are also clearly shown in several recent high-resolution cloud-representing numerical model simulations [5, 6], as well as in two-dimensional turbulence-resolving numerical models [7, 8]. The mesovortices are tied to the combined baratropic-baroclinic instability that is associated with the annulus of high potential vorticity near and within the eyewall clouds. Eyewall mesovortices have horizontal scales smaller than the diameter of the hurricane eye, but generally larger than the scales of individual cumulus clouds that constitute the eyewall.

Well-organized bands of convection and precipitation are commonly observed in and around a hurricane in satellite and radar images, which are usually called "spiral bands". Through careful analysis of land-based Doppler radar data of landfalling hurricanes, Gall et al. [9] identified a new class of spiral bands which are called "fine-scale spiral bands", and are significantly smaller than the rainbands (with $\sim$100 km scale) typically visible in satellite images. The radial wavelengths of the fine-scale spiral bands are between 5–10 km. Numerical simulations of landfalling hurricanes also identified features with similar wavelengths extending through the depth of the troposphere [10–12]. Katsaros et al. [13] identified roll-like structures with similar horizontal wavelengths (4–6 km) in the convection-free areas between the larger rainbands of several hurricanes through analysis of synthetic aperture radar (SAR) images. Nolan [14] investigated the mechanisms for the fine-scale spiral bands in the hurricane boundary layer using idealized numerical simulations and showed that an instability associated with the inflection points of the swirling boundary layers is a good candidate mechanism.

Boundary layer secondary-circulations or 'roll vortices' are known to be a common feature in the atmospheric boundary layer which have a significant influence on turbulent exchange of momentum, sensible heat and latent heat. Satellite remote sensing and Doppler radar observation indicate that rolls frequently form in the hurricane boundary layer. Wurman and Winslow [15] provided the first evidence of intense subkilometer-scale horizontal roll vortices in the high resolution Doppler radar wind retrievals in the boundary layer of Hurricane Fran (1996) during landfall. Morrison et al. [16] analyzed a large quantity of Weather Surveillance Radar-1998 Doppler (WSR-88D) data during landfalls of three hurricanes. They found rolls in 35–69% radar volumes which have a typical wavelength of 1450 m. They also found that the rolls were nearly aligned with the mean boundary layer wind. Lorsolo et al. [17] analyzed the data from the Shared Mobile Atmospheric Research and Teaching (SMART) radars in Hurricane Frances (2004) and found that the majority of wavelengths of rolls that are omnipresent in their data are between 200 and 650 in agreement with the study of Wurman and Winslow [15]. Ellis and Businger [18] analyzed WSR-88D data in two Typhoons and investigated a total of 99 cases of coherent structures and confirmed prior findings of roll observations. Foster [19] has developed a theory for roll vortices in curved flow at high wind speeds, such as in hurricanes, suggesting that hurricane boundary layer rolls transport high-momentum air from the upper boundary layer downward and enhance the transport of air-sea flux.

Since the first spaceborne microwave SAR onboard the National Aeronautics and Space Administration (NASA) SEASAT was launched in 1978 [20], tropical cyclones have also been observed by SAR images. Prior to the 2002 hurricane season, the "Hurricane Watch" program was established by the Canadian Space Agency (CSA) and Canadian Centre for Remote Sensing (CCRS) in collaboration with the National Oceanic and Atmospheric Administration (NOAA) Hurricane Research Division (HRD) flight planning team and the NOAA Aircraft Operations Center (AOC). This program helped collecting a large number of RADARSAT SAR images in and around hurricanes. SAR images were also collected in hurricanes when collocated in-situ aircraft observations were available [21]. A SAR image shows the sea surface imprint of a tropical cyclone, allowing us to extract multiscale wind features mentioned above.

2.2 Eyewall Mesovortices

It is often observed in mature hurricanes the low-level stratus or stratocumulus cloud decks in their eyes with vortical or "swirling" features embedded in these clouds. The swirling patterns in these eye clouds have different shapes ranging from circles centered in the eyewall to convoluted patterns with multiple swirls at various locations in the eye [22]. These swirling features usually observed at the inner edge of the eyewall are called "eyewall mesovortices" and are believed to form through Kelvin–Helmholtz instability [23]. Because of the concentrated angular momentum of the parent vortex into a small area, eyewall mesovortices can produce swaths of enhanced destruction for landfalling hurricanes [24]. They can also threat the operational safety of reconnaissance aircraft penetrating the hurricane eyewall [25]. Montgomery et al. [26] presented observational evidence that high-entropy air inside the low-level eye can sustain the storm at intensity above that predicted by the maximum potential intensity theory of Emanuel [27]. They suggested that the eye-eyewall mesovortices may be responsible for transferring high entropy air from the low-level eye to the eyewall.

SAR imagery has the advantage of observing features through the clouds at very high spatial resolution, thus the eyewall mesovortices if detected in a SAR image would provide extra evidence that such features do form below the clouds as well. An example of the eyewall mesovortex signature taken on 1 August 2007 at 21:57 UTC by RADARSAT-1 is shown in Fig. 2.1. The vortical structure can be seen at the upper left corner (labeled 'A') of the image which is also located at the inner edge of the eyewall. It is noticed also the filamentary features at the bottom of the eye-eyewall interface in this image (labeled 'B'). Similar features with fingers of high reflectivity extending from the eyewall to the eye have been detected by low-fuselage

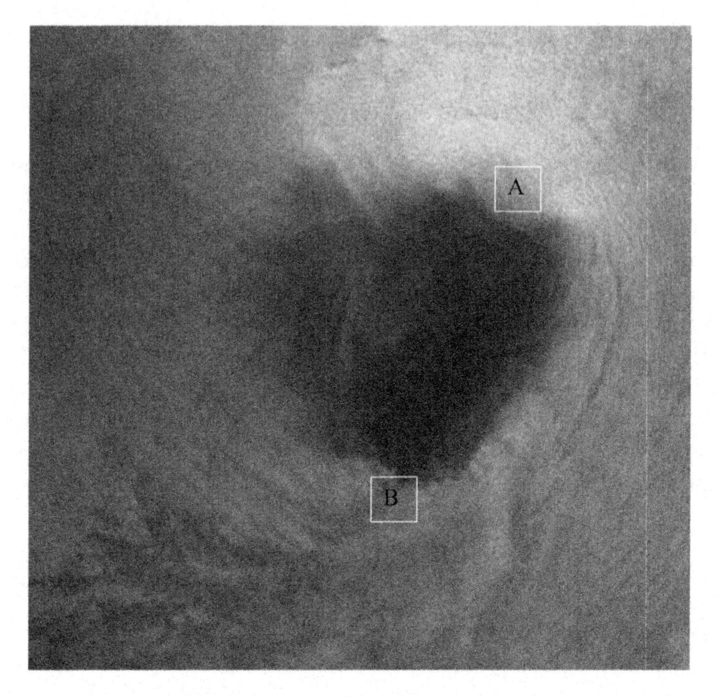

Fig. 2.1 Tropical cyclone eyewall mesovotices observed in a RADARSAT-1 synthetic aperture radar image (copyright Canadian Space Agency) taken in Typhoon Usagi at 20:57:57 UTC on 1 August 2007

radar in Hurricane Isabel on 13 September 2003 as reported by Aberson et al. [28]. They also tracked these features in subsequent radar sweeps and found that they rotate roughly 70–80 m/s. Similar features are also clearly seen in photographs of the eyewall of Hurricane Isabel on 12 September 2003 [28] as well as Hurricane Hugo on 15 September 1989 [25, 29].

Eyewall mesovotices are observed in 3 other SAR images taken in Hurricane Erin on 3 different days (11, 13 and 17 September 2001) (Fig. 2.2a, b, and c). Similar eyewall mesovortex features in Hurricane Erin have been seen in the visible satellite images by Kossin et al. [22]. They suggested a dynamic mechanism to explain the presence of mesovortices in the hurricane eye based on barotropic stability arguments. It is also noticed that the shapes of the eyes in these three images are different, ranging from triangular, square to rectangular. These shapes are associated with the asymmetric dynamics of a hurricane vortex in terms of the azimuthal wavenumbers [30]. Besides the relative strong hurricanes, we also find mesovotices in much weaker cyclones. An example showing mesovortices near the center of Tropical storm Bilis is presented in Fig. 2.3. These mesovortices are believed to be associated with eye-eyewall mixing processes.

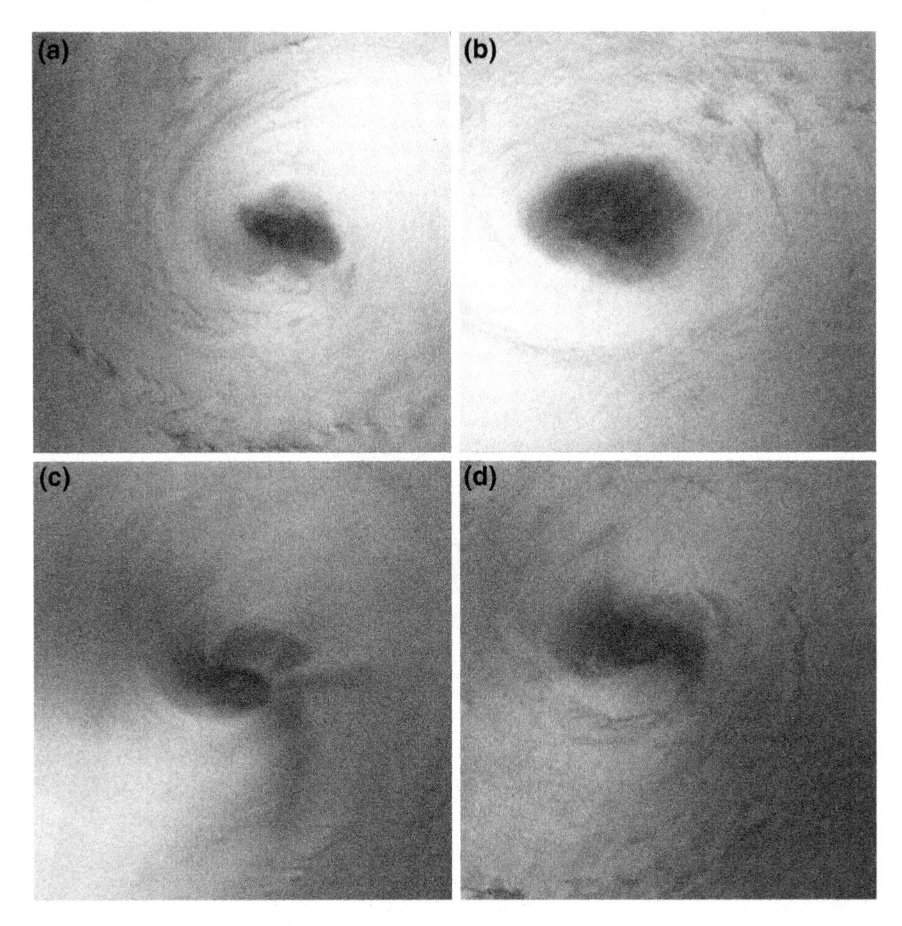

Fig. 2.2 Tropical cyclone eyewall mesovotices observed in RADARSAT-1 synthetic aperture radar images (copyright Canadian Space Agency) taken in **a** Hurricane Erin at 22:19:07 UTC on 11 September 2001; **b** Hurricane Erin at 10:03:10 UTC on 11 September 2001; **c** Hurricane Erin at 8:07:41 UTC on 11 September 2001; and **d** Hurricane Dean at 23:17:51 UTC on 19 August 2007

2.3 Spiral Rainbands, Fine-Scale Bands and Arc Clouds

Spiral band signatures have been often seen in SAR images taken in tropical cyclones. Figure 2.4 show examples of different types of spiral bands not only in their shapes but also in their degree of brightness. Li et al. [30] pointed out that the bright and dark patterns of the spiral bands are associated different physical mechanisms that change the sea surface roughness. These mechanisms are attenuation due to heavy rain; backscattering from rain drops in the air and ice particles; sea surface capillary waves induced by rain; damping of sea surface waves by rain-induced turbulence; and wind gusts [31, 32]. These spiral bands produce large amounts of rainfall and often lead to very costly and potential flooding for landfalling storms.

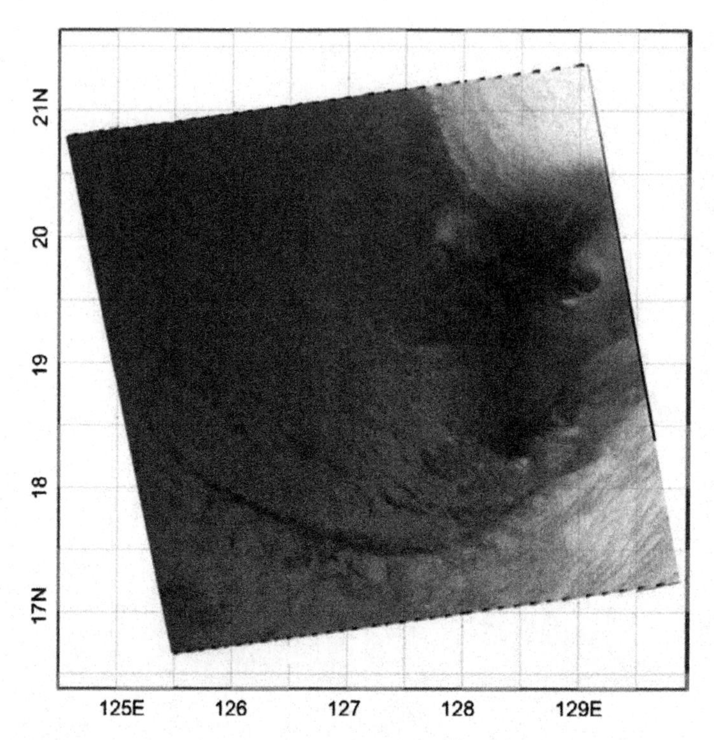

Fig. 2.3 Hurricane eyewall mesovotices observed in a RADARSAT-1 synthetic aperture radar image (copyright Canadian Space Agency) taken in Tropical storm Bilis at 09:35:42 UTC on 11 July 2006

Most tropical cyclones exhibit a set of spiral bands, so called rainbands [33, 34]. The rainbands exhibit a variety of internal structures with deep convective cores embedded in stratiform precipitation [35–37]. The pattern of rainbands in a tropical cyclone is always evolving with different shapes and wavelengths as indicated by Fig. 2.4. Some of the rainbands (Fig. 2.4d) are large with connection to the eyewall. Previous studies suggest that in some intense storms, principal and/or secondary rainbands may evolve into a secondary eyewall triggering eyewall replacement cycle and influencing the storm intensity [37]. Many previous observational studies [35, 36, 38, 39] found decreasing hurricane inflow and decreasing equivalent potential temperature in conjunction with the hurricane rainbands, which are related to the amount of convective activity and thus may have a direct impact of the overall intensity of the parent hurricane.

Arc clouds have been reported to consistently form in the periphery of tropical cyclones as noticed in visible satellite images [40]. Like a primary rainbands, an arc cloud can have a length on the order of several hundred kilometers. Evidence of arc cloud features can be seen in SAR images. For instance, Fig. 2.4b shows a well-organized arc cloud (labeled by "B") in Hurricane Maria on 5 September 2005 at 21:37:58 UTC. It is thought that the presence of arc clouds in the SAR images

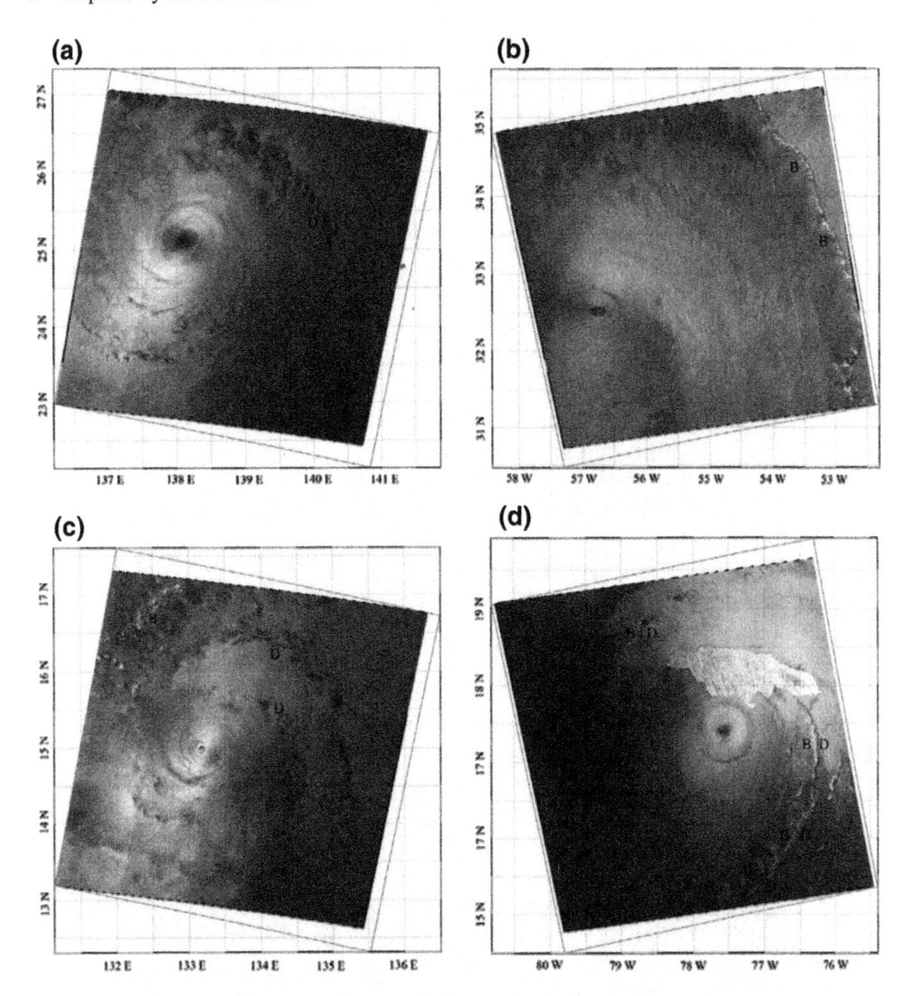

Fig. 2.4 Tropical cyclone spiral band patterns observed in SAR images (All RADARSAT-1 images copyright Canadian Space Agency). Panel **a** shows *dark* rain band pattern in Typhoon Guchol at 20:38:49 UTC on 22 August 2005; Panel **b** shows *bright* rain band pattern in Hurricane Maria at 21:37:58 UTC on 5 September 2005; Panel **c** shows *dark* pattern in inner rain band and *bright* pattern in outer rain band in Typhoon Ewiniar at 20:53:52 UTC on 3 July 2006; Panel **d** shows *bright-dark* pattern in the same rain band in Hurricane Dean at 23:16:40 UTC on 9 August 19 2007. Panel **b** also clearly reveals the signature of arc clouds. Letters "*B*" and "*D*" stand for "*Bright*" and "*Dark*", respectively. The figure is adapted from Li et al. [30]

is due to rain effects. Since we can estimate the surface wind speed from the SAR images, how the arc clouds affect surface winds can be quantified in future work. It is hypothesized by Dunion et al. [40] that arc clouds created low-level outflow may counter the typical low-level inflow that is vital for tropical cyclone formation and maintenance.

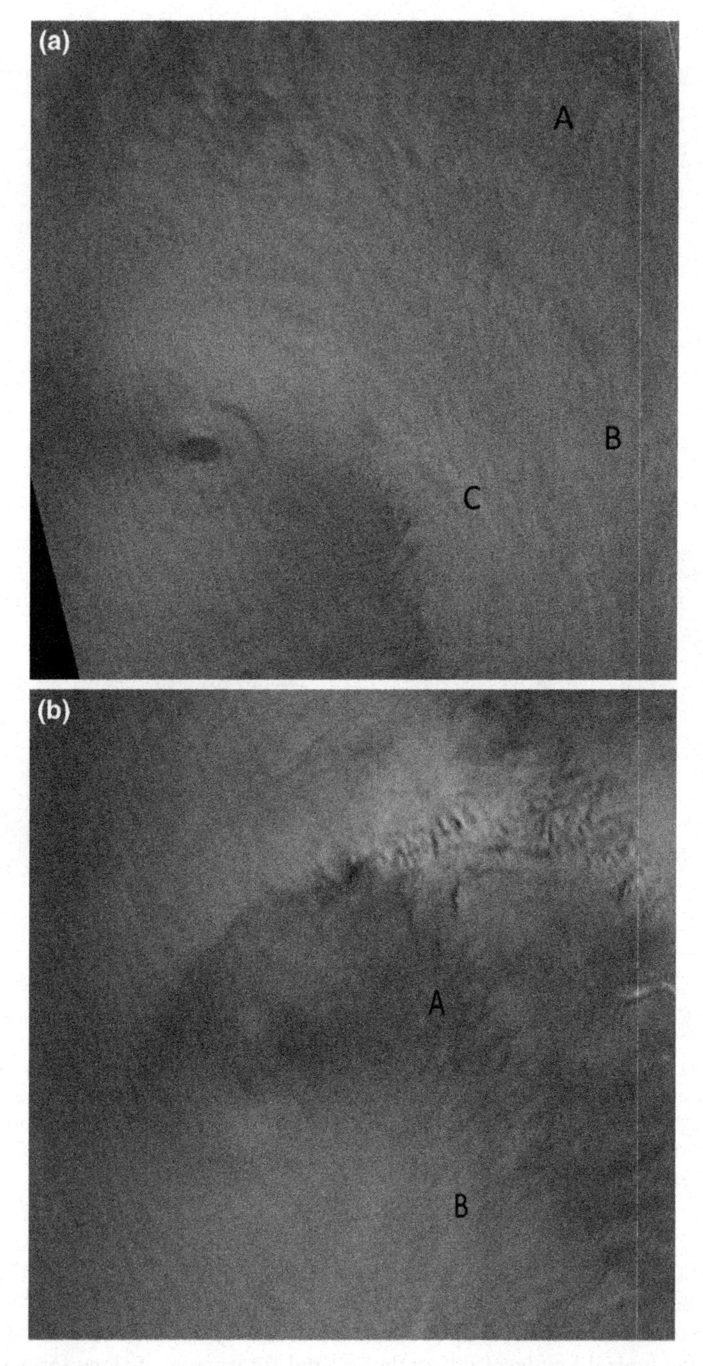

Fig. 2.5 Tropical cyclone fine-scale bands observed in RADARSAT-1 synthetic aperture radar images (copyright Canadian Space Agency) taken in **a** Hurricane Maria at 21:37:58 UTC on 5 September 2005; and **b** Hurricane Maria at 21:21:42 UTC on 9 September 2005. Letters "*A*", "*B*", and "*C*" indicate locations of fine scale bands

As mentioned earlier, another type of spiral bands, "fine-scale spiral bands", with wavelengths of 5–10 km have been identified using Doppler radar data [9]. Using SAR images, Katsaros et al. [13] identified roll-like structures with similar and slightly smaller wavelengths. We also detected fine-scale band signatures in several SAR images in hurricanes. An example is shown in Fig. 2.5 with two SAR images taken in Hurricane Maria on 5 and 9 September 2005, respectively. Features with scales larger than boundary layer rolls but much smaller than a typical primary rainband can be seen (labeled by "*A*", "*B*", "*C*") in these two images. These fine-scale spiral bands are believed to be related to the boundary layer inflection instability [14].

2.4 Boundary Layer Rolls

Boundary layer (BL) rolls or 'roll vortices' can have a significant influence on turbulent exchange of momentum, sensible heat and latent heat in the tropical cyclone boundary layer which are essential for its maintenance and intensification [21]. Foster [19] has developed a theory for roll vortices in curved flow at high wind

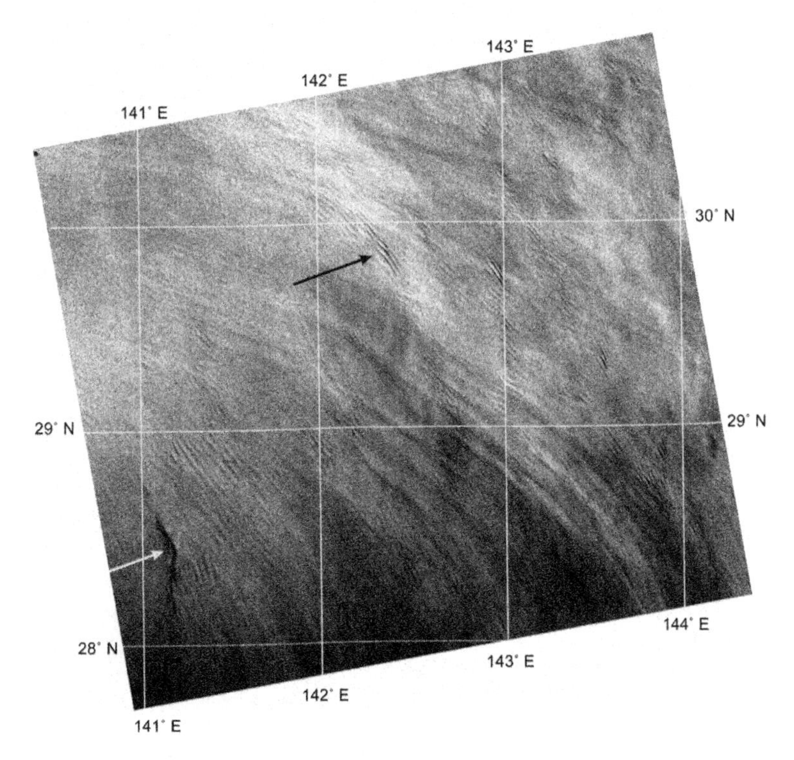

Fig. 2.6 Tropical cyclone boundary layer rolls observed in a RADARSAT-1 synthetic aperture radar image (copyright Canadian Space Agency) taken in Typhoon Fengshen at 08:32:28 UTC on 24 Jul 2002. The figure is adapted from Morrison et al. [16]

speeds, such as in hurricanes, suggesting that tropical cyclone boundary layer is a favorable environment for roll formation.

SAR can provide useful information for identifying tropical cyclone boundary layer rolls, because streak patterns in sea surface roughness can be explained by change in surface wind speed due to the formation of the rolls. Previous studies [13, 16, 19] have presented SAR images clearly showing the roll signatures. Figure 2.6 shows an example of roll signatures in a SAR image taken in Typhoon Fengshen (2002).

Fig. 2.7 Tropical cyclone boundary layer rolls observed in a RADARSAT-1 synthetic aperture radar image (copyright Canadian Space Agency) taken in Hurricane Isidore at 23:57 UTC on 23 September 2002 after landfall on the Yucatan Peninsula. Analyses of this image and the two boxes are shown in Fig. 2.8. The figure is adapted from Zhang et al. [21]

A RADARSAT-1 SAR image over Hurricane Isidore taken at 23:57 UTC on 23 September 2002 was analyzed by Zhang et al. [21]. At that time, the storm center was nearly stationary over the Yucatan. The SAR image, shown in Fig. 2.7, was first split into subscenes of 20 km (256 pixels of 50 m length and width). A two dimensional Fast Fourier Transform (2D-FFT) algorithm was then used to calculate the directional image spectra for each subscene [41]. The maximum of the 2D spectrum is due to wind streaks or rolls. The direction of the rolls is then taken to be the wind direction.

The analysis result of this SAR image in terms of the wavelength of the rolls is shown in Fig. 2.8. The upper left panel shows the wavelength and direction of the wind streaks. The histogram of the wavelengths of the rolls for the entire image is shown in the upper right panel of Fig. 2.8, indicating a maximum in the distribution of ~850 m. The lower two panels in Fig. 2.8 demonstrate the analyses of the two boxes of interest as indicated in Fig. 2.7. The wavelengths of the rolls in these two boxes vary from around 400–1100 m, suggesting how the scales of rolls may vary

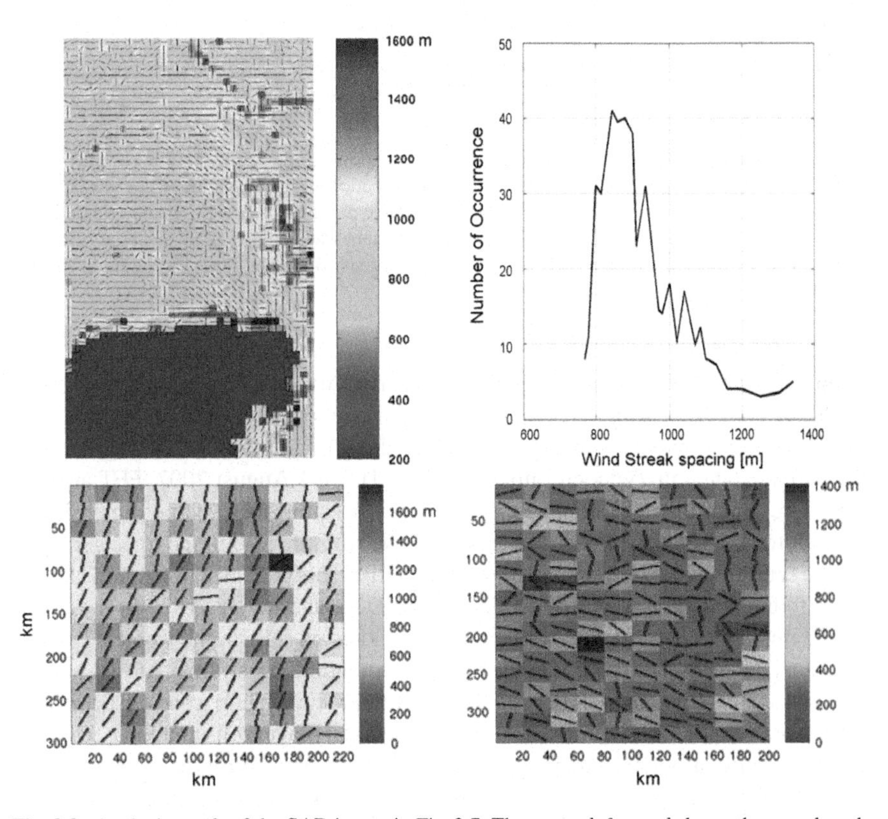

Fig. 2.8 Analysis result of the SAR image in Fig. 2.7. The *upper left panel* shows the wavelength (*color bar code*, m) and direction of the wind rolls. The *upper right panel* shows the histogram of the roll wavelengths of the entire image at *left*. The lower two (*left/right*) *panels* show the wavelength and direction analysis of the (*lower/upper*) *boxes* in Fig. 2.7. The figure is adapted from Zhang et al. [21]

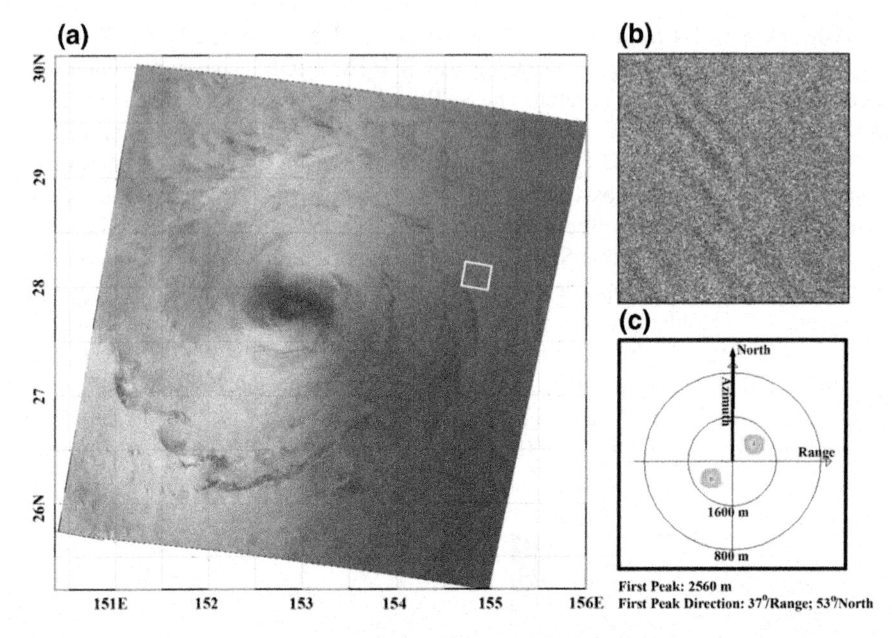

Fig. 2.9 Analysis of boundary layer rolls within Typhoon Fitow. Panel **a** shows a SAR image acquired at 19:42:51 UTC on 1 August, 2007. Full resolution subimage in the *white box* is shown in **b** and its two dimensional FFT analysis in **c**, shows the dominant wavelength and orientation of the boundary layer rolls with 180° ambiguity. The dominant wavelength is 2560 m and its direction is 37° (53°) with respect to the satellite flying range (true North) direction. The figure is adapted from Li et al. [30]

within a storm. The average wavelength of the rolls from the SAR image analysis is generally consistent with that estimated using in-situ aircraft flight-level data.

Another example of boundary layer roll analysis is demonstrated in Fig. 2.9 for a SAR image taken in Typhoon Fitow at 1942 UTC on 1 August 2007. FFT analysis (Fig. 2.9c) on a full-resolution subimage (Fig. 2.9b) shows the spatial dimension of 2–3 km for the wind streaks. The boundary layer rolls are also found to be generally in line with the wind direction. These characteristics of rolls are in agreement with those found from previous studies using Doppler radar data [16, 18].

We note that most previous observational studies on rolls in the tropical cyclone boundary layer to date have mainly focused on landfalling storms. At this point it remains unclear how frequently boundary layer rolls occur in hurricanes, especially in open-ocean conditions. How tropical cyclone boundary layer rolls modulate the mean and turbulence structure remains poorly understood either. With the extensive SAR images taken in tropical cyclones, quantifying the characteristics and mechanisms of tropical cyclone boundary layer rolls becomes promising [30].

2.5 Summary

It is a relatively new tool for tropical cyclone research and forecasting to use a SAR imagery to study its structure and dynamics. This study demonstrates the advantage of SAR sensors for imaging multiscale wind features on the sea surface beneath the storm clouds. Examples of eyewall mesovotices, spiral rainbands, fine-scale bands, arc clouds, and boundary layer rolls are clearly presented. Since SAR images view the actual ocean surface responses to the storm-forced winds with high spatial resolution ($<100\,\text{m}$), they provide evidence for the interaction between the multiscale wind features and the surface. These wind features may enhance damages when observed during tropical cyclone landfalls. Qualitative analyses of the characteristics of boundary layer rolls show consistency with previous observations using Doppler radar and/or aircraft flight-level data. This result suggests that a SAR image has potential to be used to improve our understanding of such wind features.

Since the number of spaceborne SAR satellites will increase in the coming years, more collocated observations of tropical cyclones from different spaceborne, airborne and in situ sensors will be obtained. These observations will provide guidance for mapping tropical cyclone surface wind vector in high resolution [42]. They will also help improve our understanding of the mechanisms of the formation of the multiscale wind features. How the tropical cyclone intensity is linked to these wind features can also be investigated with collocated SAR images and other in-situ observations in the future.

References

1. Pielke Jr., R.A., and C.W. Landsea. 1998. Normalized hurricane damages in the United States: 1925–1995. *Weather and Forecasting* 13 (3): 621–631.
2. Marks Jr., F.D., and R.A. Houze Jr. 1984. Airborne doppler radar observations in Hurricane Debby. *Bulletin of the American Meteorological Society* 65 (6): 569–582.
3. Kossin, J.P., and W.H. Schubert. 2004. Mesovortices in Hurricane Isabel. *Bulletin of the American Meteorological Society* 85 (2): 151–153.
4. Reasor, P.D., M.D. Eastin, and J.F. Gamache. 2009. Rapidly intensifying Hurricane Guillermo (1997). Part I: Low-wavenumber structure and evolution. *Monthly Weather Review* 137 (2): 603–631.
5. Braun, S.A., M.T. Montgomery, and Z. Pu. 2006. High-resolution simulation of Hurricane Bonnie (1998). Part I: The organization of eyewall vertical motion. *Journal of the Atmospheric Sciences* 63 (1): 19–42.
6. Nolan, D.S., J.A. Zhang, and D.P. Stern. 2009. Evaluation of planetary boundary layer parameterizations in tropical cyclones by comparison of in situ observations and high-resolution simulations of Hurricane Isabel (2003). Part I: Initialization, maximum winds, and the outer-core boundary layer. *Monthly Weather Review* 137 (11): 3651–3674.
7. Schubert, W.H., M.T. Montgomery, R.K. Taft, T.A. Guinn, S.R. Fulton, J.P. Kossin, and J.P. Edwards. 1999. Polygonal eyewalls, asymmetric eye contraction, and potential vorticity mixing in hurricanes. *Journal of the Atmospheric Sciences* 56 (9): 1197–1223.
8. Kossin, J.P., and W.H. Schubert. 2001. Mesovortices, polygonal flow patterns, and rapid pressure falls in hurricane-like vortices. *Journal of the Atmospheric Sciences* 58 (15): 2196–2209.

9. Gall, R., J. Tuttle, and P. Hildebrand. 1998. Small-scale spiral bands observed in Hurricanes Andrew, Hugo, and Erin. *Monthly Weather Review* 126 (7): 1749–1766.
10. Chen, Y., and M. Yau. 2001. Spiral bands in a simulated hurricane. Part I: Vortex Rossby wave verification. *Journal of the Atmospheric Sciences* 58 (15): 2128–2145.
11. Wang, Y. 2002a. Vortex Rossby waves in a numerically simulated tropical cyclone. Part I: Overall structure, potential vorticity, and kinetic energy budgets. *Journal of the Atmospheric Sciences* 59 (7): 1213–1238.
12. Wang, Y. 2002b. Vortex Rossby waves in a numerically simulated tropical cyclone. Part II: The role in tropical cyclone structure and intensity changes. *Journal of the Atmospheric Sciences* 59 (7): 1239–1262.
13. Katsaros, K.B., P.W. Vachon, W.T. Liu, and P.G. Black. 2002. Microwave remote sensing of tropical cyclones from space. *Journal of Oceanography* 58 (1): 137–151.
14. Nolan, D.S. 2005. Instabilities in hurricane-like boundary layers. *Dynamics of Atmospheres and Oceans* 40 (3): 209–236.
15. Wurman, J., and J. Winslow. 1998. Intense sub-kilometer-scale boundary layer rolls observed in Hurricane Fran. *Science* 280 (5363): 555–557.
16. Morrison, I., S. Businger, F. Marks, P. Dodge, and J.A. Businger. 2005. An observational case for the prevalence of roll vortices in the hurricane boundary layer. *Journal of the Atmospheric Sciences* 62 (8): 2662–2673.
17. Lorsolo, S., J.L. Schroeder, P. Dodge, and F. Marks Jr. 2008. An observational study of hurricane boundary layer small-scale coherent structures. *Monthly Weather Review* 136 (8): 2871–2893.
18. R. Ellis, and S. Businger. Helical circulations in the typhoon boundary layer. *Journal of Geophysical Research: Atmospheres*, 115(D6), 2010.
19. Foster, R.C. 2005. Why rolls are prevalent in the hurricane boundary layer. *Journal of the Atmospheric Sciences* 62 (8): 2647–2661.
20. Fu, L.L., and B. Holt. 1982. Seasat views oceans and sea ice with synthetic aperture radar. *JPL Publication* 81: 120.
21. Zhang, J.A., K.B. Katsaros, P.G. Black, S. Lehner, J.R. French, and W.M. Drennan. 2008. Effects of roll vortices on turbulent fluxes in the hurricane boundary layer. *Boundary-Layer Meteorology* 128 (2): 173–189.
22. Kossin, J.P., B.D. McNoldy, and W.H. Schubert. 2002. Vortical swirls in hurricane eye clouds. *Monthly Weather Review* 130 (12): 3144–3149.
23. Montgomery, M.T., V.A. Vladimirov, and P.V. Denissenko. 2002. An experimental study on hurricane mesovortices. *Journal of Fluid Mechanics* 471: 1–32.
24. Wakimoto, R.M., and P.G. Black. 1994. Damage survey of Hurricane Andrew and its relationship to the eyewall. *Bulletin of the American Meteorological Society* 75 (2): 189–200.
25. Marks, F.D., P.G. Black, M.T. Montgomery, and R.W. Burpee. 2008. Structure of the eye and eyewall of Hurricane Hugo (1989). *Monthly Weather Review* 136 (4): 1237–1259.
26. Montgomery, M.T., M.M. Bell, M.L. Black, and S.D. Aberson. 2006. *Hurricane Isabel (2003): New insights into the physics of intense storms, Part I: Mean vortex structure and maximum intensity estimates.*
27. Emanuel, K.A. 1986. An air-sea interaction theory for tropical cyclones. Part I: Steady-state maintenance. *Journal of the Atmospheric Sciences* 43 (6): 585–605.
28. Aberson, S.D., M.T. Montgomery, M. Bell, and M. Black. 2006. Hurricane Isabel (2003): New insights into the physics of intense storms. Part II. *Bulletin of the American Meteorological Society* 87 (10): 1349.
29. Zhang, J.A., F.D. Marks, M.T. Montgomery, and S. Lorsolo. 2011. An estimation of turbulent characteristics in the low-level region of intense Hurricanes Allen (1980) and Hugo (1989). *Monthly Weather Review* 139 (5): 1447–1462.
30. Li, X., J.A. Zhang, X. Yang, W.G. Pichel, M. DeMaria, D. Long, and Z. Li. 2013. Tropical cyclone morphology from spaceborne synthetic aperture radar. *Bulletin of the American Meteorological Society* 94 (2): 215–230.
31. Bliven, L.F., and J.P. Giovanangeli. 1993. An experimental study of microwave scattering from rain-and wind-roughened seas. *International Journal of Remote Sensing* 14 (5): 855–869.

32. Lin, I.I., W. Alpers, V. Khoo, H. Lim, T.K. Lim, and D. Kasilingam. 2001. An ERS-1 synthetic aperture radar image of a tropical squall line compared with weather radar data. *IEEE Transactions on Geoscience and Remote Sensing* 39 (5): 937–945.
33. Willoughby, H. 1978. A possible mechanism for the formation of hurricane rainbands. *Journal of the Atmospheric Sciences* 35 (5): 838–848.
34. Willoughby. H. 1988. The dynamics of the tropical cyclone core. *Australian Meteorological Magazine* 36 (3).
35. Barnes, G., E.J. Zipser, D. Jorgensen, and F. Marks Jr. 1983. Mesoscale and convective structure of a hurricane rainband. *Journal of the Atmospheric Sciences* 40 (9): 2125–2137.
36. Barnes, G., J. Gamache, M. LeMone, and G. Stossmeister. 1991. A convective cell in a hurricane rainband. *Monthly Weather Review* 119 (3): 776–794.
37. Houze Jr., R.A., S.S. Chen, L. Wen-Chau, R.F. Rogers, et al. 2006. The hurricane rainband and intensity change experiment: Observations and modeling of Hurricanes Katrina, Ophelia, and Rita. *Bulletin of the American Meteorological Society* 87 (11): 1503.
38. Powell, M.D. 1990a. Boundary layer structure and dynamics in outer hurricane rainbands. Part I: Mesoscale rainfall and kinematic structure. *Monthly Weather Review* 118 (4): 891–917.
39. Powell, M.D. 1990b. Boundary layer structure and dynamics in outer hurricane rainbands. Part II: Downdraft modification and mixed layer recovery. *Monthly Weather Review* 118 (4): 918–938.
40. Dunion, J., M. Eastin, D. Nolan, J. Hawkins, and C. Velden. 2010. Arc clouds in the tropical cyclone environment: Implications for TC intensity change. In *Preprints, 29th conference on hurricanes and tropical meteorology, Tucson, AZ, American Meteorological Society C*, Vol. 6.
41. Lehner, S., J. Schulz-Stellenfleth, B. Schattler, H. Breit, and J. Horstmann. 2000. Wind and wave measurements using complex ERS-2 SAR wave mode data. *IEEE Transactions on Geoscience and Remote Sensing* 38 (5): 2246–2257.
42. Zhang, B., W. Perrie, J.A. Zhang, E.W. Uhlhorn, and Y. He. 2014. High-resolution hurricane vector winds from C-band dual-polarization SAR observations. *Journal of Atmospheric and Oceanic Technology* 31 (2): 272–286.

Chapter 3
Observations of Typhoon Eye on Ocean Surface Using SAR and Other Satellite Sensors

Antony K. Liu, Yu-Hsin Cheng and Jingsong Yang

Abstract In this study, typhoon eyes have been delineated using wavelet analysis from the synthetic aperture radar (SAR) images of ocean surface roughness and from the warmer area at the cloud top in the infrared (IR) images, respectively. RADARSAT and ENVISAT SAR imagery, and multi-functional transport satellite (MTSAT) and Feng Yun (FY)-2 Chinese meteorological satellite IR imagery were used to examine the typhoons in the western North Pacific from 2005 to 2011. Nine cases of various typhoons in different years, locations, and conditions have been used to compare the typhoon eyes by SAR (on the ocean surface) with IR (at the cloud-top level) images. Furthermore, the best track data getting from the Joint Typhoon Warning Center (JTWC), Chinese Meteorological Administration (CMA), and the Japan Meteorological Agency (JMA) are checked for the calibration and validation along with the Moderate Resolution Imaging Spectroradiometer (MODIS) image. Because of the vertical wind shear, which acts as an upright tilt, the location of the typhoon eye on the ocean surface differs from that at the top of the clouds. Consequently, the large horizontal distance between typhoon eyes on the ocean surface and on the cloud top implies that the associated vertical wind shear profile is considerably more complex than generally expected. The upright tilt structure may be caused by the ocean's feedback or the effect of island obstruction. This result demonstrates that SAR can be a useful tool for typhoon monitoring study over the ocean surface.

A.K. Liu (✉)
Ocean College, Zhejiang University, Zhoushan, China
e-mail: tonyakliu@gmail.com

A.K. Liu
NASA Goddard Space Flight Center (Emeritus), Greenbelt, MD, USA

Y.-H. Cheng
Department of Marine Environmental Informatics, National Taiwan Ocean University, Keelung, Taiwan
e-mail: galaxysmail@gmail.com

J. Yang
State Key Laboratory of Satellite Ocean Environment Dynamics, Second Institute of Oceanography, State Oceanic Administration, Hangzhou 310012, China
e-mail: syang@sio.org.cn

X. Li (ed.), *Hurricane Monitoring With Spaceborne Synthetic Aperture Radar*, Springer Natural Hazards, DOI 10.1007/978-981-10-2893-9_3

3.1 Introduction

Satellite remote sensing with repeated coverage is the most efficient method to monitor and study ocean environment, marine productivity, and oil spills pollution. The ability of meteorological satellite remote sensing for monitoring clouds, fronts, and typhoon has been amply demonstrated. One of the first and most important applications of satellite observations of the cyclone has been the estimation of associated intensity of cyclones with the temperature of the eye, cloud organization, and surrounding environment. For example, meteorological satellite image shows changes in the cloud organization, central area of the cyclone, and rain bands. The synthetic aperture radar (SAR) images of wind-related ocean features, such as surface waves, rain cells, tropical cyclones, internal lee waves, and katabatic wind have been studied recently [1]. The combined use of infrared, optical sensors and SAR can provide frequent high-resolution coverage of the typhoon evolution.

The western North Pacific is the most frequent area throughout the world where tropical cyclones strike. Typhoons, in particular, wreak devastation along coastal areas through powerful winds and torrential rain [2–4], and play a key role in influencing the upper layer marine ecosystem, churning the ocean surface and causing upper ocean response along the storm path [5, 6]. Because the land and sea surface environments most heavily influence human lives, tracking and predicting the typhoon's operation by extracting sea-surface information is critical. Satellite remote sensing with repeated coverage has provided efficient monitoring and study of tropical cyclones [7–12]. The ability of meteorological satellite remote sensing to monitor clouds, fronts, and typhoons has been amply demonstrated. Limited to providing visible (VIS) and infrared (IR) images, the geostationary satellites can only frequently provide information on the top of clouds from per hour to per 15 min. By contrast, microwave sensors can penetrate clouds, thus improving the detection of the internal cyclone structure. Because microwave radiometers and scatterometers are limited to observing localized phenomena using lower spatial resolution (for Advanced Microwave Scanning Radiometer 89 GHz with 6.25 km resolution), that synthetic aperture radar (SAR) can penetrate clouds to observe sea surface roughness with high resolution (for ScanSAR Wide mode with 100-m resolution) [13]. In recent years, SAR has been a powerful tool to provide a new vision of the atmospheric phenomena, such as typhoon and convective cells, imprinted on the ocean surface [14–17].

Although SAR imagery yields rich spatial resolution, successively observing a typhoon is difficult because of the poor temporal resolution of SAR. The "eye" of a typhoon is a zone of weak winds that exists in the center of a swirling vortex. The weather is normally calm and free of clouds inside the eye of a typhoon. The eye zone is the warmest at the upper cloud levels, is normally circular in shape, and may range in size from 8 to 200 km in diameter [18]. A conventional definition of a typhoon eye is the cloudless area (VIS imagery) or the warmer area (IR imagery), where it is observed at the height of the tropopause. SAR images provide visual evidence of precipitation, boundary layer rolls, and typhoon eyes within these typhoons [19]. The vertical wind shear, which acts to tilt the typhoon eye's position, causes the upper and lower eyes' location to be uncoordinated [20–22]. Thus, tracking the calm areas on the sea surface using SAR imagery may be more effective in detecting the typhoon eye on the ocean surface.

Weatherford and Gray [2] used the aircraft dataset, collected by the United States Air Force WC-130 aircraft, to analyzing the relationship between cyclone inner-core intensity and outer-core wind strength. Weatherford and Gray [3] found that outer-radius wind strength and inner-core intensity can vary greatly. But they dont have enough information to measure the size of typhoon eye. Brueske and Velden [23] estimated the cyclone intensity and eye size from the NOAA-KLM series Advanced Microwave Sounding Unit with resolution of 16–48 km. The eyewall, a ring of cumulonimbus convection, surrounds the eye and contains the sharpest radial pressure gradient nearly coincident with strongest wind. Because the eyewall slopes outward, the eye is approximately an inverted, truncated cone. The air aloft in the eye is clear, warm, and dry, separated by an inversion from more moist, usually cloudy air near the ocean surface. Willoughby [4] reported that typhoon eye contains two air masses separated by an inversion and the air below the inversion exchanges momentum and moisture with the ocean and mixes in complicated ways with air from the eyewall.

In addition to the cooling response induced by a typhoon, the passage of typhoon also plays a key role in influencing the upper layer marine ecosystem [5, 6, 24]. Thus, more understanding the behavior of typhoon proves to be the key for further improving the prediction of typhoon intensity change. Conventionally, the typhoon eye observation is from weather satellite using IR or VIS wavelengths, which is the observation at cloud height about 13 km. However, because of the vertical wind shear tilt, the location of typhoon eye on the ocean surface may be quite different from that on the top of clouds. In general, the impact of typhoon on coast community is either near the land or on the ocean surface. Therefore to extract typhoon information near the ocean surface is very critical for the typhoon tracking operation and forecast. Owing to its warm-core structure, the tropical cyclones strongest winds are, in fact, located in the lowermost troposphere, typically near the top of a shallow boundary layer 500–1000 m deep.

Microwave sensors have improved the detection of internal cyclone structure, such as the location of the eye, because those wavelengths are sensed through high clouds that sometimes obscure the eye in IR and VIS images. However, the passive microwave sensors does not have enough high-resolution to estimate the size and structure of typhoon eye area. Therefore, tracking the calm areas on the ocean surface from high-resolution SAR imagery (with 25–50 m resolution) or the warmest areas on the cloud top from thermal IR imagery (with 1–4 km resolution) may have a better detection and clear description of a typhoons eye. Especially, SAR has the day-and-night all weather capability to "see" through the clouds for the signature and footprint of typhoon on the ocean surface. The objective of this study is to compare the results observed using SAR imagery of typhoon eyes located near the ocean surface with those using IR imagery of the cloud top. Consequently, this study demonstrates that SAR could be a potential tool in monitoring typhoons over the ocean surface.

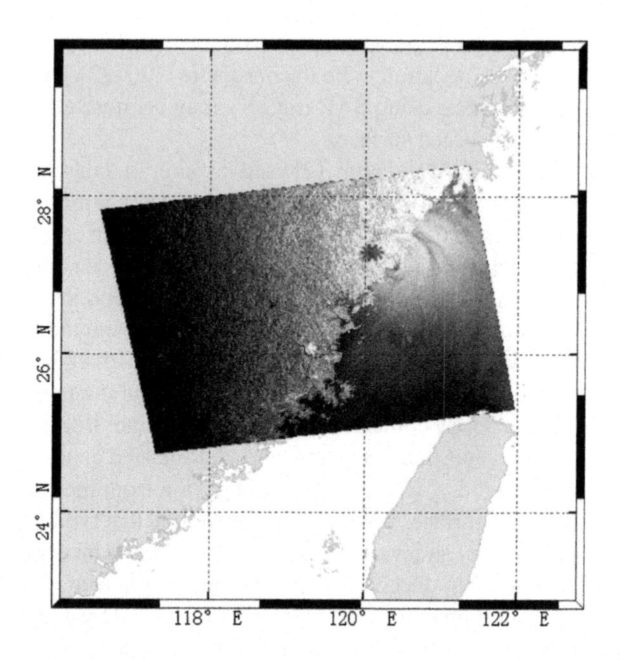

Fig. 3.1 RADARSAT SAR image of Typhoon Saomai landed along the east coast of China on August 10, 2006 at 10:02 UTC with eye over the coastline, and the typhoon eye centers estimated from MTSAT (*red dot*) and FY-2 (*green dot*) IR images

The typhoon monitoring program in Asia using SAR has been carried out at the National Taiwan Ocean University (NTOU) recently. The objectives are to study typhoon characteristics and to retrieve wind field by wavelet tracking technique using multiple sensors including SAR. The first project is focused on the tracking of typhoons eye using satellite images of SAR from RADARSAT of Canadian Space Agency (CSA), and MTSAT from JMA. Cheng et al. [25] have reported that the horizontal distances between typhoon eyes on the cloud top and on the ocean surface of the abovementioned cases are significantly large from 9 to 22 km. This implies that the vertical wind shear profile might be much more complex than generally expected. For the purpose of demonstration, Fig. 3.1 shows a RADARSAT SAR Image of Typhoon Saomai landed along the east coast of China on August 10, 2006 at 10:02 UTC with eye right on the coastline. Typhoon Saomai had become a Category 5-equivalent super typhoon by August 9, and it made landfall in Zhejiang, China on August 10 with maximum sustained winds of 115 knots. The red and green dots represent the typhoon eye centers estimated from MTSAT and FY-2, respectively. As expected, the typhoon eye on the cloud top from MTSAT and FY-2 IR data was already inland while the eye on the ocean surface from SAR image was dragging behind on the coastline due to the blocking effect of island and land.

In this chapter, first the SAR, FY-2/IR, and MTSAT/IR satellite data used are listed for reference. Then, the typhoon eyes are delineated from the smoother area in SAR images and from the warmer area in IR images by using wavelet analysis. Nine case studies for different typhoons and environment have been investigated to measure the center distance between the typhoon eye derived from SAR data and that from IR data. As demonstrated by these case studies, SAR can be a powerful tool to help on typhoon tracking and prediction especially on the ocean surface. SAR images, however, are not completely immune from precipitation attenuation and surface splash effects, but since the typhoon eye is generally free of such effects, SAR imagery is particularly useful for studies of the eye. Finally, some of the issues concerned on the definition of typhoons eye and typhoon tracking/prediction have been identified. Also, the highly tilted vertical wind shear, especially during typhoon turning and staggering has been examined and discussed.

3.2 Multiple Sensors Data

In this study, all typhoon images collected by RADARSAT and ENVISAT in conjunction with MTSAT and FY-2 with a relatively strong typhoon eye in the Western Pacific Ocean from 2003 to 2011 have been examined. In addition, JTWC, CMA, and JMAs archived best track data have been used to check the consistency of typhoons location. Also, the Moderate Resolution Imaging Spectrometer (MODIS) are viewing the entire Earth's surface every 1–2 days, and its image with 1.1 km spatial resolution collected during typhoon area has been used for calibration and comparison with SAR and IR data. The detailed image processing method is outlined in the next section for the case studies of typhoon eyes.

3.2.1 Infrared Images

The Multi-functional Transport Satellite (MTSAT) series fulfills a meteorological function for the JMA. MTSAT is a geostationary meteorological satellite and provides imagery for the northern hemisphere and its coverage includes East Asia and the Western Pacific area. The orbit of MTSAT-1R satellite is 35,800 km above the equator at 140° East longitude. The MTSAT-1R is observing every hour in normal condition and half hour during the severe weather. The MTSAT-1R meteorological sensor contains one visible and four infrared channels. The wavelength of visible channel is 0.55–0.90 μm and its resolution is 1 km at the sub-satellite point. The infrared channels include IR1 (10.3–11.3 μm), IR2 (11.5–12.5 μm), IR3 (6.5–7.0 μm), and IR4 (3.5-4.0 μm) and its resolution is 4 km. IR imagery can be employed day and night and indicates the severity of typhoons, such as revealing cold, high clouds that suggest severe convection [13]. In this study, IR images were used for analyzing typhoon eyes at the cloud top.

China began its geostationary meteorological satellite FengYun-2 in 1980 [26]. FY-n is the meteorological satellite series, including polar-orbiting and geostationary meteorological satellites, whose serial numbers n are odd and even ones respectively. FY-1A and FY-1B are experimental satellites carrying 5-channel Multichannel Visible Infrared Scanning Radiometers (MVISR-1). FY-1C and FY-1D are operational satellites carrying 10-channel Multichannel Visible Infrared Scanning Radiometers

(MVISR-2) similar to channels of NOAA/AVHRR and CZCS. FY-3 is the second generation of polar-orbiting meteorological satellites of China carrying 11 sensors. Generally, FY-3 is similar to the American National Polar-orbiting Operational Environmental Satellite System (NPOESS) and the European Meteorological Operational satellite programme (MetOp).

FY-2A and FY-2B are also experimental satellites, however, they operated over eight and five years respectively and exceed the two-year design life. The 3-channel Visible and Infrared Spin Scan Radiometers (VISSR-1) are onboard FY-2A and FY-2B. FY-2C and FY-2D are operational satellites carrying Visible and Infrared Spin Scan Radiometer (VISSR-2). The number of spectral channels of VISSR-2 is increased to five, and the infrared wave window 10.5–12.5 μm is split into two channels for the retrieval of Sea Surface Temperature (SST). FY-2E is a substitute satellite for FY-2C, carrying the VISSR-2. FY-2F, FY-2G, and FY-2H are the follow-on satellites to FY-2C, FY-2D, and FY-2E, and will carry VISSR-n which is the improved version of VISSR-2. The FY-2 satellites are positioned at 105° East longitude, and its data used in this study are operational products similar to MTSAT data with the resolution of 1 km for VIS channel and 4 km for IR channel.

3.2.2 SAR Images

The Center for Southeastern Tropical Advanced Remote Sensing (CSTARS) at the University of Miami has provided access for the scientific community to real-time, high-resolution satellite imagery since 2003. Currently, existing SAR satellites orbiting in space and accessible to CSTARS are CSAs RADARSAT-1, the ESAs ERS-2 and ENVISAT Advanced SAR. Other SAR satellites now in orbit include CSAs RADARSAT-2 (with additional beam and polarization capabilities), PALSAR (Japanese Space Agency), TerraSAR-X (Germany/DLR), CRS (Chinese Remote Sensing Satellite) and Cosmo Sky-Med (Italian Space Agency). RADARSAT-1 is a C-Band SAR that penetrates clouds and can image day-and-night and in all weather condition, and its launched in November 1995. Both RADARSAT and ENVISAT SAR images were used to examine the typhoons in the western North Pacific from 2005 to 2011. The ScanSAR Wide product used in this study is a geo-referenced ground coordinate multi-look image. The ScanSAR Wide product has a pixel size of 50 m 50 m with nominal image coverage of 500 km [27]. Because SAR has the capability to penetrate the clouds, it provides a new method to study typhoons with high-resolution observation near the ocean surface.

In order to continue the Sino-European cooperation on the coastal zone monitoring, particularly by exploiting ESA and China satellite data, a continuous project of monitoring marine environmental safety and protection has been proposed. The objective of this Dragon programme is to measure elemental marine-meteo parameters such as sea surface wind, wave and salinity, and to monitor coastal algae blooms by utilizing the space borne microwave radar to support marine environmental safety in the China and European seas, especially by using the spaceborne SAR.

Fig. 3.2 Satellite images of Typhoon Megi collected over the southeast of Taiwan on October 17, 2010 from **a** MODIS, **b** MTSAT, and **c** ENVISAT, respectively. The typhoon eye as a dark circle can be easily detected in **d** the zoom-in ENVISAT SAR subscene

All ENVISAT Advanced SAR images with wide swath mode of 150 m resolution and 400 km swath are provided by ESA for this typhoon eye observation study.

For the purpose of demonstration, Fig. 3.2a–c show the satellite images of Typhoon Megi collected over the southeast of Taiwan on October 17, 2010 from MODIS, MTSAT, and ENVISAT, respectively. The typhoon eye with minimum wind as a dark circle and the rain bands around the eye in dark due to the damping effect on the ocean surface, can be easily detected in the zoom-in ENVISAT SAR image shown in Fig. 3.2d. In IR image, the formation of a small eye (with warmer temperatures relative to the eyewall) and the expanded area of very cold cloud in bands around the eye indicate the increased intensity. The Typhoon Megi will be investigated in detail as a case study in the later section.

3.3 Approach

The eyewall of a typhoon is a zone of super gradient wind and appears extremely bright in contrast to the eye zone of a typhoon in SAR images. Therefore, the typhoon eye can be delineated from a darker area surrounded by the eyewall using wavelet analysis [28, 29]. Wavelet transform is analogous to the Fourier transform but is more localized both in frequency and time. A two dimensional wavelet transform is a highly efficient band-pass data filter, which can be used to separate various scales of processes [30]. For effective identification and tracking of common features in a chosen image, such as typhoon eyes, a two dimensional Mexican hat wavelet transform is applied to extract features with corresponding spatial scales from the image and to filter out noise [11, 28].

The Mexican hat wavelet is the second derivative of a Gaussian function, and the resulting transform is the Laplacian of a Gaussian smoothed function. Thus, zeros correspond to the inflection points of the original function. The contours of zero crossing indicate the edges in the pattern of the input function. To perform the wavelet transformation, first a suitable length scale value is chosen, which corresponds to the horizontal scale of the Gaussian function. Then an edge detection procedure is carried out to determine the pixel locations of significant differentials so that the feature of interest can be delineated from the background. However, in a noisy image, there is a tradeoff between (i) missing valid edges and (ii) over-detection: that is, detecting false edges induced by noise. Several thresholds can be explored via trial and error and a final choice selected based on observation and experience [31].

A typhoon eye is a quasi-circular zone of relatively weak winds that exists in the center of the swirling vortex of a typhoon. For the SAR imaging mechanism, a weak wind area with less surface roughness has relatively low radar backscatter compared to the adjacent eyewall. Thus, the typhoon eye can be delineated in a SAR image as the darker (less rough) area with relatively low radar backscatter, and its location pinpointed by using wavelet analysis. A typhoon eye has also the warmest temperatures at the upper cloud levels. So, in a IR image, the warmer area also can be delineated by using wavelet analysis, and its position may be regarded as the typhoons eye observed in the upper cloud field.

Once the typhoon eyes have been extracted from the images, the central location of each image is calculated as the geometric center from each delineated eye contour. The center distance between the central locations of the typhoon eyes detected using SAR and IR images is then estimated. Because the typhoon eyes observed using SAR and IR data are located at different heights, to estimate the center distance, the distortion from satellite incidence angle of MTSAT must be evaluated. In estimating distance, a spheroid coordinate was used for higher accuracy. First, the relative positions have been defined among MTSAT (S), the nadir point of MTSAT (SE), and the projected point on the surface (P) of the observed typhoon eye at the cloud top in spheroid coordinates. The altitude of MTSAT from the earth surface is approximately 35,800 km. The angle between PSE and PS can be expressed by a simple geometric equation as shown in [25]. Then, the distortion between the projected point and the nadir point of the typhoon eye at the cloud top from the satellite incident angle can be calculated. The height of the cloud top is estimated from the temperature difference of the cloud top with an average of 13 km. In summary, the estimated distortion distances of the eyes between the projected point and the nadir point at the cloud tops of all typhoons are found to be 4–5 km for MTSAT and 6–7 km for FY-2. However, the distortion of eye locations derived from two IR sensors are likely in the opposite sides of the nadir point of the typhoon eye since the typhoon eye is probably located in between MTSAT (140° E) and FY-2 (105° E) satellites.

Once the center distance between the central locations of the typhoon eyes detected using SAR and IR images has been estimated. Because IR imagery is taken every half an hour, the location of the eye in an IR image can then be linearly interpolated to the same time as that of an SAR image taken for comparison. The identical time locations of eyes of MTSAT (red dot), FY-2 (green dot), and SAR (blue dot) are plotted in the following figures. Therefore, the displacement of lower (SAR) and upper (IR) eye centers represents the vertical tilt of the typhoons eyewall shaft with increasing height.

3.4 Case Studies for RADARSAT SAR Images

Total of 6 cases of various typhoons in different years, locations, and condition have been examined for the comparison of typhoon eye between RADARSAT SAR images, MTSAT and FY-2 IR data collected at the same time. Also, the typhoon location from the archived best tracks of JTWC, CMA and JMA has been checked along with MODIS image for the calibration and validation.

3.4.1 Typhoon Usagi in August 2007

Typhoon Usagi with Category 3 intensity was headed-on towards Japan on August 1, 2007. Figures 3.3a–c show the RADARSAT-1 SAR image, MTSAT and FY-2 IR images of typhoon Usagi with the eye boundaries (red contours) delineated using wavelet analysis, respectively. In MTSAT and FY-2 IR images, the formation of a small eye, with warmer temperatures relative to the eyewall, and the expanded area of extremely cold cloud in bands around the eye indicate the increased intensity. The regional typhoon track map and the extracted typhoon eyes' center from SAR, FY-2, and MTSAT images when Typhoon Usagi was approaching Japan on August 1, 2007 are shown in Fig. 3.4, respectively. The green line, blue line, and black line are the best tracks of typhoon from JTWC, CMA, and JMA, respectively. In the track map, the blue box is the approximate boundary of SAR subscene, and the red dots indicate the dates from July 31 to August 3. The identical time locations of eyes from MTSAT (red dot), FY-2 (green dot), and SAR (blue dot) data are plotted in the

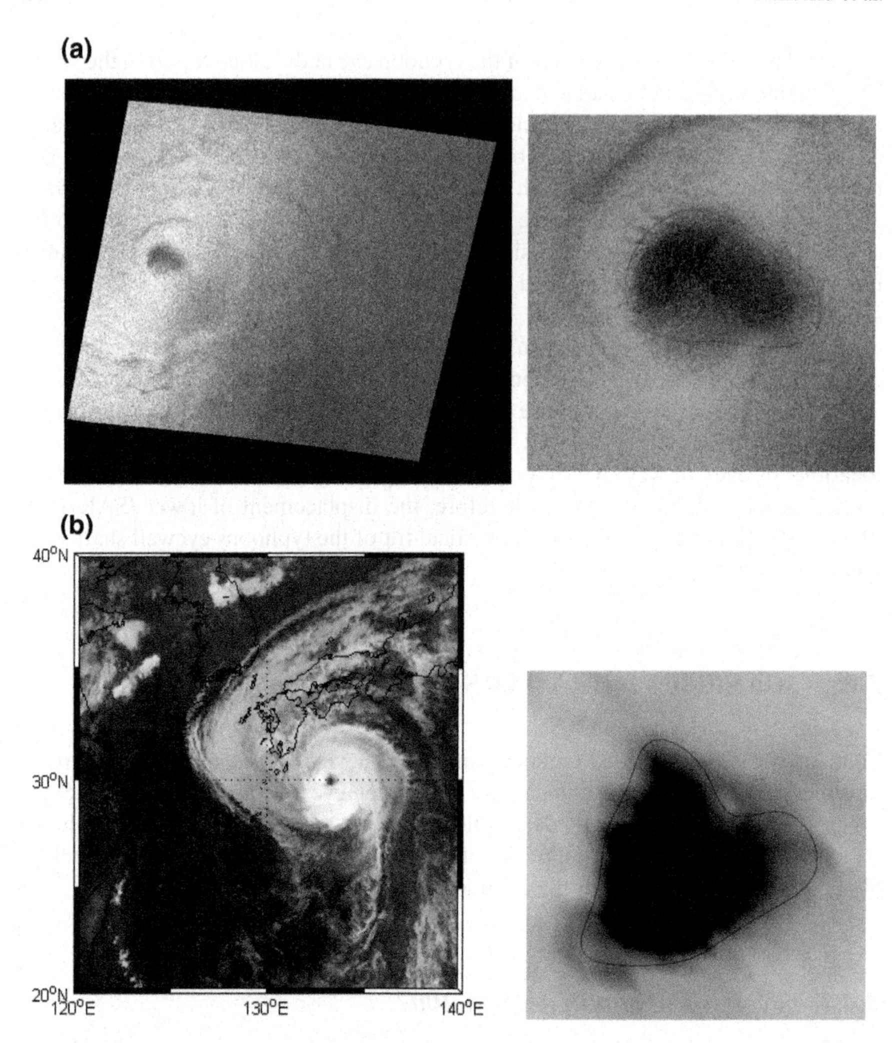

Fig. 3.3 Images and zoom-in subscenes of **a** RADARSAT SAR (*upper panel*) and **b** MTSAT/IR (*lower panel*) collected on August 1, 2007 when Typhoon Usagi headed-on to Japan with the typhoon eye boundary (*red contours*) delineated by wavelet transform

following figures. In addition, the eye tracks observed using SAR, FY-2 and MTSAT IR data are close to those of the JTWC, CMA, and JMA post-analysis. The center distance between typhoon eyes derived from SAR and MTSAT data is more than 10 km and is about 24 km between that from SAR and FY-2 data. At this stage, the typhoon intensity was quite steady based on the ocean surface pressure field. The pressure kept in 945 hPa, the maximum wind speed of typhoon was 35 km/h at this time. In this case, the typhoon eye derived from SAR image is close to the JTWC, CMA, and JMA tracks than that from MTSAT and FY-2 data. Notice that the typhoon

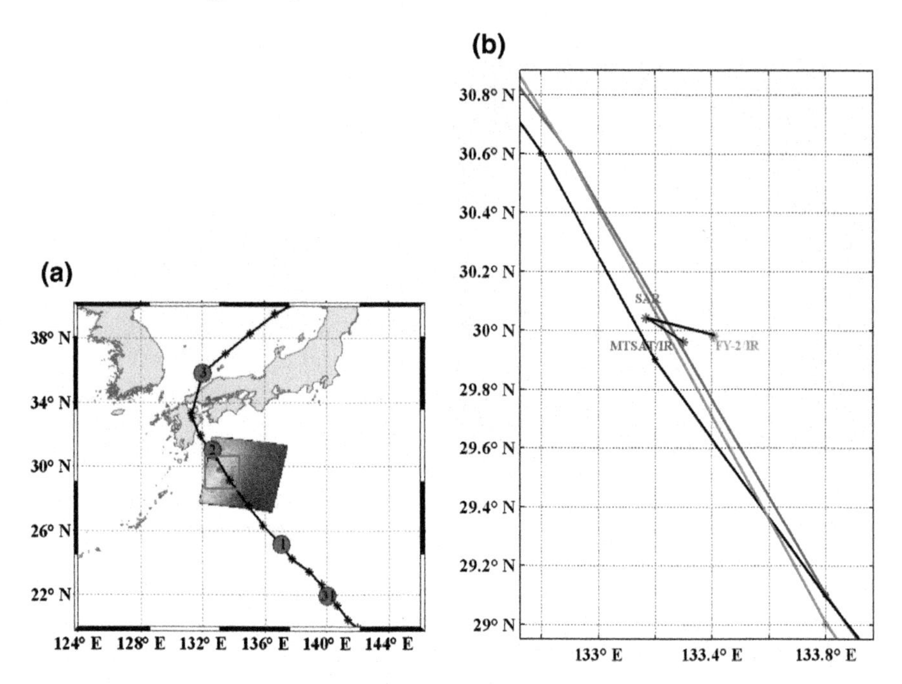

Fig. 3.4 **a** Regional typhoon track map, and **b** extracted typhoon eyes' center from SAR, FY-2, and MTSAT images for Typhoon Usagi on August 1, 2007. The *green line*, *blue line*, and *black line* are the best tracks of typhoon from JTWC, CMA, and JMA, respectively. The *blue box* in the track map is the approximate boundary of SAR subscene, and the red dots indicate the dates from July 31 to August 3. The identical time locations of eyes from MTSAT (*red dot*), FY-2 (*green dot*), and SAR (*blue dot*) data are plotted in the following figures

eye at the cloud level derived from MTSAT and FY-2 data is lagging behind that from SAR data near the ocean surface. This is probably due to the blocking effect caused by island and mountains in the wind field at the cloud level, however, the local wind field was not much affected near the ocean surface.

3.4.2 Typhoon Man-yi in July 2007

Typhoon Man-yi was in the open ocean on July11, 2007. Its intensity was increasing from 935 hPa at 1800 UTC into 930 hPa at 2400 UTC but the translation speed was decreasing from 31 to 25 km/h. On July 13, Typhoon Man-yi was turning and approaching Japan and its moving direction was changing from northwest to north-northeast. Figure 3.5 shows the RADARSAT SAR images on July 13, and the zoom-in subscene with the eye boundary (red contour) delineated by wavelet analysis. At this stage, the typhoon intensity was decreasing from 940 hPa at 0900 UTC into 945 hPa at 1200 UTC, and typhoon was also changing the direction to north approaching

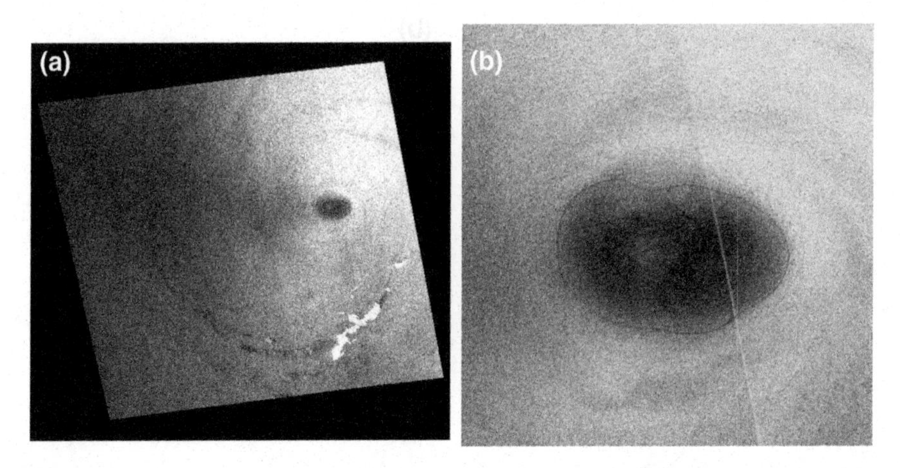

Fig. 3.5 **a** RADARSAT SAR Image and **b** zoom-in subscene collected on July 13, 2007 with the typhoon eye boundary (*red contours*) delineated using wavelet analysis for Typhoon Man-yi

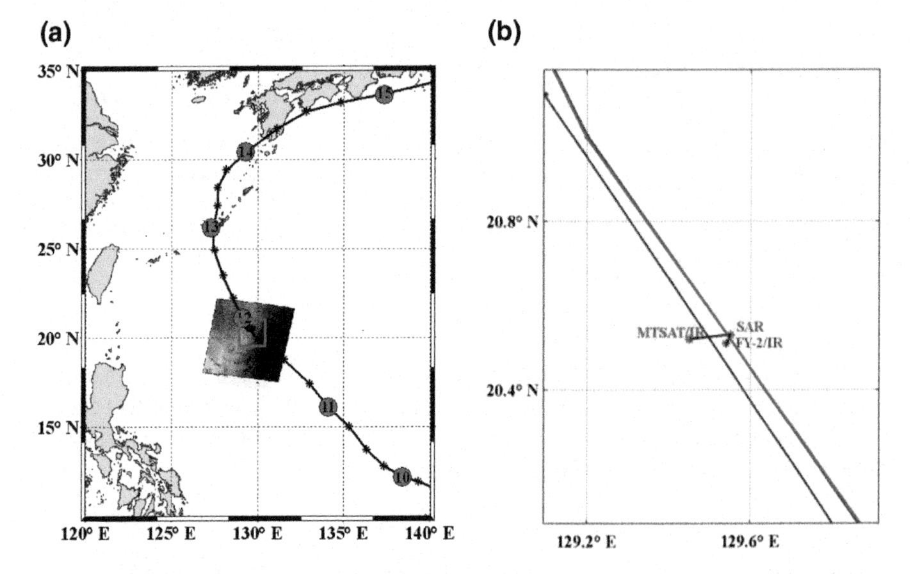

Fig. 3.6 **a** Regional typhoon track map from July 10–15, and **b** extracted typhoon eyes center from RADARSAT SAR, FY-2 and MTSAT IR images when Typhoon Man-yi turned and approached to Japan on July 11, 2007

Japan with a translation speed of 19 km/h. After Man-yi lashed the Okinawa island chain with heavy rain and fierce winds, the maximum wind speed near the center was decreasing from 48 to 43 m/s. The typhoon regional track map with SAR image overlaid is shown in Figs. 3.6a and 3.7a for July 11 and 13, respectively. Figures 3.6b and 3.7b show the extracted typhoon eyes' center from SAR, FY-2 and MTSAT images for Typhoon Man-yi turned and staggered on July 11 and approached Japan

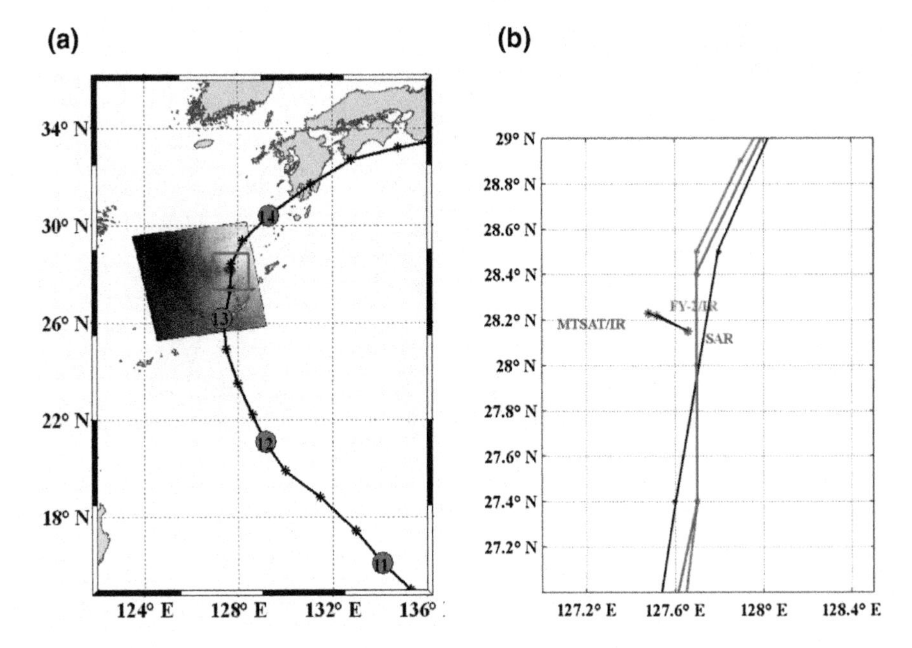

Fig. 3.7 **a** Regional typhoon track map from July 11–14, and **b** extracted typhoon eyes center from SAR and IR images when for Typhoon Man-yi on July 13, 2007

on July 13, respectively. Notably, the eyes derived from the MTSAT data are quite far from the eye derived from SAR data, and the separation has a center distance of approximately 10 and 22 km for July 11 and 13, respectively, almost perpendicular to the track direction. However, the eye derived from FY-2 is quite close to the eye derived from SAR on July 11, and close to eye derived from MTSAT on July 13. Also, the best typhoon tracks from the JTWC (green line), CMA (blue line) and JMA (black line) post-analysis are overlaid in Figs. 3.6b and 3.7b for reference. In this case, CMA tracks follow JMA track almost exactly on 11th but follow JWTC track on 13th, and the eyes derived from SAR images agree with these tracks well for both July 11 and 13.

3.4.3 *Typhoon Kirogi in October 2005*

In this special case, Typhoon Kirogi was turning and staggering on October 14 and then speeding up on October 15. Figure 3.8 shows the RADARSAT SAR image on October 14, and the zoom-in subscene with the eye boundary (red contour) delineated by wavelet analysis. The typhoon regional track map from October 9–18 with SAR image and subscene box on October 14, and the centers of typhoon eye derived from SAR image, FY-2 and MTSAT IR data are shown in Fig. 3.9. Both typhoon eyes

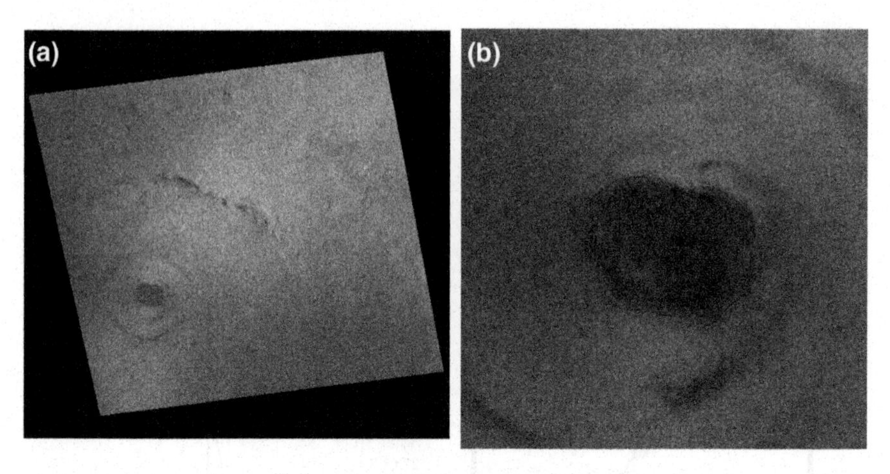

Fig. 3.8 **a** RADARSAT SAR Image and **b** zoom-in subscene collected on October 14, 2005 for Typhoon Kirogi with the eye boundaries delineated using wavelet analysis

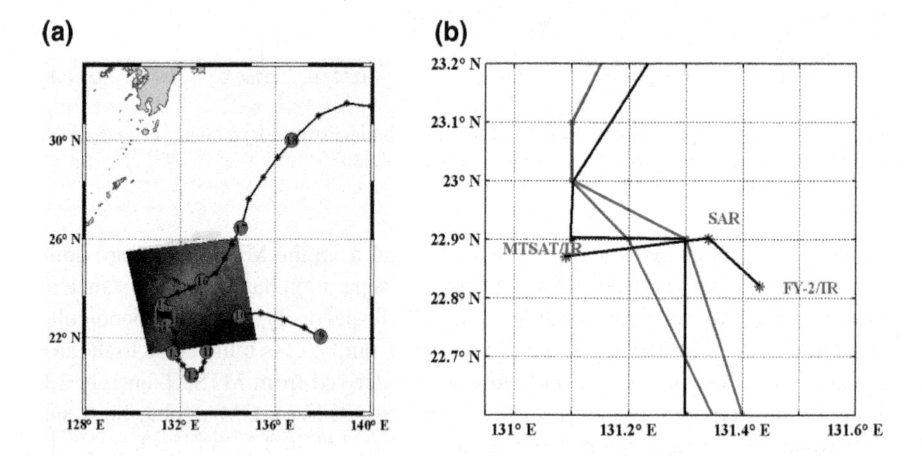

Fig. 3.9 **a** Regional typhoon track map from October 9–18, and **b** extracted typhoon eyes center from SAR, and IR images when Typhoon Kirogi turned and staggered on October 14, 2005

derived from SAR and MTSAT data follow the track from JMA post-analysis really well, but not so well from the JTWC and CMA tracks on the 14th. At this stage from October 14–15, the typhoon intensity was quite steady, the pressure remained at 940 hPa, and the maximum wind speed near the center was 43 m/s. In addition, the eye tracks observed using SAR and MTSAT data are identical to those of the JTWC (green line) and JMA (black line) post-analysis, while the eye center derived from FY-2 data is not so close to JTWC and JMA tracks but is close to CMA track. When the typhoon was turning and staggering on 14th, the SAR eye may be dragging behind the IR eye derived from MTSAT data with a distance of 20 km approximately. At this time, the translation speed of the typhoon was 6 km/h. However, on October

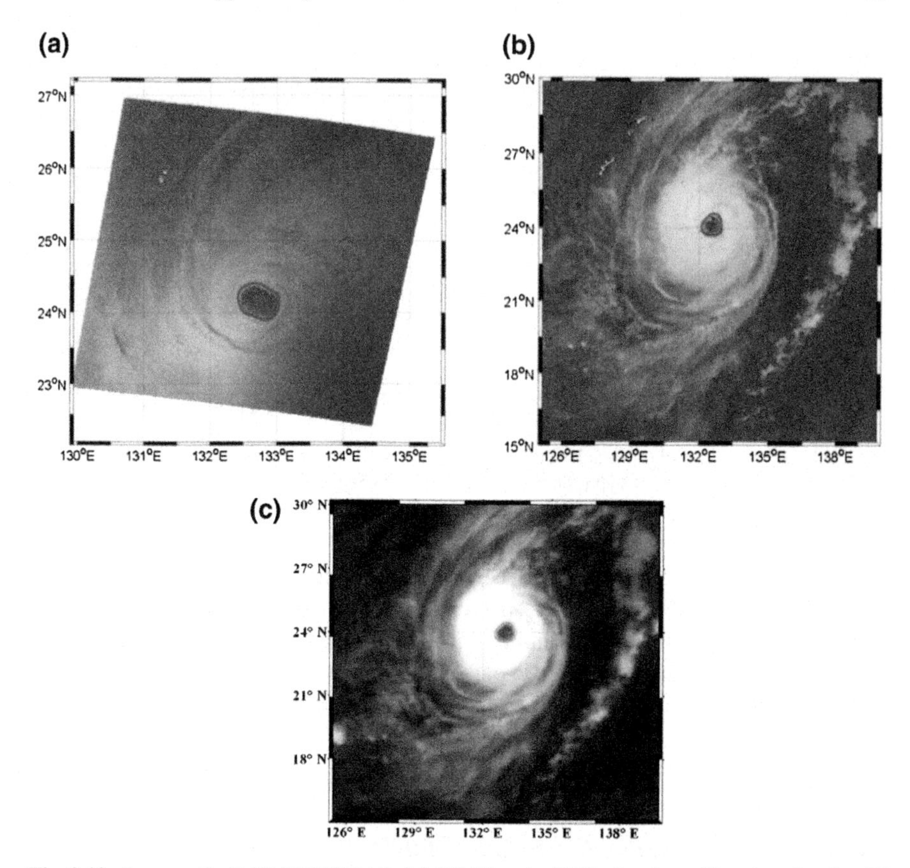

Fig. 3.10 Images of **a** RADARSAT SAR, **b** MTSAT, and **c** FY-2 of typhoon Kirogi on October 15 with the eye boundaries delineated using wavelet analysis

15, when the typhoon was increasing the translation speed to 9 km/h, the SAR eye was revealed to be suddenly ahead of the IR eye derived from MTSAT by 16 km. Figure 3.10 shows the images of RADARSAT SAR, MTSAT, and FY-2 of Typhoon Kirogi on October 15 with the eye boundaries delineated using wavelet analysis, respectively. The typhoon regional track map from October 9–18, and the derived centers of typhoon eye from SAR, FY-2, and MTSAT images on October 15 with all tracks from post-analysis are shown in Fig. 3.11. It is possible that the eye near the ocean surface was dragging behind on October 14, then took a short cut, bypassed the turning, and moved directly to the new track ahead on 15th. As mentioned before, two IR sensors are in the opposite sides of the nadir point of the typhoon eye. So, the eye locations derived from FY-2 and MTSAT data are spread to the east and west sides of that derived from SAR image with a distance of 26 and 16 km, respectively. This large separation distance may be partially caused by the distortion error due to its position at 105° East Longitude for FY-2 and 140° East Longitude for MTSAT, far from the typhoons location around 133°. If the projected distortion is corrected

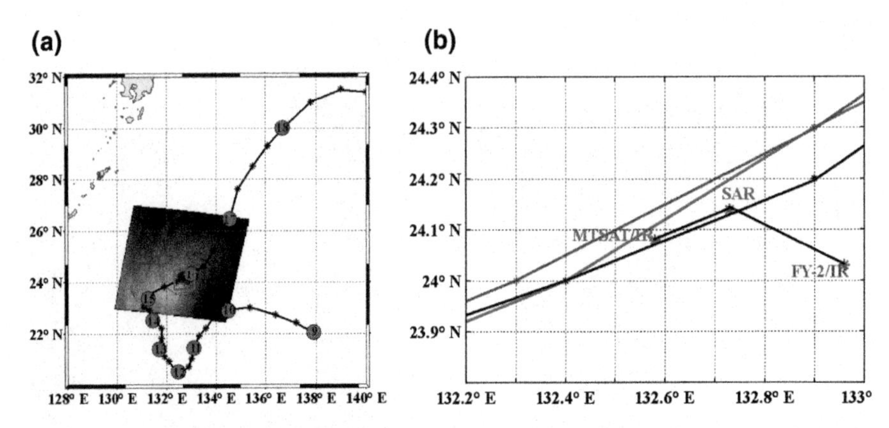

Fig. 3.11 **a** Regional typhoon track map from October 9–18, and **b** extracted typhoon eyes center from SAR and IR images for Typhoon Kirogi on October 15, 2005

for this case, then the distance between eye centers derived from SAR and FY-2, MTSAT data were 19 and 6 km, respectively. This phenomenon also happened on October 14 with a distance of 11 and 20 km between eye locations derived from SAR and that from FY-2 and MTSAT IR data. However, with distortion correction that distance between eye locations derived from SAR and FY-2, MTSAT were 7 and 15 km, respectively.

3.4.4 Typhoon Krosa in October 2007

Typhoon Krosa continued to rapidly intensify on October 4 before leveling off as a Category-4 typhoon. Fluctuations in intensity soon followed as Krosa approached Taiwan, as the JMA upgraded it to 194 km/h and the JTWC to a super typhoon early on October 5. It slowly weakened afterward before making landfall in northeastern Taiwan on October 6. The pressure of Typhoon Krosa was decreasing from 935 hPa on October 4 to 925 hPa on October 5, and the wind speed was increasing from 46 to 51 m/s. At this stage, the translation speed was decreasing from 17 to 9 km/h, and the moving direction was changing from northwest to north-northwest; its translation speed was then increasing slightly to 12 km/h. Its regional track map from October 2–9, and the eyes center using SAR, FY-2 and MTSAT images on October 4 are shown in Fig. 3.12. The distance between the eyes derived from SAR data and FY-2 and MTSAT IR images are approximately 6 and 15 km, respectively. The eye derived from FY-2 image was closer to the JMA track, but that which derived from the MTSAT image was closer to the JTWC track. However, the eye derived from SAR data was closer to CMA track and right in between all tracks. The spreading of eyes derived from MTSAT and FY-2 IR data on opposite sides of that from SAR image may be also partially caused by the distortion error. Again in this case, two

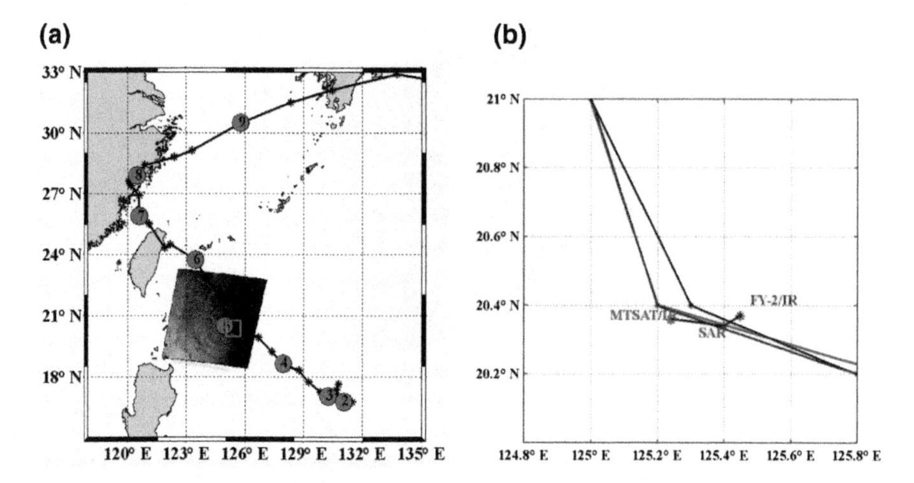

Fig. 3.12 **a** Regional typhoon track map from October 2–9, and **b** extracted typhoon eyes center from SAR, and IR images when Typhoon Krosa being in the western Pacific Ocean on October 4, 2007

IR sensors are located in the opposite sides of the nadir point of the typhoon eye (125°). Thus, with distortion correction that distance between eye locations derived from SAR and FY-2, MTSAT data were 1 and 11 km, respectively.

3.5 Case Studies for ENVISAT SAR Images

Total of 3 cases of various typhoons with clearly identified eye in different years, locations, and condition have been examined to compare the typhoon eyes separation distance between ENVISAT SAR images, MTSAT and FY-2 IR data are described as follows. The typhoon location from the post-analysis for the best tracks of JTWC, CMA and JMA has also been checked for the calibration and comparison.

3.5.1 Typhoon Megi in October 2010

Typhoon Megi formed as a tropical depression over the western Pacific Ocean on October 13, 2010. It quickly strengthened to a named storm, and three days after forming, had grown to a super typhoon. In general, rainfall roughly matches the storm track, especially west and northwest of the Philippines. The heavy rain Super-typhoon Megi unleashed as it tracked west across the Pacific between October 13 and October 23, 2010 with the heaviest rainfallmore than 600 mm. Away from the storm track, areas of heavy rainfall appear east of Taiwan, where torrential rains led

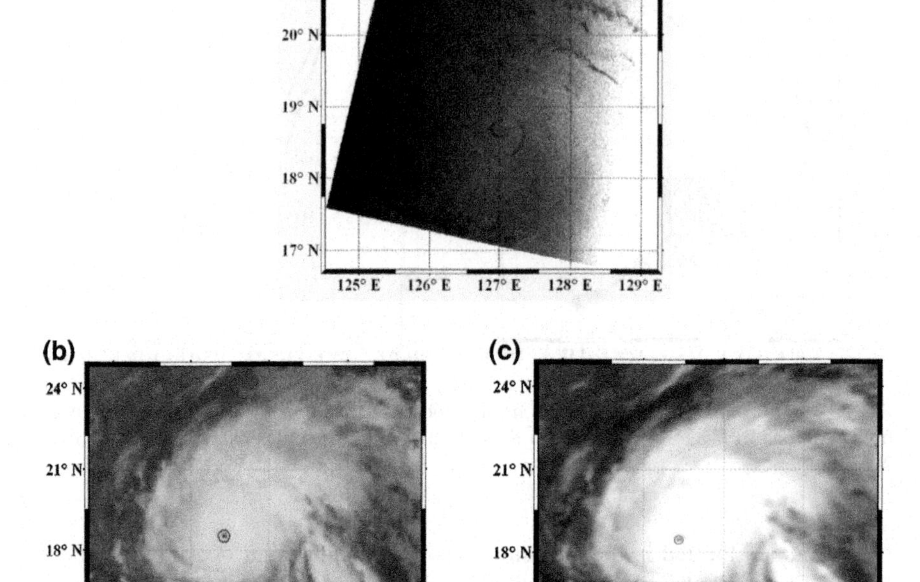

Fig. 3.13 **a** ENVISAT-1 SAR image, **b** MTSAT and **c** FY-2 IR images of typhoon Megi on October 17, 2010 with the eye boundaries delineated using wavelet analysis

to deadly landslides. As the storm track indicates, Megi reached its greatest intensity immediately east of the Philippines. The storm weakened slightly after October 17, but remained powerful across the northern Philippines.

Figures 3.13a–c show the ENVISAT-1 SAR image, MTSAT and FY-2 IR images of typhoon Megi on October 17, 2010 with the eye boundaries delineated using wavelet analysis, respectively. In MTSAT and FY-2 IR images, the formation of a clearly visible eye and the expanded area of extremely cold cloud with rain bands around the eye indicate the increased intensity (Fig. 3.2d). The regional typhoon track map from October 15–23, and the extracted typhoon eyes' center from SAR data, FY-2 and MTSAT IR images when Typhoon Megi was approaching Philippines on October 17, 2007 are shown in Fig. 3.14. The distance between eyes center derived from SAR and IR images is around 23–24 km since both eyes derived from FY-2 and MTSAT agree with each other. Notice that the eye derived from SAR data is very close to all collided tracks of JTWC, CMA, and JMA than that from IR data.

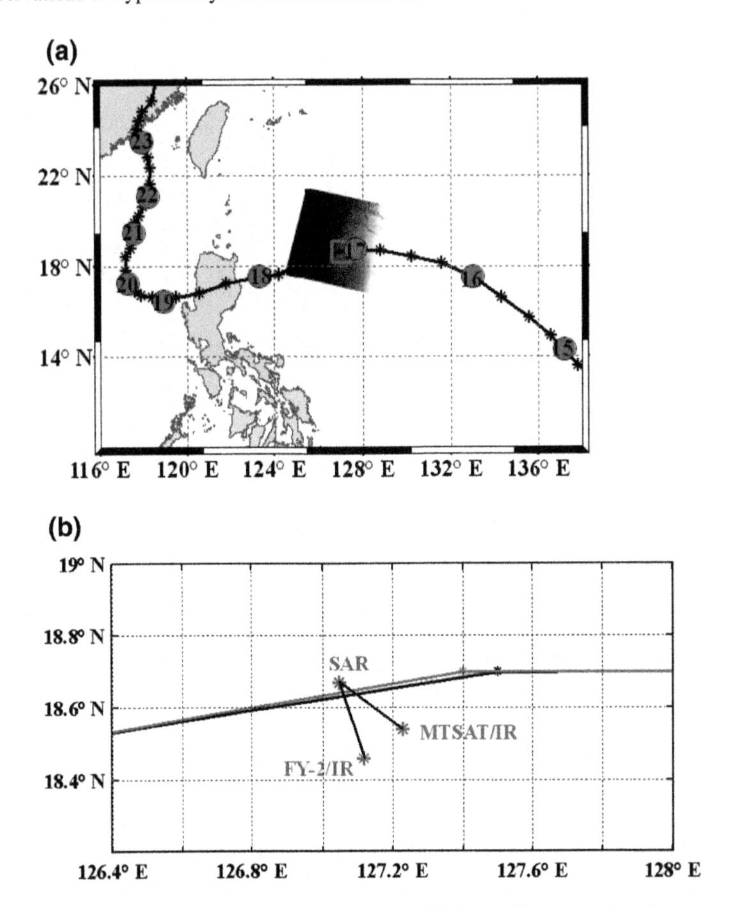

Fig. 3.14 **a** Regional typhoon track map from October 15–23, and **b** extracted typhoon eyes' center from SAR data, FY-2 and MTSAT IR images when Typhoon Megi was approaching Philippines on October 17, 2010

3.5.2 Typhoon Talim in Augusr 2005

Typhoon Talim was a building storm in the western Pacific several hundred kilometers south of Japan on August 29, 2005. As of August 30, it started to gather up stronger winds and to strike Taiwan on September 2, and then continued across the Taiwan Strait to make landfall again in Fujian Province. The typhoon reaches Category-3 strength with maximum sustained winds estimated at 121 mph by the time it strikes Taiwan. Figure 3.15 shows the ENVISAT SAR image with the eye boundaries delineated using wavelet analysis for Typhoon Talim on August 30, 2005. Notice that the eyewall was not very clearly identified as a perfect circle since it was still building its strength as a Category-2 storm on August 30. The regional typhoon track map from August 29 to September 1, and the extracted typhoon eyes' center from SAR

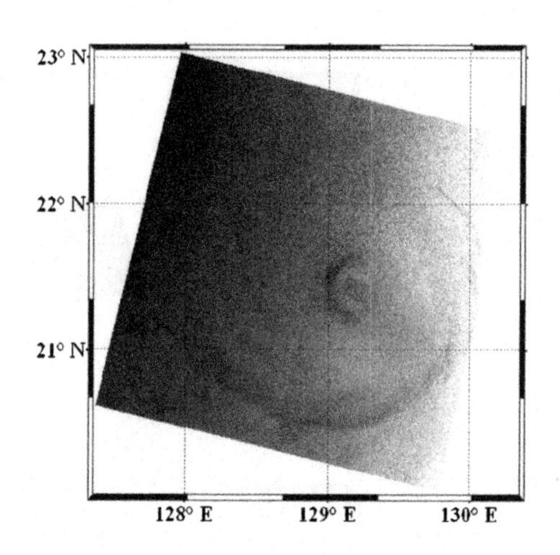

Fig. 3.15 ENVISAT SAR image with the eye boundaries delineated using wavelet analysis for Typhoon Talim on August 30, 2005

data, FY-2 and MTSAT IR images when Typhoon Talim was approaching Taiwan on August 30 are shown in Fig. 3.16. Notice that the eye center derived from SAR data is way ahead of that from IR images with large separation distances of 22 km for MTSAT and 37 km for FY-2, probably caused by its relatively weak storm intensity during formation with poorly defined eye location. These large distances between eyes may be also partially caused by the distortion error, especially for FY-2, since its position at 105° East Longitude is far from the typhoon eyes projected location around 129°. The highly tilted vertical wind shear may be considerably more complex than expected. Thus, with distortion correction that distance between eye locations derived from SAR and FY-2 data were 30 km, respectively.

3.5.3 Typhoon Khanun in September 2005

As Typhoon Khanun moved in towards China on September 11, it was moving northwest at a speed of 25 km/h with winds gusting as high as 40 m/s. It was also bringing heavy rain and made landfall near Shanghai in eastern China on September 11 at 06:00 UTC. Then, Typhoon Khanun dissipated the next day, on September 12. China's eastern coast and Japans southern islands suffered strong winds, high waves and heavy rains as the storm passed. Figure 3.17 shows the ENVISAT SAR image with the eye boundaries delineated using wavelet analysis for Typhoon Khanun on September 11, 2005. Notice that the eyewall was clearly identifiable as a perfect circle since it was moving with full strength of a Category 4 intensity with sustained winds of 59 m/s. The regional typhoon track map from September 7–12, and the extracted typhoon eyes' center from SAR data, FY-2 and MTSAT IR images when

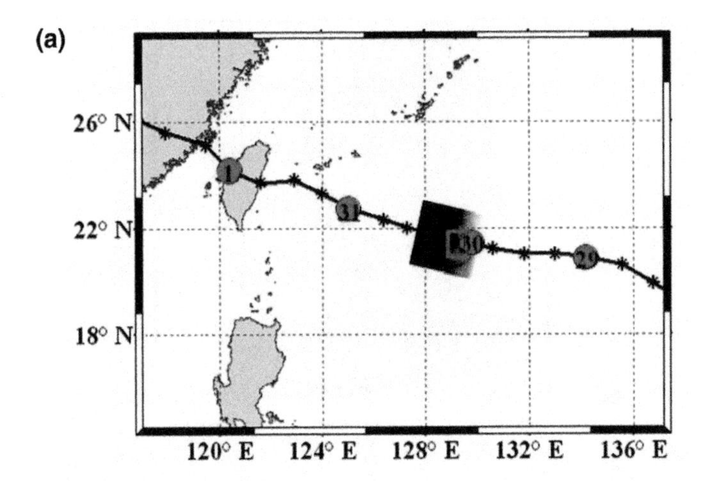

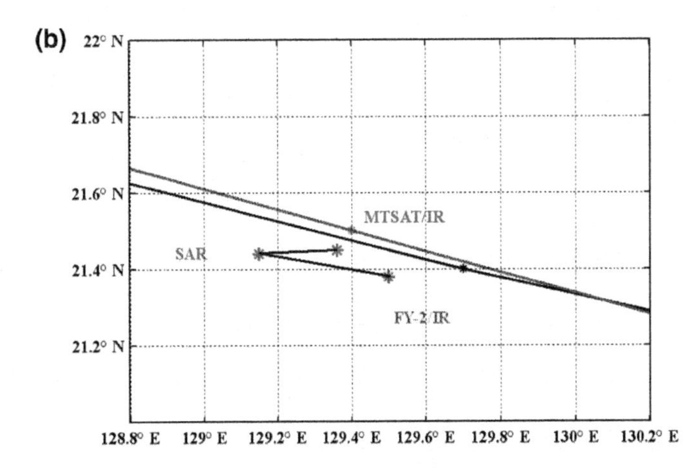

Fig. 3.16 **a** Regional typhoon track map from August 29 to September 1, and **b** extracted typhoon eyes' center from ENVISAT SAR data, FY-2 (*black dot*) and MTSAT IR images when Typhoon Talim was approaching Taiwan on August 30, 2005

Typhoon Khanun was passing through north of Taiwan on September 11 are shown in Fig. 3.18. Notice that the eye center derived from SAR is dragging behind the eyes center derived from IR images with separation distances of 26 km for MTSAT and 13 km for FY-2 as expected in the open ocean, but is further away from all post-analysis tracks than that derived from IR images.

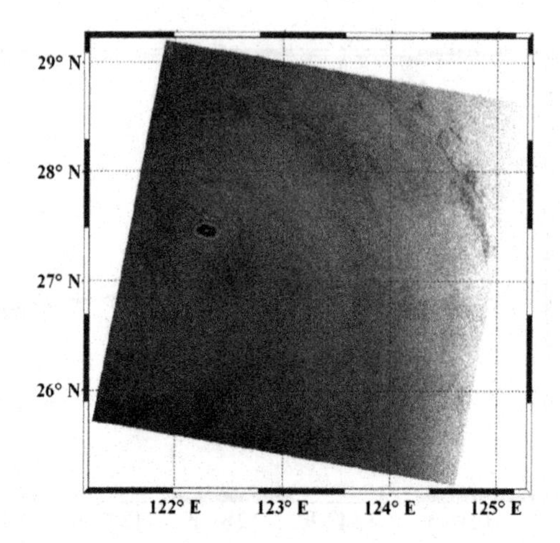

Fig. 3.17 ENVISAT SAR image with the eye boundaries delineated using wavelet analysis for Typhoon Khanun on September 11, 2005

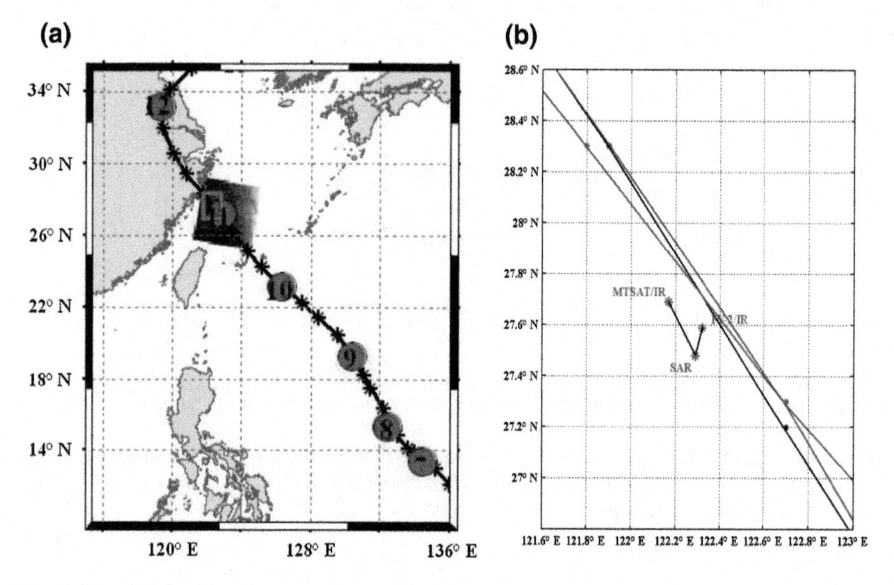

Fig. 3.18 **a** Regional typhoon track map from September 7–12, and **b** extracted typhoon eyes' center from SAR data and IR images when Typhoon Khanun was passing through north of Taiwan on September 11

3.6 Issues and Discussion

The definition of typhoon eye derived from SAR imagery is the smoother area near storm center with darker shading, representing lesser backscattering from radar relative to that underneath the adjacent eyewall. For IR data, the typhoon eye is the warmer area with higher brightness temperature. The general consensus is that a sensor for typhoon intensity studies should be one which measures the wind field directly. Lacking direct measurements of surface wind, the next best alternative is the high-resolution SAR imagery based on the surface roughness inferred from radar backscattering of capillary waves on the ocean surface. Roughness increases with wind intensity, and the retrieval of wind speed requires a suitable geophysical model function. In a heavily precipitating area such as the hurricane eyewall, SAR images may contain artifacts owing to attenuation and surface splash; nevertheless, these images are well-suited to studies of the hurricane eye [32] which is relatively calm and free of such effects, as demonstrated here.

Based on these case studies and observation, the tilted inner-core structure and associated vertical wind shear may be much more complex than expected. All these case studies of typhoon eye and comparisons of eye center distances using FY-2 and MTSAT IR images with RADARSAT SAR data are summarized in Table 3.1 and with ENVISAT SAR data in Table 3.2 for reference. It is found the distances between the center locations of these typhoon eyes are quite significant from 2 to 26 km except a special case of 37 km for Typhoon Talim with relatively weak intensity. But not all typhoons have large horizontal distance between eyes on the cloud top and on the ocean surface; such as Typhoon Kajiki case on October 20, 2007 has only a 9 km separation which is in the uncertainty and error range [25]. It is found that the typhoon eye derived from SAR always follow the typhoon tracks from the JTWC and JMA post-analysis really well. It is conceivable that these best tracks have used

Table 3.1 Summary of all case studies of typhoon "eye" and center distance using RADARSAT SAR data, FY-2 and MTSAT IR images

Typhoon name	Date	Time	RADARSAT SAR		FY-2/IR (4 km) MTSAT/IR (4 km)		Distance btw Eyes (km)	Direction (Degree from N)
	dd/mm/yyyy		Lat (°N)	Lon (°E)	Lat (°N)	Lon (°E)		
Kirogi	14/10/2005	9:12	22.9	131.34	22.82	131.43	13/7*	131
					22.87	131.09	20/15*	264
Kirogi	15/10/2005	21:04	24.14	132.73	24.03	132.96	26/20*	115
					24.08	132.58	16/11*	248
Man_yi	11/07/2007	21:13	20.53	129.55	20.51	129.54	2	207
					20.52	129.45	10	265
Man_yi	13/07/2007	9:33	28.15	127.66	28.22	127.52	16	298
					128.23	127.48	122	1295
Usagi	01/08/2007	20:57	30.04	133.17	29.98	133.41	124	103
					29.96	1133.3	10	122
Krosa	04/10/2007	21:33	20.34	125.4	20.37	125.45	6/1*	59
					20.36	125.24	15/9*	277

*With distortion correction

Table 3.2 Summary of all case studies of typhoon "eye" and center distance using ENVISAT SAR data, FY-2 and MTSAT IR images

Typhoon name	Date	Time	RADARSAT SAR		FY-2/IR (4 km) / MTSAT/IR (4 km)		Distance btw Eyes (km)	Direction (Degree from N)
	dd/mm/yyyy		Lat (°N)	Lon (°E)	Lat (°N)	Lon (°E)		
Talim	30/08/2005	1:23	21.44	129.15	21.38	129.5	37/30*	100
					21.45	129.36	22	88
Khanun	11/09/2005	1:45	27.48	122.29	27.59	122.32	13	18
					27.69	122.17	26	329
Megi	17/10/2010	1:24	18.67	127.05	18.46	127.12	24	162
					18.54	127.23	23	126

*With distortion correction

some other satellite data (e.g. MODIS, AMSR, TRMM, . . .) and even ship/buoy data over the ocean for post-analysis. As demonstrated here that SAR can be a powerful tool for typhoon tracking and prediction especially near the ocean surface.

The accuracy of this wavelet typhoon eye extraction method is only limited by the persistence of the features, the spatial resolution, and navigational accuracy of satellite data. The geometric accuracy of the MTSAT, FY-2, and SAR data are less than 1.97 km, 2 km, and 100 m, respectively [33–35]. Because the remapped SAR image has pixel size of 50 m, it has no serious accuracy issue. However, the MTSAT and FY-2 IR image has spatial resolution of 4 km; the derived eye may have geo-location error of approximately 5 km due to sensor resolution [25]. Because the most of eye contours are relatively axis-symmetric, the central location of eye will not be shifted significantly using the wavelet detection method and the threshold technique to delineate eye contour. The distortion distance of these cases caused by the satellite incidence angle of MTSAT and FY-2 is approximately 47 km. For an average eye size of 50 km in diameter, the maximum error is estimated to be around 8 km (2 pixels). As the average size of the eye is usually much larger than the height of the typhoon, the slight shift of the typhoon eyewall shaft would be hard to detect by human eyes. Therefore, only high-resolution instrument and objective analysis could offer the true observations of typhoon eye.

However, the accuracy of eye location from MTSAT IR data may still be a concern as compared with that derived from MODIS/IR data with 10–16 km eye center separation [25]. Also, the typhoon eye separation between FY-2 and MTSAT at the same time brings the concern on the navigation issues, especially for the case with distance between eyes of 37 km. Determining the distortion distance of typhoon eyes between the projected point and the nadir point at the cloud top from the satellite incident angle using MTSAT and FY-2 images has helped to justify the resolution and uncertainty of this study. More detailed comparison study between MTSAT, FY-2, and MODIS images is warranted to calibrate the accuracy of MTSAT and FY-2 data.

The large horizontal distance between typhoon eyes on the cloud top and on the ocean surface of these case studies are from 10–26 km. This implies that the eyewall shaft may be highly tilted and the vertical wind shear profile is much more complex than generally expected. Compared to the eye diameter of 50 km, the horizontal distance of typhoon eyes at various heights is insignificant; but compared to the average height of clouds 13 km, the horizontal distance of 1026 km seems significantly large. For a small island such as Taiwan, 22 km found in the study of Typhoon Talim passed through Taiwan in August 2005 could be the distance between two cities. As shown in the case of Typhoon Usagi, the typhoon eye derived from IR data is lagging behind that derived from SAR data by 17 km in average probably due to island/mountain blocking effect. Then in this case, the typhoon could land on the coast almost half an hour earlier than the prediction based on the IR data alone.

Definitely, the high-resolution observation by SAR (100 m) could be complementary to the IR frequently observations (every 30 min) from meteorological satellite. Although lacking the temporal continuity of geostationary imagery, a solitary SAR image may afford some advance warning that the MTSAT/FY-2 eye location requires correction to account for position errors of order 10 km. As SAR data become increasingly abundant with computer working efficiency, they should be used when combining multiple sensors, such as MTSAT, FY-2, and MODIS, for typhoon tracking in near real-time applications. High-resolution SAR could definitely facilitate operating MTSAT for higher accuracy in prediction and model assimilation of typhoon tracking over the open ocean, such as storm surges along coastal communities where typhoons have most serious impact. In addition, highly tilted and twisted vertical wind shear profiles must be considered in the mesoscale typhoon model for more accurate prediction and more effective operation. Based on these case studies and observations, results from the ocean upwelling might be significant, especially when a typhoon is turning and staggering such as in the case study of typhoon Kirogi. Therefore, understanding the behavior of typhoons more clearly is critical for predicting more accurately a typhoon's track and intensity.

References

1. Hsu, M.K., C.T. Wang, A.K. Liu, and K.S. Chen. 2010. *Satellite Remote Sensing of Southeast Asia-SAR Atlas*, 133. Taipei: Tingmao Publish Co.
2. Weatherford, C.L., and W.M. Gray. 1988. Typhoon structure as revealed by aircraft reconnaissance. Part I: Data analysis and climatology. *Monthly Weather Review* 116 (5): 1032–1043.
3. Weatherford, C.L., and W.M. Gray. 1988. Typhoon structure as revealed by aircraft reconnaissance. Part II: Structural variability. *Monthly Weather Review* 116 (5): 1044–1056.
4. Willoughby, H.E. 1998. Tropical cyclone eye thermodynamics. *Monthly Weather Review* 126 (12): 3053–3067.
5. Lin, I., W.T. Liu, C.C. Wu, G.T.F. Wong, C. Hu, Z. Chen, W.D. Liang, Y. Yang, and K.K. Liu. 2003. New evidence for enhanced ocean primary production triggered by tropical cyclone. *Geophysical Research Letters*, 30 (13).

6. Walker, N.D., R.R. Leben, and S. Balasubramanian. 2005. Hurricane-forced upwelling and Chlorophyll-a enhancement within cold-core cyclones in the Gulf of Mexico. *Geophysical Research Letters*, 32 (18).
7. Bell, M.M., and M.T. Montgomery. 2008. Observed structure, evolution and potential intensity of category five Hurricane Isabel (2003) from 12–14 September. *Monthly Weather Review* 136: 2023–2046.
8. Hasler, A.F., K. Palaniappan, C. Kambhammetu, P. Black, et al. 1998. High-resolution wind fields within the inner core and eye of a mature tropical cyclone from GOES 1-min images. *Bulletin of the American Meteorological Society* 79 (11): 2483.
9. Liu, C.C., G.R. Liu, T.H. Lin, and C.C. Chao. 2010. Accumulated rainfall forecast of Typhoon Morakot (2009) in Taiwan using satellite data. *Journal of the Meteorological Society of Japan* 88 (5): 785–798.
10. Wu, C.C., K.H. Chou, H.J. Cheng, and Y. Wang. 2003. Eyewall contraction, breakdown and reformation in a landfalling typhoon. *Geophysical Research Letters*, 30 (17).
11. Wu, S.Y., and A.K. Liu. 2003. Towards an automated ocean feature detection, extraction and classification scheme for SAR imagery. *International Journal of Remote Sensing* 24 (5): 935–951.
12. Zhu, T., D.L. Zhang, and F. Weng. 2004. Numerical simulation of Hurricane Bonnie (1998). Part I: Eyewall evolution and intensity changes. *Monthly Weather Review* 132 (1): 225–241.
13. Katsaros, K.B., P.W. Vachon, W.T. Liu, and P.G. Black. 2002. Microwave remote sensing of tropical cyclones from space. *Journal of Oceanography* 58 (1): 137–151.
14. Friedman, K., and X. Li. 2000. Storm patterns over the ocean with wide swath SAR. *Johns Hopkins University APL Technical Digest* 21: 80–85.
15. Liu, K.S., and J.C.L. Chan. 1999. Size of tropical cyclones as inferred from ERS-1 and ERS-2 data. *Monthly Weather Review* 127 (12): 2992–3001.
16. Mourad, P.D. 1999. Footprints of atmospheric phenomena in synthetic aperture radar images of the ocean surface: A review. In *Air-Sea Exchange: Physics, Chemistry and Dynamics*, ed. G.L. Geernaert, 269–290. Berlin: Springer.
17. Sikora, T.D., K.S. Friedman, W.G. Pichel, and P. Clemente-Colon. 2000. Synthetic aperture radar as a tool for investigating polar mesoscale cyclones. *Weather and Forecasting* 15 (6): 745–758.
18. Hilton, R.G., A. Galy, N. Hovius, M.C. Chen, M.J. Horng, and H. Chen. 2008. Tropical-cyclone-driven erosion of the terrestrial biosphere from mountains. *Nature Geoscience* 1 (11): 759–762.
19. Friedman, K.S., P.W. Vachon, and K. Katsaros. 2004. Mesoscale storm systems. In *Synthetic Aperture Radar Marine User's Manual*, 331–340. US Department of Commerce, National Oceanic and Atmospheric Administration, National Environmental Satellite, Data, and Information Serve, Office of Research and Applications.
20. Elsner, J.B., J.P. Kossin, and T.H. Jagger. 2008. The increasing intensity of the strongest tropical cyclones. *Nature* 455 (7209): 92–95.
21. Jordan, C.L. 1952. On the low-level structure of the typhoon eye. *Journal of Meteorology* 9 (4): 285–290.
22. Shea, D.J., and W.M. Gray. 1973. The hurricane's inner core region. Part I. Symmetric and asymmetric structure. *Journal of the Atmospheric Sciences* 30 (8): 1544–1564.
23. Brueske, K.F., and C.S. Velden. 2003. Satellite-based tropical cyclone intensity estimation using the NOAA-KLM series Advanced Microwave Sounding Unit (AMSU). *Monthly Weather Review* 131 (4): 687–697.
24. Babin, S.M., J.A. Carton, T.D. Dickey, and J.D. Wiggert. 2004. Satellite evidence of hurricane-induced phytoplankton blooms in an oceanic desert. *Journal of Geophysical Research: Oceans*, 109(C3).
25. Cheng, Y.H., S.J. Huang, A.K. Liu, C.R. Ho, and N.J. Kuo. 2012. Observation of typhoon eyes on the sea surface using multi-sensors. *Remote Sensing of Environment* 123: 434–442.
26. He, M.X., S.Y. He, Y.F. Wang, Q. Yang, J.W. Tang, and C.M. Hu. 2008. Chinese Spaceborne Ocean Observing Systems and Onboard Sensors (1988–2025). EC DRAGONESS Project WP2 First Annual Report.

27. Vachon, P.W., P. Adlakha, H. Edel, M. Henschel, B. Ramsay, D. Flett, M. Rey, G. Staples, and S. Thomas. 2000. Canadian progress toward marine and coastal applications of synthetic aperture radar. *Johns Hopkins APL Technical Digest* 21 (1): 33–40.
28. Du, Y., and P.W. Vachon. 2003. Characterization of hurricane eyes in RADARSAT-1 images with wavelet analysis. *Canadian Journal of Remote Sensing* 29 (4): 491–498.
29. Houze Jr., R.A. 2010. Clouds in tropical cyclones. *Monthly Weather Review* 138 (2): 293–344.
30. Liu, A.K., C.Y. Peng, and S.S. Chang. 1997. Wavelet analysis of satellite images for coastal watch. *IEEE Journal of Oceanic Engineering* 22 (1): 9–17.
31. Liu, A.K., S.Y. Wu, and Y. Zhao. 2003. Wavelet analysis of satellite images in ocean applications. In *Frontiers of Remote Sensing Information Processing*, 141–162.
32. Dunkerton, T.J., and W.A. Redmond. 2010. Microwave imagery and in situ validation of eye mesovortex structure in Hurricane Katrina (2005) at peak intensity. In *Proceedings of AMS 29th Conference on Hurricanes and Tropical Meteorology, Tucson, AZ, USA*, vol. 10.
33. Horstmann, J., W. Koch, S. Lehner, and R. Tonboe. 2000. Wind retrieval over the ocean using synthetic aperture radar with C-band HH polarization. *IEEE Transactions on Geoscience and Remote Sensing* 38 (5): 2122–2131.
34. Takeuchi, W., and Y. Yasuoka. 2007. Precise geometric correction of MTSAT imagery. In *Proceedings of the ACRS*.
35. Wolfe, R.E., M. Nishihama, A.J. Fleig, J.A. Kuyper, D.P. Roy, J.C. Storey, and F.S. Patt. 2002. Achieving sub-pixel geolocation accuracy in support of MODIS land science. *Remote Sensing of Environment* 83 (1): 31–49.

Chapter 4
Tropical Cyclone Wind Field Reconstruction from SAR and Analytical Model

Xiaofeng Yang, Xuan Zhou, Xiaofeng Li and Ziwei Li

Abstract Sea surface wind field retrieval from Synthetic Aperture Radar (SAR) imagery is based on geophysical model functions (GMFs), which describe the relationship between the near surface winds and the normalized radar backscatter cross section. However, existing GMFs will saturate under tropical cyclone conditions, and cause huge wind retrieval errors. This study develops an approach to estimate tropical cyclone parameters and wind fields based on an improved Holland model and the SAR images. To evaluate its accuracy, three case studies of Typhoon Aere, Typhoon Khanun and Hurricane Ophelia are presented. Estimated results are validated by the best track data of the Joint Typhoon Warning Center (JTWC) and reanalyzed H*wind fields from the Hurricane Research Division (HRD). These results indicate that the tropical cyclone center, maximum wind speed and central pressure are generally consistent with the best track data, and wind fields agree well with reanalyzed data.

4.1 Introduction

Tropical cyclones, which are locally known by the name typhoons in the northwest Pacific Ocean and by hurricanes in the Atlantic Ocean and northeast Pacific Ocean, are the most intense air-sea interaction processes on a synoptic scale. They can produce extremely powerful winds and torrential rain so that marine activities, human life

X. Yang (✉) · X. Zhou · Z. Li
Institute of Remote Sensing and Digital Earth, Chinese Academy of Sciences, Beijing, China
e-mail: yangxf@radi.ac.cn

X. Zhou
e-mail: zhouxuan@radi.ac.cn

Z. Li
e-mail: lizw@radi.ac.cn

X. Li
GST, National Oceanic and Atmospheric Administration (NOAA)/NESDIS, College Park, MD, USA
e-mail: Xiaofeng.Li@noaa.gov

X. Li (ed.), *Hurricane Monitoring With Spaceborne Synthetic Aperture Radar*, Springer Natural Hazards, DOI 10.1007/978-981-10-2893-9_4

and property are seriously threatened. Therefore, the observation of tropical cyclones plays an important role in understanding the air-sea interaction mechanism, forecasting and early warning tropical cyclone track and intensity [1]. Traditional observation methods, such as buoy, ocean station, and drilling platform, are generally limited to its sparse distribution, so the wide range of measurement is hardly achieved. On the contrary, spaceborne microwave sensors such as scatterometer, radiometer, altimeter and SAR offer all-weather, day-night and wide-area coverage so as to become increasingly vital in monitoring tropical cyclones [2]. Unfortunately, scatterometer and radiometer, which are restricted by its low spatial resolution, have no ability to give reliable wind fields near the coast (up to 50 km off coast) and measure the small-scale ocean phenomena. In addition, altimeter only operates in a nadir mode and cant be used to retrieve wind direction. Compared to the above-mentioned sensors, SAR has an ability of high spatial resolution to offer the fine-scale wind field, which can provide a powerful means to monitor tropical cyclones [3].

Wind fields exert an influence on sea surface through wind stress so as to produce wind-generated surface waves, which can alter sea surface roughness and radar backscatter cross section. Therefore, SAR can extract wind fields by measuring the backscattered intensity. An empirical relation between them is called GMF. It is well known that the normalized radar backscatter cross section increases monotonically with wind speeds so the method of SAR wind field retrieval based on GMF have the high accuracy under the no-rain and low-to-moderate conditions [4, 5]. However, for high wind speed measurements, the accuracy rapidly decreases due to saturation of the backscattered intensity and reduction of friction between the sea surface and the wind [6].

The central and outer sea of tropical cyclone have the low wind speed, but the tropical cyclone eye wall have the strongest wind speed, which represents the tropical cyclone intensity. Reference [7] retrieve wind speeds of tropical cyclones from ENVISAT ASAR using a GMF, which are compared with HRD hurricane wind analysis. The results show a significant underestimate exists near the tropical cyclone eye wall where high wind speeds are frequently observed. Horstmann et al. [8] develop an algorithm for extracting automatically wind information under typhoon conditions from SAR images in near real time, which is based on the projection and local gradient method. But this algorithm does not improve the accuracy of high wind speed retrieval. Reference [9, 10] propose a method to estimate the tropical cyclone intensity using SAR images with the help of a parametric model of tropical cyclone. However, the Holland model they used is an axisymmetric model, meaning that the asymmetric structure of tropical cyclone cannot be represented by it [11]. It is well known that an actual tropical cyclone is rarely axisymmetric [12]. Within the same tropical cyclone, the differences in wind speeds at different azimuthal directions can be substantial. Highly asymmetric structures often lead to large errors in tropical cyclone research. In this study, an improved Holland model is introduced which take into account the asymmetry of cyclone eye.

The wind speed near the tropical cyclone eye wall is often more than 20 m/s, and the saturation of the backscattered intensity will cause error using a GMF to retrieve wind speed. In order to solve this problem, an approach to reconstruct tropical cyclone wind fields from SAR images and this model is proposed. Methods to obtain all the coefficients and variables needed to run this model are also presented. Three cases of tropical cyclone are examined and the reconstructed winds are compared with reanalyzed best track data.

4.2 Data

4.2.1 SAR Data

For the following investigations SAR images acquired by the European satellite ENVISAT and the Canadian satellite RADARSAT-1 are used. Both satellites operated a SAR with a frequency of 5.3 GHz (C band). In this study the ScanSAR wide-swath mode is chosen, which offers the widest possible coverage, up to 450 km with a spatial resolution of 100 m for ENVISAT (e.g. Fig. 4.1) and up to 500 km with a spatial resolution of 150 m for RADARSAT-1. Many GMFs have been established for C-band VV polarized radar, but RADARSAT-1 operates at HH polarization. Therefore, a polarization ratio describing the relationship between VV and HH polarization NRCS is needed [13–15].

$$\sigma_0^{HH} = \sigma_0^{VV} \frac{(1 + \alpha \tan^2 \phi)^2}{(1 + 2tan^2\phi)^2} \tag{4.1}$$

where σ_0^{HH} and σ_0^{VV} are the HH- and VV-polarized NRCS; ϕ is incidence angle; α is constant and set to 0.6 following Unal et al. [13].

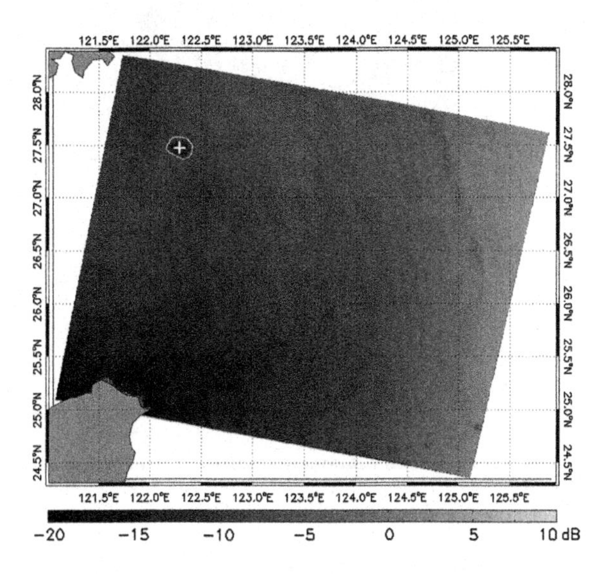

Fig. 4.1 ENVISAT ASAR image observed at 01:46, 11 September, 2005

4.2.2 Validation Tropical Cyclone Winds Data

The validation data contains the best track data and H*Wind products. The Joint Typhoon Warning Center (JTWC) provides the best track data of the western North Pacific from 1945, which contains tropical cyclone center locations and intensities (i.e., the maximum 1-min mean sustained 10-m wind speed) at six-hour intervals.

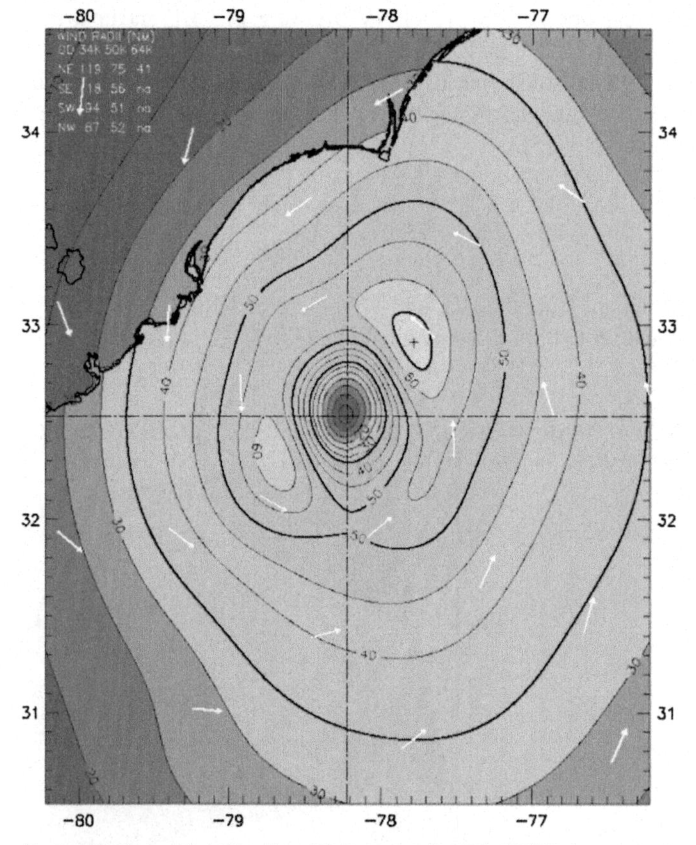

Fig. 4.2 H*Wind product of Hurricane Ophelia

The Hurricane Research Division provides H*Wind products from 1996, which is designed to improve understanding of the extent and strength of the wind field, and to improve the assessment of hurricane intensity. This product is based on actual wind measurements collected in the storm over a 4–6 h period. Figure 4.2 shows H*Wind products of Hurricane Ophelia on September 14 at 01:30 UTC.

4.3 Estimation of Tropical Cyclone Parameters and Wind Fields

Firstly, the wind direction, tropical cyclone center and radius of maximum wind are derived from the SAR image, and the low-to-moderate wind speed is retrieved using the wind direction as the input of CMOD5 model. Furthermore, we use these parameters to fit the improved Holland model to the low-to-moderate wind speed (<20 m/s) by the least squares method in order to calculate maximum wind and central pressure of tropical cyclone. Finally, wind speed distributions are reconstructed by putting these parameters into the improved Holland model. The flowchart of our approach is shown in Fig. 4.3.

4.3.1 Wind Direction Extraction

The periodic streaks parallel to wind direction often exist in the SAR images, which are caused by sources such as atmospheric boundary layer rolls and Lingmuer

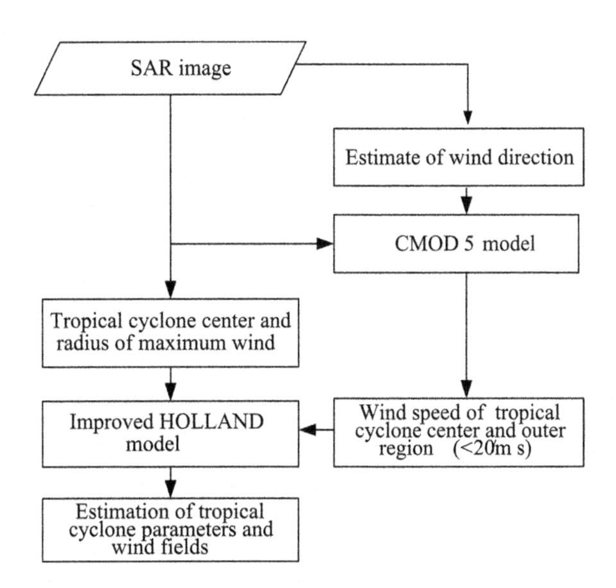

Fig. 4.3 Flowchart of estimation of tropical cyclone parameters and wind fields from SAR images

circulation [16]. The SAR wind direction retrieval is based on these phenomena. For the tropical cyclone wind field, there are the streaks of different spatial scales in an SAR image. The Discrete Wavelet Transform (DWT) has the ability to multi-scale analysis, which is suitable for the detection of texture feature at different spatial scales [17]. In order to derive the multi-scale wind direction of tropical cyclone, the algorithm based on DWT is used for the wind direction retrieval.

To begin with, the vertical detail of an image after DWT is rotated through 180° with a given rotational interval 180/n ($1 < n < 180$) and n images are produced. Then for each image, the standard deviation of the mean value of each column in a vertical direction is calculated. Finally the wind direction is considered as the rotation angle of the image for the maximum of the standard deviation.

However, the wind direction based on this algorithm has a 180 ambiguity and the periodic streaks do not always exist in each sub region of SAR image. So a simple typhoon wind direction model [18] is used to resolve the 180° wind direction ambiguity and provide the wind direction of no wind streaks region. In the northern hemisphere, typhoons rotate counterclockwise. At a given radial distance from the typhoon center, the wind direction chosen is 75° from the radius vector in this model. Figure 4.4 shows the reference wind direction of typhoon, which can remove the ambiguity and provide the wind direction in no wind streaks region. Figure 4.5 shows the wind direction of Typhoon Khanun estimated by this algorithm.

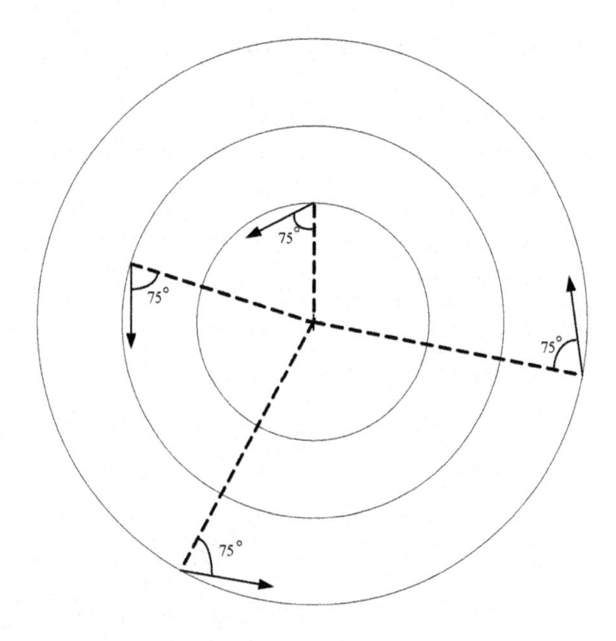

Fig. 4.4 A simple typhoon wind direction model

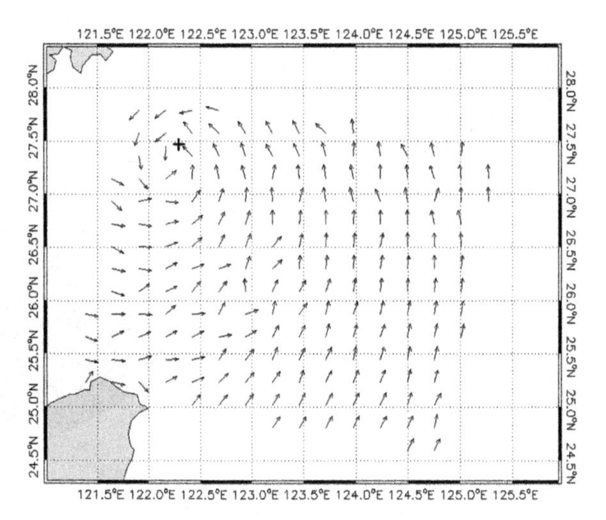

Fig. 4.5 The wind direction of Typhoon Khanun

4.3.2 *Wind Speed Retrieval*

The SAR wind field retrieval is based on the GMF in operation, which establishes the quantitative relationship between wind vector, angle of incidence and normalized radar cross section. The CMOD5 model is developed by ECMWF and KNMI for the C-band radar with vertical polarization [19], whose form is

$$\sigma_{wind} = B_0(1 + B_1 \cos\phi + B_2 \cos 2\phi)^{1.6} \tag{4.2}$$

where B_0, B_1 and B_2 are functions of wind speed v and incidence angle, and is the relative wind direction with respect to the radar look direction.

The NRCS calibration error of spaceborne SAR is about 1.4 dB [20]. This calibration error can cause great wind speed retrieve error in high wind conditions. Figure 4.6 gives the wind speed retrieval error distribution varying with the wind speed and direction at an incidence angle of 35°. When wind speeds are below 20 m/s, the accuracy is rather high and wind speed errors retrieved by CMOD5 model are less than 2 m/s. But when high winds are over 30 m/s, the errors rapidly increase with wind speeds, which are up to 10 m/s for wind speeds of 30 m/s. Obviously, the NRCS calibration accuracy of spaceborne SAR have a big impact on the wind speed retrieval. For the center and outer sea of tropical cyclones, wind speeds are low and the spaceborne SAR wind speed retrieval can achieve the high accuracy using CMOD5 model. However, for the sea of high wind speeds near tropical cyclone center, the errors are rather high.

For the above-mentioned characteristics of CMOD5 model, we only use SAR derived wind speed to fit the coefficients of the improved Holland model. And use Holland model to reconstruct the high wind speed distribution of tropical cyclones.

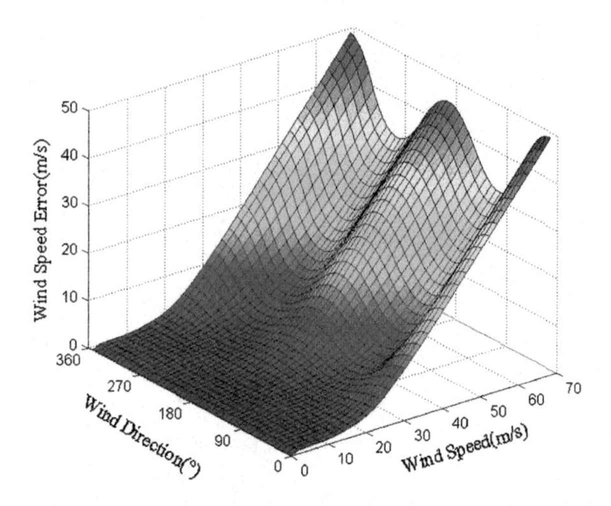

Fig. 4.6 Plot of the wind speed retrieval error for spaceborne SAR depending on wind speed and directions at a fixed incidence angle 35°

4.3.3 Tropical Cyclone Center and Radius of Maximum Wind

ENVISAT ASAR and RADARSAT-1 SAR which both operate at C-band can measure the normalized radar cross section of sea surface through clouds. For the mature tropical cyclone, its center would have been developed into a large eye, which is normally calm and free of clouds. So the reflected radar signal is relatively weak in the eye compared with other areas of tropical cyclone [21]. For the tropical cyclone eye, the dark region is surrounded by eye wall, which contains the strongest wind speeds (see Fig. 4.1).

We estimates the tropical cyclone center based on these inherent characteristics of tropical cyclone. To begin with, the SAR sub-image containing the tropical cyclone eye is extracted from the SAR image and is smoothed with a db4 wavelet analysis. Then the darker region of the eye is first roughly detected using a threshold calculated as 80% of the mean value of the sub-image, and the average values of latitude and longitude within the detected darker area are used as the initial tropical cyclone center. Furthermore, the maximum gradient points are then detected along polar directions for all integer angles from 0° to 360°. These maximum gradient points provide a discontinuous boundary. The average value of these points provides a threshold for determination of the extent of the tropical cyclone eye. Finally, the tropical cyclone center is calculated by the average value of latitude and longitude within the tropical cyclone eye.

The radius of maximum wind is assumed to be nearby the tropical cyclone eye and it can be estimated by extracting the maximum value of NRCS along each polar direction. Figure 4.1 shows the center and the radius of maximum wind of Typhoon Khanun, which are extracted from the ENVISAT ASAR image. The white plus sign is the tropical cyclone center and the white irregular circle is the radius of maximum wind.

4.3.4 Tropical Cyclone Wind Field Reconstruction Model

A variety of tropical cyclone wind field models have been developed for describing the distribution of wind fields. Depperman et al. in early studies used a modified Rankine vortex model to approximate the structure of a generic tropical cyclone wind field, which deficiency is that it requires accurate measurements of the radius of maximum wind speed [22]. Schleomer proposed a parametric model that relates the wind field to the pressure field [23]. However, it obviously underestimates the radial extent of tropical cyclone wind speed. Holland presented an analytic model of radial profiles of sea level pressure and winds for tropical cyclones, which resemble a family of rectangular hyperbolas [11]. Holland's model is relatively simple parametric model and generally consistent with tropical cyclone wind speed. So It has been used to estimated tropical cyclone wind speed from SAR imagery [9, 10, 24]. Unfortunately, this model only fits an axisymmetric structure of tropical cyclone, which account for less than 20% of all tropical cyclones. According to the characteristics of aforementioned models, we use an improved Holland model to approximate the distribution of tropical cyclone wind field [25]. The formula is

$$V = \left[\frac{R_{max}(\theta)}{R_{max}}\right]^{B/2}\left\{\left[\frac{B}{\rho}\left(\frac{R_{max}}{r}\right)^{B}(P_n - P_c)\,exp\left(-\frac{R_{max}^{B}}{r^{B}}\right) + \frac{r^2 f^2}{4}\right] - \frac{rf}{2}\right\} \tag{4.3}$$

where V is wind speed at a distance r from tropical cyclone center; $R_{max}(\theta)$ is the radius of maximum wind at θ direction and represents the asymmetric wind structure; R_{max} is the average of $R_{max}(\theta)$; $\rho = 1.15\,\text{kg}\cdot\text{m}^{-3}$ is air density; $P_n = 1015\,\text{mb}$ is the ambient pressure; P_c is the central pressure; $B = \rho e \frac{v_M^2}{P_n - P_c}$ is a shape parameter of tropical cyclone, where v_M is maximum wind; f is the Coriolis parameter.

4.3.5 Model Parameters Fitting

Parameters in Eq. (4.3) such as $R_{max}(\theta)$, R_{max} and r can be estimated from method in Sect. 4.1; ρ, P_n, f and are already known; the low wind speed V at the central and outer region of tropical cyclone is retrieved by putting wind direction into CMOD5 model; However, P_c and v_M remains unsolved. So we enter $R_{max}(\theta)$, R_{max}, r, ρ, P_n, f into Eq. (4.3) and then fit the improved Holland model to the low-to-moderate wind speed V. Finally, P_c and v_M are calculated by the least squares method.

The central pressure and maximum wind of Typhoon Khanun are 938 mb and 59 m/s, which are derived from ENVISAT ASAR image using our method. Wind fields are estimated by putting tropical cyclone parameters into Eq. (4.3). Figure 4.7 shows that the distribution of Typhoon Khanun wind speed is estimated from

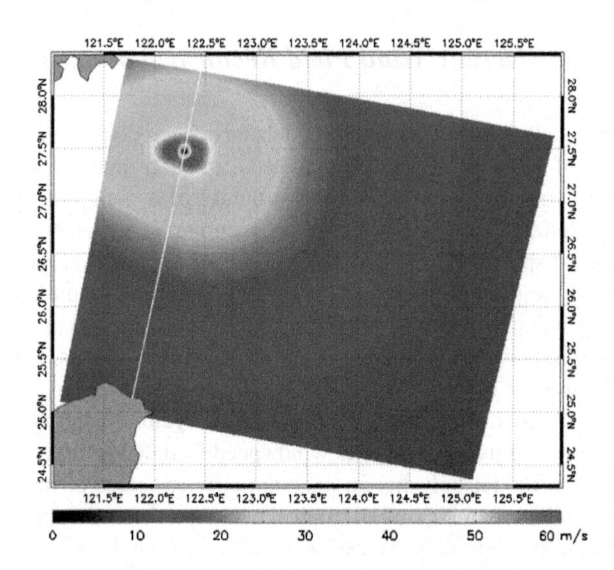

Fig. 4.7 Wind speed from ENVISAT ASAR image based on the improved HOLLAND model

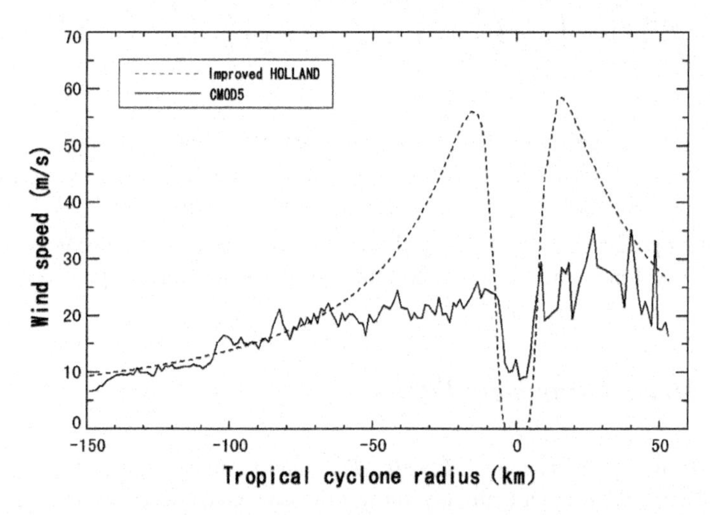

Fig. 4.8 Comparison of wind speeds based on CMOD5 model and the improved HOLLAND model

ENVISAT ASAR based on the improved HOLLAND model, where the wind speed profile represented by the white solid line is shown in Fig. 4.8.

Figure 4.8 shows the comparison of wind speeds based on CMOD5 model (solid line) and the improved HOLLAND model (dash-dotted line). Wind speeds of the tropical cyclone eye wall are remarkably underestimated by CMOD5 model. In addition, the asymmetry of the wind speed profile based on the improved HOLLAND

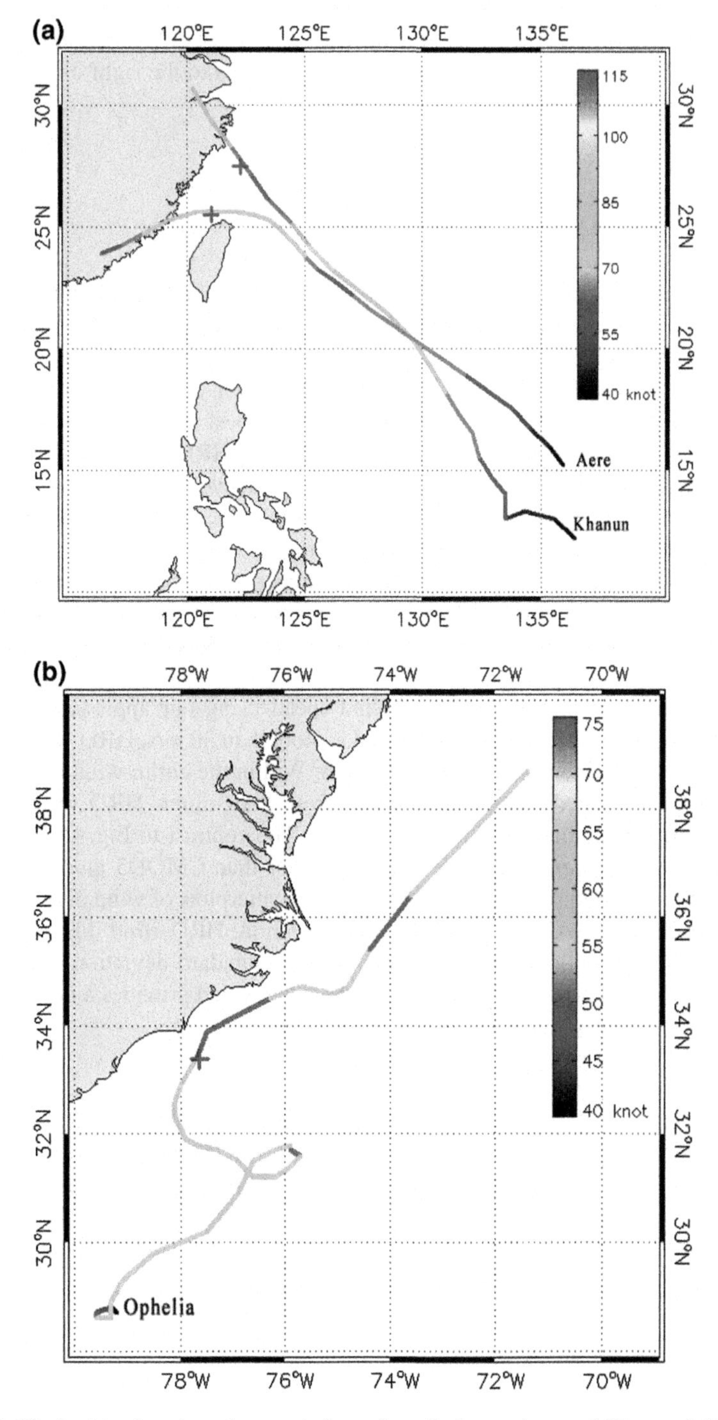

Fig. 4.9 The best track and maximum wind speeds. **a** Typhoons Aere and Khanun, **b** Hurricane Ophelia

model is shown, which is in good agreement with the result of Willoughby et al. [26]. They demonstrated that the stronger wind speed occurs to the right of direction of movement in tropical cyclones.

4.4 Validation

Our approach to the estimation of tropical cyclone parameters and wind fields from SAR images is validated by three tropical cyclone SAR images: the ENVISAT ASAR image of Typhoon Aere (25 August, 2004), the ENVISAT ASAR image of Typhoon Khanun (11 September, 2005) and the RADARSAT-1 SAR image of Hurricane Ophelia (14 September, 2005). Figure 4.9 shows the best track and maximum wind speeds from JTWC (Typhoons Aere and Khanun) and HRD (Hurricane Ophelia).

The comparison between the estimation of tropical cyclone parameters using our method and JTWC or HRD data is shown in Table 4.1, and the blue plus sign in Fig. 4.9 is the cyclone center at SAR overpass time.

Table 4.1 and Fig. 4.9 show that tropical cyclone parameters estimated from SAR is generally consistent with JTWC or HRD data. In addition, the comparison between the estimation of wind fields using our method and H*Wind product for Hurricane Ophelia is shown in Fig. 4.10.

Figure 4.10a shows scatterplots of wind speed using our approach from SAR versus H*Wind product. When wind speed is more than 30 m/s, HRD wind speed is slightly less than wind speed of our approach. Within the entire wind speed range, the bias of wind speed using our approach from SAR minus HRD wind speed is 0.3 m/s, and the standard deviation is 1.4 m/s. The contrast to Fig. 4.5 shows that the accuracy of our approach is remarkably better than CMOD5 model under the high wind speed conditions. Figure 4.10b shows scatterplots of wind direction using our approach from SAR versus HRD wind direction. HRD wind direction agrees well with wind direction of our approach and the standard deviation is only 2.1°. In summary, our approach achieves the high accuracy and provides a new technical means to monitor tropical cyclone.

Table 4.1 Tropical cyclone parameters estimated from SAR versus JTWC or HRD data

Tropical cyclone	Spaceborne SAR		JTWC or HRD	
	Central pressure (mb)	Maximum wind (m/s)	Central pressure (mb)	Maximum wind (m/s)
Aere	955	50	954	46
Aland Islands	938	59	933	57
Ophelia	976 mb	42	980	39

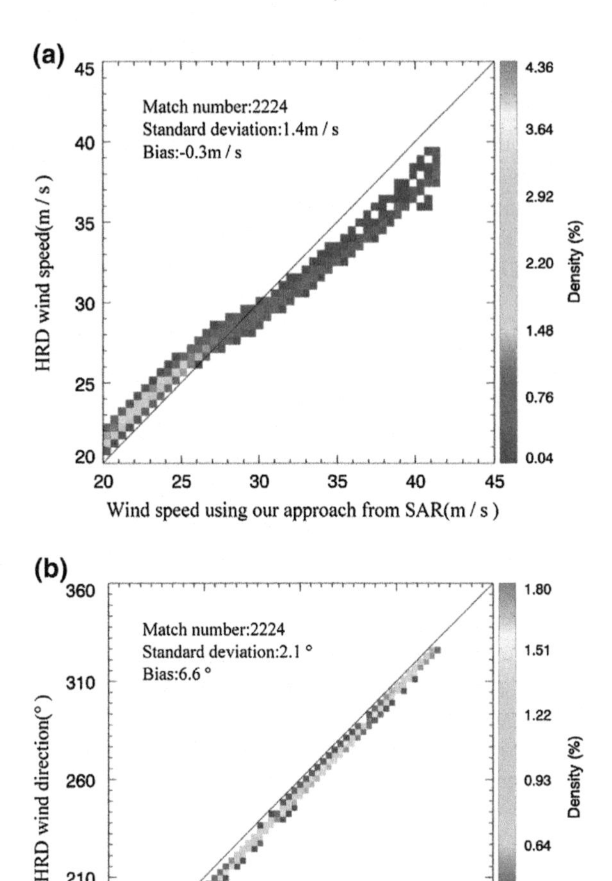

Fig. 4.10 Wind fields estimated from SAR versus H*Wind product **a** Wind speed, **b** Wind direction

4.5 Discussion and Conclusion

In this study, an approach to estimate tropical cyclone parameters and reconstruct its wind fields from SAR images and tropical cyclone model has been presented. Three tropical cyclones located in west Pacific and Atlantic are analyzed using the proposed method. Moreover, the best track and reanalyzed data of tropical cyclone are used to validate the reconstructed wind field.

Although the comparison results show that reconstructed wind field are close to reanalyzed data, there still are some aspects need to take into account when implement it.

(1) The proposed method is based on tropical cyclone eye recognition and center location. However, tropical cyclones dont always have a clear round/ellipse shape eye in SAR images. Actually, only less than 30% of SAR observed tropical cyclones has round/ellipse shape eye [12]. Even if there is a clear eye in SAR image, its not easy to automated extract the eye area and exactly locate its center. This eye center location error cannot be ignored when the eye is asymmetric.

(2) Tropical cyclones always accompanied by heavy rains. Rain is an important factor to affect the satellite SAR received backscatter signals. The impact of rainfall and the saturation of backscattered intensity under the condition of tropical cyclones should be considered for the estimation of tropical cyclone parameters and wind fields.

(3) This study assumes that the radius of maximum wind is in the vicinity of cyclone eye and the maximum value of NRCS corresponds to maximum wind speed in each azimuth direction. However, when the wind speed is greater than 20 m/s, the NRCS is gradually become saturated. Thus, the radius of maximum wind does not always correspond to the maximum value of NRCS. Recent studies show that cross-pol observations can overcome NRCS saturation in high wind conditions [27, 28]. More accurate maximum wind extraction may be obtained from cross-pol SAR image in the future.

In summary, the combination of satellite SAR imagery and tropical cyclone model can provide tropical cyclone center, maximum wind speed and central pressure as well as sea surface wind field information. These extracted location and shape parameters are in good agreement with the best track data. Besides, the reconstructed winds are more accurate than the directly retrieved SAR wind products, especially where wind speed is greater than 25 m/s.

References

1. Zhang, Q.H., Q. Wei, and L.S. Chen. 2010. Impact of landfalling tropical cyclones in mainland China. *Science China Earth Sciences* 50 (10): 1559–1564.
2. Katsaros, K.B., P.W. Vachon, W.T. Liu, and P.G. Black. 2002. Microwave remote sensing of tropical cyclones from space. *Journal of Oceanography* 58 (1): 137–151.
3. Horstmann, J., W. Koch, S. Lehner, and R. Tonboe. 2002. Ocean winds from RADARSAT-1 ScanSAR. *Canadian Journal of Remote Sensing* 28 (3): 524–533.
4. Donelan, M.A., B.K. Haus, N. Reul, W.J. Plant, M. Stiassnie, H.C. Graber, O.B. Brown, and E.S. Saltzman. 2004. On the limiting aerodynamic roughness of the ocean in very strong winds. *Geophysical Research Letters*, 31 (18).
5. Fernandez, D.E., J.R. Carswell, S. Frasier, P.S. Chang, P.G. Black, and F.D. Marks. 2006. Dual-polarized C- and Ku-band ocean backscatter response to hurricane-force winds. *Journal of Geophysical Research: Oceans*, 111 (C8).
6. Powell, M.D., P.J. Vickery, and T.A. Reinhold. 2003. Reduced drag coefficient for high wind speeds in tropical cyclones. *Nature* 422 (6929): 279–283.
7. Pichel, W.G., X.F. Li, F. Monaldo, C. Wackerman, C. Jackson, C.Z. Zou, et al. 2007. ENVISAT ASAR applications demonstrations: Alaska SAR demonstration and Gulf of Mexico hurricane studies. In *Proceedings of Envisat Symposium 2007, Montreux, Switzerland.*

8. Horstmann, J., C. Wackerman, R. Forster, M. Caruso, and H.C. Graber. 2012. Estimating winds from synthetic aperture radar in typhoon conditions. *IEEE Geoscience and Remote Sensing Society*.
9. Reppucci, A., S. Lehner, J. Schulz-Stellenfleth, and C.S. Yang. 2008. Extreme wind conditions observed by satellite synthetic aperture radar in the North West Pacific. *International Journal of Remote Sensing* 29 (21): 6129–6144.
10. Reppucci, A., S. Lehner, J. Schulz-Stellenfleth, and S. Brusch. 2010. Tropical cyclone intensity estimated from wide-swath SAR images. *IEEE Transactions on Geoscience and Remote Sensing* 48 (4): 1639–1649.
11. Holland, G.J. 1980. An Analytic Model of the Wind and Pressure Profiles in Hurricanes. *Monthly Weather Review* 108 (8): 1212–1218.
12. Li, X., J.A. Zhang, X. Yang, W.G. Pichel, M. DeMaria, D. Long, and Z. Li. 2013. Tropical cyclone morphology from spaceborne synthetic aperture radar. *Bulletin of the American Meteorological Society* 94 (2): 215–230.
13. Unal, C.M.H., P. Snoeij, and P.J.F. Swart. 1991. The Polarization-Dependent relation between radar backscatter from the ocean surface and surface wind vector at frequencies between 1 Ghz and 18 Ghz. *IEEE Transactions on Geoscience and Remote Sensing* 29 (4): 621–626.
14. Thompson, D.R., T.M. Elfouhaily, and B. Chapron. 1998. Polarization ratio for microwave backscattering from the ocean surface at low to moderate incidence angles. In *1998 IEEE International Geoscience and Remote Sensing Symposium Proceedings, 1998. IGARSS'98*, vol. 3, 1671–1673.
15. Horstmann, J., W. Koch, S. Lehner, and R. Tonboe. 2000. Wind retrieval over the ocean using synthetic aperture radar with C-band HH polarization. *IEEE Transactions on Geoscience and Remote Sensing* 38 (5): 2122–2131.
16. Alpers, W., and B. Brummer. 1994. Atmospheric boundary-layer rolls observed by the synthetic-aperture radar aboard the ERS-1 satellite. *Journal of Geophysical Research* 99 (C6): 12613–12621.
17. Du, Y., P.W. Vachon, and J. Wolfe. 2002. Wind direction estimation from SAR images of the ocean using wavelet analysis. *Canadian Journal of Remote Sensing* 28 (3): 498–509.
18. Jones, W.L., V.J. Cardone, W.J. Pierson, J. Zec, L.P. Rice, A. Cox, and W.B. Sylvester. 1999. NSCAT high-resolution surface wind measurements in Typhoon Violet. *Journal of Geophysical Research: Oceans* 104 (C5): 11247–11259.
19. Hersbach, H., A. Stoffelen, and S. de Haan. 2007. An improved C-band scatterometer ocean geophysical model function: CMOD5. *Journal of Geophysical Research: Oceans*, 112 (C3).
20. Srivastava, S.K., S. Cote, P.L. Dantec, R.K. Hawkins, and K. Murnaghan. 2007. RADARSAT-1 calibration and image quality evolution to the extended mission. *Advances in Space Research* 39 (1): 7–12.
21. Du, Y., and P.W. Vachon. 2003. Characterization of hurricane eyes in RADARSAT-1 images with wavelet analysis. *Canadian Journal of Remote Sensing* 29 (4): 491–498.
22. Depperman, R.C. 1947. Notes on the origin and structures of Philippine typhoons. *Bulletin of the American Meteorological Society* 28: 399–404.
23. Schloemer, R.W. 1954. Analysis and synthesis of hurricane wind patterns over Lake Okeechobee. *NOAA Hydrometeorological Report* 31: 49.
24. Young, I.R. 1993. An estimate of the geosat altimeter wind-speed algorithm at high wind speeds. *Journal of Geophysical Research: Oceans* 98 (C11): 20275–20285.
25. Xie, L., S.W. Bao, L.J. Pietrafesa, K. Foley, and M. Fuentes. 2006. A real-time hurricane surface wind forecasting model: formulation and verification. *Monthly Weather Review* 134 (5): 1355–1370.
26. Willoughby, H.E., and M.E. Rahn. 2004. Parametric representation of the primary hurricane vortex. Part I: Observations and evaluation of the Holland (1980) model. *Monthly Weather Review* 132 (12): 3033–3048.
27. Zhang, B., and W. Perrie. 2012. Cross-polarized synthetic aperture radar a new potential measurement technique for hurricanes. *Bulletin of the American Meteorological Society* 93 (4): 531–541.

28. Zhang, B., W. Perrie, P.W. Vachon, X.F. Li, W.G. Pichel, G. Jie, and Y.J. He. 2012. Ocean vector winds retrieval from C-band fully polarimetric SAR measurements. *IEEE Transactions on Geoscience and Remote Sensing* 50 (11): 4252–4261.

Chapter 5
High Wind Speed Retrieval from Multi-polarization SAR

Biao Zhang and William Perrie

Abstract Synthetic aperture radar (SAR) has capability to observe tropical cyclones (TCs) with high-resolution and large coverage under all weather conditions. The intensity parameters and two-dimensional fine structure characteristics for TCs can be derived from SAR observations. In this chapter, we make an overview of high wind retrieval from C-band multi-polarization SAR. Co- and cross-polarized geophysical model functions (GMFs) for high wind speed derivation are summarized. The validation results are presented using various GMFs and data sources. The intense effect of rainfall on SAR high winds retrieval is emphasized. We summarize potential challenges and provide possible solutions for high wind retrieval in the future.

5.1 Introduction

Surface winds over the ocean are an essential meteorological variable to determine many climate variables, such as moisture, heat, and momentum fluxes between the atmosphere and ocean, mixed layer depth, and Ekman transports, etc. Surface winds directly influence ocean circulation, water mass formations, and energy transports between ocean basins. The destructive forces of cyclonic winds periodically produce devastating weather-related natural disasters, such as intense rainfall and storm surges, as well as serious flooding. As a result, the investigation of high winds is of benefit to tropical cyclone (TC) intensity and track forecasting, which is extremely important for the protection of coastal residents and infrastructure.

Optical and microwave satellite sensors are important remote sensing instruments for synoptic TC monitoring from space. Moderate Resolution Imaging Spectrora-

B. Zhang (✉)
Nanjing University of Information Science & Technology, Nanjing, Jiangsu, China
e-mail: zhangbiao@nuist.edu.cn

W. Perrie
Fisheries and Oceans Canada, Bedford Institute of Oceanography, Dartmouth, NS, Canada
e-mail: William.Perrie@dfo-mpo.gc.ca

X. Li (ed.), *Hurricane Monitoring With Spaceborne Synthetic Aperture Radar*, Springer Natural Hazards, DOI 10.1007/978-981-10-2893-9_5

diometer (MODIS), Advanced Very High Resolution Radiometer (AVHRR), and the Geostationary Orbiting Environmental Satellite (GOES) imager are capable of acquiring cloud properties at the top of TC, but they cannot measure the size and shape as well as storm intensity, due to the dense cloud coverage [1]. However, SAR transmits electromagnetic signals which are able to penetrate clouds, and measure sea surface roughness associated with local winds with high-resolution and large coverage under all weather conditions. The first applications of microwave remote sensing technology to hurricanes involved an airborne active scatterometer and passive radiometer during Hurricane Allen, demonstrating their potential for reliable measurements [2]. Although scatterometers and radiometers can also observe ocean surface conditions all day or night, they can only provide measurements with coarse resolutions, and thus their observations are not very suitable to probe high-resolution wind gradient variations between the TC eye and eyewall. Compared to other existing sensors, the significant advantage of SAR is that it can contribute fine-resolution sea surface wind field information that can not otherwise be obtained below the cloud deck.

Wind speed measurements from SAR imagery are routinely estimated utilizing various empirical geophysical model functions (GMFs) [3–6]. These GMFs are derived from C-band VV-polarized scatterometer measurements, and they relate wind vectors to the normalized radar cross section (NRCS) and the radar incidence angle. Using wind direction and incidence angle, along with the NRCS at each pixel, the associated wind speed can be computed with these GMFs. Thus far, studies on wind retrieval from SAR data have shown that measurements can achieve an accuracy of about 2 m/s in wind speed (at 10-m reference height) for moderate winds [7–10]. Even though SAR is able to provide high-resolution sea surface wind fields under moderate sea states, the oceanic and meteorological communities are also interested in the potential ability of SAR data to retrieve hurricane force winds.

5.2 High Winds from Co-polarized SAR

Radar backscatter saturation, wind direction, polarization ratio (PR) and intense rainfall are the four important factors to affect high wind retrievals using co-polarized SAR. Airborne scatterometer measurements under high wind conditions has demonstrated that the GMFs mentioned previously were found to overpredict the NRCS for winds above 15 m/s, and that the NRCS starts to saturate for wind speed above 20 m/s [11]. Global Positioning System (GPS) dropwindsonde wind observations in tropical cyclones revealed that drag coefficient and sea surface roughness do not continuously increase, and that surface momentum flux values level off as wind speeds increase above hurricane force (>33 m/s) [12]. Laboratory experiments involving extreme winds showed that the aerodynamic roughness approaches a limiting value in high winds [13]. A unified boundary layer model, taking account of both the wave field and sea spray production, also suggested drag coefficient reduction at very high wind speeds [14]. Moreover, both airborne C- and Ku-band ocean surface NRCS measurements of hurricane-force winds clearly showed that the NRCS in VV polarization does not continue to increase as winds increase beyond marginal hurricane strength [15]. Thus, NRCS experiences saturation under high wind conditions which creates a wind speed ambiguity problem for SAR wind speed retrieval [16].

A typical C-band RADARSAT-1 SAR image of Hurricane Ivan on September 10, 2004 was used to investigate the feasibility of retrieval of hurricane force winds utilizing conventional GMFs, such as CMOD4 and CMOD5 [17]. Since two unknown parameters, i.e., wind speed and wind direction, simultaneously exist in the GMF, one must inevitably obtain the wind direction from external sources, prior to wind speed retrieval. Fast Fourier transform (FFT) and Local Gradient (LG) methods [7, 18, 19] were employed to first retrieve the wind direction using the linear features associated with wind streaks and marine atmospheric boundary layer (MABL) rolls visible in the SAR image. However, it should be noted that the use of linear features in the SAR image to estimate wind direction can sometimes lead to erroneous results. Although wind-induced signatures, such as wind rows, are most conspicuous under unstable atmospheric conditions, however, in some cases, especially in neutral or stable conditions, they are not present at all [20]. In addition, there can be other features, such as oceanic or atmospheric internal waves, that produce linear features on the same spatial scale as wind rows. These nonwind streak features are not generally aligned with the local wind vectors and can contaminate wind direction retrievals directly inferred from the SAR image. Moreover, wind directions from the two methods mentioned above have a 180° ambiguity problem, which needs to be resolved using ancillary information from meteorological weather forecasts or scatterometer observations. In hurricane wind regimes, although the wind direction dependence of the NRCS for CMOD5 is much weaker than that of CMOD4 [17], inaccurate wind directions are still able to cause serious errors in wind speed retrievals.

To avoid the influence of inaccurate wind directions on wind speed inversion, a new algorithm was proposed to retrieve hurricane wind speed and wind direction simultaneously, based on the primary hurricane structure characteristics [21]. As an example, this methodology was used to retrieve high wind speeds from a RADARSAT-1 SAR image of Hurricane Isabel acquired on September 17, 2003, and results were validated using QuikSCAT scatterometer measurements and in situ buoy observations. This method assumes constant wind speeds and equal-distant wind directions in three neighboring sub-image blocks of any specific concentric circle around the hurricane eye. No external information is needed to resolve wind vectors. By comparison, the PR model [22] involving the hurricane winds retrieval algorithms mentioned above, is only associated with radar incidence angle in order to transform NRCS in HH polarization into VV polarization, and thus provide input to the GMF model for wind retrievals. In fact, the PR model is not only dependent on incidence angle, but also on wind speed and wind direction, as well sea states (significant wave height and wave steepness). However, PR research is an area of rapid development, and several theoretical and empirical PR models have been proposed [23–27], which relate PR, and radar and sea state parameters for NRCS models. These PR models can be further employed for possible improvement to the accuracy of hurricane wind retrievals from SAR imagery acquired at HH-polarization.

Compared to scatterometer and radiometer data, SAR can provide high resolution measurements of tropical cyclones, which can be used to infer information on the fine-scale storm structure, for example, eye size, radius of maximum wind speed, and rain band structure. Recently, 83 hurricanes imaged by spaceborne SAR have been used to investigate morphology of hurricane eye structure, meso-vortices, rain bands and unusual observations, i.e., high winds within the eye [28]. A parametric radial wind profile model for hurricanes [29], in combination with ENVISAT ASAR imagery of Hurricane Katrina and Typhoon Kiko were acquired at VV-polarization, and used to estimate the tropical cyclone intensity (maximum wind speed and radius of maximum wind speed) [30]. Although the C-band microwave signal is much less affected by rain than that of Ku-band, dark spiral shaped features associated with the rain bands are clearly visible in the SAR image of Hurricane Katrina. Unfortunately, the available GMFs for wind field retrieval do not include a rain rate parameter, and thus cannot account for the rain contamination on the NRCS under high winds, especially in the intense rainfall environment. Thus, strong NRCS dampening due to heavy rain could possibly lead to serious underestimation of wind speeds in hurricane conditions when standard wind retrieval algorithms are applied. In addition to C-band SAR, L- and X-band GMFs have also been developed to retrieve ocean surface wind speeds under moderate sea states [31–33], rather than for extreme weather events. Moreover, as a challenge for future research, it is interesting to investigate if SAR-measured ocean surface backscatters at these two bands are saturated, or not, under high wind conditions.

5.3 High Winds from Cross-Polarized SAR

For moderate radar incidence angles ($20° \sim 60°$) and wind speeds ($5 \sim 20$ m/s), ocean surface Bragg scattering is the dominant scattering mechanism; the radar backscatter is proportional to the spectral density of the surface roughness on scales comparable to the radar wavelengths. The NRCS in co-polarizations (HH, VV) are not only related to radar incidence angle and wind speed, but also dependent on wind direction. In general, because radar backscatter in cross-polarizations (HV, VH) are induced by volume scattering or by surface tilts (i.e., polarization orientation angle shift), thus cross-polarizations are less correlated with co-polarizations [34].

Recently, C-band cross-polarized ocean backscatter has been documented as being insensitive to the wind direction and the radar incidence angle, and quite linear with respect to the wind speed, and thus can be used to directly retrieve wind speed [26, 35]. The cross-polarized SAR return exhibits the typical double structure; its wind speed dependence increases with wind speed from linear to cubic. The signature of the double structure is in the wind speed dependence of the radar returns: linear for scattering from gentle waves and cubic for wave-breaking contributions. The enhanced sensitivity of cross-polarization returns in the higher wind data of available measurements suggests that it may be ideal for hurricane wind retrievals [25, 36].

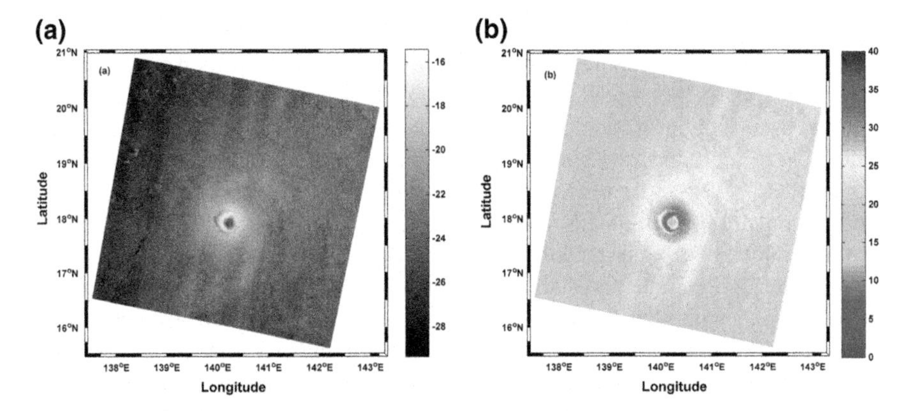

Fig. 5.1 **a** RADARSAT-2 VH-polarization SAR image acquired over Typhoon Soudelor at 20:31 UTC on August 3, 2015; *Colorbar* shows sigma-naught VH polarization (σ^0_{VH}) in dB, **b** SAR-retrieved wind speeds from the C-2PO model and σ^0_{VH}. *Colorbar* shows wind speeds at 10-m height (U_{10}) in m/s. RADARSAT-2 Data and Product MacDonald, Dettwiler and Associates Ltd, -All Rights Reserved

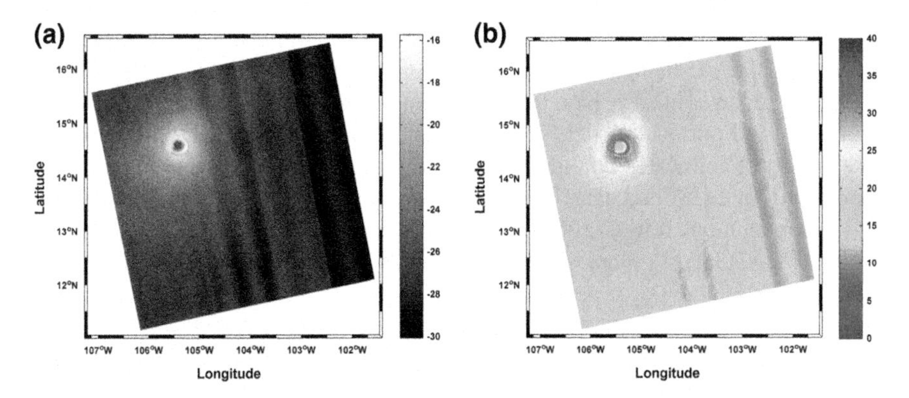

Fig. 5.2 **a** RADARSAT-2 VH-polarization SAR image acquired over Hurricane Adrian at 01:18 UTC on June 10, 2011; Colorbar shows sigma-naught VH polarization (σ^0_{VH}) in dB, **b** SAR-retrieved wind speeds from the C-2PO model and σ^0_{VH}. *Colorbar* shows wind speeds at 10-m height (U_{10}) in m/s. RADARSAT-2 Data and Product MacDonald, Dettwiler and Associates Ltd, -All Rights Reserved

In presently available SAR data, the observed NRCS in cross-polarizations do not seem saturated under high winds, and increase linearly with wind speed up to 26 m/s, which indicates that they can potentially be used to retrieve hurricane winds, including eye structure observations. An empirical C-band Cross-Polarization Ocean (C-2PO) backscatter model was proposed to retrieve high winds from satellite SAR data, which eliminates the need for external wind direction and radar incidence angle inputs and provides a linear response at high wind speeds for hurricanes [37]. The advantage of C-2PO is that it avoids the errors in wind speed retrievals that can occur in CMOD5.N due to errors in wind directions. It also simplifies the standard SAR wind retrieval procedure. Figures 5.1 and 5.2 show the winds on scales of 1 km resolution from C-2PO model, using two RADARSAT-2 VV-polarized SAR images acquired over Typhoon Soudelor and Hurricane Adrian. The results from C-2PO are better able to reproduce the hurricane eye structure than those of CMOD5.N [37]. Moreover, C-2PO is a more straightforward mapping of observed NRCS to wind speed, compared to CMOD5.N. Wind speed retrieval errors present in CMOD5.N results, induced by inaccurate wind directions, can be avoided by C-2PO.

Since the C-2PO model is established from RADARSAT-2 fine quad-polarization mode SAR data with a low noise floor, one should exercise care when using this model to retrieve relatively high winds from dual-polarization SAR imagery, because the noise floor for dual-polarization data is higher than for quad-polarization data. Moreover, quad-polarization allows the retrieval of cross-channel leakage for antennas of both low and high cross-polarization isolation. However, the cross-channel leakage cannot be retrieved if only single- or dual-polarization measurements are collocated, as is the case with the single- or dual-polarization modes of ENVISAT ASAR, Advanced Land Observing Satellite (ALOS) Phased-Array L-band Synthetic Aperture Radar (PALSAR), RADARSAT-2 and TerraSAR-X [38]. Although the wind speed retrieval accuracy of C-2PO is better than that of CMOD5.N, it still needs to be tested using other C-band SARs. More high wind observations are needed to improve the C-2PO model. Thus, we have developed a new C-band cross-polarized ocean surface high wind retrieval model for dual-polarization SAR (C-2POD) model, based on a dataset of collocated dual-polarized SAR imagery and high wind observations from SFMR and H*Wind data. Figures 5.3 and 5.4 show the winds on spatial scales of 1 km resolution from C-2POD model, using two RADARSAT-2 VV-polarized (VV, VH) SAR images acquired of Typhoon Nuri and Hurricane Kennenth.

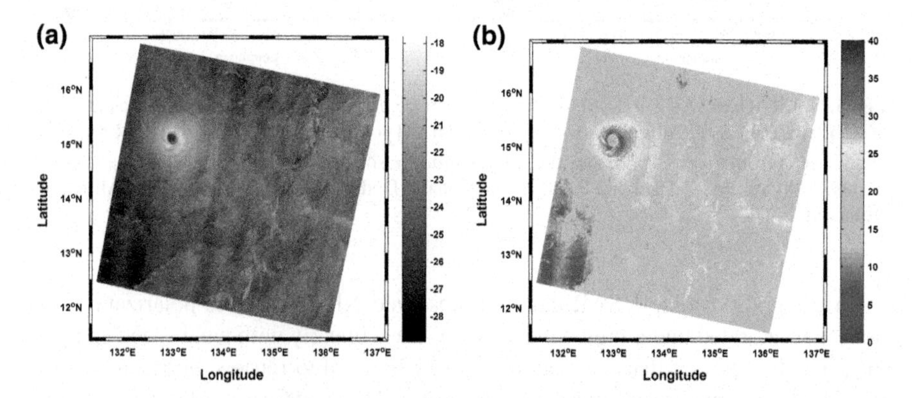

Fig. 5.3 **a** RADARSAT-2 VH-polarization SAR image acquired over Typhoon Nuri at 20:53 UTC on November 1, 2014; Colorbar shows sigma-naught VH polarization (σ^0_{VH}) in dB, **b** SAR-retrieved wind speeds from the C-2POD model and σ^0_{VH}. *Colorbar* shows wind speeds at 10-m height (U_{10}) in m/s. RADARSAT-2 Data and Product MacDonald, Dettwiler and Associates Ltd, -All Rights Reserved

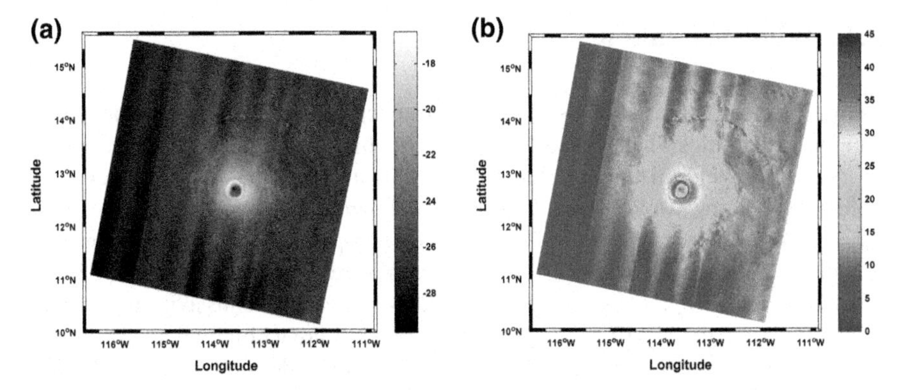

Fig. 5.4 **a** RADARSAT-2 VH-polarization SAR image acquired over Hurricane Kennenth at 13:24 UTC on November 1, 2011; Colorbar shows sigma-naught VH polarization (σ^0_{VH}) in dB, **b** SAR-retrieved wind speeds from the C-2POD model and σ^0_{VH}. *Colorbar* shows wind speeds at 10-m height (U_{10}) in m/s. RADARSAT-2 Data and Product MacDonald, Dettwiler and Associates Ltd, -All Rights Reserved

Recently, an algorithm has been proposed to retrieve ocean vector winds utilizing C-band fully polarimetric SAR measurements [39], thus, both wind speeds and wind directions. C-2PO model was firstly used to directly retrieve wind speeds without any external wind direction and radar incidence angle inputs. Subsequently, the retrieved wind speeds, along with incidence angles and CMOD5.N, are employed to invert the wind direction with ambiguities. Finally, the odd symmetry characteristic of polarimetric correlation coefficient (PCC) between co- and cross-polarization channels is applied to remove wind direction ambiguities. However, this method is not able to be extended to dual-polarization SAR because it only provides backscatter intensity, not scattering matrix measurements and thus cannot estimate PCC.

Due to the friction between hurricanes and the ocean surface, the winds inside the hurricane rotate counter-clockwise and the inflow angle is towards the storms center, in the northern hemisphere. The average inflow angle is approximately 20° [40]. Recent investigation by [41] has shown that the mean inflow angle in the hurricane is in the range $-22.6 \pm 2.2°$ (95% confidence) from analysis of near-surface (10-m) inflow angles using wind vector data from over 1600 quality-controlled Global Positioning System dropwindsondes deployed by aircraft on 187 flights in 18 hurricanes. They proposed an analytical parametric model for the surface inflow angle which requires, as input, the storm motion speed, maximum wind speed, and radius of maximum wind. For the tropical cyclone wind field mapping, using dual-polarization SAR measurements and the inflow angle derived from the parametric inflow angle model, the co- and cross-polarized high wind speed retrieval models mentioned above can potentially be used to simultaneously retrieve high wind speed and wind direction, without ambiguity.

5.4 Wind Validation

Buoys can measure ocean winds under low to moderate sea states, but they lose observational capabilities when wind speed approaches hurricane force. During the past 15 years, surface winds in hurricanes have been estimated remotely using the Steppped-Frequency Microwave Radiometer (SFMR) on board the NOAA hurricane research aircraft [42]. The advantage of SFMR is that it can potentially provide along-track mapping of surface wind speeds and rain rates in relatively high spatial (1.5 km) and temporal (1 Hz) resolutions. Using the latest microwave emissivity-wind speed model function, the SFMR wind speed measurements are within ~4 m/s RMS error of the dropwindsonde-measured surface wind speeds and within ~5 m/s of direct 10-m wind speed measurements [43]. The SFMR-derived surface winds are among the most important observations of direct hurricane inner-core surface wind speeds available for Tropical Prediction Center forecasters. Although the Ku-band scatterometer QuikSCAT can provide tropical cyclone wind fields with large coverage [44, 45], it can be influenced by precipitation and thus result in underestimations of high wind speeds [46]. Moreover, scatterometer observations are contaminated by land signals near coastal regions. In addition, the coarse resolution is not suitable for observations of the fine structure of tropical cyclones. The NOAA Hurricane Research Division's Hurricane Wind Analysis System (H*Wind) [47, 48] produces a gridded analysis by interpolating and smoothing wind speed observations from a vast array of marine, land, aircraft, and satellite platforms. H*Wind assimilates all of the available wind observations from a specific time period into a moving "storm-relative" coordinate system that allows for the production of an objectively blended wind field. However, there are some sources of uncertainty existing in the H*Wind fields, for example, observation errors and errors introduced by the data assimilation techniques [49].

For validation of co-polarized SAR high winds, [21] compared high winds from a RADARSAT-1 SAR image of Hurricane Isabel, with collocated QuikSCAT measurements. The results showed that the wind bias and root-mean-square (RMS) error of their method are −4.26 m/s and 12.07°, 6.05 m/s and 29.02°, respectively. They suggested that the negative bias between SAR retrieved wind speed and the QuikSCAT measurement is possibly associated with a precipitation effect, which can lead to high or low biases in wind speeds, depending on the size of rain drops [50]. Moreover, the maximum wind speeds estimated utilizing five ENVISAT ASAR tropical cyclone images, were also validated using NOAA Hurricane Research Division (HRD) and Japan Meteorological Agency (JMA) in situ observations, with the bias and RMS error of 3.0 and 3.9 m/s, respectively [30].

To evaluate the cross-polarized SAR high winds retrieval accuracy, the wind speeds were retrieved using C-2PO and CMOD5.N with RADARSAT-2 Hurricane SAR images of Earl and Ike, and compared with wind estimates from in situ buoy observations and SFMR measurements, as well as H*Wind analysis [37]. It is shown that winds from C-2PO are quite consistent with SFMR measurements with a correlation coefficient of 0.80, whereas the CMOD5.N wind speeds are underestimated

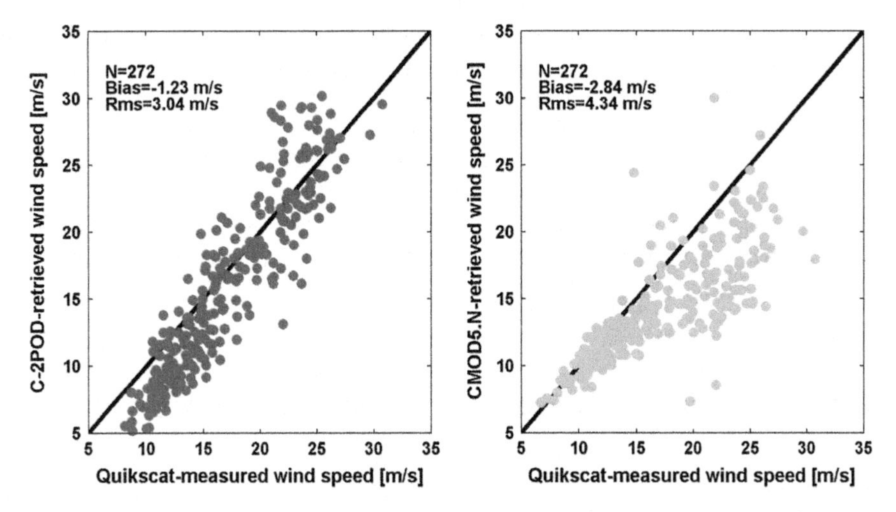

Fig. 5.5 **a** Comparisons of C-2POD and CMOD5.N SAR-retrieved wind speeds with RADARSAT-2 acquired dual-polarization SAR image of Hurricane Bill at 22:26 UTC on August 22, 2009, with collocated QuikSCAT-measured wind speeds at 22:54 UTC on August 22, 2009

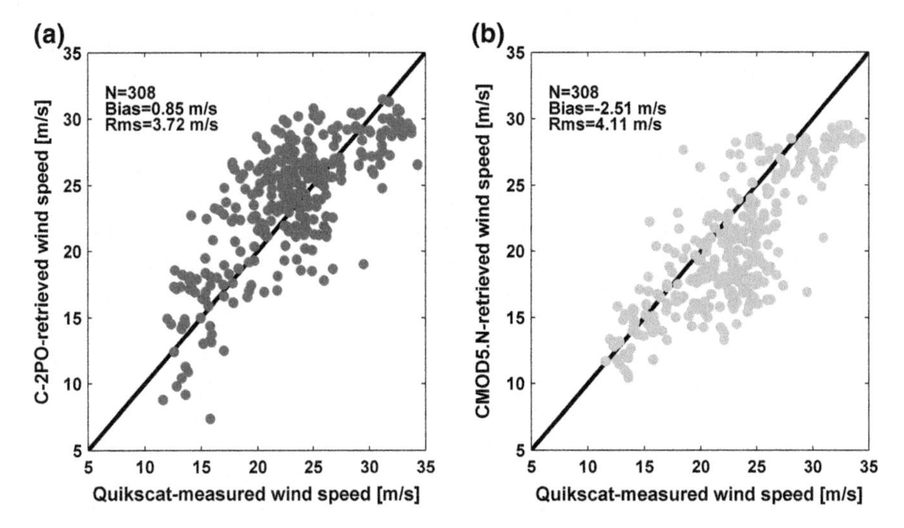

Fig. 5.6 **a** Comparisons of C-2POD and CMOD5.N SAR-retrieved wind speeds with RADARSAT-2 acquired dual-polarization SAR image of Hurricane Bertha at 10:14 UTC on July 12, 2008, with collocated QuikSCAT-measured wind speeds at 09:42 UTC on August 12, 2008

compared to SFMR values, with a correlation coefficient of 0.52 [37]. Winds from CMOD5.N have a bias of −4.14 m/s and a RMS error of 6.24 m/s; whereas for C-2PO winds, the bias and RMS difference are −0.89 and 3.23 m/s, respectively, which are significant improvements [37]. Wind speeds from CMOD5.N and C-2PO are also compared with the collocated NDBC buoy 41001 observations. The buoy measured

(10-m) wind speed is 18.1 m/s, whereas the CMOD5.N and C-2PO estimates are 19.5 and 18.7 m/s, respectively. To make a quantitative comparison of relatively high (up to 38 m/s) wind retrievals, H*wind fields from Hurricane Ike were compared with the wind speed results from C-2PO and CMOD5.N. It is shown that winds from CMOD5.N have a bias of −4.89 m/s and an RMS error of 6.51 m/s; for C-2PO winds, the bias and RMS difference are −0.88 and 4.47 m/s, respectively [37]. Moreover, QuikSCAT data were also used to validate the SAR retrieved high winds from C-2POD, using SAR images of Hurricanes Bill and Bertha [51], as presented in Figs. 5.5 and 5.6, showing similar results.

5.5 Intense Rainfall Effect on SAR Wind Retrievals

Challenges to satellite retrievals of ocean winds related to precipitation effects have been elaborated [52]. Although atmospheric attenuation and volume scatter induced by rain are markedly less at C-band based on straightforward electromagnetic theory, it is more difficult to determine the change in the surface splash backscatter [53]. An inevitable limitation to microwave remote sensing of the TC is the presence of extensive intense rain across most of the storm. Theoretical simulation investigations showed that for high winds and NRCS in VV polarization, attenuation is dominant until the rain rate reaches 15 mm/hr, which then leads to underestimates in the wind speeds [46]. For higher rain rates, for example, exceeding 20 mm/hr, the contribution to the effective NRCS from surface backscatter is much smaller than that due to volume scattering by rain. Thus, retrievals of surface wind vectors using NRCS under intense rain conditions are almost impossible without precise knowledge of the rain rate. Some studies using scatterometer data have shown that for wind above 30 m/s and rain rate exceeding 15 mm/hr, the error in the winds can be more than 10 m/s [54]. Moreover, the influence of heavy rain on the C-band ocean backscatter has been estimated from an existing radiative transfer model [55]. It was shown that the NRCS attenuation can be more than -1 dB for rain rates of 30 mm/hr, and even larger damping of up to −2 dB for rain rates of 50 mm/hr. Recent investigations show that for winds of more than 20 m/s, in storm or hurricane conditions, the 0.5 $\sim$ 1 dB NRCS calibration error will induce 3 $\sim$ 8 m/s errors using standard wind retrieval algorithms [10]. For the example of Hurricane Earl mentioned above, the heaviest rain rate (34.9 mm/hr) occurred at 23:12 UTC, corresponding to the largest wind speed difference between the CMOD5.N estimate (16.6 m/s) and the SFMR measurement (35.0 m/s) [37]. These large differences might be induced by dampened NRCS values, caused by heavy rain contamination, as well as wave effects and high sea states generated by winds. Using studies of three storms, recent investigation has demonstrated that the highest wind regions within these hurricanes are usually accompanied by significant rain [56]. This means that a major fraction of the TC's air-sea interface, at the highest winds, is affected by modifications to roughness induced by rain impacts. Therefore, a GMF is really needed that not only includes radar and wind vector parameters but also involves rain rate, in order to quantitatively estimate the heavy rain influence on radar backscatter. This suggested GMF can then be employed to implement rain effect modifications on NRCS values in heavy rainfall regions, which has potential to improve the high winds retrieval accuracy especially under intense rainfall environments. Further research in this direction should focus on simultaneous wind and rain retrievals with SAR.

5.6 Summary

High winds from multi-polarization SAR imagery face some challenges: (1) VV-polarized normalized radar cross section (NRCS) is saturated as the wind speeds approach and exceed hurricane force, (2) heavy rain contamination and additional effects associated with severe sea states can strongly dampen the co- (HH or VV) and cross-polarized (HV or VH) NRCS, (3) validation of SAR high wind retrievals is needed with in situ buoy and high-resolution research aircraft measurements, as well as observed Global Positioning System dropwind-sonde (GPS sonde) observations. Ongoing research will include additional high wind observations to improve the C-2PO model. Investigations have suggested that C-2PO represents a potential technique for hurricane observation from space. The future C-band RADARSAT Constellation Mission (RCM) SAR satellites will provide SAR data in compact polarization (CP) mode with large swath (350 km) and medium resolution (50 m), which can be used to develop a CP-SAR vector wind retrieval algorithm. Moreover, an empirical or theoretical GMF is expected to be developed for correcting the heavy rain effect on NRCS values, utilizing collocated SAR hurricane imagery with high resolution ground-based Next-Generation Radar (NEXRAD), and Tropical Rainfall Measuring Mission (TRMM) precipitation radar rain rate observations. Currently, dual-polarization (VV, VH) SAR has potential to simultaneously retrieve wind speed and wind direction. Wind speeds are directly inverted with C-2PO and HV- or VH-polarized NRCS. The resulting wind speeds, along with radar incidence angle and VV-polarized NRCS are imported into the CMOD5.N to retrieve wind directions with ambiguities. For the tropical cyclone with quasi-circular rotational wind structures centered on their eyes, the wind directions near the sea surface generally have an inflow angle relative to the tangential angle. This inflow angle can be used as a criterion to remove the directional ambiguities. The wind alias nearest to the counter-clockwise tangential direction, less than estimated inflow angle (toward the storm center), is chosen as the correct wind direction. Moreover, an approach needs to be developed for measuring TC intensity using multi-polarization SAR, in combination with the revised model for radial profiles of hurricane winds [57].

References

1. Friedman, K.S., and X. Li. 2000. Monitoring hurricanes over the ocean with wide swath SAR. *Johns Hopkins APL Technical Digest* 21 (1): 80–85.
2. Jones, W.L., P.G. Black, V.E. Delnore, and C.T. Swift. 1981. Airborne microwave remote-sensing measurements of Hurricane Allen. *Science* 214 (4518): 274–280.
3. Stoffelen, A., and D. Anderson. 1997. Scatterometer data interpretation: Estimation and validation of the transfer function CMOD4. *Journal of Geophysical Research: Oceans* 102 (C3): 5767–5780.
4. Quilfen, Y., B. Chapron, T. Elfouhaily, K. Katsaros, and J. Tournadre. 2003. Observation of tropical cyclones by high-resolution scatterometry. *Journal of Geophysical Research: Oceans* 103 (C4): 7767–7786.
5. Hersbach, H., A. Stoffelen, and S. de Haan. 2007. An improved C-band scatterometer ocean geophysical model function: CMOD5. *Journal of Geophysical Research: Oceans* 112 (C3).
6. Hersbach, H. 2010. Comparison of C-band scatterometer CMOD5.N equivalent neural winds with ECMWF. *Journal of Atmospheric and Oceanic Technology* 27 (4): 721–736.
7. Lehner, S., J. Horstmann, W. Koch, and W. Rosenthal. 1998. Mesoscale wind measurements using recalibrated ERS SAR image. *Journal of Geophysical Research: Oceans* 103 (C4): 7847–7856.
8. Horstman, J., H. Schiller, J. Schulz-Stellenfleth, and S. Lehner. 2003. Global wind speed retrieval from SAR. *IEEE Transactions on Geoscience and Remote Sensing* 41 (10): 2277–2286.
9. Monaldo, F.M., D.R. Thompson, R.C. Beal, W.G. Pichel, and P. Clemente-Colon. 2004. Comparison of SAR-derived wind speed with model predictions and buoy comparisons. *IEEE Transactions on Geoscience and Remote Sensing* 42 (2): 283–291.
10. Yang, X., X. Li, W.G. Pichel, and Z. Li. 2011. Comparison of ocean surface winds from ENVISAT ASAR, MetOp ASCAT scatterometer, buoy measurements, and NOGAPS model. *IEEE Transactions on Geoscience and Remote Sensing* 49 (12): 4743–4750.
11. Donnelly, W.J., J.R. Carswell, R.E. McIntosh, P.S. Chang, J. Wilkerson, F. Marks, and P.G. Black. 1999. Revised ocean backscatter model at C and Ku band under high-wind conditions. *Journal of Geophysical Research* 104 (C5): 11485–11497.
12. Powell, M.D., P.J. Vickery, and T.A. Reinhold. 2003. Reduced drag coefficient for high wind speeds in tropical cyclones. *Nature* 422 (6929): 279–283.
13. Donelan, M.A., B.K. Haus, N. Reul, W.J. Plant, M. Stiassnie, H.C. Graber, O.B. Brown, and E.S. Saltzman. 2004. On the limiting aerodynamic roughness of the ocean in very strong winds. *Geophysical Research Letters* 31 (18).
14. Bye, J.A.T., and A.D. Jenkins. 2006. Drag coefficient reduction at very high wind speeds. *Journal of Geophysical Research: Oceans* 11 (C3).
15. Fernandez, D.E., J.R. Carswell, S. Frasier, P.S. Chang, P.G. Black, and F.D. Marks. 2006. Dual-polarized C- and Ku-band ocean backscatter response to hurricane-force winds. *Journal of Geophysical Research: Oceans* 111 (C8).
16. Shen, H., Y. He, and W. Perrie. 2009. Speed ambiguity in hurricane wind retrieval from SAR imagery. *International Journal of Remote Sensing* 30 (11): 2827–2836.
17. Horstmann, J., D.R. Thompson, F. Monaldo, S. Iris, and H.C. Graber. 2005. Can synthetic aperture radars be used to estimate hurricane force winds? *Geophysical Research Letters* 32 (22).
18. Horstmann, J., W. Koch, S. Lehner, and R. Tonboe. 2002. Ocean winds from RADARSAT-1 ScanSAR. *Canadian Journal of Remote Sensing* 28 (3): 524–533.
19. Koch, W. 2004. Directional analysis of SAR images aiming at wind direction. *IEEE Transactions on Geoscience and Remote Sensing* 42 (4): 702–710.
20. Monaldo, F.M., D.R. Thompson, R.C. Beal, W.G. Pichel, and P. Clemente-Colon. 2001. Comparison of SAR-derived wind speed with model predictions and buoy comparisons. *IEEE Transactions on Geoscience and Remote Sensing* 39 (12): 2587–2600.

21. Shen, H., W. Perrie, and Y. He. 2006. A new hurricane wind retrieval algorithm for SAR images. *Geophysical Research Letters* 33 (21).
22. Thompson, D., T. Elfouhaily, and B. Chapron. 1998. Polarization ratio for microwave backscattering from the ocean surface at low to moderate incidence angles. *1998 IEEE International Geoscience and Remote Sensing Symposium Proceedings, IGARSS'98*. 3:1671–1673.
23. Mouche, A., D. Hauser, J.F. Daloze, and C. Guerin. 2005. Comparison of SAR-derived wind speed with model predictions and buoy comparisons. *IEEE Transactions on Geoscience and Remote Sensing* 43 (4): 753–769.
24. Johnsen, H., G. Engen, and G. Guitton. 2008. Sea-surface polarization ratio from ENVISAT ASAR AP data. *IEEE Transactions on Geoscience and Remote Sensing* 46 (11): 3637–3646.
25. Hwang, P.A., B. Zhang, and W. Perrie. 2010a. Depolarized radar return for breaking wave measurement and hurricane wind retrieval. *Geophysical Research Letters* 37 (1).
26. Zhang, B., W. Perrie, and Y. He. 2011. Wind speed retrieval from RADARSAT-2 quad-polarization images using a new polarization ratio model. *Journal of Geophysical Research: Oceans* 116(C8).
27. Bergeron, T., M. Bernier, K. Chokmani, A. Lessard-Fontaine, G. Lafrance, and P. Beaucage. 2011. Wind speed estimation using polarimetric RADARSAT-2 images: finding the best polarization and polarization ratio. *IEEE Journal of Selected Topics in Applied Earth Observations and Remote Sensing* 4 (4): 896–904.
28. Li, X., J.A. Zhang, X. Yang, W.G. Pichel, M. DeMaria, D. Long, and Z. Li. 2013. Tropical cyclone morphology from spaceborne synthetic aperture radar. *Bulletin of the American Meteorological Society* 94 (2): 215–230.
29. Holland, G.J. 1980. An analytic model of the wind and pressure profiles in hurricanes. *Monthly Weather Review* 108 (8): 1212–1218.
30. Reppucci, A., S. Lehner, J. Schulz-Stellenfleth, and S. Brusch. 2010. Tropical cyclone intensity estimated from wide-swath SAR images. *IEEE Transactions on Geoscience and Remote Sensing* 48 (4): 1639–1649.
31. Isoguchi, O., and M. Shimada. 2009. An L-band ocean geophysical model function derived from PALSAR. *IEEE Transactions on Geoscience and Remote Sensing* 47 (7): 1925–1936.
32. Ren, Y., S. Lehner, S. Brusch, X. Li, and M. He. 2012. An algorithm for the retrieval of sea surface wind field using X-band TerraSAR-X data. *International Journal of Remote Sensing* 33 (23): 7310–7336.
33. Thompson, D.R., J. Horstmann, A. Mouche, N.S. Winstead, R. Sterner, and F.M. Monaldo. 2012. Comparison of high-resolution wind fields extracted from TerraSAR-X SAR imagery with predictions from the WRF mesoscale model. *Journal of Geophysical Research: Oceans* 117(C2).
34. Lee, J.S., and E. Pottier. 2009. *Polarimetric radar imaging: from basics to applications*. Boca Raton: CRC Press.
35. Vachon, P.W., and J. Wolfe. 2011. C-band cross-polarization wind speed retrieval. *IEEE Geoscience and Remote Sensing Letters* 8 (3): 456–459.
36. Hwang, P.A., B. Zhang, J.V. Toporkov, and W. Perrie. 2010b. Comparison of composite Bragg and quad-polarization radar backscatter from RADARSAT-2: with applications to wave breaking and high wind retrieval. *Journal of Geophysical Research: Oceans* 115(C8).
37. Zhang, B., and W. Perrie. 2012. Cross-polarized synthetic aperture radar: a new potential technique for hurricanes. *Bulletin of the American Meteorological Society* 93 (4): 531–541.
38. Touzi, R., P.W. Vachon, and J. Wolfe. 2010. Requirement on antenna cross-polarization isolation for the operational use of C-band SAR constellations in maritime surveillance. *IEEE Geoscience and Remote Sensing Letters* 7 (4): 861–865.
39. Zhang, B., W. Perrie, P.W. Vachon, X. Li, W.G. Pichel, J. Guo, and Y. He. 2012. Ocean vector winds retrieval from C-band fully polarimetric SAR measurements. *IEEE Transactions on Geoscience and Remote Sensing* 50 (11): 4252–4261.
40. Powell, M.D., S.H. Houston, and T.A. Reinhold. 1996. Hurricane Andrew's landfall in south Florida. Part I: standardizing measurements for documentation of sea wind fields. *Weather Forecasting* 11 (3): 304–328.

41. Zhang, J., and E. Uhlhorn. 2012. Hurricane sea-surface inflow angle and observation-based parametric model. *Monthly Weather Review* 140 (11): 3587–3605.
42. Uhlhorn, E.W., and P.G. Black. 2003. Verification of remotely sensed sea surface winds in hurricanes. *Journal of Atmospheric and Oceanic Technology* 20 (1): 99–116.
43. Uhlhorn, E.W., P.G. Black, J.L. Franklin, M. Goodberlet, J. Carswell, and A.S. Goldstein. 2007. Hurricane surface wind measurements from an operational stepped frequency microwave radiometer. *Monthly Weather Review* 135 (9): 3070–3085.
44. Yueh, S.H., B.W. Stiles, and W.T. Liu. 2003. QuikSCAT wind retrievals for tropical cyclones. *IEEE Transactions on Geoscience and Remote Sensing* 41 (11): 2616–2628.
45. Laupattarakasem, P., and W.L. Jones. 2010. C. C. H, J. R. Allard, A. R. Harless, and P. G. Black. Improved hurricane ocean vector winds using SeaWinds active/passive retrievals. *IEEE Transactions on Geoscience and Remote Sensing* 48 (7): 2909–2923.
46. Tournadre, J., and Y. Quilfen. 2003. Impact of rain cell on scatterometer data: 1. theory and modeling. *Journal of Geophysical Research: Oceans* 108(C7).
47. Powell, M.D., S.H. Houston, L.R. Amat, and N. Morisseau-Leroy. 1998. The HRD real-time hurricane wind analysis system. *Journal of Wind Engineering and Industrial Aerodynamics* 77: 53–64.
48. Powell, M.D., S. Murillo, P. Dodge, E. Uhlhorn, J. Gamache, V. Cardone, A. Cox, S. Otero, N. Carrasco, B. Annane, and R.S. Fleur. 2010. Reconstruction of Hurricane Katrina's wind fields for storm surge and wave hindcasting. *Ocean Engineering* 37 (1): 26–36.
49. DiNapoli, S.M., M. Bourassa, and M.D. Powell. 2012. Uncertainty and intercalibration analysis of H*Wind. *Journal of Atmospheric and Oceanic Technology* 29 (6): 822–833.
50. Portabella, M. 2002. *Wind field retrieval from satellite radar systems*. Ph. D. thesis, University of Barcelona, Barcelona, Spain.
51. Zhang, B., W. Perrie, J.A. Zhang, E. Uhlhorn, and Y. He. 2014. High resolution hurricane vector winds from C-band dual-polarization SAR observations. *Journal of Atmospheric and Oceanic Technology* 31 (2): 272–286.
52. Weissman, D.E., B.W. Stiles, S.M. Hristova-Veleva, D.G. Long, D.K. Smith, K.A. Hilbum, and W.L. Jones. 2012. Challenges to satellite sensors of ocean winds: addressing precipitation effects. *Journal of Atmospheric and Oceanic Technology* 29 (3): 356–374.
53. Stiles, B.W., and R.S. Dunbar. 2010. A neural network technique for improving the accuracy of scatterometer winds in rainy condition. *IEEE Transactions on Geoscience and Remote Sensing* 48 (8): 3114–3122.
54. Yang, J., J.A. Zhang, X. Chen, Y. Ke, D. Esteban, J.R. Carswell, S. Frasier, D.J. Mclaughlin, P. Chang, P.G. Black, and F. Marks. 2004. Effect of precipitation of ocean wind scatterometry. In *2004 Proceedings of IEEE International Geoscience and Remote Sensing Symposium, IGARSS'04*, 20–24, Anchorage, Alaska.
55. Reppucci, A., S. Lehner, J. Schulz-Stellenfleth, and C.S. Yang. 2008. Extreme wind conditions observed by satellite synthetic aperture radar in the North West Pacific. *Nature* 29 (21): 6129–6144.
56. Weissman, D.E., and M.A. Bourassa. 2011. The influence of rainfall on scatterometer backscatter within tropical cyclone environments-implications on parameterization of sea-surface stress. *IEEE Transactions on Geoscience and Remote Sensing* 49 (12): 4805–4814.
57. Holland, G.J., J.I. Belanger, and A. Fritz. 2010. A revised model for radial profiles of hurricane winds. *Monthly Weather Review* 138 (12): 4393–4401.

Chapter 6
Observation of Sea Surface Wind and Wave in X-Band TerraSAR-X and TanDEM-X Over Hurricane Sandy

XiaoMing Li and Susanne Lehner

Abstract Several TerraSAR-X and TanDEM-X ScanSAR images are acquired in October, 2012 to track the Hurricane Sandy. Three of the images are acquired in the open sea, which are presented in this chapter to demonstrate observations of sea surface wind and wave extracted from X-band ScanSAR image with high spatial resolution of 17 m in the hurricane. In the case of the TerraSAR-X image acquired on October 26, 2012, we analyze the peak wave direction and length of swell generated by Hurricane Sandy, as well as interaction of swell with the Abaco Island, Bahamas. In the other two cases, sea surface wind field derived from the TerraSAR-X and TanDEM-X acquired on October 27 and 28 are presented. The sea surface wind speed retrieved by the X-band Geophysical Model Function (GMF) XMOD2 using wind direction derived from SAR images and the NOAA Hurricane Research Division (HRD) wind analyses are both presented for comparisons. We also compare the retrieved sea surface wind speed with Stepped Frequency Microwave Radiometer (SFMR) to quantify effect of rainfall on X-band SAR images.

6.1 Introduction

Tropical cyclones (TCs) are storm systems characterized by a large low pressure center that normally produce strong winds and heavy rainfall, as well as able to produce high waves and damaging storm surge when the TCs make their landfall over coastal regions. Thanks to spaceborne observations, TCs can be frequently

X. Li (✉)
Institute of Remote Sensing and Digital Earth, Chinese Academy of Sciences,
Beijing, People's Republic of China
e-mail: lixm@radi.ac.cn

X. Li
Hainan Key Laboratory of Earth Observation, Sanya, People's Republic of China

S. Lehner
Remote Sensing Technology Institute, German Aerospace Center (DLR),
Oberpfaffenhofen, Germany
e-mail: Susanne.Lehner@dlr.de

X. Li (ed.), *Hurricane Monitoring With Spaceborne Synthetic Aperture Radar*, Springer Natural Hazards, DOI 10.1007/978-981-10-2893-9_6

observed from space for analysis of wind, wave, rainfall, temperature and cloud-top properties. The qualitative and quantitative information of TC derived from satellite observations have been recognized as a crucial for improving numerical predictions of TC in short term and hindcast for climatology studies, e.g., as demonstrated by Velden et al. [1] for hurricane tracking forecast and by Zhang et al. [2] for Pacific typhoon reanalysis.

The spaceborne active remote sensing have the unique capability of observing sea surface through cloud, which has been playing an important role of monitoring response of sea surface under extreme weather situations. Scatterometers on board the European Remote Sensing (ERS), the Quick Scatterometer (QuikSCAT), and the Meteorological Operational (MetOp) satellites are particularly suitable for measurements of sea surface wind field, as both wind direction and wind speed can be derived without needing external information. Another active remote sensing instrument, spaceborne Synthetic Aperture Radar (SAR), e.g., the ERS-1/2 SAR, ENVISAT/ASAR, RADARSAT-1/2, TS-X/TD-X and Cosmo-Skymed, can not only provide sea surface backscatter intensity like scatterometer, but also image the sea surface in two-dimension with large spatial coverage and high spatial resolution, which provides abundant oceanic and atmospheric information of TCs, such as hurricane-generated long swell waves in small scales [3, 4], hurricane/typhoon eye morphology [5] and roll vortices occurred in marine boundary layer [6] in meso-scale.

With respect to measurements of sea surface wind field in TCs using scatterometer or SAR, two major sources may limit the accuracy of retrieval for high winds: (1) deficiencies of the Geophysical Model Function (GMF) for high winds, as presented in [7]. Improvement of GMF, such as CMOD5 [8] is dedicated for retrieval of high wind using scatterometer or SAR data, has somewhat reduces this error sources for inversion of sea surface wind field in hurricane scale [9]. However, one still faces the problem of speed ambiguity when applying CMOD5 for retrieving high winds [10] and saturation or damping of radar signal under severe weather conditions [11]. (2) Effects of heavy rains on radar signal. Microwave signals are likely to suffer effect of heavy rains which are permanent features in TCs and therefore errors are induced of deriving sea surface wind speed, e.g., studies presented by Quilfen et al. [7] and Weissman et al. [12]. Yueh et al. [13] proposed an updated GMF for retrieval of sea surface wind field considering the rain rate as a parameter, which is applied to the hurricane Floyd with maximum wind speed reaching 60 m/s showing a good agreement NOAA Hurricane Research Division (HRD) wind reanalysis.

The new generation spaceborne SAR sensors, such as represented by RADARSAT-2, TerraSAR-X/TanDEM-X (TS-X/TD-X) and Cosmo-Skymed, are highlighted by their high spatial resolution (down to 1 m), flexible imaging modes and polarimetric capabilities. Some preliminary studies related to TCs are presented for sea surface wind field retrieval for Typhoon Megi [14] using TS-X data and Hurricane Earl [15] using RASARSAT-2 data. In this chapter, we present detailed analysis of sea surface wind and wave field derived from three TS-X and TD-X images acquired over Hurricane Sandy.

6.2 TerraSAR-X and Tandem-X Data

The German X-band SAR TS-X was launched successfully on 15 June 2007 from Baikonur, Kazakhstan. The satellite is in a near-polar orbit around the Earth, at an altitude of 514 km. Using its active radar antenna, TS-X is able to produce image data with a resolution down to one meter, independent of weather conditions and daylight. It has been fully operational since January 7, 2008. Main technical parameters of TS-X are briefed in Table 6.1. The detailed information of TS-X mission, design, as well as ground segment is available in [16, 17]. Figure 6.1 illuminates three different imaging modes of TS-X, i.e., Spotlight, Stripmap and ScanSAR modes. For both Stripmap and ScanSAR modes, the radar beam can be electronically tilted within a range of 20–45° perpendicular to the flight direction without having to move the satellite itself. For Spotlight mode, the radar beam can be further tilted to 55°. Scenes sizes and resolutions of the three imaging modes of TS-X are listed in Table 6.2.

Table 6.1 Main TerraSAR-X system parameters

Height	4.88 m
Width	2.4 m
Payload mass	About 400 kg
Radar frequency	9.65 GHz
Power consumption	800 watt (on average)
Resolution	1, 3 m, or 16 m depending on image size
Polarization	HH/VV/HV/VH
Orbit altitude	514 km
Inclination	97.4°, sun-synchronous
Mission life time	At least 5 years

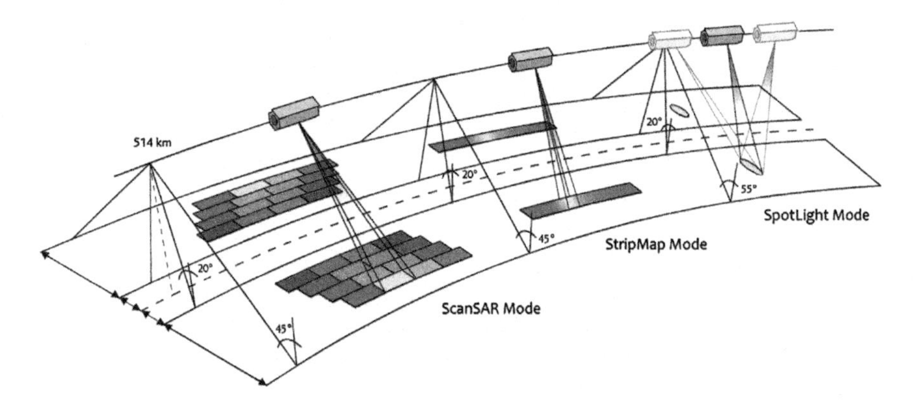

Fig. 6.1 Illumination of TS-X imaging modes (© DLR)

Table 6.2 Features of TS-X imaging modes

Imaging mode	Resolution (Range × Azimuth)	Scene size (Range × Azimuth)
Spotlight	1 m × 1.5 m ... 3.5 m	10 km × 5 km (variable)
cStripmap	3 m × 3 m	30 km × 50 km (variable)
ScanSAR	16 m × 16 m	100 km × 150 km (variable)

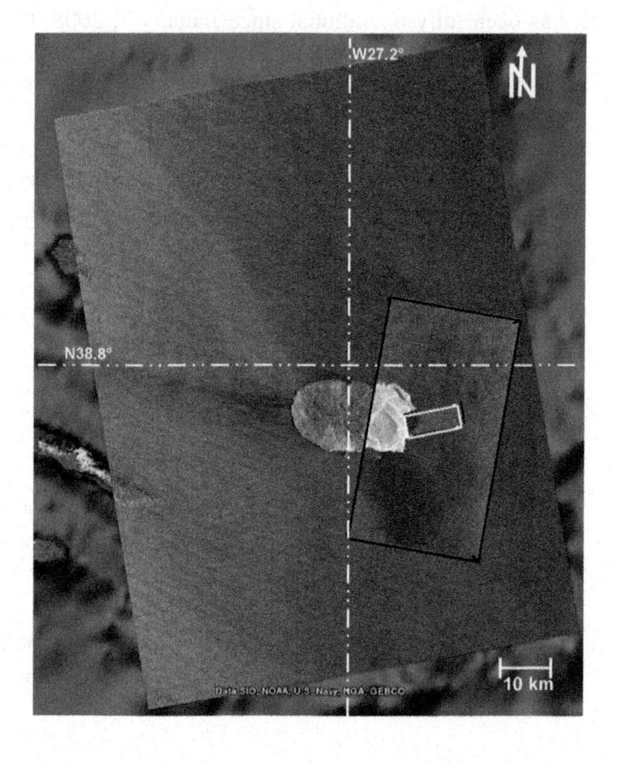

Fig. 6.2 Overlay of TSX ScanSAR (largest one) image acquired on March 20, 2008, StripMap image (in the *black rectangle*) acquired on January 15, 2009 and Spotlight mode image (in the *white rectangle*) acquired on March 26, 2008 over Terceira island (map in background © Google earth)

To demonstrate variable applications of TS-X in different imaging modes in oceanography, three quick looks of TS-X images over the Terceira Island in the North Atlantic operated respectively in ScanSAR, StripMap and Spotlight mode overlaid on top of each other are shown in Fig. 6.2 [18]. The ScanSAR image has the largest area coverage of 140 and 100 km in azimuth and range direction, respectively, with a pixel size of 8.25 m.

Wind streaks are visible in the ScanSAR image, and surface wind blowing toward northwest is inferred by the shadow zone behind the Santa Barbara volcano (1021 m) at the NW end of the island. The ScanSAR image yields an overview of sea state and wind field for the entire oceanic region around the island. The StripMap image (inside the black rectangle) acquired in the eastern coast of Terceira island shows spatial variations of wave refraction when approaching to the coasts. The Spotlight image inside the white rectangle has the smallest coverage of 5 km in azimuth and

12 km in range, which is are particularly suitable for investigation of near-shore processes, harbor monitoring, and targets detection.

The TanDEM-X (TerraSAR-X add-on for Digital Elevation Measurement, TD-X) mission was launched successfully as well on June 21, 2010. The first bistatic SAR mission is formed by adding a second while almost identical X-band SAR to TS-X, which opens a new era in spaceborne radar remote sensing. The twins can operate either collaboratively, e.g., in bistatic and monostatic modes, or independently.

6.3 X-Band GMF for Sea Surface Wind Retrieval from TS-X and TD-X

The X-band Geophysical Model Function (GMF) called XMOD2 [5] is used for retrieval of the sea surface wind field from TS-X and TD-X. Brief description of XMOD2 is given in following.

XMOD2 is written as

$$z(v, \phi, \theta) = B_0^p(v, \theta)(1 + B_1(v, \theta)cos\phi + B_2(v, \theta)cos2\phi) \tag{6.1}$$

where B_0, B_1, and B_2 are functions of incidence angle θ and sea surface wind speed v at 10 m height. Relative direction ϕ is the angle between wind direction ϕ and radar look direction α, i.e. $\phi = \varphi - \alpha$. The constant p has value of 0.625. In the proposed GMF XMOD2 for X-band SAR, transfer functions used to depict B_0 and B_2 are adopted from the CMOD5, while a second-order polynomial function is used to describe the dependence of B_1 on the sea surface wind speed and incidence angle as given in Eq. (6.2).

$$B_1 = \sum_{j=0}^{2} \sum_{i=0}^{2} \alpha_{ij} \theta^i v^i \tag{6.2}$$

In total, 32 coefficients are included in XMOD2, which are determined by tuning dataset consisting of in situ buoy measurements (371 collocations) and DWD (German Weather Service) analysis atmospheric model (639 collocations). Figure 6.3 shows the simulated for X-band TS-X using the XMOD2 for incidence angles of 20°, 30° and 40° against the relative sea surface wind direction.

The simulation shows that the XMOD2 can represent properly the anisotropic effect of wind direction on the sea surface backscatter. Moreover, the incidence angle effect on the difference between upwind and crosswind, as well as on the difference between upwind and downwind, are also distinct. One can find that the higher incidence angle, the more sensitive is on the sea surface wind direction. Figure 6.4a and b show the simulated sea surface backscatter using XMOD2 for incidence angles of 20°, 30° and 40° in upwind and cross wind situations. The same simulations but using the CMOD5 are also presented for comparison. It is interesting

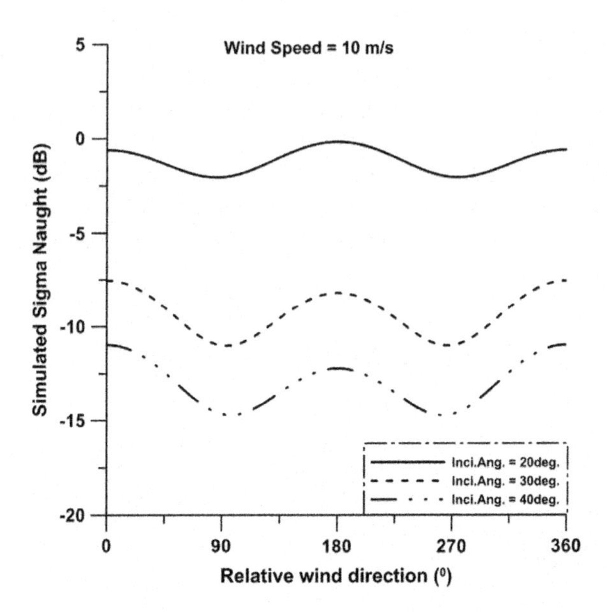

Fig. 6.3 Simulated sea surface backscatter in X-band SAR using the XMOD2 for incidence angles of 20°, 30°, and 40° against the relative sea surface wind direction in the sea surface wind speed of 10 m/s

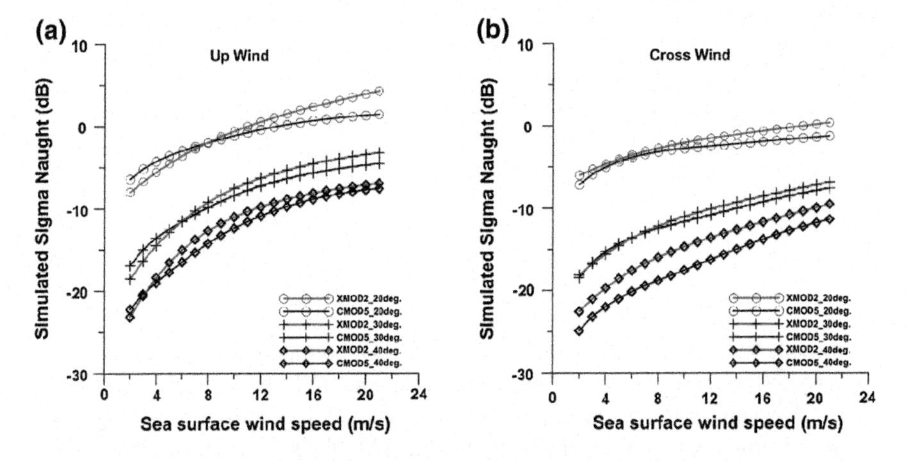

Fig. 6.4 Simulated σ_0 in X-band SAR using the XMOD2 for incidence angles of 20°, 30°, and 40° against the sea surface wind speed in upwind (**a**) and crosswind (**b**), as represented by *black lines*. The *red lines* show the simulation of σ_0 in C-band SAR using the CMOD5 for comparison

to notice that, for upwind, the transition of the difference between X-band and C-band σ_0 shows a dependence on incidence angle. For incidence of 20°, 30°, and 40°, the transition appears at around 8, 6 and 4 m/s.

In order to reduce irregular distributions of wind speed and wind direction in the tuning dataset, large amount of DWD wind model results, which is around two

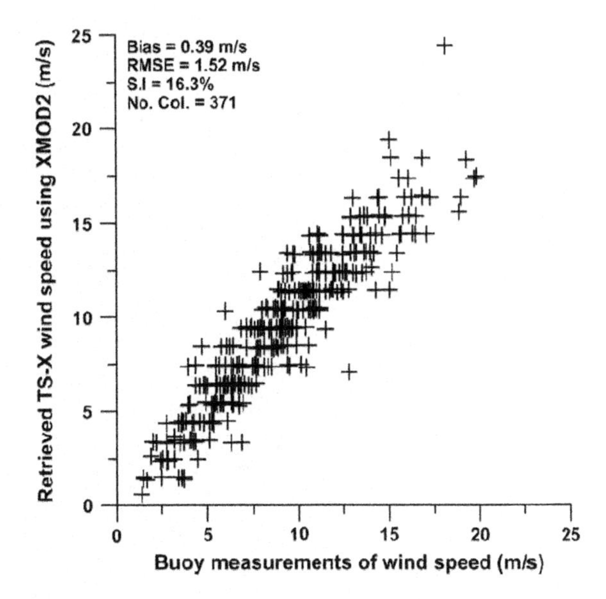

Fig. 6.5 Comparison of the retrieved TS-X/TD-X U_{10} using the XMOD2 against in situ buoy measurements. Buoy measurements of the sea surface wind direction is used as input for the retrieval

times of the collocated buoy amount, is also added in the tuning dataset. Therefore, comparison of the retrieved sea surface wind speeds to buoy measurements is still necessary to verify XMOD2. The collocated buoy measurements of wind direction, incidence angle and averaged sea surface backscatter σ_0 of the TS-X subscenes are used as input to XMOD2 to retrieve the sea surface wind speed. The comparison is shown in Fig. 6.5.

6.4 Analysis of TS-X and TD-X Data for Hurricane Sandy

When the two X-band SAR sensors operate independently, it can increase opportunity to track oceanic or atmospheric phenomena with significant temporal and spatial variations, such as tropical cyclones analyzed in this chapter. The three colorful rectangles shown in Fig. 6.6 represent three ScanSAR images acquired by TS-X or TD-X in VV polarizations on October 26, 27, and 28, respectively. Acquisition details of the three X-band ScanSAR images over the Hurricane Sandy are given in Table 6.3.

6.4.1 Observation on October, 26

Two continuous TS-X ScanSAR scenes are acquired in descending orbit on October, 26 at 11:17 UTC, as indicated by the two pink rectangles in Fig. 6.6. The first scene

Fig. 6.6 TS-X and TD-X acquisitions over the Hurricane Sandy. The *pink*, *yellow* and *green rectangles* represent the X-band ScanSAR images on October 26, 27, and 28, respectively. Best track of the Hurricane Sandy issued by NOAA is overplayed on the Google Earth map

Table 6.3 Acquisitions of TS-X and TD-X ScanSAR images over the Hurricane Sandy

Mission	Time	Incidence angle	Distance to Hurricane center
TS-X	2012-10-26T 11:17 UTC	19.6° – 30.3°	~36 km
TS-X	2012-10-27T 23:05 UTC	33.9° – 42.3°	~170 km
TD-X	2012-10-28T 22:50 UTC	31.7° – 40.5°	~190 km

acquired in open sea, about 70 km away to the north of the Abaco island, Bahamas, is shown in right panel of Fig. 6.7.

The key to derive surface wave information, e.g., peak wavelength and direction, directly from SAR image spectra without through nonlinear inversion [19] is that SAR should image linearly the sea surface waves under certain conditions. The velocity-bunching mechanism [20] plays an important role on whether surface waves are imaged linearly or not by SAR. To quantify whether SAR images surface waves in linear or nonlinear regime, a critical velocity-bunching cutoff significant wave height (H_s) given in [21] is represented here.

$$H_s = \frac{1}{\pi^{3/2}\sqrt{g}} \frac{v}{h} \frac{\lambda^{3/2}}{\sqrt{1 + tan^2\theta sin^2\phi cos\phi}} \tag{6.3}$$

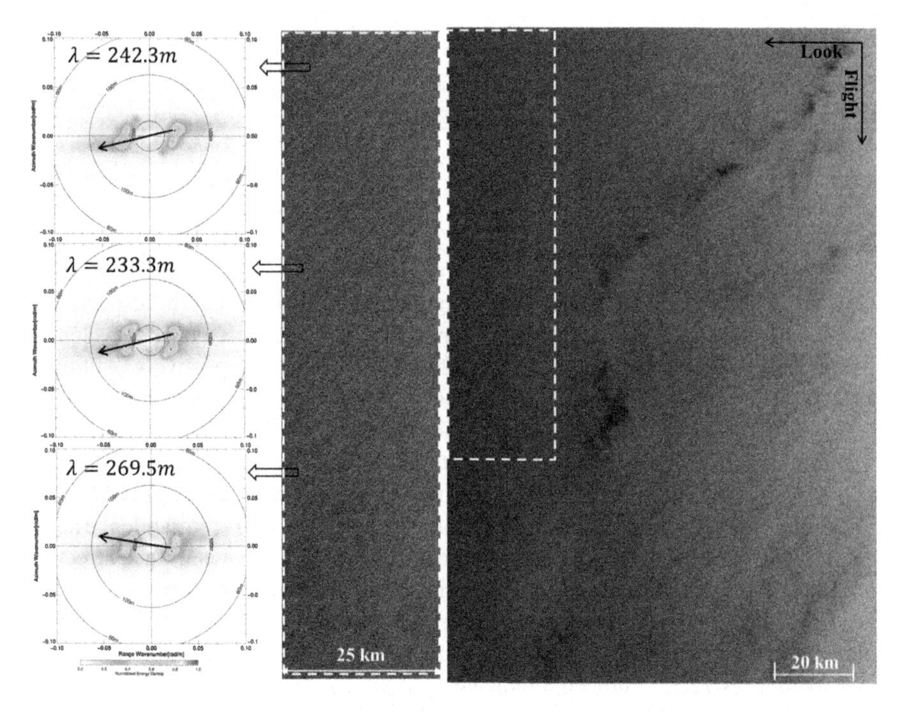

Fig. 6.7 The TS-X ScanSAR first scene (*right panel*) acquired on October 26, 2012 at 11:17 UTC near to Hurricane Sandy. The *right panel* shows the sub-image in the *white rectangle* and image spectra

where v is satellite velocity and h is satellite orbit altitude. For TS-X and TD-X, v has a value of 7.6 km/s which is similar with the previous spaceborne SAR sensors such as ERS/SAR, RADARSAT and ENVISAT/ASAR. However, orbit altitude of TS-X and TD-X is 514 km, which is much lower than those platforms. λ is cutoff wavelength in unit of meter. θ is incidence angle in degree. The relative angle between surface wave propagation and flight (azimuth) direction is defined as ϕ. For azimuth traveling waves, $\phi = 0°$ and for range traveling waves, $\phi = 90°$. Prior to deriving wave parameters directly from TS-X image spectra, we firstly analyze whether sea state conditions are favorable for TS-X imaging sea surface linearly.

In this case, the NOAA/NDBC buoy 41010 locates around 60 km west of the TS-X first scene, as indicated in Fig. 6.6. The buoy measurements at 11:20 UTC on October 26, 2012 are listed in Table 6.4. As the mean wave direction is not available in buoy 41010, we further add measurements of buoy 41009 for analysis, which is 220 km away from the TS-X image.

The dispersion relation of surface waves in finite depth water is:

$$\omega^2 = gktan(kH) \tag{6.4}$$

Table 6.4 TS-X Collocated Buoy 41010 and 41009 measurements

Buoy ID	Time	Hs	Dominant Peak period	Mean wave direction	Wind speed	Water depth
41010	11:20 UTC	7.18 m	12.12 s	N/A	18.7 m/s	872.6 m
41009	11:50 UTC	5.79 m	12.12 s	83°	15.1 m/s	40.5 m

where ω is wave frequency, k is wave number and H is water depth, respectively. By knowing the measured dominant peak period T, the calculated peak wavelength in location of buoy 41010 and 41009 is 229.5 and 196.5 m, respectively. Although buoy 41009 is about 160 km away from buoy 41010, we notice that the dominant peak periods measured by both buoys are identical.

Buoy 41010 locates near to far range of the TS-X image, where incidence angle is 30.3°. Therefore, estimated velocity-bunching cutoff significant wave height using Eq. (6.3) against wave propagation direction between 0° and 80° for incidence angle of θ 30° and cutoff wavelength λ is plotted in Fig. 6.8. Although mean wave direction is not available in buoy 41010, we can infer that it should be near to 83° from measurements of buoy 41009. Azimuth angle of the TS-X image is 190.8° clockwise relative to North. Therefore, angle ϕ in Eq. (6.3), i.e. wave propagation relative to SAR azimuth is 72.2°. Figure 6.8 indicates that the cutoff H_s for $\phi = 72°$ and 73° is 6.8 and 7.2 m, respectively, which is near to H_s of 7.18 m measured by buoy 41010. Therefore, it is inferred that the TS-X under this condition can image swell waves linearly, which ensures that we can use SAR image spectra derived from TS-X to analyze swell wave propagation in this case.

Three image spectra derived from sub-image are shown in left panel of Fig. 6.7. The image spectra are derived from subscene with pixel size of 1024 x 1024, which

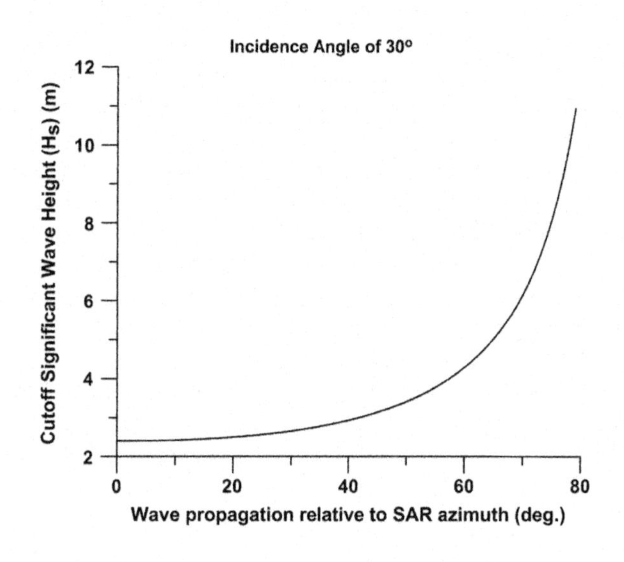

Fig. 6.8 Estimated cutoff H_s of TS-X as a function of wave propagation direction relative to SAR azimuth for incidence angle of 30° and cutoff wavelength of 250 m

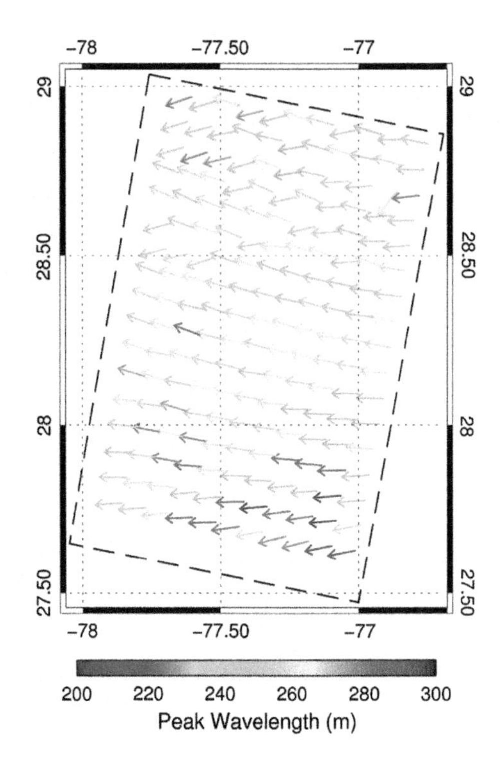

Fig. 6.9 Peak wavelength and direction from SAR image spectra for the TS-X scene shown in *right panel* of Fig. 6.7

corresponds to a spatial size of 8.5 km × 8.5 km. Each subscene is further divided into four small subsets with pixel size of 256 × 256 (around 2 km × 2 km). Then the four corresponding FFT spectra are summed and averaged as the image spectrum for a subscene. The second image spectra shows peak wavelength is 233.3 m which agrees well with buoy 41010 measurement of 229.5 m. The peak wave direction has a slight change of around 20° from north to south within the sub-image. Figure 6.9 shows the peak wavelength and direction from SAR image spectra for the whole TS-X scene to demonstrate spatial variations of swell generated by Hurricane Sandy. 180° ambiguity of wave propagation direction derived from SAR image spectrum is removed based on buoy measurement. Over the image sea surface, swell wavelength is longer than 220 m. The peak wavelength increases homogeneously from around 220 to 300 m from northwest to southeast. Over the sea above 28.5°N, the peak wave direction varies between northeast and east. It should indicate a crossing swell sea state, as also observed in the second image spectra shown in Fig. 6.7. However, the southern water region shows that swell comes homogeneously from east with peak wavelength near to 300 m.

The second TS-X scene acquired on October 26 is shown in Fig. 6.10. It covers the northern and southern coastal region of the Abaco island. Land in the lower right part of the image is the Great Abaco and the Little Abaco is in the middle of the image. On "top" of Great and Little Abaco, there is a long chain of islands consist

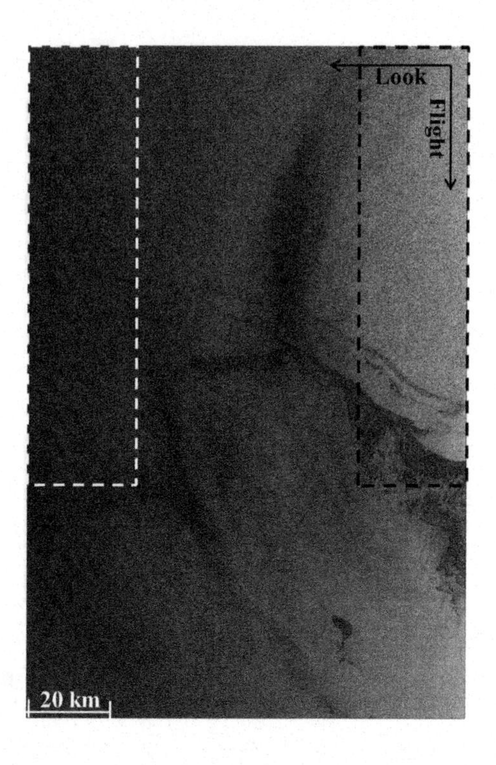

Fig. 6.10 The second TS-X scene acquired on October 26, 2012 at 11:17 UTC over the Abaco islands, Bahamas

of a large amount of sand bank and coral reefs in shallow water, which block the hurricane generated waves propagating further to south. Therefore, in the second TS-X scenes, we can clearly observe that sea state exhibits significant spatial variations in the northern and southern water regions of the long chain of islands.

Figure 6.11 shows the two sub-images inside the rectangles shown in Fig. 6.10. Two SAR image spectra are derived from subscenes in the two areas are shown as well. The two image spectra derived from the sub-image in right panel shows consistent peak wavelength and direction with those derived in the first TS-X scene shown in Fig. 6.9. The long island chain in the north of the Abaco island plays a role as a barrier which blocks swell further propagates toward to the island coast. In the left panel, the first SAR image spectrum is derived from subscene in the north of the island chain, which also shows consistent swell peak wavelength and direction with those in the near range sub-image (right panel). However, in the second SAR image spectrum which is derived from subscene locates in the south of the island chain, we observe a peak locates in the inner circle in order of kilometers, which is much longer than wavelength of swell observed in the north. It is inferred as the peak of wind streaks. Wind direction is perpendicular to connected line of the inner peaks, as indicated by dash line in the image spectrum. The other peak, however, only has wavelength of 209 m, which is 105 m shorter than swell wavelength in the north of the island chain. Further, we also notice that peak wave direction (solid line) of this wave system is close to wind direction. Thus, it should be the young swell

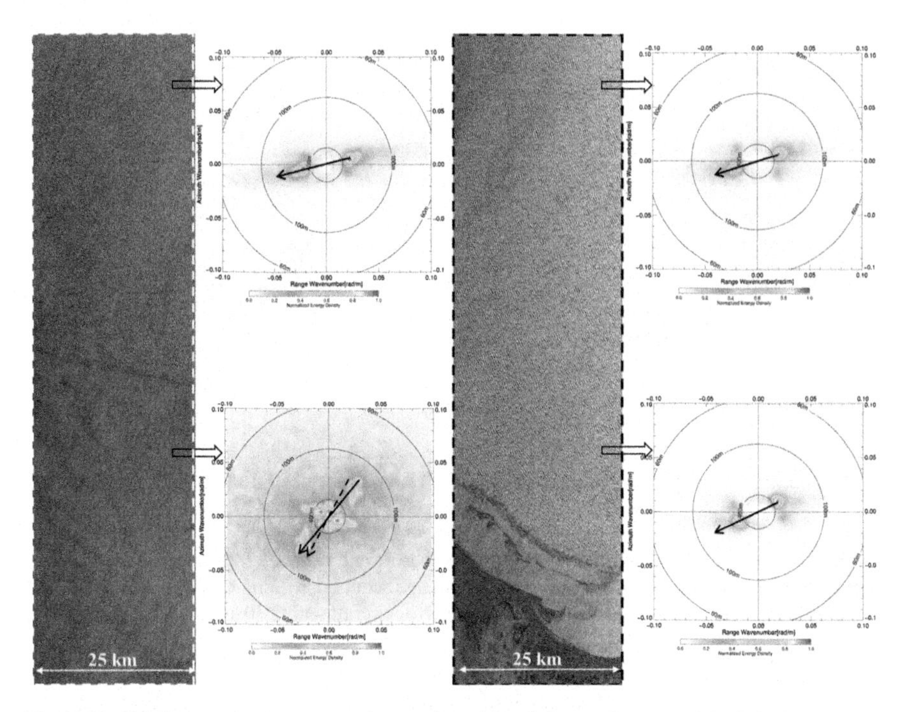

Fig. 6.11 Sub-images in the near (*right panel*) and far (*left panel*) range of the TS-X image and SAR image spectra. *Solid lines* indicate wave direction. *Dash line* indicates wind direction

generated by local hurricane wind, as most incoming long swells are blocked by the island chain. Therefore, this special situation provides us an opportunity for further investigation of wave growth in hurricane scale wind field.

6.4.2 *Observations on October, 27 and 28*

TS-X and TD-X acquired another two images over Hurricane Sandy on October 27 and 28, respectively, as represented by the yellow and green rectangles in Fig. 6.6. With respect to the two images, we focus on sea surface wind field analysis, particularly to verify performance of the XMOD2 for sea surface wind retrieval in extreme weather situations.

6.4.2.1 TS-X/TD-X Images and Retrieved Sea Surface Wind Field

Figure 6.12a and b show the HRD sea surface wind field on October 27 and 28 at 22:30 UTC, respectively. Spatial coverage of the TS-X images represented by white rectangles is overlaid on the HRD wind field map. The black lines show track

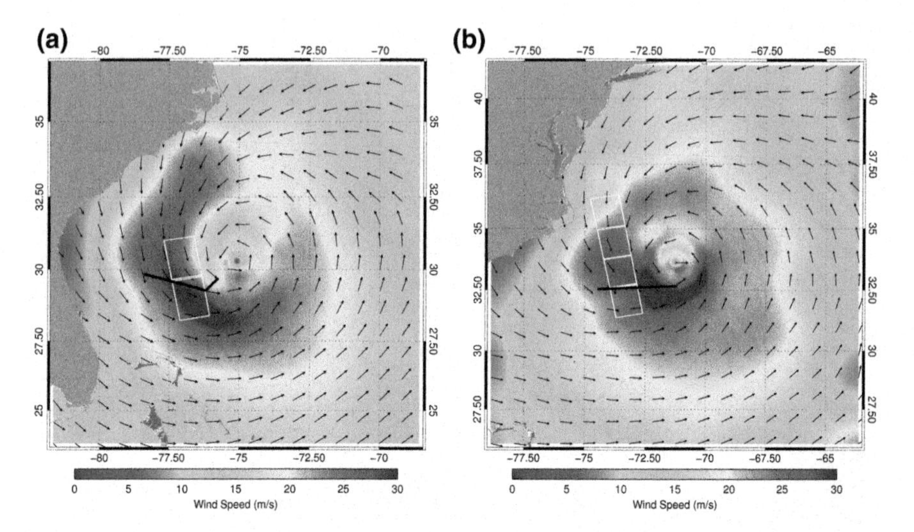

Fig. 6.12 HRD sea surface wind field at **a** on October 27 and October 28 **b** at 22:30 UTC. *White rectangles* represent spatial coverage of the TS-X image acquired at 23:05 UTC on October 27 and 22:50 UTC on October 28. *Block line* shows the track of aircraft within 30 min of SAR acquisitions

of the US aircraft which obtained the Stepped Frequency Microwave Radiometer (SFMR) measurements at close time (within 30 min) to SAR acquisitions. The TS-X calibrated image on October 27 at 23:05 UTC and October 28 at 22:50 UTC is shown in Figs. 6.13a and 6.14a, respectively. In both cases, we dont observe complete eye structure in the SAR images. However, from characters of the image, e.g., large contrast of sea surface backscatter, we may refer that both are acquired near to center of Sandy, which on the other hand indicates that Sandy has a big size considering that swath width of the TS-X/TD-X ScanSAR is 100 km.

As SAR has only one antenna, sea surface wind direction has to be obtained firstly in order to retrieve sea surface wind speed from SAR image by exploiting GMFs. In both cases, wind direction obtained from HRD analysis wind field are used a priori to retrieve the sea surface wind speed by XMOD2. The results are shown in Figs. 6.13b and 6.14b, respectively. The other way to obtain sea surface wind direction is to use FFT method (e.g., [22]) or Local Gradient (LG) method [23] if wind streaks are visible in SAR images. Figures 6.13c and 6.14c are the retrieved sea surface wind speed using wind direction derived from SAR images through the FFT method for the TS-X and TD-X cases, respectively. The HRD wind direction generally agrees well with that derived from SAR images directly. With respect to the case on October 27, we find that the major differences between HRD and TS-X wind directions are in lower left part of the image, i.e. the southwest quadrant. The HRD wind direction has a bias of around 15° compared to wind streaks visible in the TS-X image, which in turn induces the retrieved wind speed around 5 m/s lower than that by using wind direction derived from wind streaks in the TS-X image. We also observe that the sea

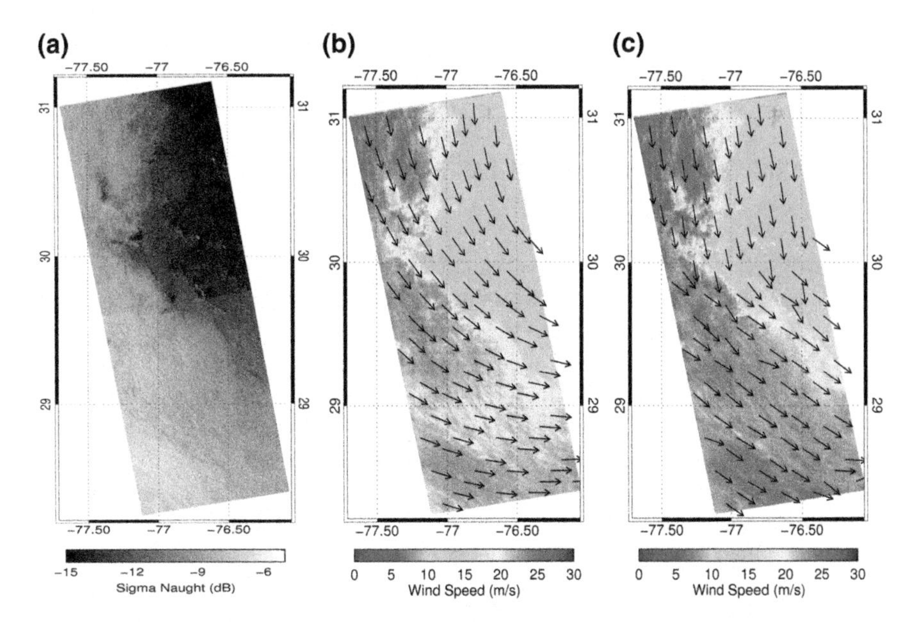

Fig. 6.13 **a** TS-X calibrated image acquired on October 27 at 23:05 UTC. **b** Retrieved TS-X wind speed using HRD wind direction; **c** Retrieved TS-X wind speed using TS-X wind direction

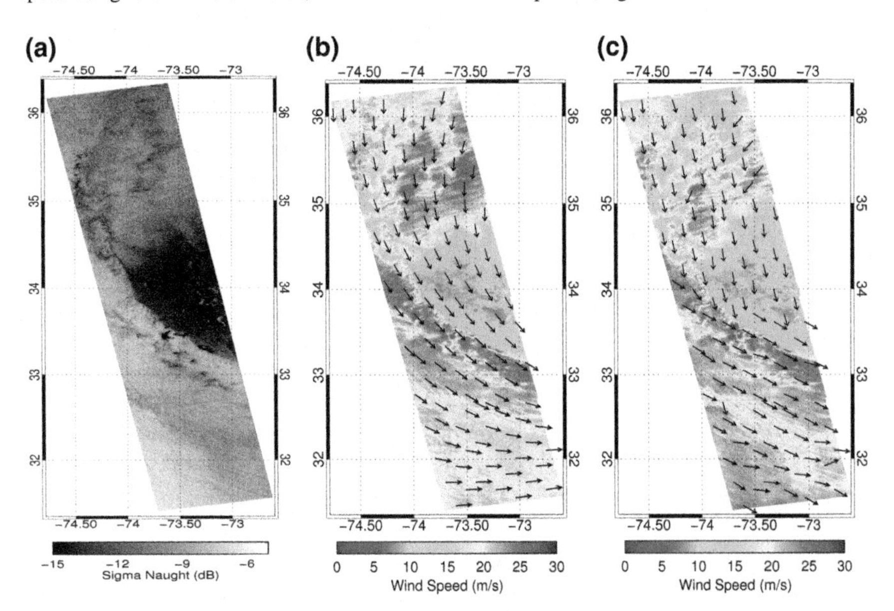

Fig. 6.14 **a** TS-X calibrated image acquired on October 28 at 22:50 UTC. **b** Retrieved TS-X wind speed using HRD wind direction; **c** Retrieved TS-X wind speed using TS-X wind direction

surface wind speed retrieved by using TD-X wind direction is higher than that by using the HRD wind direction in the southwest quadrant in the second case.

6.4.2.2 Influence of Rainfall on X-Band SAR Data and Its Retrieved Sea Surface Wind Field

On both TS-X and TD-X images, we can observe effect of rainfall on X-band SAR data, for instance, one can find some bright slicks in the large dark pattern area in both images. The rain band of Sandy is also clearly visible in the TD-X image (Fig. 6.14 (a)). In a previous study, Melsheimer et al. [24] investigate effects of rain cell on SAR imagery using SIRC/X-SAR data in multifrequency (L-, C-, and X-band) and multipolarization (HH, VV and HV). They summarized that the radar backscatter over ocean in the presence of rain cells is mainly associated with three processes: (1) scattering and attenuation of radar microwaves by hydrometeors in atmosphere, (2) the modification (enhancement or reduction) of sea surface roughness by rain drops, and 3) the enhanced sea surface roughness by wind gust. Generally, the latter two processes are easily identified in SAR imagery as both can significantly change the sea surface roughness. However, if the radar microwaves are attenuated by rain cell in atmosphere, it maybe not as clearly manifested in SAR image as other processes, particularly when rain fall has a large spatial coverage such as in hurricanes or typhoons. To quantify attenuation of radar backscatter induced by rain fall, Urlaby [25] proposed a radiative transfer model for C-band radar. Danklmayer and Chandra [26] present a model to quantify the attenuation for Ka- and X- band SAR. This two-way path attenuation model is given as:

$$A(t) = r(t) \cdot 2 \cdot \frac{H}{cos\theta}\,[dB] \tag{6.5}$$

where, H is rain layer height, θ is local incidence angle of SAR. $r(t)$ is the attenuation coefficient in unit of dB/km, which is calculated via rain rate:

$$r(t) = a \cdot R^b \tag{6.6}$$

in which, R is rain rate in unit of mm/hr. The regression coefficients a and b is 0.0136 and 1.15 for X-band microwave. We use this model and rain rate measured by SFRM to quantify attenuation of X-band SAR radar backscatter induced by rain fall in Hurricane Sandy.

Figure 6.15a and b show the retrieved SAR sea surface wind speed (red line) and collocated SFMR measurements of sea surface wind speed (blue line) and rain rate (black line) on October 27 and October 28, respectively. The retrieved SAR sea surface wind speed shows a trend of underestimation compared to SFMR measurements, which tends to be consistent with rainfall rate measured as well by the SFMR.

We calculate X-band radar backscatter using the SFMR measurements of the sea surface wind speed, wind direction and incidence angle derived from SAR, which

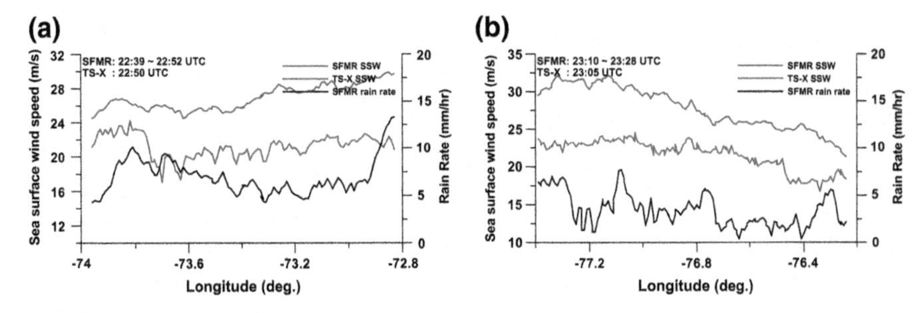

Fig. 6.15 Comparisons of SAR retrieved sea surface wind speed and SFMR measurements on October 27 (**a**) and October 28 (**b**)

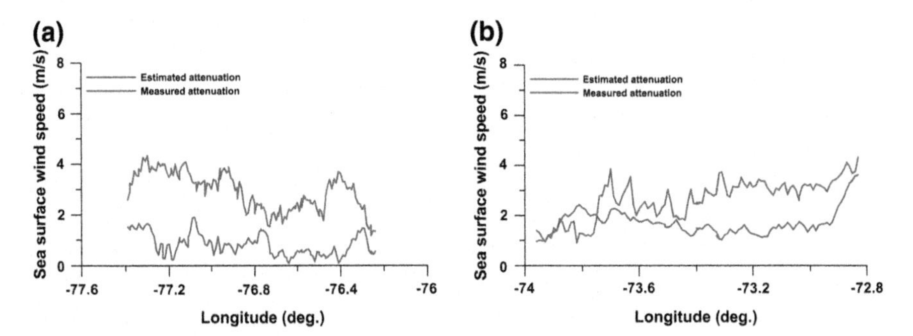

Fig. 6.16 Comparisons of estimated X-band SAR attenuation induced by rain fall to measured attenuation on October 27 (**a**) and October 28 (**b**)

is denoted σ_0^{sim}. It is assumed that the differences between observed TS-X/TD-X radar backscatter σ_0^{SAR} and the simulation σ_0^{sim} is the attenuation $\Delta\sigma$ induced by rain fall. By applying SFMR measurements of rain fall rate to Eqs. (6.5) and (6.6), the X-band SAR attenuation $\Delta\tilde{\sigma}$ is estimated. Figure 6.16a and b shows the comparisons between $\Delta\sigma$ and $\Delta\tilde{\sigma}$ for the TS-X and TD-X case, respectively. However, as shown in Fig. 6.15a and b, the estimated attenuation $\Delta\tilde{\sigma}$ (blue line) is still lower than the calculated $\Delta\sigma$ (red line) with mean value of 2 dB and 1 dB for the case on October 27 and October 28, respectively. This indicates that attenuation of radar backscatter induced by rain fall in X-band SAR should be stronger than that estimated by Eq. (6.5). Comparing Figs. 6.15 with 6.14, we can find that if rain fall rate is larger than 5 mm/hr, e.g., in the case on October 28, the estimated attenuation $\Delta\tilde{\sigma}$ by Eq. (6.5) shows better agreement with $\Delta\sigma$ than that in the case on October 27 which has rain rate lower than 5 mm/hr. Nevertheless, we need further investigation to investigate rain fall attenuation on X-band SAR data more accurately.

6.5 Discussion and Conclusion

Three X-band TS-X and TD-X ScanSAR images with swath width of 100 km and spatial resolution of 17 m acquired over Hurricane Sandy are presented. For the TS-X image acquired on October 26, 2012 over the Abaco island, we focus on analyzing spatial variations of hurricane generated swell. In order to derive peak wave parameters directly from the SAR image spectra, the so-called cutoff H_s is calculated using buoy measurements. Although sea state of this case is very rough with H_s of 7.18 m, the calculated cutoff H_s indicates that imaging of sea surface waves by TS-X under this situation is still a linear process. This should attribute to the orbit altitude of TS-X and TD-X (514 km) is much lower than that of the conventional spaceborne SAR sensors, such as RADARSAT and ENVISAT/ASAR (790 km). The TS-X image spectra in this case show that the long island chain of the Abaco island blocks swell propagates further towards to coastal area. Therefore, only windsea and young swell waves are observed behind the island, which are generated purely by local hurricane wind.

The sea surface wind fields are derived from the TS-X and TD-X images acquired on October 27 and 28, respectively by using XMOD2. To retrieve sea surface wind speed from SAR by applying GMF, the sea surface wind direction has to be obtained firstly. In both cases, the HRD wind direction is used for sea surface wind speed retrieval from TS-X and TD-X data. However, we found that HRD wind direction has a bias of around 15° compared with wind direction derived from SAR wind streaks in the southwest quadrant, which in turn induces a bias of around 5 m/s for the retrieved wind speed. This finding indicates that wind direction derived from SAR images may improvement the analysis wind field in TCs. The comparisons of the SAR retrieved wind speed with SFMR measurements show a negative bias of 6.2 and 5.8 m/s. It is noticed that the underestimation of SAR wind speed tends to be related with rainfall rate measured by SFMR as well. Although we try to simulate the attenuation induced by rain fall in X-band SAR using rain rate measured by SFMR, it is found that the simulated attenuation is still much lower than that estimated using TS-X/TD-X and SFMR measurements. The interesting finding is that the simulated attenuation shows better agreement with estimated one in the case on October 28 with rain rate above 5 mm/hr than that in the case on October 27 with rain rate lower than 5 mm/hr. Therefore, we consider that on the one hand XMOD2 needs to be further improvement, particularly for high winds. On the other hand, X-band SAR tends be rather sensitive to rain fall even for rain rate lower than 5 mm/hr. Thus, we also need improved model or method to relate rain rate with attenuation of radar backscatter in X-band SAR.

References

1. Velden, C.S., C.M. Hayden, W.P. Menzel, J.L. Franklin, and J.S. Lynch. 1992. The impact of satellite-derived winds on numerical hurricane track forecasting, weather and forecast. *Weather and Forecast* 7 (1): 107–118.
2. Zhang, X., T. Li, F. Weng, C.C. Wu, and L. Xu. 2007. Reanalysis of western Pacific typhoons in 2004 with multi-satellite observations. *Meteorology and Atmospheric Physics* 97 (1): 3–18.
3. Holt, B., A.K. Liu, D.W. Wang, A. Gnanadesikan, and H.S. Chen. 1998. Tracking storm-generated waves in the Northeast Pacific Ocean with ERS-1 synthetic aperture radar imagery and buoys. *Journal of Geophysical Research* 103 (C4): 7917–7929.
4. Li, X., W.G. Pichel, M. He, S. Wu, K.S. Friedman, P. Clemente-Colon, and C. Zhao. 2002. Observation of hurricane-generated ocean swell refraction at the Gulf Stream North Wall with the RADARSAT-1 synthetic aperture radar. *IEEE Transactions on Geoscience and Remote Sensing* 40 (10): 2131–2142.
5. Li, X.M., and S. Lehner. 2014. Algorithm for sea surface wind retrieval from TerraSAR-X and TanDEM-X data. *IEEE Transactions on Geoscience and Remote Sensing* 52 (5): 2928–2939.
6. Katsaros, K.B., P.W. Vachon, P.G. Black, P.P. Dodge, and E.W. Uhlhorn. 2000. Wind fields from SAR: Could they improve our understanding of storm dynamics? *Johns Hopkings APL Technical Digest* 21 (1): 86–93.
7. Quilfen, Y., B. Chapron, T. Elfouhaily, K. Katsaros, and J. Tournadre. 2003. Observation of tropical cyclones by high-resolution scatterometry. *Journal of Geophysical Research: Oceans* 103 (C4): 7767–7786.
8. Hersbach, H., A. Stoffelen, and S. de Haan. 2007. An improved C-band scatterometer ocean geophysical model function: CMOD5. *Journal of Geophysical Research: Oceans* 112 (C3).
9. Horstmann, J., D.R. Thompson, F. Monaldo, S. Iris, and H.C. Graber. 2005. Can synthetic aperture radars be used to estimate hurricane force winds? *Geophysical Research Letters* 32 (22).
10. Shen, H., W. Perrie, and Y. He. 2006. A new hurricane wind retrieval algorithm for SAR images. *Geophysical Research Letters* 33 (21).
11. Reppucci, A., S. Lehner, J. Schulz-Stellenfleth, and S. Brusch. 2010. Tropical cyclone intensity estimated from wide-swath SAR images. *IEEE Transactions on Geoscience and Remote Sensing* 48 (4): 1639–1649.
12. Weissman, D.E., M.A. Bourassa, and J. Tongue. 2002. Effects of rain rate and wind magnitude on Sea Winds scatterometer wind speed errors. *Weather and Forecast* 19 (5): 738–746.
13. Yueh, S.H., B.W. Stiles, W.Y. Tsai, H. Hu, and W.T. Liu. 2001. QuikSCAT geophysical model function for tropical cyclones and applications to Hurricane Floyd. *Meteorology and Atmospheric Physics* 39 (12): 2601–2612.
14. Li, X.M., and S. Lehner. 2001. Observation of typhoon Megi using TerraSAR-X data. Oberpfaffenhofen, Germany. TerraSAR-X Science Team Meeting.
15. Zhang, B., and W. Perrie. 2011. Cross-polarized synthetic aperture radar: A new potential measurement technique for hurricanes. *Bulletin of the American Meteorological Society* 93 (4): 531–541.
16. Werninghaus, R., and S. Buckreuss. 2010. The TerraSAR-X mission and system design. *IEEE Transactions on Geoscience and Remote Sensing* 48 (2): 606–614.
17. Bruckreuss, S., and B. Schttlerh. 2010. The TerraSAR-X ground segment. *IEEE Transactions on Geoscience and Remote Sensing* 48 (2): 623–631.
18. Li, X.M., S. Lehner, and W. Rosenthal. 2010. Investigation of ocean surface wave refraction using TerraSAR-X data. *IEEE Transactions on Geoscience and Remote Sensing* 48 (2): 830–840.
19. Hasselmann, K., and S. Hasselmann. 1991. On the nonlinear mapping of an ocean wave spectrum into a synthetic aperture radar image spectrum and its inversion. *Journal of Geophysical Research: Oceans* 96 (C6): 10713–10729.
20. Alpers, W.R., and C.L. Rufenach. 1979. The effect of orbital motions on synthetic aperture radar imagery of ocean waves. *IEEE Transactions on Antennas and Propagation* 27 (5): 685–690.

21. Raney, R.K., P.W. Vachon, R.A.D. Abreu, and A.S. Bhogal. 1989. Airborne SAR observations of ocean surface waves penetrating floating ice. *IEEE Transactions on Geoscience and Remote Sensing* 27 (5): 492–499.
22. Lehner, S., J. Horstmann, W. Koch, and W. Rosenthal. 1998. Mesoscale wind measurements using recalibrated ERS SAR image. *Journal of Geophysical Research: Oceans* 103 (C4): 7847–7856.
23. Koch, W. 2004. Directional analysis of SAR images aiming at wind direction. *IEEE Transactions on Geoscience and Remote Sensing* 42 (4): 702–710.
24. Melsheimer, C., W. Alpers, and M. Gade. 1998. Investigation of multifrequency/multipolarization radar signatures of rain cells over the ocean using SIR-C/X-SAR data. *Journal of Geophysical Research: Oceans* 103 (C9): 18867–18884.
25. Ulaby, F.T., R.K. More, and A.K. Fung. 1981. *Microwave remote sensing: Active and passive.* Massachusetts: Addison-Wesley.
26. Danklmayer, A., and M. Chandra. 2010. Precipitation Effects for X- and Ka-band SAR. Aachen, Germany. 8th European Conference on Synthetic Aperture Radar (EUSAR).

Chapter 7
Extracting Hurricane Eye Morphology from Spaceborne SAR Images Using Morphological Analysis

Isabella K. Lee, Ali Shamsoddini, Xiaofeng Li, John C. Trinder and Zeyu Li

Abstract Hurricanes are among the most destructive global natural disasters. Thus recognizing and extracting their morphology is important for understanding their dynamics. Conventional optical sensors, due to cloud cover associated with hurricanes, cannot reveal the intense air-sea interaction occurring at the sea surface. In contrast, the unique capabilities of spaceborne synthetic aperture radar (SAR) data for cloud penetration, and its backscattering signal characteristics enable the extraction of the sea surface roughness. Therefore, SAR images enable the measurement of the size and shape of hurricane eyes, which reveal their evolution and strength. In this study, using six SAR hurricane images, we have developed a mathematical morphology method for automatically extracting the hurricane eyes from C-band SAR data. Skeleton pruning based on discrete skeleton evolution (DSE) was used to ensure global and local preservation of the hurricane eye shape. This distance weighted algorithm applied in a hierarchical structure for extraction of the edges of the hurricane eyes, can effectively avoid segmentation errors by reducing redundant skeletons attributed to speckle noise along the edges of the hurricane eye. As a consequence, the skeleton pruning has been accomplished without deficiencies in the key hurricane eye skeletons. The subsequent reconstructed of the hurricane eyes thereby proves the morphology-based analyses results in a high degree of agreement with the hurricane eye areas derived from reference data based on NOAA manual work.

I.K. Lee · J.C. Trinder · Z. Li
School of Civil & Environmental Engineering, The University of New South Wales, Sydney, NSW, Australia

A. Shamsoddini
Tarbiat Modares University, Tehran, Iran

X. Li (✉)
GST, National Oceanic and Atmospheric Administration (NOAA)/NESDIS, College Park, MD 20740, USA
e-mail: xiaofeng.li@noaa.gov

X. Li (ed.), *Hurricane Monitoring With Spaceborne Synthetic Aperture Radar*, Springer Natural Hazards, DOI 10.1007/978-981-10-2893-9_7

7.1 Introduction

Hurricanes with maximum winds speeds exceeding 64 knots occur in the Atlantic Ocean, the Pacific Ocean (referred to as typhoons or cyclones), and in the Indian Ocean (also referred to as cyclones) [1–5]. They usually generate damaging storm surges and heavy rain, resulting in coastal residents being particularly vulnerable due to extensive coastal and inland flooding. The difficulties of recognizing and locating the eyes of hurricanes on optical satellite images, as described in [6, 7] are caused by cloud obstructions, especially during the formation stage of a hurricane. According to the previous studies, visible and infrared (IR) images with 1-km spatial resolution such as those acquired by HURSAT/GOES-12 (Geostationary Operational Environmental Satellite) and MODIS (Moderate Resolution Imaging Spectroradiometer) /AVHRR (Advanced Very High Resolution Radiometer) satellites have been studied by [8], as they provide information, such as structure and intensity of the top of the clouds associated with hurricanes.

As opposed to cloud-top information acquired by optical sensors, microwave sensors, e.g. SAR data, have the unique capabilities of imaging the Earth surface with a high spatial resolution of 100 m, such as images obtained by RADARSAT-1, in almost all weather conditions, day or night. The Advanced Land Observation Satellite (ALOS) Phased Array L-Band SAR (PALSAR) operated at a wavelength of 23.62 cm is commonly used in locations that are subject to tropical rainfall [9, 10]. ERS-1,-2, RADARSAT-1,-2 spacecraft and Environmental Satellite (ENVISAT) Advanced SAR (ASAR) are C-Band Radar Systems (4–8 cm wavelength), and have distinct advantages not only for the analysis of the structure of hurricanes in diverse and dynamic weather conditions, but also have been well positioned for imaging hurricane events in the Gulf of Mexico. Moreover, Tropical Rainfall Measuring Mission (TRMM) precipitation radar and microwave radiometers have assisted in answering some questions in terms of climate model estimation as well as projection of rainfall from U.S. landing hurricanes [11].

The dataset utilized in this study are derived from C-band RADARSAT-1 and ENVISAT satellites, which have design characteristics for the retrieval of ocean surface wind fields and for the development of geophysical model functions (GMF) for studying hurricanes [12]. High resolution X-Band SAR data, such as from COSMO-SkyMed and TerraSAR-X provides a unique opportunity of measuring precipitation effects and rainfall rate over land [13–15], but this research concentrates on the analysis of SAR images over the oceans.

In addition, the ScanSAR mode or wide swath mode SAR images from RADAR SAT-1, -2 and ENVISAT with >450 km swath width can cover the entire area of a hurricane. The core value of the broad coverage SAR data also enables ocean climate researchers to track hurricanes by analysing hurricane eye areas. Therefore SAR images have recently become increasingly important for hurricane studies. SAR images can reveal the structure on the sea surface at the centre of the hurricane, which helps scientists improve their understanding of hurricane morphology, including its shape, size and area of the eye, wind speed, surface waves and rain cells, and wind related ocean features which optical images do not reveal [6, 7]. Thus, hurricane studies using SAR images have the potential to play a vital role in monitoring the evolution and trajectory of hurricanes [16, 17].

The swirling vortex at the eye of a hurricane is typically almost calm and cloud-free with relatively weak winds. According to the previous quantitative investigations [18], the normalized radar cross section (NRCS), is a backscattering physical parameter that SAR measures in decibel dB units, for which pixel intensity values, range from +5 dB (very bright) to −40 dB (very dark). The variation of the NRCS is related to the wind speed and direction. Therefore, the relatively smooth ocean surface of hurricane eyes corresponds to low NRCS, which appears dark in the display of an ocean SAR image surrounded by a bright area for the spiral rain bands and cumulus clouds. Conversely, NRCS values extracted at the outer ring of the eye area will significantly increase as wind speeds increase. Thus, extracting hurricane eye areas automatically involves locating areas of localised low NRCS. The dynamic behaviour of a hurricane eye has a strong correlation with its intensity [19, 20]. The eye will become more pronounced if the intensity increases. The intensity can be determined from the maximum wind speed near the surface boundary layer around the eyewall [21, 22]. This information is critical for determining the hurricane wavenumber or scale category, from the most devastating Category 5 to the weakest Category 1. The intensity values, in evaluating the destructive effect of hurricane event, are represented in 5-knot (kt) increments at landfall based on Dvorak enhanced infrared (EIR) hurricane techniques and hurricane intensity estimation (HIE) algorithm [23, 24], and Saffir-Simpson Hurricane Wind Scale [25]. The unit 'knot' is used to avoid the ambiguity if strength of hurricanes is estimated in mph and km/h, since both metrics have been utilised to define the category or intensity of hurricanes in historical records and may be used for future hurricanes.

Fundamentally, mathematical morphology has been used extensively for image analysis for computing the geometric characteristics of shape and structure of features [26]. One of the popular approaches of morphological shape representation is a morphological based skeleton [27] and pruning variants. During skeleton transformation, the outcomes usually yield redundant skeleton branches caused by noise along the edges of the features in the images. Introducing an appropriate skeleton pruning technique can overcome the instability of the skeleton by eliminating these redundant segments. Then, the computation of shape representation can be considerably reduced with more efficient programming [28].

In this article, we have implemented morphological skeletonizing, pruning with the Discrete Skeleton Evolution (DSE) algorithm and shape reconstruction, to extract hurricane eye areas from SAR images. The flowchart of our algorithm is shown in Fig. 7.1. The results are compared with those derived using a wavelet-based algorithm for edge extraction described in [29].

This paper is organized as follows: we first describe the image pre-processing, including noise reduction by employing Adaptive Filters UNSW (University of New South Wales) Adaptive Filter (UAF) and image enhancement in Sect. 7.2. Then, the mathematical morphology methods for extracting hurricane eye areas including the skeleton extraction and pruning are described in Sect. 7.3. In Sect. 7.4, background

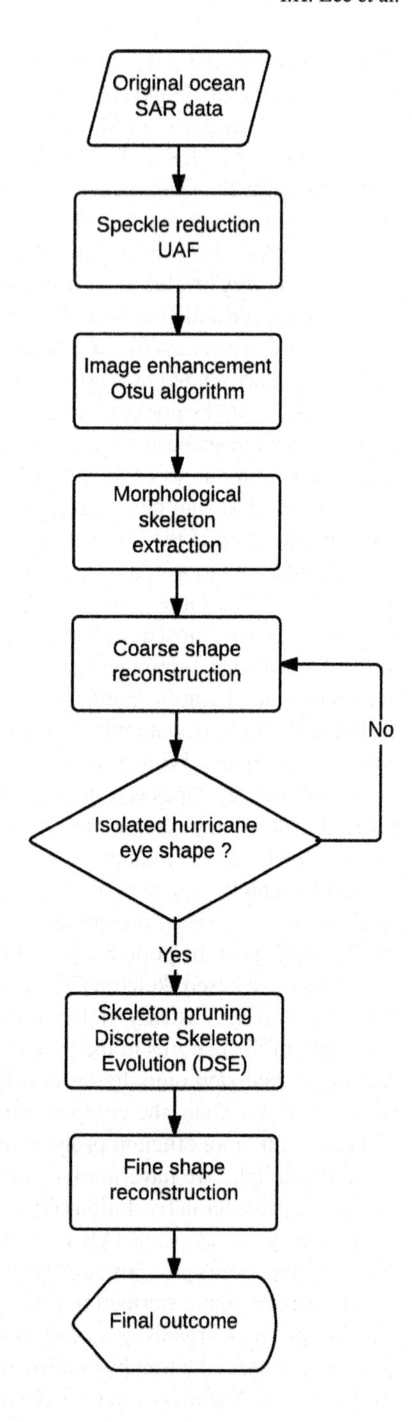

Fig. 7.1 Processing flow chart of hurricane eye areas extraction

information of six hurricane events is introduced. In Sect. 7.5, we have discussed the experimental results for determining the shape and size of the hurricane eye and compared them with reference data of manually measured shape and size. Concluding remarks are given in Sect. 7.6.

7.2 SAR Image Pre-processing

Identifying hurricane eyes on a SAR image is a challenging task as the SAR imagery is subject to speckle noise. In order to improve the interpretation ability of the target images, robust pre-processing of SAR data should be undertaken, including an adaptive speckle noise filter as well as image enhancement to identify the significant parts of the hurricane eye areas.

7.2.1 Applying Adaptive Filters UAF for SAR Speckle Reduction

Generally, speckle noise is common to all SAR imagery acquired by coherent systems, resulting in reduced potential for recognition of objects/targets [30]. The classification accuracy and edge detection can be dramatically influenced by speckle noise, so de-noising becomes one of the imperatives in pre-processing [31]. However, some adaptive filters, such as the Kuan [32], Lee, minimum mean square error (MMSE) [33] and Frost Filters [34] use a priori statistical knowledge in the model to suppress speckle noise at the expense of detail degradation. [35] employed a gamma maximum a posteriori (MAP) Bayes criterion while [36] employed the enhanced Lee filter that uses the level of homogeneity in the image [37], to determine how the pixels are affected by the noise and the preservation of textural information, yet with residual noise. In these filters, the homogeneity level was determined by calculation of two thresholds including C_u and C_{max} to differentiate the pixels which are affected by fully-developed speckle model from those which are not [36].

The common statistical filters such as Kuan, Lee and Frost Filters pay less attention to the homogeneity of intensity levels of pixels over a SAR image, leading to a loss of edge details and hence a degradation of spatial resolution, whereas filters such as Enhanced Lee inefficiently reduce the speckle noise level. SAR speckle reduction is essential for the study of hurricane eyes, since speckle will cause redundant sub-branches during the skeletonizing process in the morphological analysis [38]. The University of New South Wales (UNSW) adaptive filter (UAF) [31] was designed to preserve edges and textural information and simultaneously reduce the noise to an acceptable level. In addition to the provision of a trade-off between speckle noise reduction and texture preservation, UAF is able to efficiently filter the noisy pixels which were affected by neighbouring point scatter pixels. Moreover, this filter efficiently uses a damping factor to increase the capability of UAF to adjust between speckle noise reduction and texture preservation. Finally, according to Shamsoddini and Trinder (2012) [31], this filter performed better than the other common filters to reduce the noise level and preserve the edges and texture information. Due to the complex texture of hurricane-affected areas, including bare islands, noisy inland areas, the relatively calm ocean surface at hurricane eye areas, and rough ocean surface away from the eye caused by rainbands, UAF filter was found to perform better than the traditional; because the traditional filters focus on either noise reduction or edge information preservation but not both. Since textual information has been well maintained after applying the UAF filter, more effective and robust interpretation could be achieved in the later procedures. The result of the application of UAF with a moderate damping factor of 5 on a sample SAR image of all hurricane events are shown in Table 7.1 (Column B).

7.2.2 *Image Enhancement*

Image contrast enhancement is required for better target extraction. Lowering the grey-level threshold values for separating the hurricane eye area from surrounds maintains global representation of the eye area at the expense of increasing noise effects on the skeleton definition. However, a relatively high threshold can produce a more effective skeleton [39], yet homotopy may be reduced, but this effect would normally be small. The classic Otsu automatic threshold selection method is practical and efficient for dealing with image thresholds [40], in order to enhance the hurricane eye with respect to its background. Otsu's thresholding technique aims at finding the threshold value t by considering the sum of the foreground pixel values in the hurricane eye, and the background pixel values in the non-eye area. The process is iterated through possible threshold values by calculating a measure of the distribution of the pixel levels each side of the threshold. In order to evaluate the optimal threshold in the image, a discriminant criterion to measure the class separability was introduced by [41]:

$$\sigma^2_{Within}(t) = \omega_0(t)\sigma^0_2(t) + \omega_B(t)\sigma^2_B(t) \tag{7.1}$$

$$\omega_0 = \sum_{i=t}^{n-1} p(i) \tag{7.2}$$

$$\omega_B = \sum_{i=0}^{t-1} p(i) \tag{7.3}$$

$$\sigma^2_{Between}(t) = \sigma^2 - \sigma^2_{Within}(t) = \omega_B(t)\omega_0(t)[\mu_B(t) - \mu_0(t)]^2 \tag{7.4}$$

Table 7.1 Demonstration of original SAR image (Column A) of 6 hurricanes, after speckle noise reduction by UAF adaptive filter (Column B), followed by image enhancement based on Ostus thresholding method (Column C)

Column A	Column B	Column C
Original	UAF noise reduction	Ostu enhanced
	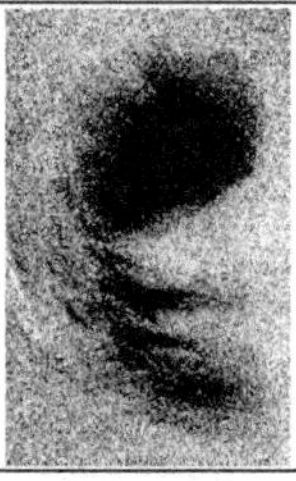	
Hurricane Katrina 27/08/2005 (threshold t=0.68)		
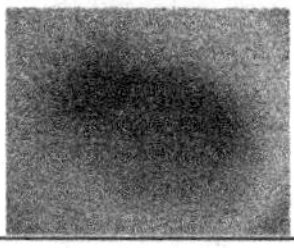	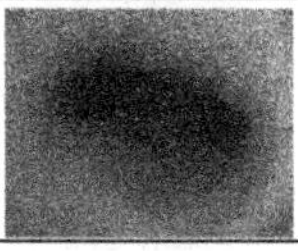	
Hurricane Katrina 28/08/2005 (threshold t=0.16)		
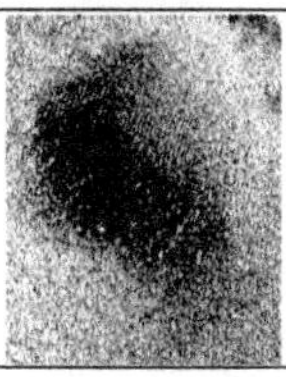	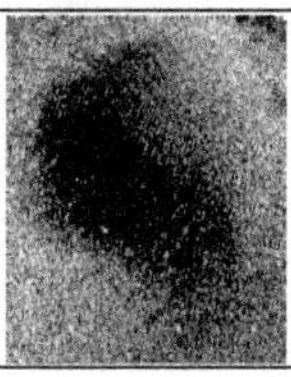	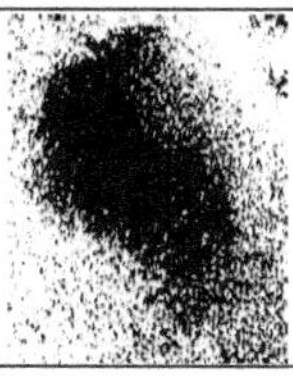
Hurricane Dean 17/08/2007 (threshold t=0.52)		
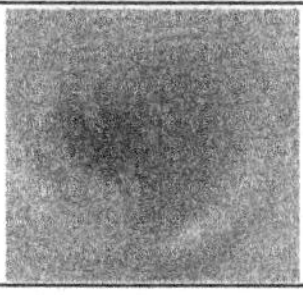	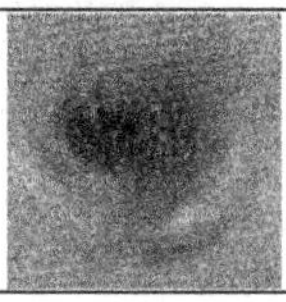	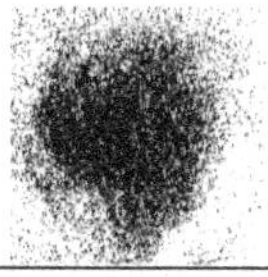
Hurricane Dean 19/08/2007 (threshold t=0.63)		
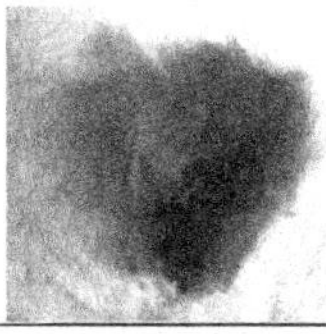		
Hurricane Rita 22/09/2005 (threshold t=0.19)		
	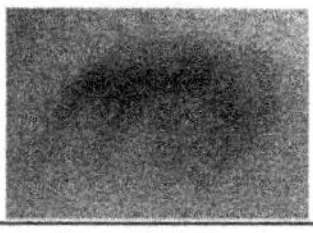	
Hurricane Earl 02/09/2010 (threshold t=0.16)		

Table 7.2 Demonstration of initial morphological reconstruction with original skeleton compared to the second reconstruction based on DSE skeleton pruning method

Column A	Column B	Column C
Original data	Original skeleton	Pruned skeleton based on DSE

Hurricane Katrina 27/08/2005 (*PPS: 60.8%)	Sum of original skeleton pixels:2880	Sum of pruned skeleton pixels:1750
	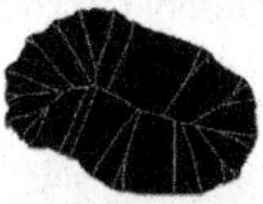	
Hurricane Katrina 28/08/2005 (*PPS: 16.0%)	Sum of original skeleton pixels:2845	Sum of pruned skeleton pixels:455
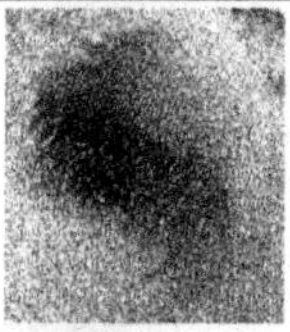		
Hurricane Dean 17/08/2007 (*PPS: 20.7%)	Sum of original skeleton pixels:3004	Sum of pruned skeleton pixels:621
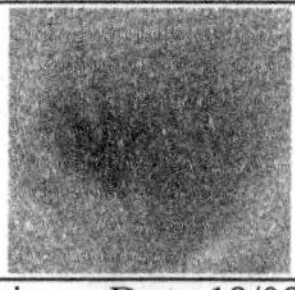		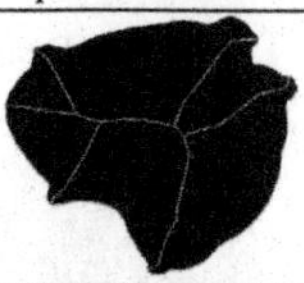
Hurricane Dean 19/08/2007 (*PPS: 40.0%)	Sum of original skeleton pixels:2642	Sum of pruned skeleton pixels:1055
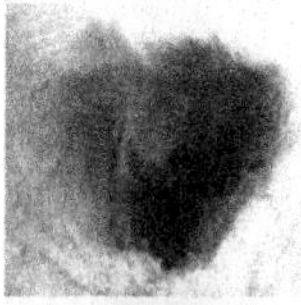		
Hurricane Rita 22/09/2005 (*PPS: 42.1%)	Sum of original skeleton pixels:810	Sum of pruned skeleton pixels:341

Hurricane Earl 02/09/2010 (*PPS:30.8%)	Sum of original skeleton pixels:2586	Sum of pruned skeleton pixels:796
The mean of relative skeleton comparison after pruning : 35%		

*PPS: Percentage of pruned skeleton

where σ^2_{Within} is defined as the within-class (background) variance as the weighted sum of the variances of both background and foreground $\sigma^2_{Between}$ is defined as between-class (foreground) variance, which is derived by subtracting the within-class variance from the total variance of the combined distribution. In Eq. (7.4), $\mu_0(t)$ and $\mu_B(t)$ denote the average grey scale of foreground and the average grey scale of background respectively. $p(i)$ represents the probabilities of the foreground and background separated by a threshold t, [0, n-1] is the range of intensity levels, $\sigma^2_B(t)$ is the variance of the pixels in the background (below threshold), and $\sigma^2_0(t)$ is the variance of the pixels in the foreground (above threshold), where also ω_0 and ω_B denote the weights, which are the probabilities of the two classes being separated by a threshold t.

This algorithm can be implemented in three steps:

- Separating the pixels into two clusters according to the threshold,
- Finding the mean of each cluster and squaring the differences from the means,
- Then multiplying the result from the previous step by the number of pixels in each cluster.

The method can also be used for finding the best threshold to efficiently separate the foreground and background based on variances for each threshold t. The results of the processing are displayed in Table 7.1 (Column C). By implementing the Ostu threshold that is calculated based on the filtered image, the hurricane eye pattern is unambiguously identifiable from the speckle noise in the original SAR data, displayed in Table 7.1 (Column C).

However, in the case of Hurricane Dean (17/08/2007), some noise in the lower right corner is likely to affect the hurricane eye pattern recognition even after applying the UAF. This will be solved by morphological transformation in Sect. 7.3.3 (Table 7.2). For instance, a closing morphological operator is applied to remedy some new hollows which may appear during the image enhancement [38]. This is because such hollows might cause new complex skeleton structures and wrong connections in the following step. As a result, a binary image is obtained for the skeleton extraction.

7.3 Mathematical Morphology of Hurricane Eyes Extraction

Primarily, the proposed algorithm has three major steps which will be detailed in the following: morphological skeleton computation, morphological reconstruction, and skeleton pruning based on Discrete Skeleton Evolution (DSE).

7.3.1 Morphological Skeleton

The concept of skeletonization was initially proposed by Blum [42, 43], who emphasized certain properties of images as a result of the Medial Axis Transform (MAT) procedures. Skeletonization is often used to describe a binary image for presenting the simple skeleton of the object in terms of the distance to its edges. It reduces the foreground of regions in a binary image by means of morphological operators, to produce a skeletal remnant that largely preserves the extent and connectivity of the original region, while abandoning most of the original foreground pixels. For instance, a contour on an object corresponds to the topological properties of the skeleton. It determines the closest boundary points to each inner point within an object, which has at least two nearest boundary points belonging to the skeleton as shown in Fig. 7.2.

The skeleton approach is designed to simplify the patterns of the hurricane eyes into a series of thin lines, called Euclidean skeletons. Euclidean distance transformation and maximum value tracking was proposed by Shih and Mitchell [44], which is a nonlinear measure, and which can ensure that the localization of skeleton points in n-dimensional objects is well-defined and robust enough to preserve the homotopy with the original object. According to the definitions of binary operations, for instance morphological opening and closing, the morphological skeleton is denoted by the following formulas [26]:

$$SK(X) = U_{n=0}^{N} SK_n(X) \tag{7.5}$$

$$SK_n(X) = (X \ominus nB) - (X \ominus nB) \circ B, 0 \leq n \leq N \tag{7.6}$$

$$(X \ominus nB) = (...(X \ominus B) \ominus B) \ominus ...) \ominus B, 0 \leq n \leq N \tag{7.7}$$

$$N = Max\,\{n|(X \ominus nB) \neq \phi\} \tag{7.8}$$

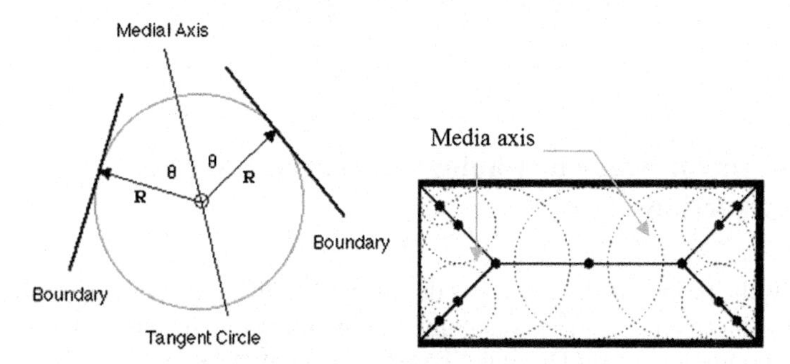

Fig. 7.2 Skeleton of a rectangle defined in terms of 2D Medial Axis Indication of the geometric relationship between a point on a 2-D medial axis and its corresponding boundary points. The tangent circle grazes the boundary, so the boundary is perpendicular to its radius vector. Also, the medial axis bisects the two radius vectors. The media axis indicated in the rectangle defines the centres of maximal discs

A single maximal disc structuring element B exists in the input image X. Skeleton $SK(X)$ of set X is the locus of the centres of the maximal discs such that the disc is tangent to the boundary of the area in at least two distinct points of set X. In Eq. (7.5), $SK_n(X)$ represents the union of the skeleton of original image X that has been skeletonized n times by the disc or structuring element B. Discs B are represented as a set of pixels with size 5 in this study. A series of N erosion operations are performed by the structuring element B by Eqs. (7.6) and (7.7) on set X for the skeletonizing processing until the set is empty, and the skeleton is reduced to one pixel wide.

Extracting a robust skeleton is essential for identifying and detecting endpoints and closed loops, based on Euclidean distance functions [27], which can retain homotopy of the initial objects and ensure the extracted skeleton is satisfactory for the reconstruction of the original image.

7.3.2 Reconstruction Algorithm

Reconstruction can be explained as an iteration of dilations, to create padding with the purpose of recreating the original binary image after skeletonization. Maragos and Schafer [45] advanced a more overall-awareness reconstruction method, due to the associative and distributive properties of set union. According to the definitions of Blum (1967) and Lantuejoul (1978) [46] for medial axis and Euclidean distance, let each skeleton point $s \in SK$ and $r(s)$ indicate the radius of the maximal disc $B(s, r(s))$ of an original image set X. An adjacent point is a bifurcation of the branches. A further definition introduced by [47] is presented in Fig. 7.3:

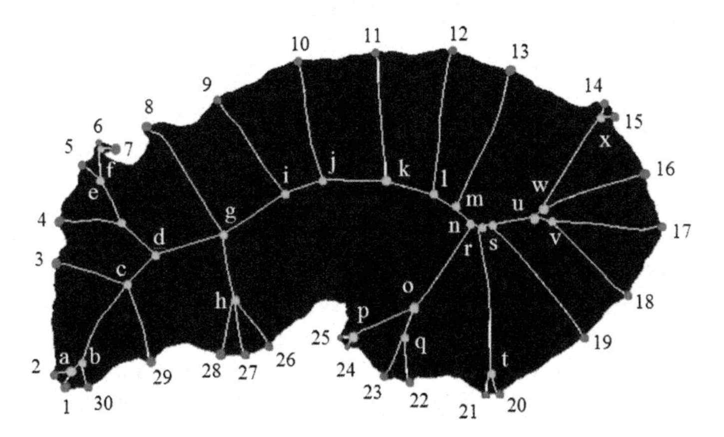

Fig. 7.3 Demonstration of the structure of skeleton endpoints and junction points. It shows the skeleton endpoints (denoted by number) and junction points (denoted by letter) on the skeleton of Hurricane Earl eye (09/02/2010) after first reconstruction without skeleton pruning

(1) If the skeleton point only has one adjacent point, it is a skeleton endpoint, denoted by points from 1 to 30 in Fig. 7.3;
(2) a junction point (denoted by letter) is defined as skeleton point containing more than two adjacent points;
(3) a connection point denoted using letters from a to x in Fig. 7.3 is defined for a skeleton point which is neither a junction point (denoted by letter) nor an endpoint, denoted by points from 1 to 30.

Thus, the skeleton end branch $l_i (i = 1, 2, \ldots, N;)$ is the distance between a skeleton endpoint (denoted by number) and the nearest junction point $f(l_i)$, and the minimum distance between l_i and $f(l_i)$ is denoted as $P(l_i, f(l_i))$. The reconstruction of a skeleton SK is denoted as $R(SK)$ shown in Eq. (7.9) below:

$$R(SK) = \bigcup_{s \in SK} B(s, r(s)) \tag{7.9}$$

In this study, we required the shape reconstruction to be done twice because the first reconstruction failed to yield an accurate result of the actual hurricane eye area due to messy skeleton branches. However, following skeleton pruning to remove spurious branches with DSE, which is described in next section (Sect. 7.3.3), the trimmed skeleton ensures that the computation of all the sub-skeletons after the second iteration, are sufficient to derive a thoroughly reconstructed object. Thus, a more accurate hurricane eye area can eventually be preserved based on the contributions of only the significant branches.

7.3.3 Skeleton Pruning with Discrete Skeleton Evolution

A sound presentation of skeleton pruning comprises topological information preservation, accuracy of extraction of the skeletons and stability of the transformations [48]. Bai et al. (2007) [47] proposed the DSE pruning method based on a distance transform method, which yields outstanding results in the fields of human visual perception, by iteratively removing skeleton end branches which are irrelevant for the latter fine shape reconstruction, as well as suppressing spurious branches or biased skeleton points. In addition, the stability of pruning with DSE produces a hierarchical skeleton structure while simplifying key skeletons, which relies on the centres of the maximal discs in the skeletons. This process cannot be achieved by other combined pruning methods [47]. The DSE method guarantees the elimination of multiple distortions and simplifies the skeleton shape while preserving the significant visual segments. It is more stable than traditional Discrete Curve Evolution (DCE) based on a contour partitioning method [28, 49]. Additionally, it can produce a satisfactory result even under noise effects and shape variations without dislocating the centres of maximal discs based on Euclidean distance functions for the skeleton points [27].

Skeleton pruning based on DSE indicates a process of simplification for a messy skeleton with many unnecessary branches. According to Bais algorithm, the key iterative process is to remove the end branch of skeleton with the lowest weight w_i for each end branch $P(l_i, f(l_i))$ for the latter fine reconstruction in Eq. (7.10), where $Area()$ is the set area function, referred to in Sect. 3.2, Eq. (7.9).

$$w_i = 1 - \frac{Area(R(SK - P(l_i, fl_i)))}{Area(R(SK))} \tag{7.10}$$

Using $w_i^k (i = 1, 2, \ldots, N^k)$ to calculate the weight for each end branch w_i based on the input skeleton SK^k, $k = 0$ or SK^0, where k denotes the iteration in Eq. (7.11).

$$w_i^{(k)} = 1 - \frac{Area(R(SK^k - P(l_i^{(k)}, fl_i^{(k)})))}{Area(R(SK^{(k)}))} \tag{7.11}$$

If the smallest weight $w_{min}^{(k)}$ is lower than the threshold $t = 0.005$, the end branch $P_{min}^{(k)}$ should be removed, else the evolution can be terminated and the output for $SK^{(k)}$ is the final result, shown in Eq. (7.12).

$$SK^{k+1} = SK^k - P_{min}^{(k)} \tag{7.12}$$

After skeleton pruning, the robust skeleton can be used to reconstruct the targeted shape more accurately. Examples of the extraction of the hurricane eyes will be described in the next section.

7.4 Processing of RADARSAT-1 and ENVISAT SAR Images

According to the hurricane intensity and azimuthal wavenumber defined by numerous studies, the most destructive Category 4 and 5 hurricanes are the main focus in this study. Table 7.3 shows a dataset containing details of six RADARSAT-1 images acquired for hurricane eye shape analysis, each of them with distinguishable eye areas. Two pairs of significant hurricane events with 1–2 days interval, Hurricane Katrina and Dean, were carefully selected over numerous hurricane events in NOAA's dataset. Both hurricane events rapidly intensified, growing from a Category 2 to a Category 5 in just a few days. Moreover, the famous and notorious Hurricane Rita was chosen because it broke the record with nine counties in the USA suffering significant damages. This makes Category 5 Hurricane Rita the most intense tropical cyclone ever observed in the Gulf of Mexico. Since 1991 Hurricane Earl was the first major hurricane to batter the northeast Caribbean, which became a Category 4 hurricane on 2 Sep 2010.

Table 7.3 SAR imagery information of hurricane events

Hurricane event	Satellite/ sensor	Acquisitions time	Event time (UTC)	Spatial resolution (m)	Swath (km)	SAR image mode (Li et al. 2013)	Wave number (Li et al. 2013)
Katrina	RADARSAT-1	27/08/2005	11:28	48.94, 53.98	500	SCW	Cat 2
Katrina	ENVISAT/ ASAR	28/08/2005	15:51	79.52, 88.19	405	WSM	Cat 5
Rita	RADARSAT-1	22/09/2005	03:44	49.02, 50.43	500	SCW	Cat 5
Dean	RADARSAT-1	17/08/2007	09:50	50, 50	500	SCW	Cat 2
Dean	RADARSAT-1	19/08/2007	23:17	48.60, 50.57	500	SCW	Cat 4
Earl	ENVISAT/ ASAR	02/09/2010	02:44	82.06, 95.18	405	WSM	Cat 4

*SCW (ScanSAR wide beam); WSM (Wide swath mode);
*Cat denotes intensity Category of hurricane; Cat 5 wind speed more than 252 km/hr (157 mph); Cat 4 wind speed more than 209–251 km/hr (130–156 mph); Cat 2 wind speed more than 157–177 km/hr (96–110 mph) (Siegel 2015) [50]

The methodology of hurricane eye measurement proposed in this study aims at providing a close insight into the related ocean conditions and to uncover patterns of hurricane eye evolution through this analysis. The original SAR images, either SCW (ScanSAR wide beam), and WSM (Wide swath mode), used for these investigations are displayed in Fig. 7.4a–f.

7.5 Experimental Results and Discussions

To assess the robustness of the methods described in this study for determining the shape of the hurricane eye, two components were evaluated:

(1) Effectiveness and completeness of the skeleton pruning by comparison of pre- and post-pruning areas,
(2) the accuracy of the extracted areas of the hurricane eye compared with reference data based on manually extracted areas [7].

A comparison of morphological skeleton and skeleton pruning based on DSE for the morphological development of hurricane eyes is shown in Table 7.2 (Column B and C) for all SAR hurricane images. The ridged edges of the hurricane eyes usually yield redundant information that affects the local contributions to the skeletons. In the case of Hurricanes Katrina (28/08/2005), Earl (02/09/2010) and Dean (17/08/2007 and 19/08/2007), the total number of pruned pixels in the skeleton is significantly decreased. In Table 7.2 (Columns A and C), the outcome shows approximately 16.0,

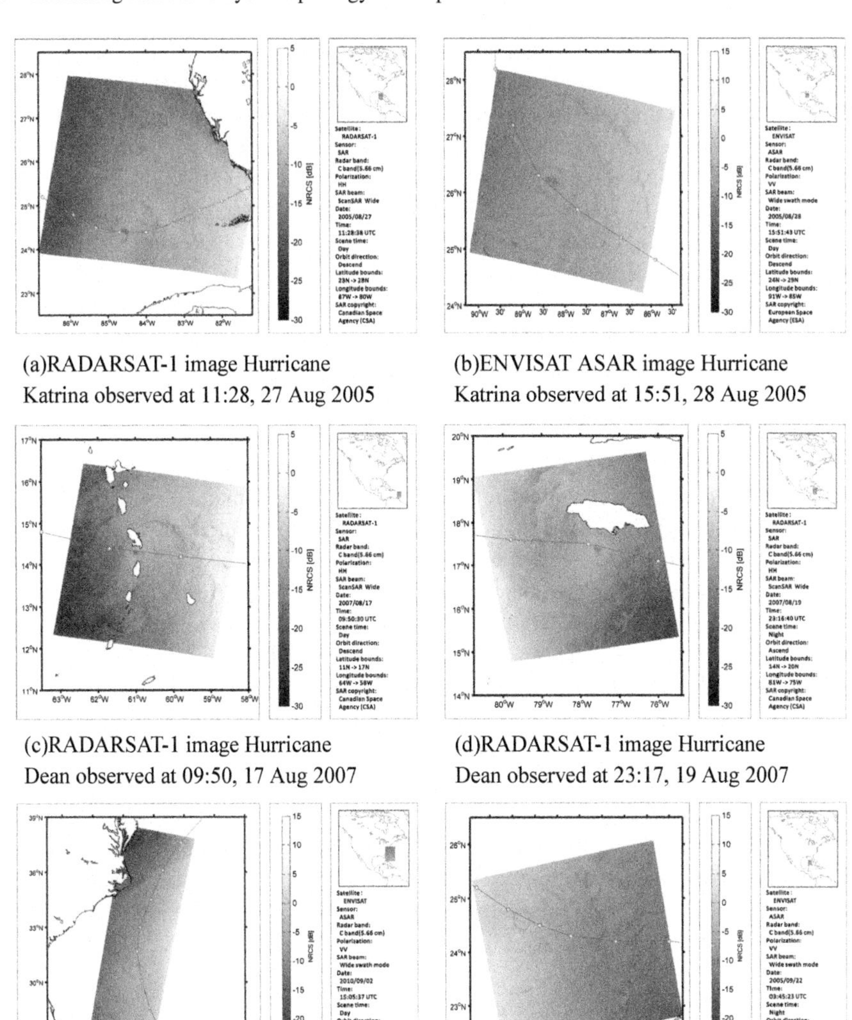

(a)RADARSAT-1 image Hurricane Katrina observed at 11:28, 27 Aug 2005

(b)ENVISAT ASAR image Hurricane Katrina observed at 15:51, 28 Aug 2005

(c)RADARSAT-1 image Hurricane Dean observed at 09:50, 17 Aug 2007

(d)RADARSAT-1 image Hurricane Dean observed at 23:17, 19 Aug 2007

(e)ENVISAT ASAR image Hurricane Earl observed at 02:44, 2 Sep 2010

(f)RADARSAT-1 image Hurricane Rita observed at 03:44, 22 Sep 2005

Fig. 7.4 Original hurricane SAR images used in this study: **a** Hurricane Katrina 27/08/2005; **b** Hurricane Katrina 28/08/2005; **c** Hurricane Dean 17/08/2007; **d** Hurricane Dean 19/08/2007; **e** Hurricane Earl 02/09/2010; **f** Hurricane Rita, 22/09/2005

30.8, 20.7 and 40.0% of the original skeleton pixels remained, respectively, leading to a more accurate output for the latter fine reconstruction of the hurricane eye area. The fishbone shaped skeletons are simplified into a few main branches. This is due to the robustness of the DSE based skeleton pruning which eliminates the influence of the spurs in the extracted boundaries. In addition, this greatly contributes to the simplified skeletons for Hurricanes Katrina (27/08/2005) and Rita (22/09/2005), where about 39.2 and 57.9% of the redundant skeleton pixels have been trimmed by implementing the DSE algorithm.

The skeleton pruning results for the six scenes are revealed in Column C in Table 7.2, showing a comparison between the original skeleton after the former coarse hurricane eye shape reconstruction and pruned skeleton based on DSE after the latter fine reconstruction. The latter fine shape reconstructions for Katrina 28/08/2005, Dean 17/08/2007, Dean19/08/2007 and Earl 02/09/2010 are significantly smoothed. The resulting neat skeletons reduce the chaotic sub-skeletons near the boundaries of hurricane eye patterns since only the significant branches remain.

The differences between the manually derived reference hurricane eye areas, and those extracted by mathematical morphological processing before and after the application of the DSE method, are shown in Table 7.4. The mean relative differences expressed as a percentage of the area of the six hurricanes eye areas after pruning, are estimated to be approximately 4.0%, compared with the mean relative accuracy for the six hurricanes before skeleton pruning of 14.5%.

Moreover, the large relative differences for Hurricane Katrina (27/08/2005) and Hurricane Rita (22/09/2005) of about 27.5 and 18.2% respectively before pruning are much worse than the corresponding relative differences after skeleton pruning of 2.4 and 1.4%, respectively. One possible reason for the larger differences in these eye patterns as discussed above, are that they have a more irregular structural distribution than the others, and as well, the surrounding noise interference creates more challenges for morphological reconstruction. The eye areas of Hurricanes Katrina (28/08/2005), Dean (17/08/2007 and 19/08/2007), and Earl (02/09/2010) are

Table 7.4 Evaluation of hurricane eye information extraction

Hurricane event	Event date	Manual work (km^2) (Li et al. 2013)	MM* hurricane area before pruning (km^2)	Relative discrepancy before pruning (%)	MM* hurricane area after pruning (km^2)	Relative discrepancy after pruning (%)	Relative accuracy after pruning (%)
Katrina	27/08/2005	382	527	27.5	372.9	2.4	97.6
Katrina	28/08/2005	754	811	7.0	719.7	4.5	95.5
Rita	22/09/2005	828	1012	18.2	816.2	1.4	98.6
Dean	17/08/2007	278	299	7.0	264.3	4.9	95.1
Dean	19/08/2007	279	334	16.5	261.6	6.2	93.8
Earl	02/09/2010	863	968	10.8	824.3	4.5	95.5
				Mean:14.5		Mean:4.0	Mean:96.0

MM* Mathematical Morphology Method

similar in shape, being visually represented as a dark blobs. Since their eye areas are regular the relative discrepancies in the areas of Hurricanes Katrina (28/08/2005), Dean (17/08/2007 and 19/08/2007), and Earl (02/09/2010) after skeleton pruning are smaller than achieved for Hurricanes Katrina (27/08/2005) and Rita (22/09/2005).

It can therefore be concluded that wide swath SAR imagery is suitable for a broad range of applications in the future with the aim of tracking or forecasting hurricane events. The extracted skeletons together with their pruning are key steps; otherwise the results derived for specific hurricane areas will be distorted or subject to errors caused by redundant noise, which according to the outcomes of this evaluation would significantly affect the processing of the hurricane eye areas.

In the study of (Du and Vachon 2003), wavelet analysis has been discussed to estimate the scale and area of hurricane eye for eight well-defined eye areas between 1998 and 2001. The study of Du and Vachon also employed RADARSAT-1 as input data, as well as the standard reference dataset recorder produced by NOAA (Li et al. 2013). Four results of extracted hurricane eye areas by Du and Vachon from three hurricane events were compared with NOAAs reference dataset. These hurricane events included Hurricane Erin (11/09/2001 and 13/09/2001), Hurricane Humberto (26/09/2001) and Hurricane Olga (28/11/2001). After an analysis the first entry should be considered as an outlier. Therefore as shown in Table 7.5, the mean relative discrepancy between the results of the wavelet analysis, excluding the first entry and the manually extracted reference areas was approximately 19.6%, and the mean relative accuracy was 80.3%. This mean relative accuracy is considerably lower than that achieved with the proposed mathematical morphological method which has a mean relative accuracy of 96.0%.

Table 7.5 Estimation of area extraction for hurricane by wavelet analysis

Hurricane event	Satellite/sensor	Acquisitions time	Event time (UTC)	Spatial resolution (m)	Vachon's work by wavelet analysis (km^2)	Manual work (NOAA) (km^2) (Li et al. 2013)	Relative discrepancy (%)	Relative accuracy (%)
Erin	R1	11/09/2001	22:19	100	392.0	2973.0	86.8	13.2
Erin	R1	13/09/2001	10:04	100	6442.0	7368.0	12.6	87.4
Humberto	R1	26/09/2001	21:42	100	300.4	366.0	17.9	82.1
Olga	R1	28/11/2001	09:49	100	693.0	970.0	28.6	71.4
							Mean (exclude Erin 11/09/2011): 19.6	Mean(exclude Erin 11/09/2011): 80.3

R1, RADARSAT-1

7.6 Conclusion

In this paper, we have developed an efficient image processing solution that can significantly preserve the hurricane eye's shape and topological properties, while suppressing the speckle noise along the eye edges. The pruning technique based on the DSE algorithm is able to conserve connectivity of the pruned skeleton as well as efficiently trim the messy sub-branches, thereby yielding an accurate result of hurricane eye pattern that closely matches the eye area derived from NOAA's reference dataset.

In the experimental validation, the pruning algorithm together with DSE overcomes the problem of the messy and excessive skeletons and enhances the stability of the shape. In addition, the results are more stable than simple morphological solutions even though the shape undergoes rigid image transformations. This combination proved to be stronger and more flexible than using skeleton pruning alone. The extracted hurricane eye areas were generally in agreement within 4.0% with manually extracted reference data.

SAR images present unique capabilities of measuring the microwave signal backscatter response from the sea surface for characterizing hurricanes. By extracting and locating hurricane eye areas in SAR data, this method should enable meteorologists to predict or evaluate with high accuracy the hurricane movement and its landfall, based on the time series ocean SAR data when it becomes available. However, the use of RADARSAT-1 also shows the limitation of hurricane event observation with a 24-day repeat cycle. With the new ScanSAR mode (width: 100 km and length: 150 km), TerraSAR-X has been optimised for application of vessels, oil spill detection and sea ice. However, even though with a high spatial resolution up to a few tens meters and TerraSAR-X's scanSAR mode acquisition, the TerraSAR coverage is still not as wide as that of RADARSAT-1, making it difficult to cover the whole hurricane domain. Interfacing with multiple sensors, e.g. Doppler radars and TRMM satellite will spread the benefits for meteorologists and oceanologists in better understanding the paths of hurricanes.

Hurricane eyes are surrounded by hurricane eyewalls, which are associated with the maximum sustained wind speed in the hurricane. Hurricane location associated with hurricane eye analysis plays an important role on hurricane studies. In principle, as the shape and size of hurricane eye has been well extracted from the SAR image, the evolution and movement of hurricane can be estimated objectively by time series ocean SAR data if it is available from advanced satellite missions. Hurricane eye position together with a wind speed retrieval model could not only provide a promising approach to improving the estimation accuracy of the direction and velocity of hurricane, but also assist in generating a trajectory diagram to improve the prediction together with historical hurricane datasets, thus leading to possible prevention of major personnel and property losses during severe hurricane seasons.

References

1. Klotzbach, P.J., W.M. Gray, and E. Wilmsen. 2012. *Extended range forecast of Atlantic seasonal hurricane activity and landfall strike probability for 2012*. Department of Atmospheric Science: Colorado State University.
2. Katsaros, K.B., P.W. Vachon, W.T. Liu, and P.G. Black. 2002. Microwave remote sensing of tropical cyclones from space. *Journal of Oceanography* 58 (1): 137–151.
3. Pielke, R.A. 2013. *The hurricane*. London: Routledge.
4. Turk, F.J., J. Hawkins, K. Richardson, and M. Surratt. 2011. A tropical cyclone application for virtual globes. *Computers & Geosciences* 37 (1): 13–24.
5. Perez, E., and P. Thompson. 1995. Natural hazards: Causes and effects: Lesson 5tropical cyclones (hurricanes, typhoons, baguios, cordonazos, tainos). *Prehospital and Disaster Medicine* 10 (03): 202–217.
6. Zhang, B., and W. Perrie. 2012. Cross-polarized synthetic aperture radar: a new potential technique for hurricanes. *Bulletin of the American Meteorological Society* 93 (4): 531–541.
7. Li, X., J.A. Zhang, X. Yang, W.G. Pichel, M. DeMaria, D. Long, and Z. Li. 2013. Tropical cyclone morphology from spaceborne synthetic aperture radar. *Bulletin of the American Meteorological Society* 94 (2): 215–230.
8. Zehr, R. 2004. Satellite products and imagery with Hurricane Isabel. In *Proceedings 13th Conference Satellite Meteorology and Oceanography*, pp 30–37. Norfolk, VA.
9. Arnesen, A.S., T.S. Silva, L.L. Hess, E.M. Novo, C.M. Rudorff, B.D. Chapman, and K.C. Mcdonald. 2013. Monitoring flood extent in the lower Amazon River floodplain using ALOS/PALSAR ScanSAR images. *Remote Sensing of Environment* 130: 51–61.
10. Ramsey, E., A. Rangoonwala, and T. Bannister. 2013. Coastal flood inundation monitoring with satellite C-band and L-band synthetic aperture radar data. *JAWRA Journal of the American Water Resources Association* 49 (6): 1239–1260.
11. Kummerow, C., W. Barnes, T. Kozu, J. Shiue, and J. Simpson. 1998. The tropical rainfall measuring mission (TRMM) sensor packag. *Journal of Atmospheric and Oceanic Technology* 15 (3): 809–817.
12. Xu, Q., Y. Cheng, X. Li, C. Fang, and W.G. Pichel. 2011. Ocean surface wind speed of Hurricane Helene observed by SAR. *Procedia Environmental Sciences* 10: 2097–2101.
13. Marzano, F., S. Mori, M. Chini, L. Pulvirenti, N. Pierdicca, M. Montopli, and J. Weinman. 2011. Potential of high-resolution detection and retrieval of precipitation fields from X-band spaceborne synthetic aperture radar over land. *Hydrology and Earth System Science* 15: 859–875.
14. Bringi, V., and V. Chandrasekar. 2001. *Polarimetric Doppler weather radar: principles and applications*. Cambridge: Cambridge University Press.
15. Weinman, J., F. Marzano, W. Plant, A. Mugnai, and N. Pierdicca. 2009. Rainfall observation from X-band, space-borne, synthetic aperture radar. *Computers & Geosciences* 9 (1): 77–84.
16. Murali, P., P.N. Raja, I. L V, Vasavi, and K. Yacham. 2014. Real time detection of hurricanes from satellite cloud imagery using haar wavelet. In *World Congress on Engineering and Computer Science*, 1.
17. NOAA. 2013. *Coastal Remote Sensing*. US: NOAA Coastal Services Center.
18. Horstmann, J., D. Thompson, F. Monaldo, S. Iris, and H. Graber. 2005. Can synthetic aperture radars be used to estimate hurricane force winds? *Geophysical Research Letters* 32 (22): L22801.
19. Liu, K., and J.C. Chan. 1999. Size of tropical cyclones as inferred from ERS-1 and ERS-2 data. *Monthly Weather Review* 127 (12): 2992–3001.
20. Powell, M.D. 1990. Boundary layer structure and dynamics in outer hurricane rainbands. Part II: Downdraft modification and mixed layer recovery. *Prehospital and Disaster Medicine* 118 (4): 918–938.
21. Bell, M.M. 2006. *Observed structure, evolution and potential intensity of category five Hurricane Isabel (2003) from 1214 September*. Ph.D. thesis, Colorado State University, 2006.

22. Houze, R.A., S.S. Chen, B.F. Smull, W.C. Lee, and M.M. Bell. 2007. Hurricane intensity and eyewall replacement. *Science* 315 (5816): 1235–1239.
23. Dvorak, V.F. 1984. *Tropical cyclone intensity analysis using satellite data*. US Department of Commerce, National Oceanic and Atmospheric Administration, National Environmental Satellite, Data, and Information Service.
24. Goodman, S.J., J. Gurka, M. Demaria, T.J. Schmit, A. Mostek, G. Jedlovec, C. Siewert, W. Feltz, J. Gerth, and R. Brummer. 2012. The GOES-R proving ground: accelerating user readiness for the next-generation geostationary environmental satellite system. *Bulletin of the American Meteorological Society* 93 (7): 1029–1040.
25. Rogers, R., S. Aberson, A. Aksoy, B. Annane, M. Black, J. Cione, N. Dorst, J. Dunion, J. Gamache, and S. Goldenberg. 2013. NOAA's hurricane intensity forecasting experiment: A progress report. *Bulletin of the American Meteorological Society* 94 (6): 859–882.
26. Serra, J. 1982. *Image analysis and mathematical morphology*. London: Academic Press.
27. Calabi, L. 1965. *A study of the skeleton of plane figures*. Parke Mathematical Laboratories.
28. Bai, X., and L.J. Latecki. 2007. *Discrete skeleton evolution. Energy minimization methods in computer vision and pattern recognition*. Berlin: Springer.
29. Du, Y., and P.W. Vachon. 2003. Characterization of hurricane eyes in RADARSAT-1 images with wavelet analysis. *Canadian Journal of Remote Sensing* 29 (4): 491–498.
30. Bamler, R. 2000. Principles of synthetic aperture radar. *Surveys in Geophysics* 21 (2–3): 147–157.
31. Shamsoddini, A., and J.C. Trinder. 2012. Edge-detection-based filter for SAR speckle noise reduction. *International Journal of Remote Sensing* 33 (7): 2296–2320.
32. Kuan, D.T., A.A. Sawchuk, T.C. Strand, and P. Chavel. 1987. Adaptive restoration of images with speckle. *IEEE Transactions on Acoustics Speech and Signal Processing* 35 (3): 373–383.
33. Lee, J.S. 1981. Speckle analysis and smoothing of synthetic aperture radar images. *Computer Graphics and Image Processing* 17 (1): 24–32.
34. Frost, V.S., J.A. Stiles, K.S. Shanmugan, and J.C. Holtzman. 1982. A model for radar images and its application to adaptive digital filtering of multiplicative noise. *IEEE Transactions on Pattern Analysis and Machine Intelligence* 2: 157–166.
35. Lopes, A., E. Nezry, R. Touzi, and H. Laur. 1990a. Maximum a posteriori speckle filtering and 1st order texture models in SAR images. *Remote Sensing Science for the Nineties* 1–3: 2409–2412.
36. Lopes, A., R. Touzi, and E. Nezry. 1990b. Adaptive speckle filters and scene heterogeneity. *Remote Sensing Science for the Nineties* 28 (6): 992–1000.
37. Marques, R.C.P., F.N. Medeiros, and J.S. Nobre. 2012. Sar image segmentation based on level set approach and $\mathcal{G}_A^0$ model. *IEEE Transactions on Pattern Analysis and Machine Intelligence* 34 (10): 2046–2057.
38. Solilie, P. 2003. *Morphological image analysis: principles and applications*. New York: Springer.
39. Jin, S., S. Wang, and X. Li. 2014. Typhoon eye extraction with an automatic SAR image segmentation method. *International Journal of Remote Sensing* 35 (11–12): 3978–3993.
40. Otsu, N. 1975. A threshold selection method from gray-level histograms. *Automatica* 11 (285–296): 23–27.
41. Fukunaga, K. 1990. *Introduction to statistical pattern recognition*. New York: Academic Press.
42. Blum, H. 1967. A transformation for extracting new descriptors of shape. *Models for the Perception of Speech and Visual Form* 19: 362–380.
43. Blum, H. 1973. Biological shape and visual science (Part I). *Journal of theoretical Biology* 38 (2): 205–287.
44. Shin, F.Y., and O.R. Mitchell. 1989. Threshold decomposition of gray-scale morphology into binary morphology. *IEEE Transactions on Pattern Analysis and Machine Intelligence* 11 (1): 31–42.
45. Maragos, P.A., and R.W. Schafer. 1986. Morphological skeleton representation and coding of binary images. *IEEE Transactions on Acoustics Speech and Signal Processing* 34 (5): 1228–1244.

46. Lantuejoul, C. 1978. *La squelettisation et son application aux mesures topologiques des mosaiques polycristallines*. Ph.D. thesis, Ing. Ecole des Mines de Paris.
47. Bai, X., L.J. Latecki, and W.Y. Liu. 2007. Skeleton pruning by contour partitioning with discrete curve evolution. *IEEE Transactions on Pattern Analysis and Machine Intelligence* 29 (3): 449–462.
48. Montero, A.S., and J. Lang. 2012. Skeleton pruning by contour approximation and the integer medial axis transform. *Computers & Graphics* 36 (5): 477–487.
49. Latecki, L.J., and R. Lakmper. 1999. Polygon evolution by vertex deletion. *Scale-Space Theories in Computer Vision*. Berlin: Springer.
50. Siegel, F.R. 2015. Global perils that reduce the earths capacity to sustain and safeguard growing populations tactics to mitigate or suppress them. In *Countering 21st Century Social-Environmental Threats to Growing Global Populations*. Berlin: Springer

Chapter 8
Tropical Cyclone Center Location in SAR Images Based on Feature Learning and Visual Saliency

Shaohui Jin, Shuang Wang, Xiaofeng Li, Licheng Jiao and Jun A. Zhang

Abstract Synthetic aperture radar (SAR), with its high spatial resolution, large area coverage, day/night imaging capability, and penetrating cloud capability, has been used as an important tool for tropical cyclone monitoring. The accuracy of locating tropical cyclone centers has a large impact on the accuracy of tropical cyclone track prediction. This study focuses on the center location of tropical cyclones in the SAR images. Based on the analysis of the characteristics of the tropical cyclone SAR images, combined with the theory and methods of SAR image segmentation and computer vision, center location methods for both the tropical cyclones with eyes in the SAR images and the tropical cyclones without eyes in the SAR images are presented in this chapter. The main work is as follows: 1, For a tropical cyclone with its eye in the SAR image, The eye area in image appears as black or dark grey area for there being no rain and little wind in the eye area. But the gray level contrast is not always obvious. There may be no complete and clear eye when a tropical cyclone is in the development period or the recession period. The eye area in the tropical cyclone SAR image may appears as light grey area at these periods. So it is necessary to enhance the gray level contrast before image segmentation. Besides, denoising the speckle noise is also necessary for the SAR image processing. A tropical cyclone eye

S. Jin (✉) · S. Wang · L. Jiao
Key Laboratory of Intelligent Perception and Image Understanding of Ministry of Education, International Research Center for Intelligent Perception and Computation, Xidian University, Xian 710071, Shaanxi Province, China
e-mail: shaohuijin2005@163.com

S. Wang
e-mail: shwang@mail.xidian.edu.cn

L. Jiao
e-mail: lchjiao@mail.xidian.edu.cn

X. Li
GST, National Oceanic and Atmospheric Administration (NOAA)/NESDIS, College Park, MD, USA
e-mail: Xiaofeng.Li@noaa.gov

J.A. Zhang
Rosenstiel School of Marine and Atmospheric Science, University of Miami, and NOAA/AOML/Hurricane Research Division, Miami, FL, USA
e-mail: jun.zhang@noaa.gov

X. Li (ed.), *Hurricane Monitoring With Spaceborne Synthetic Aperture Radar*, Springer Natural Hazards, DOI 10.1007/978-981-10-2893-9_8

extraction method based on non-local means method and labeled watershed algorithm is given. PPB filter is used to denoise the speckle noise. Then the top-hat transform is used to enhance the contrast. At last the tropical cyclone eye is extracted labeled watershed algorithm. The eye area extracted with this method is computed to compare with the eye area extracted manually. The comparison indicates the accuracy of the extraction accuracy. 2, Generally speaking, the center of the tropical cyclone without its eye is located with template matching method for a single image. The spiral cloud band of the tropical cyclone without its eye is the information can be fully used in the tropical cyclone SAR image. Take the advantage of simple background with little texture information, a center location method of the tropical cyclone without its eye in the SAR image based on feature learning and visual saliency detection is proposed. Spiral cloud bands appear as light and dark spiral structure in the tropical cyclone SAR image, containing rich directional information. Therefore salient region map taking advantage of the gray contrast feature and orientation feature is built. The salient region map makes the spiral cloud bands outstanding and the irrelevant clouds excluded. Then the morphology method is used to extract the spiral bands in the salient region map, the skeleton lines of spiral cloud bands is extracted. At last the tropical cyclone center is estimated with the inflow angle model and the particle swarm optimization algorithm. And the estimation results are compared with the Best Track Data, confirming the validity of the algorithm.

8.1 Introduction

Tropical cyclones are a kind of disastrous weather system, which can cause extremely powerful winds and torrential rain so that marine activities, human life and property are seriously threatened [1]. The location of the center of a tropical cyclone is key information that is needed for timely and accurate tropical cyclone forecasting. It closely relates to the position and motion tendency of a tropical cyclone and is vital for prediction, to enable people to avoid or prevent disaster caused by strong winds and torrential rain.

Ever since the launch of the first polar-orbiting meteorological satellite in the early 1960s, remote sensing techniques have proved to be a useful method for tropical cyclone analyses and forecasting [2–5]. Satellite cloud images, acquired by passive remote-sensing instruments operating in the visible and infrared (IR) bands, vividly describe cloud-level tropical cyclone horizontal structures with large area coverage and frequently repeated observations [6]. However, due to cloud cover and rain effects, the inner core structures and air-sea interaction near the ocean surface cannot be directly observed with visible or IR sensors. The microwave scatterometer is an active remote-sensing sensor. It works well at night and has a wide range of observation. In addition, it can simultaneously obtain wind speed and the wind direction of a wind field on the sea surface [7]. Weather radar is a kind of active microwave radar that emits radar pulses to the sky and then receives the radar backscatter. It can identify a tropical cyclone center based on echo signal intensity [8].

Spaceborne synthetic aperture radar (SAR) has become another popular tool for tropical cyclone monitoring over recent years. This is a type of active microwave radar that emits radar pulses that can penetrate through clouds and then receive the radar backscatter from the Earths surface. Therefore, SAR can take images of the Earths surface and these images can reveal detailed structures of tropical cyclones on the ocean surface [6]. In addition, SAR has a high spatial resolution (10–100 m) and can operate day and night under almost all weather conditions. Katsaros et al. (2002) discussed the usefulness of SAR in tropical cyclone monitoring [9]. Horstmann et al. (2005) studied sea surface tropical cyclone wind retrievals with RADARSAT-1 SAR with an existing geophysical model function, i.e., CMOD5 [10]. Yang et al. (2011) assessed the impact of radar calibration accuracy on the retrieval of wind speeds with a high degree of accuracy interpretation [11]. Li et al. (2013) systematically analyzed 83 typhoons and hurricanes observed in RADARSAT-1 and ENVISAT SAR images and manually extracted tropical cyclone morphology from these images [2]. These studies illustrate that extracting quantitative tropical cyclone information from SAR images has been a focus of research.

SAR images of tropical cyclones on the sea surface show the sea surface imprint of tropical cyclones and are related to the surface roughness affected by sea surface winds, rain roughening of the surface, waves and so on [12]. A complete tropical cyclone in a SAR image appears as bright-dark spiral patterns. There is almost no rain and wind within a tropical cyclone eye area, so the radar backscattering from this part of the sea surface is relatively weak compared to signal returns from other areas in a tropical cyclone. High winds and rain result in a brighter area around the eye wall. Therefore, for the tropical cyclones with their eyes in the SAR image, researchers extract eyes to locate the centers of tropical cyclones. Du et al. (2003) proposed a wavelet analysis method to extract tropical cyclone eye shape and size [6]. Recently, Zheng et al. (2016) extracted typhoon eyes in SAR images using two newly developed algorithms and showed good results when validated against the tropical cyclone best track data sets [13]. Lee et al. (2016) presented a mathematical morphology method for automatically extracting the hurricane eyes from SAR images. And the discrete skeleton evolution algorithm is used to ensure global and local preservation of the hurricane eye shape [5].

However, not every tropical cyclone has an obvious eye feature in a SAR image. There may even be no eye where a tropical cyclone is in the developing period or the declining period. Sometimes limited by the capture range of radar, a SAR image may contain parts of a tropical cyclone without an eye. Generally speaking, methods to estimate the center of a tropical cyclone automatically or semi-automatically fall into two categories: wind field analysis and pattern matching. Wind vector fields need a sequence of images. However, SAR provides only a single snap view at a time and therefore it is difficult to determine the wind direction from a single image. Pattern matching methods can be performed using only a single image. Researchers have previously tried to locate the center of tropical cyclones using pattern matching methods in satellite cloud images that do not contain eye information. Wang et al. (2006) developed an auto-center-locating algorithm based on Hough transform [14]. They first locate the rain bands using image segmentation and mathematic morphology, and then match the skeleton lines of rain bands with a logarithm spiral using Hough transform. Segmentation methods such as threshold segmentation are used here to obtain the rain bands of a tropical cyclone from a SAR image. This is a straightforward but not very effective approach as rain bands cannot be distinguished from clouds with a similar reflectivity. Besides, the Hough transform has its limitations such as large calculations and poor detection performance when there is noise in an image. Xu et al. (2009) proposed a spiral rain band segmentation method by transforming it to the problem of classification with a support vector machine [15]. However, samples are chosen manually in this method. Wong et al. (2005) proposed a tropical cyclone eye fixing method using a genetic algorithm with temporal information [8]. Later in 2008, they developed a more automatic framework for tropical cyclone eye fixing with a method using a genetic algorithm [16]. Good results indicate that this genetic algorithm is robust and can be widely used. However, the extraction of regions of interest (the positions where the centers of the tropical cyclones may lie) needs the assistance of the previous image or forecaster. Moreover, the matching model used in the above work is a logarithm spiral, which looks similar to the geometrical structure of tropical cyclones but is not suitable for the accurate matching of diverse tropical cyclones. Besides, the matching of a logarithm spiral needs precise extraction of skeleton lines. As listed above, the pattern matching method for center location of tropical cyclones in satellite cloud images has been developed for years. Rain bands extraction, pattern matching model and fitting method are three aspects researchers pay attention to. It would be a meaningful attempt to locate tropical cyclone centers with a single SAR image using pattern matching method.

In this chapter study about center location of tropical cyclones with their eyes in the SAR images and without their eyes in the SAR images are both presented. A method based on the analysis of characteristics of tropical cyclones in SAR images and the chain of SAR image processing is given to extract tropical cyclone eyes in Sect. 8.2. A method based on the visual saliency and the pattern matching is given to estimate the centers of tropical cyclones without their eyes in the SAR images in Sect. 8.3.

8.2 Tropical Cyclone Eye Extraction with an Automatic SAR Image Segmentation Method

SAR images of tropical cyclones are mainly imaged on the sea surface and they show the sea surface imprint of tropical cyclones and are related to the surface roughness affected by sea surface winds, rain roughening of the surface, waves and so on. There is almost no rain and wind within a tropical cyclone eye area, so the radar backscattering from this part of the sea surface is relatively weak compared to signal returns from other areas in a tropical cyclone. It makes the eye area in the SAR image appears as a black or dark region. There are high winds and rain in a brighter area around the eye wall. So an image segmentation method can be used to extract the eye region in a SAR image.

However, after analysing large number of tropical cyclone SAR images, we find that the textures of some tropical cyclone eyes are not pronounced. The shape and structure of tropical cyclones are very different. Sometimes the grey level change range of a tropical cyclone SAR image is small and the grey level difference between the eye region and other regions is small. For better extraction of the eye, enhancing the image contrast is necessary.

Before developing image analysis methods for automatic tropical cyclone eye extraction, we should consider characteristics of SAR tropical cyclone images. Speckle noise in SAR is a type of multiplicative noise that reduces the actual resolution of the SAR image, and may affect target identification and sometimes cause features to disappear in the image. The existence of speckle noise makes extracting useful information directly from the SAR image difficult. So de-noising of the tropical cyclones SAR images is necessary.

All these challenges make it difficult to set up general rules for automatic tropical cyclone-eye information extraction under various conditions. To overcome these difficulties, a three-step image processing chain is developed to distinguish the tropical cyclone eye from background on SAR images. The three-step image processing procedure includes: (1) applying an extend Non-local means image denoising algorithm to reduce speckle noise, while still keeping the edges of tropical cyclone structures; (2) applying top-hat transform to denoised image to enhance the contrast. After contrast enhancement, simple segmentation method and gray level feature are used to extract tropical cyclone information; and (3) applying labelled watershed method to segment the area of interest. We implement the above-mentioned method is implemented with some SAR tropical cyclone images to demonstrate the efficiency of the proposed method. The extracted tropical cyclone-eye areas are comparable to those measured manually.

8.2.1 Image Denoising with Probabilistic Patch-Based (PPB) Filter

Affected by speckle noise, not all scattering characteristics of targets are shown clearly in SAR images. Besides, low signal-to-noise ratio makes automatic image processing difficult [17]. To identify targets in SAR images, an image denoising procedure is necessary. In general, a good denoising method can not only reduce the noise but also keep as much detailed information such as edges and shapes as possible. The denoising quality will directly affect the accuracy of the follow on image processing.

SAR image denoising has been a focus of SAR image processing research since 1980s. The early typical methods are Lee and Kuan filtering techniques [18, 19]. They are both spatial filters that assume that the mean and variance of a pixel is equal to the local mean and variance of all its surrounding pixels within a fixed range. Lee filter uses the first-order Taylor series expansion while Kuan filter uses Minimum

Mean Square Error MMSE criterion, which is a more general denoising expression. That kind of spatial filters get local statistical characteristics of an image with sliding windows, inhibiting speckle noise of homogenous regions. However, they perform not so well in non-homogenous regions such as edges and details in SAR images. If the sliding window is large, there are few samples in the window similar to the present pixel, leading to distortion after denoising. If the window is small, the speckle reduction is not enough in homogenous regions. Later improved algorithms mainly contain two types. One kind of algorithms are dividing an image into homogenous regions and non-homogenous regions, denoising with different denoising methods in different regions [20, 21]. Another kind of algorithms are constructing adaptive sliding window [22, 23].

There being strong self-similarity in an image is an important progress in image understanding in recent years. This self-similarity means pixels similar to the present pixel exist in the whole image, not just in the local neighbourhood. Based on this knowledge, another kind of filtering method, non-local filtering appears. A typical algorithm is Non-Local means (NL-means) filter proposed by Buades et al. [24]. It provides as many similar samples as possible to the denoised pixel to inhibit noise globally and make evaluated image close to the real image. For NL-means filter, image is denoised by computing similarity between pixels. The similarity depends on the Euclidean distance between pixels corresponding patches, not the spatial distance or gray value difference of pixels. For a de-noised pixel, the weights of its similar pixels are proportional to their similarity to the de-noised pixel. As shown in Fig. 8.1, the estimated value $v(p)^{NL}$ of pixel p is the weighted average of all pixel values in search region Δ_p. It is defined as:

$$v(p)^{NL} = \frac{1}{z(p)} \sum_{q \in \Delta_p} w(p,q)v(q) \tag{8.1}$$

Here Δ_p can be as large as the whole image. But actually it is smaller than the whole image to reduce computational complexity. $Z(p) = \sum_{q \in \Delta_p} w(p,q)$ is a normalized constant. $w(p,q)$ is the similarity weight of pixel q to pixel p. $w(p,q)$ is defined as

$$w(p,q) = exp(-\frac{1}{h^2} \sum_k |v(N_{p,k}) - v(N_{q,k})|^2) \tag{8.2}$$

Here h is a smoothing parameter, controlling decay of weights with the changes of Euclidean distance. $N_{q,k}$ and $N_{p,k}$ are respectively the k-th neighbour patches centered at p and q. The similarity calculation of NL-means filter breaks the limitation of neighbourhood window. It searches much more samples similar to the present pixel in a larger set, so it can get results with less distortion.

In 2009, Deledalle et al. proposed a generalized NL-means filter -Probabilistic Patch-Based (PPB) filter [25]. It designs a kind of iterative similar evaluation method and combines two kinds of similar distances. One kind is the distance between two patches in the last denoised results. Another kind is the distance between two patches

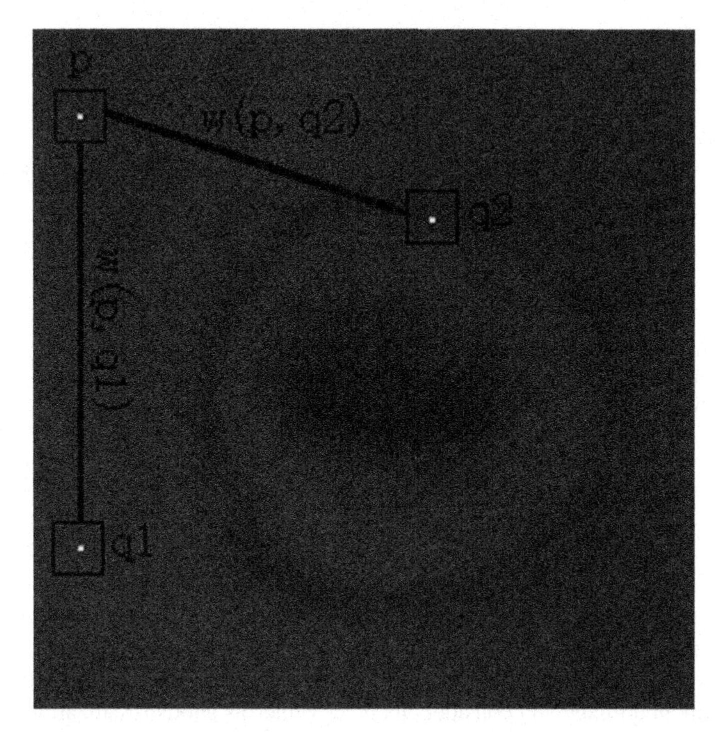

Fig. 8.1 Scheme of similarity weights of Non-local means filter. In the image, a de-noised pixel value (p) is obtained by the weighted average of all pixels in the whole image that are similar to it. The similarity between two pixels p and q1 depends on the similarity of the intensity grey level vectors $v(N_p)$ and $v(N_{q1})$, where N_k denotes a square patch of fixed size centered at a pixel k. This similarity is measured as a decreasing function of the weighted Euclidean distance $\|v(N_p) - v(N_{q1})\|_{2,a}^2$. Here pixel $q1$ with a more similar gray level in its square patch $v(N_{q1})$ have larger weights in average than pixel $q2$

in the original image. It makes use of priori knowledge in the similar evaluation, so it performs better. Inheriting NL-means filter, a de-noised pixel value in PPB filter is defined as a weighted average of the values of similar pixels. However, pixel similarity of PPB filter is not defined as the Euclidean distance between patches but probabilities instead. For a SAR image, Pixel amplitudes are modelled as independent and identically distributed according to Nakagami-Rayleigh distribution:

$$P(A_s|R_s^*) = \frac{2L^L}{\Gamma(L)R_s^{*L}} A_s^{2L-1} exp\left(-\frac{LA_s^2}{R_2^*}\right) \tag{8.3}$$

The amplitude image A is the square root of the reflectivity image R. L is the equivalent number of looks. And the denoising process of PPB filter is expressed as weighted maximum likelihood estimation (WMLE) problem in a Bayesian framework. The best estimate $\hat{R}$ is:

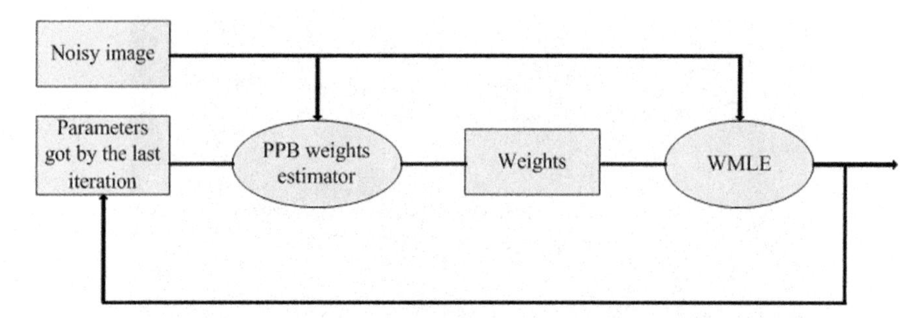

Fig. 8.2 Scheme of the iterative PPB filter. Weights are computed by PPB weights estimator using both the noisy patches of noisy image and the parameters got by the last iterative estimation. The new estimated weights and noisy image are used by the WMLE to compute the new estimate parameters. The procedure is repeated until the two consecutive estimates converge

$$\hat{R}_s^{WMLE} = \frac{\sum_t w(s,t)A_t^2}{\sum_t w(s,t)} \tag{8.4}$$

As shown in Fig. 8.2, in an iterative PPB filter, weights are defined by two terms. The first term depends on the original noisy image and considers the original pixel values as a realization of noise model. The second term is calculated from the previously estimated image and considers its pixel values as another realization of noise model. This consideration makes sure the denoising model stability over the different iterations. The weights of iteration i are defined as:

$$w(s,t)^{it,PPB} = exp\left[-\sum_k\left(\frac{1}{h}log\left(\frac{A_{s,k}}{A_{t,k}}+\frac{A_{t,k}}{A_{s,k}}\right)+\frac{L}{T}\frac{\left|\hat{R}_{s,k}^{i-1}-\hat{R}_{t,k}^{i-1}\right|^2}{hat R_{s,k}^{i-1}\hat{R}_{t,k}^{i-1}}\right)\right] \tag{8.5}$$

The parameters T and h act as dual parameters to balance the trade-off between the noise reduction and the fidelity of the estimate [26]. Experiment illustrate that appropriate T and h make good denoising result and influence later image processing. PPB filter performs well according to both objective evaluation and subjective visual effect reduce speckle noise and keep detail information such as edges at the same time. Therefore, we implement PPB filter to reduce speckle noise of tropical cyclone image.

8.2.2 *Image Enhancement with Top-Hat Transform*

After denoising and smoothing, SAR tropical cyclone image will have certain loss and fuzziness. In addition, small dynamic range of intensity of a SAR tropical cyclone image makes it hard to distinguish the tropical cyclone eye from its background at times. Therefore, before performing tropical cyclone eye extraction, we add a step to enhance the image contrast. Histogram enhancement method is commonly used. For example, Kim proposed a Bi-Histogram Equalization image enhancement method [27]. The main idea is to divide an image histogram into two sub-histograms and then equalize the two sub-histograms. However, it will enhance noise and make excessive enhancement. A mathematical morphology method named top-hat transform can be used to achieve a better performance for image enhancement [28]. Top-hat transform can be considered as a filter and its combined applications can be used to extract high frequency signals such as edges and valley values in an image. So here we implement top-hat transform to enhance contrast.

Top-hat transform is based on four basic mathematical morphology operators containing dilation, erosion, opening and closing. Mathematical morphology can be seen as a kind of nonlinear filtering. It regards an image as a set and uses another smaller set called as structuring element, basic operators and their combination to process image. Let f and b represent a grayscale image and structural element, respectively. The dilation and erosion of $f(x, y)$ by $b(u, v)$ are defined by $f \bigoplus b$ and $f \Theta b$ respectively.

$$f \bigoplus b = max\left\{f(x-u, y-v) + b(u, v) | ((x-u), (y-v)) \in D_f; (u, v) \in D_b\right\} \tag{8.6}$$

$$f \Theta b = max\left\{f(x+u, y+v) - b(u, v) | ((x+u), (y+v)) \in D_f; (u, v) \in D_b\right\} \tag{8.7}$$

D_f is the definition domain of f and D_b is the definition domain of b. Opening and closing operators are defined by $f \circ b$ and $f \cdot b$ respectively.

$$f \circ b = (f \Theta b) \bigoplus b \tag{8.8}$$

$$f \cdot b = (f \bigoplus b) \Theta b \tag{8.9}$$

Dilation operator can be used to fill hollow in an area and erosion operator can be used to clear up bright areas smaller than structural element. Openings can be used to remove small objects, protrusions from objects, and thin connections between objects, while closing eliminates small holes, smoothes concaves and fills gaps in the contour. By using opening and closing, the top-hat transform including white top-hat transform and black top-hat transform are defined as follows:

$$W_{hat}(f) = f - (f \circ b) \tag{8.10}$$

$$B_{hat}(f) = (f \cdot b) - f \tag{8.11}$$

White top-hat transform can enhance edge information for its high-pass filtering while black top-hat transform can get valley to make divide outstanding. So by adding result image of White top-hat transform to the original image and then reducing result image of black top-hat transform, the difference of gray level of tropical cyclone and background in SAR image can be enlarged and details such as boundary stand out. It enhances image contrast effectively.

8.2.3 Tropical Cyclone Eye Extraction with Labelled Watershed Algorithm

For SAR tropical cyclone images, the backscatter intensity mainly depends on the roughness of the sea surface [12]. Tropical cyclone structure appears as the changes of gray pixel values and SAR Tropical cyclone images do not contain much texture information. Therefore, we choose labelled watershed to extract tropical cyclone eyes. Watershed algorithm makes use of gradient map of an image reflecting the change tendency of the image to capture edge information and form closed segmented areas. It is very suitable for images lacking texture information.

The watershed algorithm was originally developed by Beucher and Lantuejoul [29] and its applications were extensively described by Beucher and Meyer [30]. Its intuitive idea comes from geography. As shown in Fig. 8.3, there is a landscape flooded by water, watersheds being the divide lines of the domains of attraction of rain falling over the region [31]. Considering an image as a landscape, light and dark areas of an image can be respectively considered as hills and valleys. To segment a given image the landscape is flooded, whereby catchment basins fill up with water

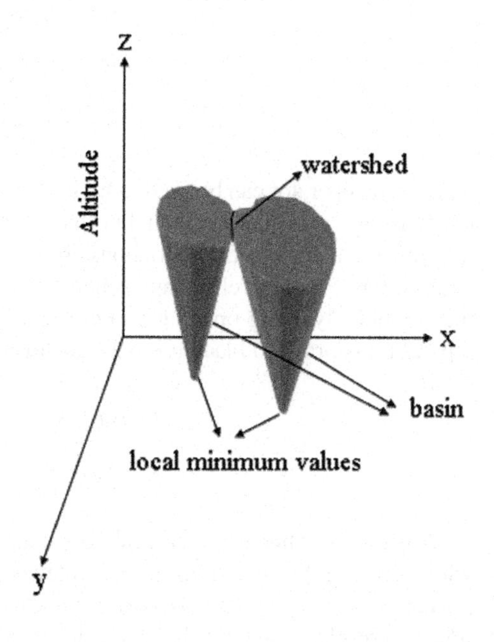

Fig. 8.3 Scheme of watershed algorithm

starting at local minima (regions with low gray values) and dams are built when water coming from different catchment basins would meet. When the water level has reached the highest peak in the landscape, the process is stopped. Therefore an image is partitioned into closed and connected regions separated by dams, called watersheds. Watershed algorithm commonly segments an image into a set of non-overlapping regions, each of which is spatially consistent and forms a closed set with boundaries (watershed). The width of their boundaries is one pixel. The union of all these regions and watershed line is the whole image. It makes use of contrast feature (the gradient) and neighbourhood spatial relationship and segments images fast and effective.

However, due to the influence of dark noise and dark detail texture, there will be a lot of false minimum values in the image, producing false segmentation areas. This fact will cause serious over-segmentation without meaningful result. So a key issue for improving the effect of applying watershed algorithm is to deal with over-segmentation. Actually over-segmentation is solved by three kinds of methods. One kind of method is pre-processing by filtering. It often reduces local minimum points of gradient map by filtering and then segments image by watershed algorithm [32]. Its improvement is limited. Another kind of method is merging small regions after watershed algorithm [33, 34]. The key is rules of merging regions and it always determine algorithm performance. However, there are so many small regions after watershed that merging computation is large. The third kind of method is adding labels to gradient map [35, 36]. It determines minimum points based on some rules and limits numbers of segmented regions, so as to avoid over-segmentation.

Here labelled watershed is used to control over-segmentation [36]. Its main idea is modifying gradient map of an image with labels to make sure that regional minima only appear in labelled positions. As a result, other regional minima are discarded and over-segmentation is weakened. Labels contain both inner labels associated with targets and outer labels associated with background. First, we assume that an inner label is a connected region surrounded by higher altitude (pixels with high gray values), and pixels in one inner label region have the same gray value. As shown in Fig. 8.4b, slight gray regions in the image meet the above assumptions and they are called inner labels. The next step, we apply watershed algorithm on the image

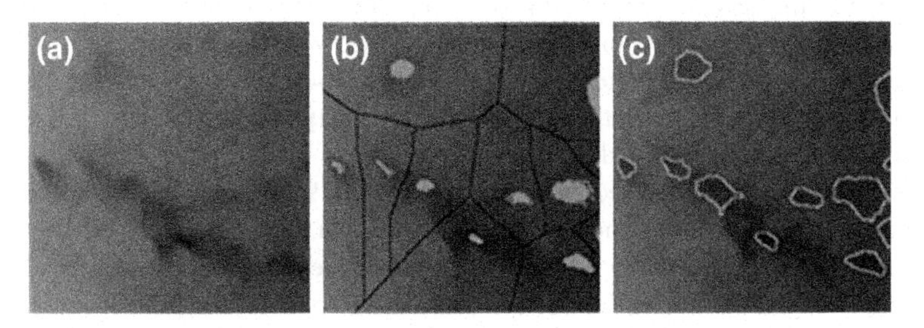

Fig. 8.4 Labelled watershed. **a** Original image; **b** Slight *gray regions* in the image are inner labels and *black lines* in the image are outer labels; **c** Watershed segmentation result

and limit inner labels only appear at some local minima. Then we get divide lines, defined as outer labels. As shown in Fig. 8.4b, black lines in the image are outer labels. Figure 8.4b shows that outer labels segment image into different regions, and each region contains only one inner label.

Steps of the above labelled watershed are as follows:

1. Get gradient map of an image.
2. Select inner labels: selecting inner labels is to find local minima. Gray values of its corresponding region are in a certain gray range and gray values of all pixels out the region are higher than that of pixels in the region.
3. Select outer labels: outer labels are obtained by applying watershed transform on inner labels.
4. Gradient modify: use force minimum technology to modified gradient, so that local minimum regions only appear in labelled positions.
5. After watershed transform for modified gradient image, we get final segmentation results.

Force minimum technology in (4) compares gradient minimum with threshold T, making sure that local minimum only appear in labelled position and deleting other local minimum. So it helps to eliminate some noise, texture details and irrelevant information, therefore effectively control over-segmentation. Here T is an experience threshold. If T is too large, the algorithm will lose edges. Otherwise, if T is too small, over-segmentation cannot be solved.

8.2.4 Experiment Results

We present the sample typhoon eye extraction experimental results using three SAR images. The detailed information of these images is given in Table 8.1. Their geographical positions are shown in Fig. 8.5.

The three SAR typhoon images are shown in Figs. 8.6, 8.7 and 8.8. What we are interested in are typhoon eyes. Typhoon eye extraction experiment results with our method are shown in Figs. 8.9 and 8.10. The three obtained denoised images are presented in the second column in Fig. 8.9 and the corresponding enhanced results are presented in the third column. Then the labelled watershed algorithm is used to segmentation enhanced images. We find closed curves in a segmented image and calculate their areas to select the biggest one, getting typhoon eye extraction results. The final results are presented in Fig. 8.10.

One can see from the second column of Fig. 8.9, image noise is effectively restrained after denoising, and the edges of typhoon eyes also are well kept. The third column of Fig. 8.9 shows that image enhancement algorithm works well, and typhoon structures are much clearer. In the third step, SAR images are segmented using labelled watershed. One can see from Fig. 8.10, the typhoon eyes in all the three images are well segmented from the complex background and their edges are well detected. In Fig. 8.10c, the left boundary of typhoon eye is a little outward extension.

Table 8.1 Information of SAR images and hurricane eye area estimates

Tropical cyclone	Satellite	Date	Time (UTC)	Beam mode	Swath (km)	Resolution (m)	Typhoon eye area estimates		
							Manual work by Li et al. (2013) (km^2)	This work (km^2)	Relative difference (%)
Aere	ENVISAT	2004-08-25	01:52:14	Wide swath	405	150	1793	1618	9.76
Khanun	ENVISAT	2005-09-11	01:46:11	Wide swath	405	150	379	345	8.97
Neoguri	ENVISAT	2008-04-18	02:31:50	Wide swath	405	150	502	522	3.98

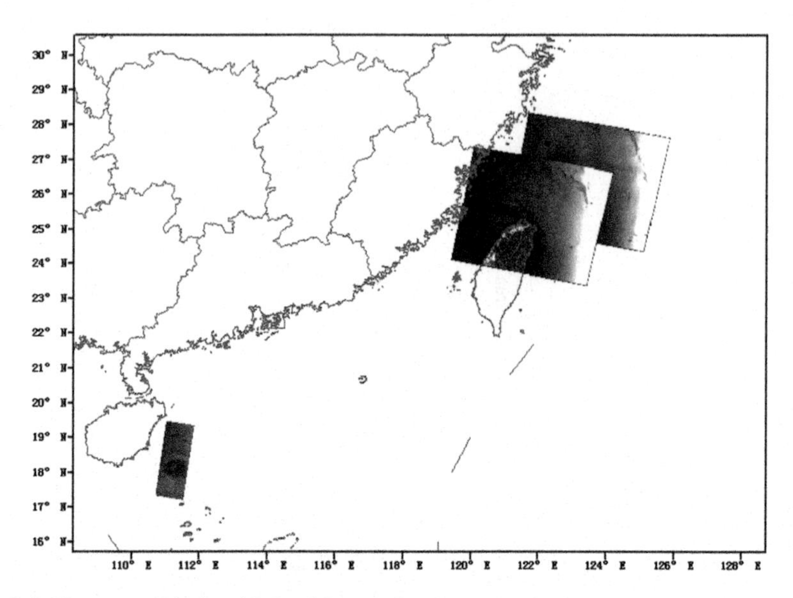

Fig. 8.5 The geographical positions of three typhoons

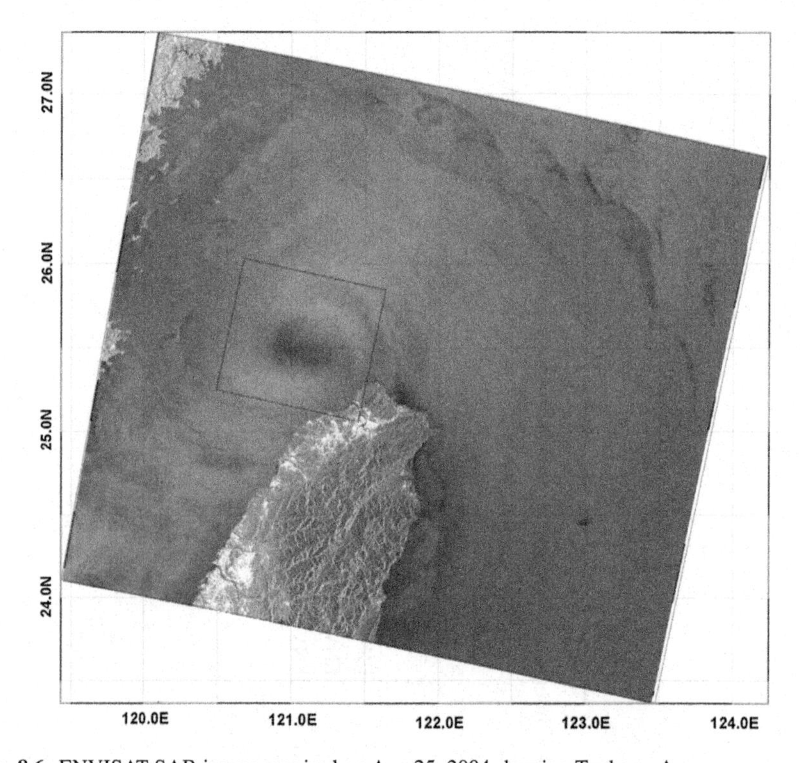

Fig. 8.6 ENVISAT SAR image acquired on Aug.25, 2004 showing Typhoon Aere

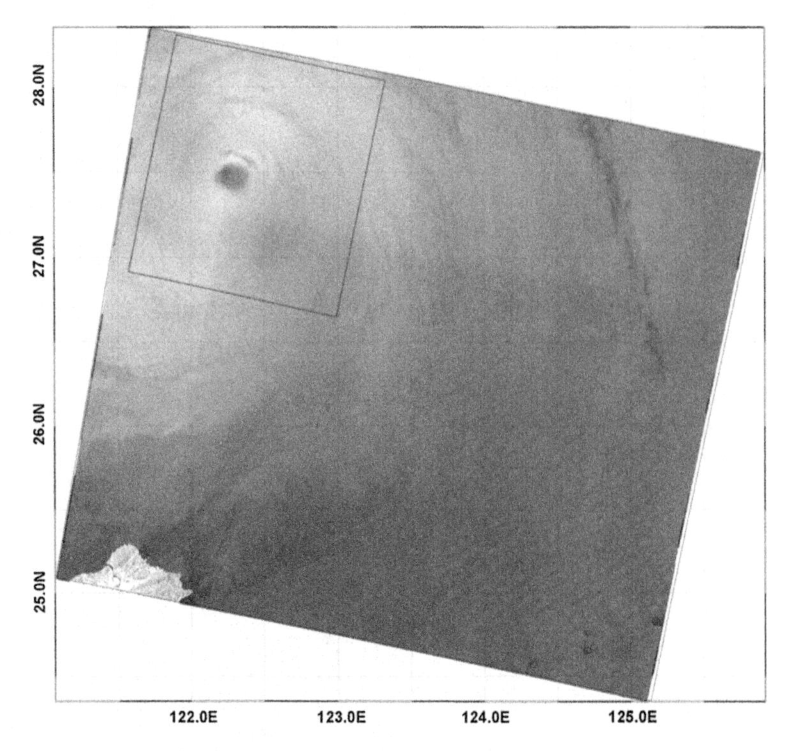

Fig. 8.7 ENVISAT SAR image acquired on Sep.11, 2005 showing Typhoon Khanun

Actually, groups of experiments indicate that segmentation results are influenced by denoised and enhanced results. If edges are not well kept or image details lose after the first two steps, typhoon eyes may not be detected accurately.

For further explanation, we calculate eye areas of the three typhoon cyclones, and then compare them with those calculated manually in Ref. [2]. As shown in Table 8.1, typhoon cyclone eye areas extracted by this technique are close to those extracted manually. We know that manual work is time consuming and labour demanding. The manual results may be different if different personnel perform the same work. Automatic procedure is robust and not influenced by peoples experience. Experiment results show that our method works well for automatically extracting typhoon cyclone eye information.

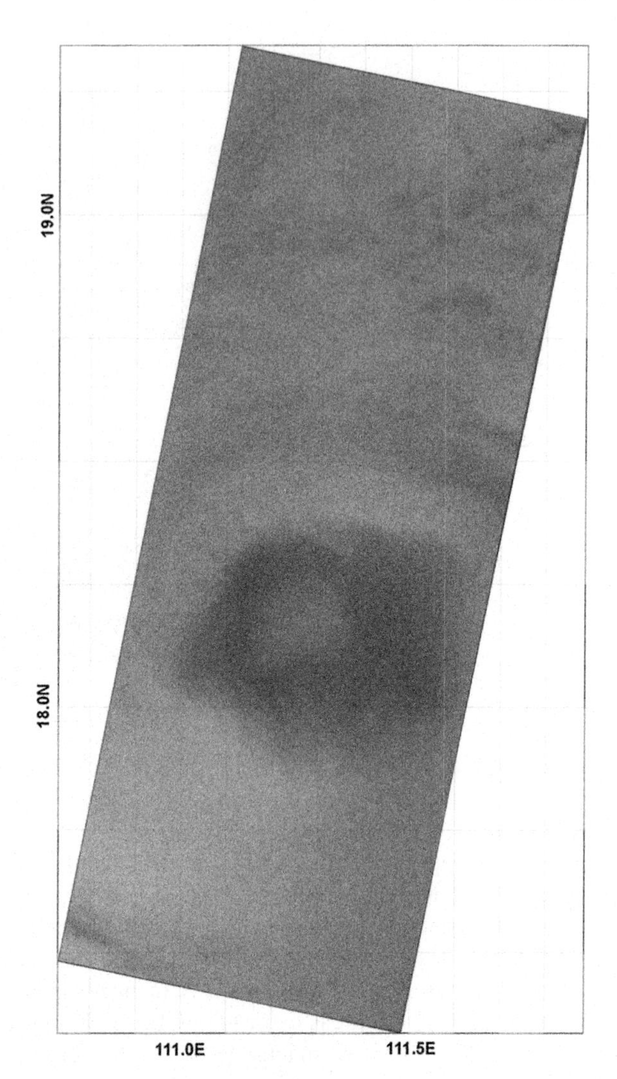

Fig. 8.8 ENVISAT SAR image acquired on Apr.18, 2008 showing Typhoon Neoguri

8.3 A Salient Region Detection and Pattern Matching Based Algorithm for Center Detection of a Partially-Covered Tropical Cyclone in a SAR Image

Not every tropical cyclone has an obvious eye feature in a SAR image. There may even be no eye where a tropical cyclone is in the developing period or the declining period. Sometimes limited by the capture range of radar, a SAR image may contain parts of a tropical cyclone without an eye. SAR provides only a single snap view at a time. So the pattern matching methods can be performed to estimate the center position of tropical cyclones without their eyes in SAR images.

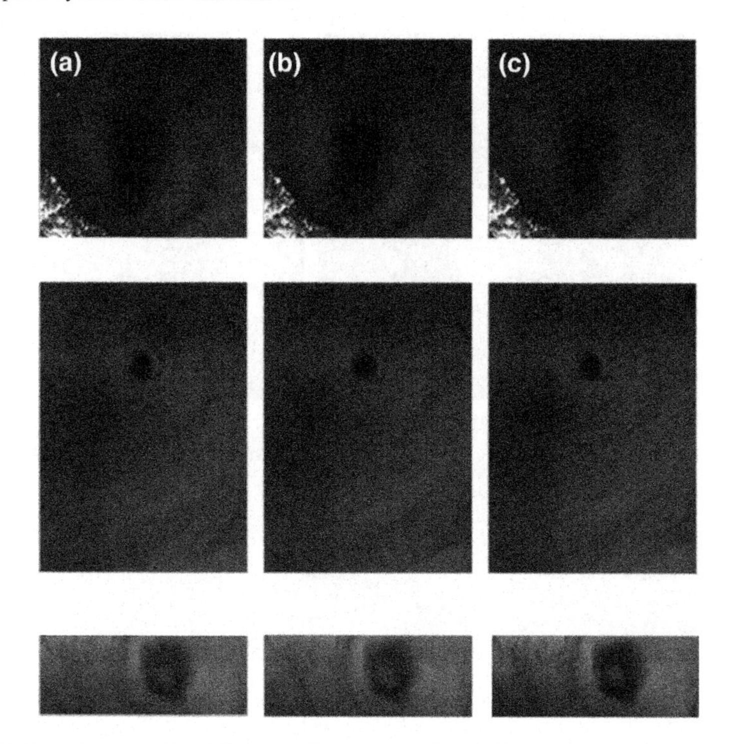

Fig. 8.9 Experiment results of three-step image processing chain to extract typhoon eye information. The Column **a** contains original images, the Column **b** contains denoised images with PPB filter and the Column **c** contains enhanced results with top-hat transform

Taking advantage of the characteristics of tropical cyclones in SAR images, a semi-automatic tropical cyclone center location method by combining a proposed salient region detection method with a particle swarm optimization algorithm (PSOA) is presented. The algorithm flowchart is shown in Fig. 8.11, and it mainly contains three steps. Step (1), rain bands are extracted using a salient region detection algorithm. Firstly, calculate the standard deviation of gray values of image patches and get a gray value contrast feature map. Secondly, calculate the Gabor features of a denoised tropical cyclone image and combine different orientation maps into a Gabor feature map. Thirdly, combine the gray value contrast feature map and the Gabor feature map by weights to get the salient region map. Rain bands are mainly contained in the salient region map. Step (2), the salient region map is segmented into a binary image and rain bands are selected by two filter criteria. Then extract the skeleton lines. To get smooth lines for better matching, apply expansion operators and pruning before and after the extraction. Step (3), transform the matching problem into an optimization problem, and then apply the PSOA to estimate the optimum solution, which corresponds to the center of the tropical cyclone. To estimate the correctness of the located center, compare the results with the National Oceanic and Atmospheric Administration National Hurricane Centers Best Track Data sets. Experiments demonstrate that the

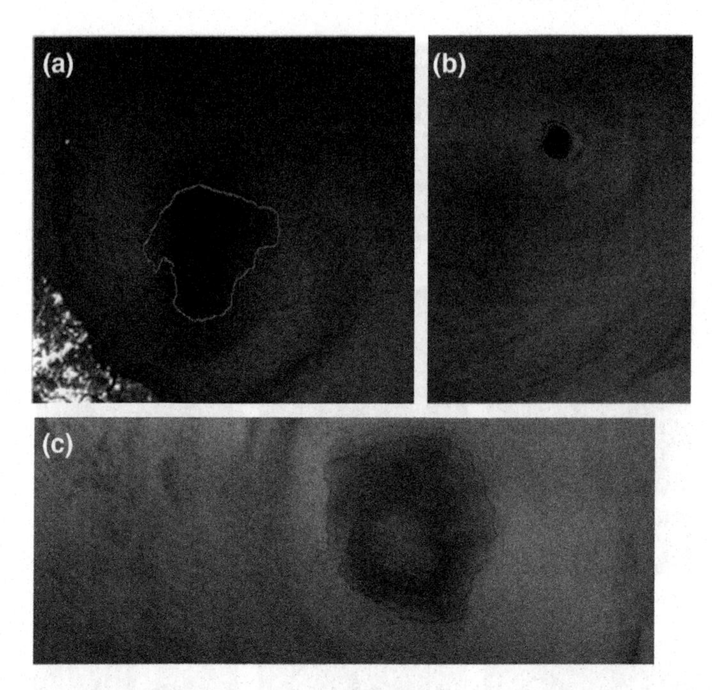

Fig. 8.10 Labelled watershed segmentation results of three images in Fig. reffig8.9a. **a** Is the result of Typhoon Aere, **b** is the result of Typhoon Khanun and **c** is the result of Typhoon Neoguri

proposed image processing method can correctly locate a tropical cyclone center in a SAR image with good accuracy.

8.3.1 Salient Region Detection Algorithm to Produce a Rain-Band Map

The automatic center location of tropical cyclones without eyes using pattern matching methods usually needs the help of rain band information. Therefore we need to obtain the spiral information of rain bands before matching. Researchers often segment satellite cloud images using a threshold segmentation method [8, 14–16]. These algorithms usually work well when the gray levels of the rain bands are obviously different from the gray levels of other regions. This is easy but not very effective when the gray levels of rain bands are more or less the same as those of other cloud clusters, which makes it hard to distinguish rain bands from other cloud clusters. Besides, influencing factors such as speckle noise, various configurations of tropical cyclone images acquired by different SAR instruments with different polarization, azimuth and spatial resolution, etc. will make automatic segmentation of large numbers of rain bands more difficult. In addition, a tropical cyclone SAR image shows

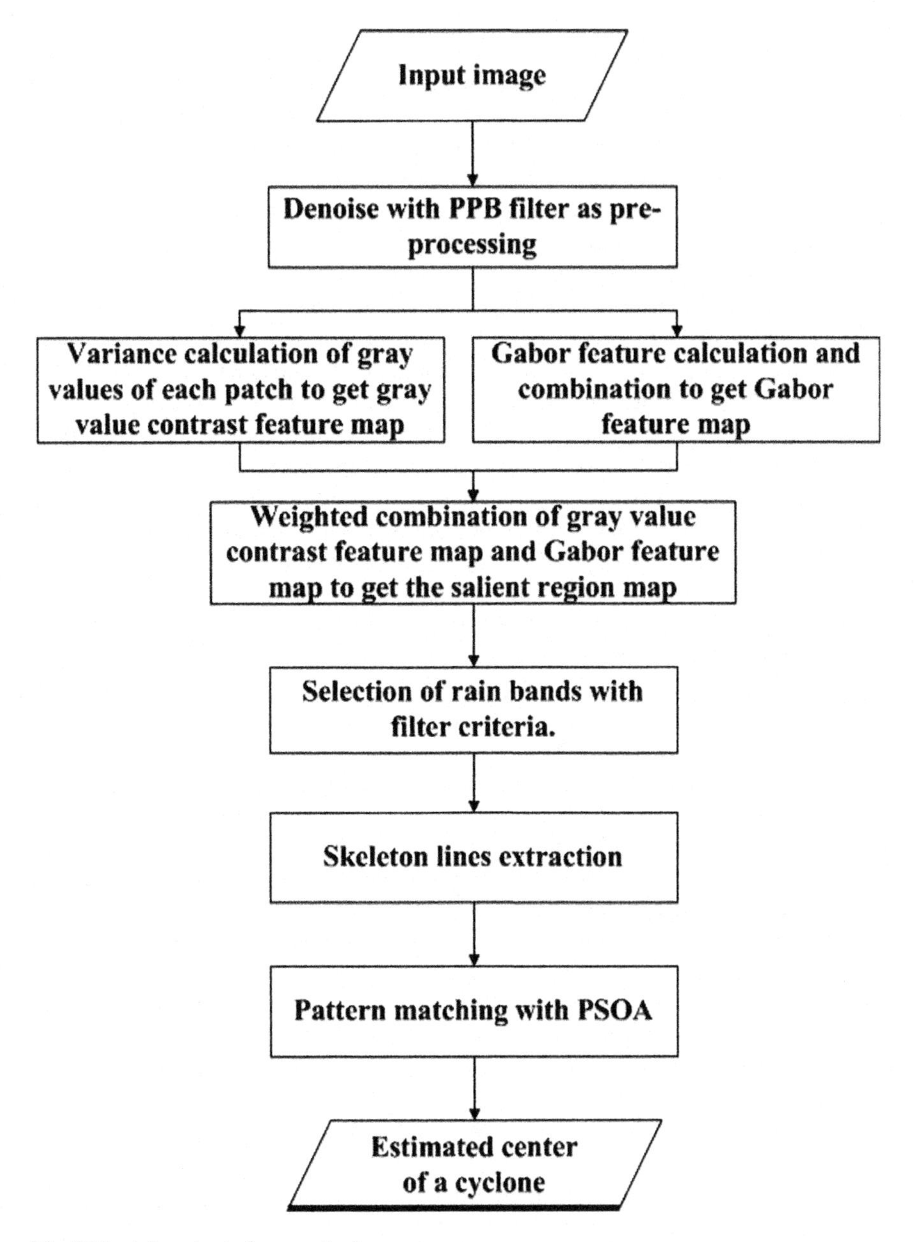

Fig. 8.11 A flowchart of our method

the sea surface imprint of a tropical cyclone with little texture information. Therefore we consider obtaining rain bands by salient region detection.

In computer vision, salient regions are defined as regions that attract human visual attention at the earliest visual processing when looking at an image. Salience is based on a variety of visual stimulation, such as color, brightness, texture, shape, edge, etc. High contrast between stimuli creates space reorganization of the receptive field cells, attracting the attention of the observer. That is the occurrence of the salience [37–39]. People tend to rapidly search for the most important parts and ignore the less important parts when they watch an image based on vision task and their prior knowledge. This selective attention mechanism enables people to efficiently capture the areas that they are interested in. These captured areas can be called the focus of attention areas or salient regions.

Since Itti and Koch proposed a model of saliency-based visual attention based on the human visual attention selective mechanism [40], visual salient region detection has become a popular research topic and many methods have been proposed [40–46]. Although different saliency detection methods are based on different hypotheses different theories, they all have one common characteristic: they focus on the center-surround difference. For a tropical cyclone SAR image, the major characteristic that makes a tropical cyclone different from its surroundings is the gray values contrast and its special spiral structure associated with rain bands, so the gray values contrast and orientation information can be used for salient region detection. Besides, little texture information in a tropical cyclone SAR image is propitious to salient region detection based contrast. Based on these advantages we propose a salient region detection method based on gray value contrast and orientation information.

Standard deviation, the average value of the distances from each data set to the mean value in a data set, reflects the degree of the discrete degree of the data set. Given an image patch, if the standard deviation of the gray values of the pixels is large, the contrast of the adjacent pixels is also large. This means the gray values change greatly and there may be context change in the image patch. As mentioned above, the gray values of clouds of a tropical cyclone are different from those of its surrounding background in a tropical cyclone SAR image. So we can consider that the salient spiral structure is next to, or contained in, the regions with a large standard deviation. We can divide an image into patches with a sliding window, whose step is half of its side length. We can then calculate the standard deviation of each patch. Generally, the standard deviation of a patch containing two or more kinds of homogenous regions is larger than that of a patch containing only one kind of homogenous region. So a patch with a larger standard deviation can be considered more salient. Figure 8.12 shows patches with different standard deviations. As shown, patch P1 is a homogenous region where the values of the pixels are more or less the same, so its standard deviation is small. But there is a context change in patch P2 and the values of the pixels in it are different, so its standard deviation is large. The larger the standard deviation, the more salient the corresponding patch. We get a gray value contrast feature map after normalization of all the standard deviations.

Since there are some irrelevant patterns whose gray values are similar to these of the rain bands, we may not extract the exact salient region only with gray value information. We further look at the spiral structure of rain bands that contain obvious orientation information. So we also make use of orientation information with Gabor features to improve the degree of saliency.

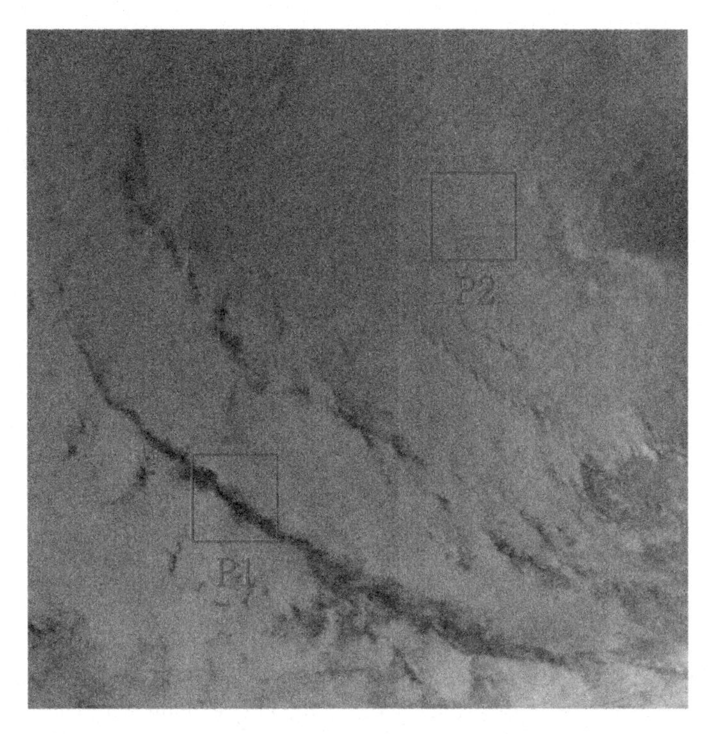

Fig. 8.12 Different patches with different standard deviation. Patch P1 passes through an edge of a rain band and contains a part of the rain band and a part of the background. Gray values of the two parts are different. Patch P2 only contains a part of the background, whose gray values are more or less the same. So Patch P1 has a bigger standard deviation than that of Patch P2

A 2D Gabor filter was first applied by Daugman [47] in the field of computer vision. The 2D Gabor filter has a visual characteristic and biology background. It can be seen as a good approximation to the sensitivity profiles of neurons found in the visual cortex of higher vertebrates [48]. The Gabor features have good directional characteristics that are sensitive to edge information. The Gabor filter will respond strongest if the filters orientation is consistent with the orientation of specific features in an image. In addition, it is good at multi-frequencies and in multi-orientations, and is non-sensitive to light change. These advantages make the Gabor filter widely used in visual information extraction, pattern recognition and image processing. The 2D Gabor filter can be written as:

$$h(x, y) = \frac{1}{2\pi\sigma^2} exp\left(-\frac{x'^2 + y'^2}{2\sigma^2}\right) exp[2\pi j(Ux' + Vy')] \tag{8.12}$$

where $x' = x\cos\theta + y\sin\theta$ and $y' = x\cos\theta - y\sin\theta$, θ is the orientation, is the scaling parameters of the filter and (U, V) is the center frequency. The Gabor filter has good direction selectivity in the spatial domain and frequency selectivity in the frequency domain by choosing different directions and adjusting the frequency. The Gabor feature of an image $I(x, y)$ is the convolution of the image and the Gabor filter $h(x, y)$.

We calculate the Gabor features of a tropical cyclone SAR image in four orientations $(0°, 45°, 90°, 135°)$. Then the Gabor feature map is obtained by combining the four orientation maps with weights. Regions with high gray values in the Gabor feature map have more directional information. Then we combine the Gabor feature map and the gray value contrast feature map with different weighting to construct the final salient region map.

8.3.2 Rain Bands Selection with Two Filter Criteria

After the salient region detection, outlined in the previous section, the salient region map of a SAR tropical cyclone image is obtained. Regions with high gray values in the salient map are mainly rain bands. The skeleton lines of rain bands make their spiral shape and rolling tendency intuitive and visual. It is easy to understand and convenient to process spiral information by computer. So in this section we consider how to extract the correct skeleton lines of rain bands for pattern matching.

First of all, we need to get rain bands which have an obvious shape characteristic. Salient region detection has eliminated most of the irrelevant parts of the image. Regions with high gray values in the salient map are mainly rain bands. However, there are other irrelevant minor features, too. The combination of several minor features may also reflect a spiral tendency, but their skeleton lines are not continuous. Therefore, they cannot be used for pattern matching. To avoid the interference of minor regions, we segment the salient region map of a SAR tropical cyclone image into a binary image by a threshold, and then remove regions with small areas in the binary image with two filter criteria. So the first filter criterion is by area of region. We calculate the area of each region in the binary image, arrange them in order of area and only keep the several largest regions. Small regions are deleted with the first filter criterion. The second filter criterion evaluates the distance of the left-top point and the right-bottom point of a remained region. Several regions with the longest distances can be remained. We remain the one region with the longest distance for automatic computation. As shown in Figs. 8.16g and 8.21g, regions with long and thin strips are kept when the second filter criterion was applied.

8.3.3 Skeleton Lines Extraction with Morphology and Pruning

The edges of the rain bands in a SAR image are irregular with protrusions and hollows. If we extract skeleton lines directly, it will contain branches. If there are holes in the rain band regions, the skeleton line will contain circular rings. All these conditions will affect the continued pattern matching. Here we take two measures to obtain smooth skeleton lines. Firstly, the rain bands are expanded using morphologic operators before extraction of the skeleton lines [49]. The hollows are filled and the edges of the connecting domains are smoothed, reducing the possibility of the skeleton lines forking. Secondly, the little branches and protrusions of the skeleton lines are pruned [50]. The two steps can be iterated if necessary. As shown in Figs. 8.16a and 8.21a, the skeleton lines are smooth without additional branches, while the spiral shape of the original connected domains is preserved.

8.3.4 Particle Swarm Optimization Algorithm Estimate Tropical Cyclone Centers Based on Pattern Matching

We focus on pattern matching to estimate tropical cyclone centers in this section. Here we choose an analytical model of two-dimensional surface storm-relative inflow angle in a tropical cyclone to estimate tropical cyclone centers. Zhang et al. (2012) presented an analysis of near-surface inflow angles using wind observation data from over 1600 quality-controlled global positioning system dropwindsondes deployed by aircraft on 187 flights into 18 hurricanes and proposed a parametric model of inflow angle based on these observations and analysis [51]. Here the inflow angle α_{SR} can be defined as the arctangent of the ratio of radial wind component v_r to tangential wind component v_t. Analysis results in the reference indicate a statistically significant dependence of inflow angle on the radial distance from the tropical cyclone center.

The model is defined as follows:

$$\alpha_{SR}(r^*, \theta, V_{max}, V_s) = A_{\alpha 0}(r^*, V_{max}) + A_{\alpha 1}(r^*, V_s, V_{max}) \cdot cos[\theta - P_{\alpha 1(r^*, V_s)}] + \varepsilon \quad (8.13)$$

where θ is the azimuth angle measured clockwise from tropical cyclone motion direction, V_{max} is the maximum wind speed, V_s is the tropical cyclone motion speed and ε is the model error. $r^* = \frac{r}{R_{max}}$, where r is the radial distance measured in a polar coordinate system and R_{max} is the radial distance of maximum wind speed. $A_{\alpha 0}$, $A_{\alpha 1}$ and $P_{\alpha 1}$ are defined as:

$$A_{\alpha 0} = a_{A0} r^* + b_{A0} V_{max} + c_{A0} \quad (8.14)$$

$$A_{\alpha 1} = -A_{\alpha 0}(a_{A1} r^* + b_{A1} V_s + c_{A1}) \quad (8.15)$$

Table 8.2 Coefficients for the inflow angle model in Eqs. (8.14)–(8.16)

Equation	Variables	a	b	c
(8.14)	$A_{\alpha 0}$	−0.90	−0.90	−14.33
(8.15)	$A_{\alpha 1}$	0.04	0.05	0.14
(8.16)	$P_{\alpha 1}$	6.88	−9.60	85.31

$$P_{\alpha 1} = a_{P1} r^* + b_{P1} V_s + c_{P1} \tag{8.16}$$

The coefficients (a, b, c) are shown in Table 8.2. The above inflow angle model is developed based on a subset of the full observation sample and he coefficients of $A_{\alpha 0}$ is the most accurately estimated quantity while the coefficients of $A_{\alpha 1}$ is the least.

Given the skeleton line l_{ske} of a rain band, we can obtain all the pixels $\{(x_i, y_i)| (x_i, y_i) \in l_{ske}\}$ on it. r_i is the Euclidean distance from one point (x_i, y_i) on the skeleton line to the center of the tropical cyclone (x_c, y_c) which we want to get. We can define the normalized distance r_i^* as:

$$r_i^* = \frac{\sqrt{(x_i - x_c)^2 - (y_i - y_c)^2}}{\sqrt{(x_{max} - x_c)^2 + (y_{max} - y_c)^2}} \tag{8.17}$$

where (x_{max}, y_{max}) is the position of the maximal wind speed. The corresponding azimuth can be defined as:

$$\theta_i = arctan\left(\frac{y_i - y_c}{x_i - x_c}\right) \tag{8.18}$$

Given a tropical cyclone wind speed data, we can get the information about the maximal wind speed V_{max} and its position (x_{max}, y_{max}) and the tropical cyclone motion speed V_s. If we also know the center (x_c, y_c) of a tropical cyclone, we can get r^* and. Then the inflow angles can be determined with parameters $(r^*, \theta, V_{max}, V_s)$.

If the inflow angle values along a skeleton line are known, the key to solve the matching problem is to find the best combination of parameters (x_c, y_c) which makes the calculated inflow angles using Eq. (8.13) best match the given inflow angles. We can then treat the matching problem as an optimization problem. A candidate combination of parameters (x_{c_i}, y_{c_i}) corresponds to an estimated inflow angle set. When we find the optimal combination of parameters $(x_{c_{best}}, y_{c_{best}})$, the estimated inflow angles are closest to the given inflow angles. It means the optimal $(x_{c_{best}}, y_{c_{best}})$ is closest to the real tropical cyclone center. So the optimization problem here is to find the optimal $(x_{c_{best}}, y_{c_{best}})$ on a skeleton-line image.

As mentioned in Sect. 8.1, the simple and easy Hough transform was often used as the matching method in previous work [14]. However, it only performs well on part of the matching pixels, and it is difficult to reach a good matching result. Wong et al. used a genetic algorithm to match skeleton lines with a logarithm spiral [8]. The genetic algorithm is robust and has global search ability. However, the crossover operator and mutation operator of the genetic algorithm guide the search iterative process randomly. So they provide the opportunity to evolve but inevitably produce the possibility of degradation at the same time. There will be a lot of redundant iteration when the solution reaches a certain range, resulting in a low efficiency of exact solutions.

Different from the genetic algorithm, the Particle Swarm Optimization algorithm (PSOA) proposed by Eberhart and Kennedy [52] is simpler and more effective with fewer input parameters required. The calculation converges to the optimal solution quicker. These advantages make the PSOA widely used in optimization problems [53–56]. We chose the PSOA to search the optimum solution to solve the matching problem.

The PSOA is based on research of birds predation. We suppose that a flock of birds is searching for food randomly. If there is only one piece of food, the easiest but most effective strategy to find the food is to search the surrounding area of the bird which is nearest to the food. This idea is used in the PSOA. Here we can consider the best center (x_c, y_c) as the food. A solution (an estimated center) for the optimization problem is considered to be the position of a bird, which is also called a particle. The method involves in inputting a binary image of skeleton lines, counting the number of lines and getting all the pixel positions of each skeleton line. Then it is necessary to initialize the position and speed of the original searching particle (parameters (x_c, y_c)). Every particle has its own position and speed, which determines its direction and distance of flight. The current position of each particle is initialized as its original best position (it can be called pbest). There is a fitness value determined by a fitness function. The corresponding fitness value z of each particle is calculated using the fitness function. The best fitness value is considered to be the global fitness value. The global best position is initialized (it can be called gbest) with the position of the particle having the best fitness value. Then each particle adjusts its speed and position according to gbest and its own speed and position. The fitness value z of each particle is calculated. If the fitness value of one particle is better than its current pbest, the current pbest moves into its position. If the best pbest of all particles is better than the current gbest, move the current gbest into the position of the particle having the best pbest. The process reiterates until the optimum solution is found or the iterative time has been reached.

A matching result can be evaluated by the degree of error between a given inflow angle α_{SR} and the estimated inflow angle α. The smaller the error is, the better the matching result is. Suppose that there are N pixels on a skeleton line and we know their inflow angles $\{\alpha_{SR_i}\}$, the fitness function can be defined as:

$$z = |\alpha_{SR_i} - \left\{A_{\alpha 0}(r_i^*, V_{max}) + A_{\alpha 1}(r_i^*, V_s, V_{max}) \cdot [\theta_i - P_{\alpha 1}(r_i^*, V_s)] + \varepsilon\right\}| \tag{8.19}$$

If we do not know the inflow angles, we can change the fitness function. The mean inflow angle in tropical cyclones $\alpha_{average}$ is found to be $-22.6° \pm 2.2°$ (95% confidence) [42], which agrees well with the previous results [48]. Therefore the fitness function can be changed as:

$$z = \left|\frac{\sum_{i=1}^{N}(r_i^*, V_{max}) + A_{\alpha 1}(r_i^*, V_s, V_{max}) \cdot cos[\theta_i - P_{\alpha 1}(r_i^*, V_s)] + \varepsilon}{N} - \alpha_{average}\right| \tag{8.20}$$

The speed v and position present of each particle can be changed using Eqs. (8.21) and (8.22).

$$v_i = w \cdot v_{i-1} + c_1 \cdot rand_1 \cdot (pbest_{i-1} - present_{i-1}) + c_2 \cdot rand_2 \cdot (gbest_{i-1} - present_{i-1}) \tag{8.21}$$

$$present_i = present_{i-1} + v_i \tag{8.22}$$

where c_1 and c_2 are acceleration coefficients, w is the weight and $rand_1$ and $rand_2$ are random numbers between 0 and 1.

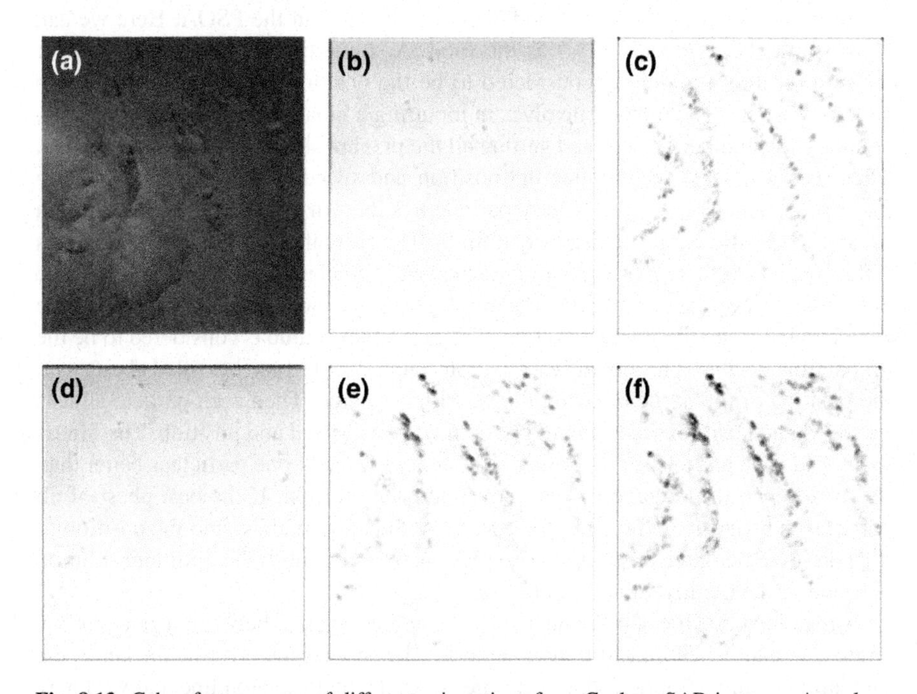

Fig. 8.13 Gabor feature maps of different orientations for a Cyclone SAR image. **a** A cyclone SAR image. **b–e** Gabor feature maps when the orientation are 0°, 45°, 90° and 135°. When the orientation are 0° and 90°, the rain bands are disappeared or fuzzy. While when the orientations are 45° and 135°, the rain bands are obvious. **f** The combined Gabor feature map

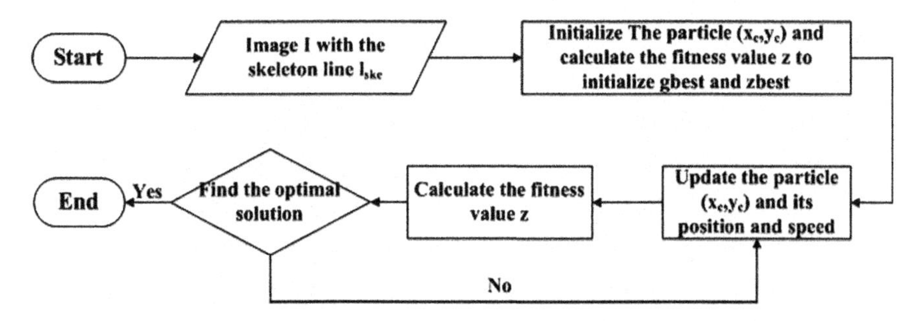

Fig. 8.14 Scheme of the PSOA

The optimum solution of (x_c, y_c) can be considered as the center of the tropical cyclone. Sometimes there may be several spiral lines after the extraction of skeleton lines. Each spiral line has an optimal solution. The optimal center achieved by one matching spiral line may not be the optimal solution of another matching spiral line. Theoretically there is a center point which is the compromised optimal solution for all the spiral lines. Hence we take the average of all optima as the final center point.

8.3.5 Experimental Results and Discussions

The data used in our experiments are RADARSAT-1 SAR tropical cyclone images. RADARSAT-1 C-band SAR images of tropical cyclones have been acquired worldwide since 1998 to support scientific research through the Hurricane Watch program which is a collaborative program between the National Oceanic and Atmospheric (NOAA), the Canada Centre for Remote Sensing (CCRS) of Natural Resources Canada and the Canadian Space Agency (CSA) [57]. These images are ScanSAR wide beam (SCW) images with a medium resolution of 100 m and a swath of 450 km. They are horizontal-transmit and horizontal-receive (HH polarization) data. The six tropical cyclones and their positions are shown in Fig. 8.15. Figure 8.15a contains the full structure of tropical cyclone Franklin while the eye area is fuzzy. Figures 8.15b and 8.15f contain the other five partially-covered tropical cyclones in SAR images.

We compute the salient region maps from the denoised images (Figs. 8.16b, 8.17b, 8.18b, 8.19b, 8.20b and 8.21b). To get a gray value contrast feature map, an image is first divided into patches using a 8×8 sliding window. Each step of the sliding window is half of its length. If the sliding window is too large, there will be a serious blocking effect, and a large number of unrelated areas will be contained in the gray value contrast feature map. If the sliding window is too small, the difference within each patch will not be well presented. In addition, the rain-band regions will be discontinuous and holes will easily appear in the connected domain. These side effects will affect the later extraction of skeleton lines. Then the standard deviation of each patch is calculated and normalized. The larger the standard deviation value, the

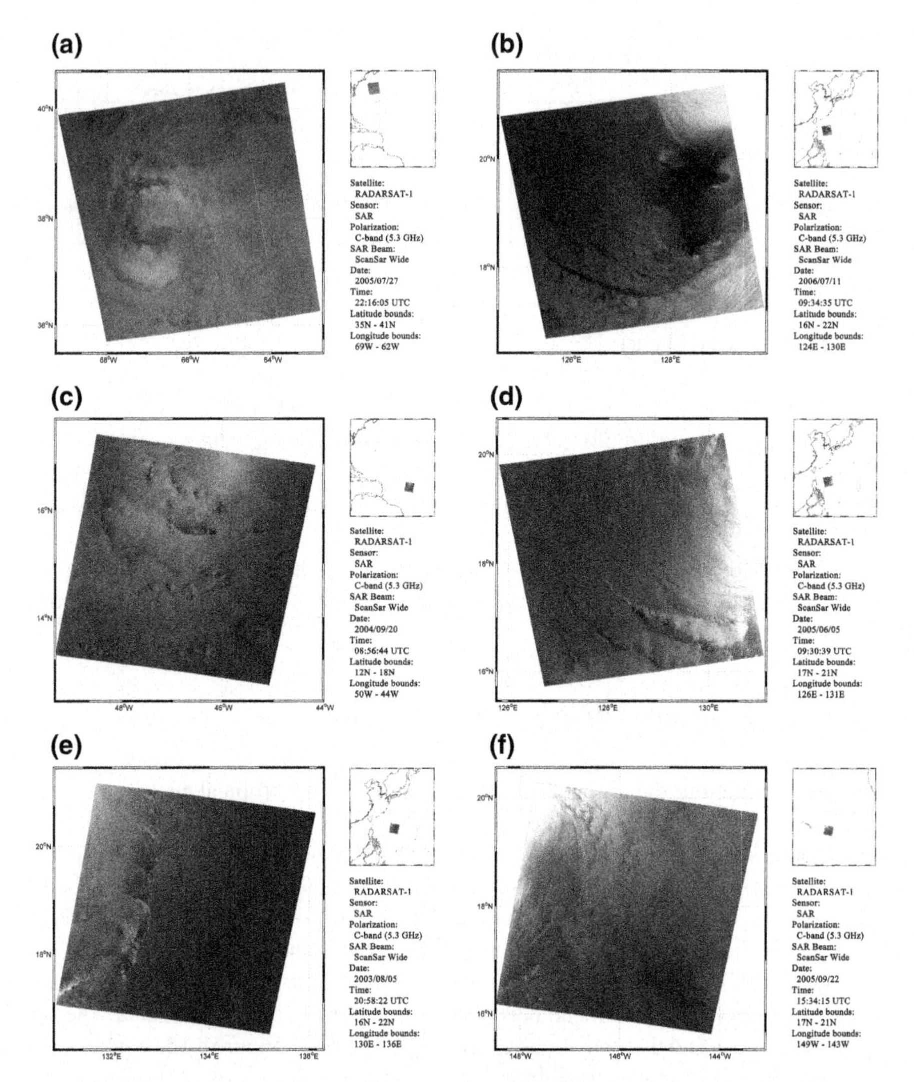

Fig. 8.15 The geographical positions of the six tropical cyclones and their wind speed. **a** Tropical cyclone Franklin captured on 22:16:05, July 28th, 2005. **b** Tropical cyclone Bilis captured on 09:34:35, July 11th, 2006. **c** Tropical cyclone Karl captured on 08:56:44, September 20th, 2004. **d** Tropical cyclone Nesat captured on 09:30:39, June 5th, 2005. **e** Tropical cyclone Etau captured on 20:58:22, August 5th, 2003. **f** Tropical cyclone Jova captured on 15:34:15, September 22th, 2005

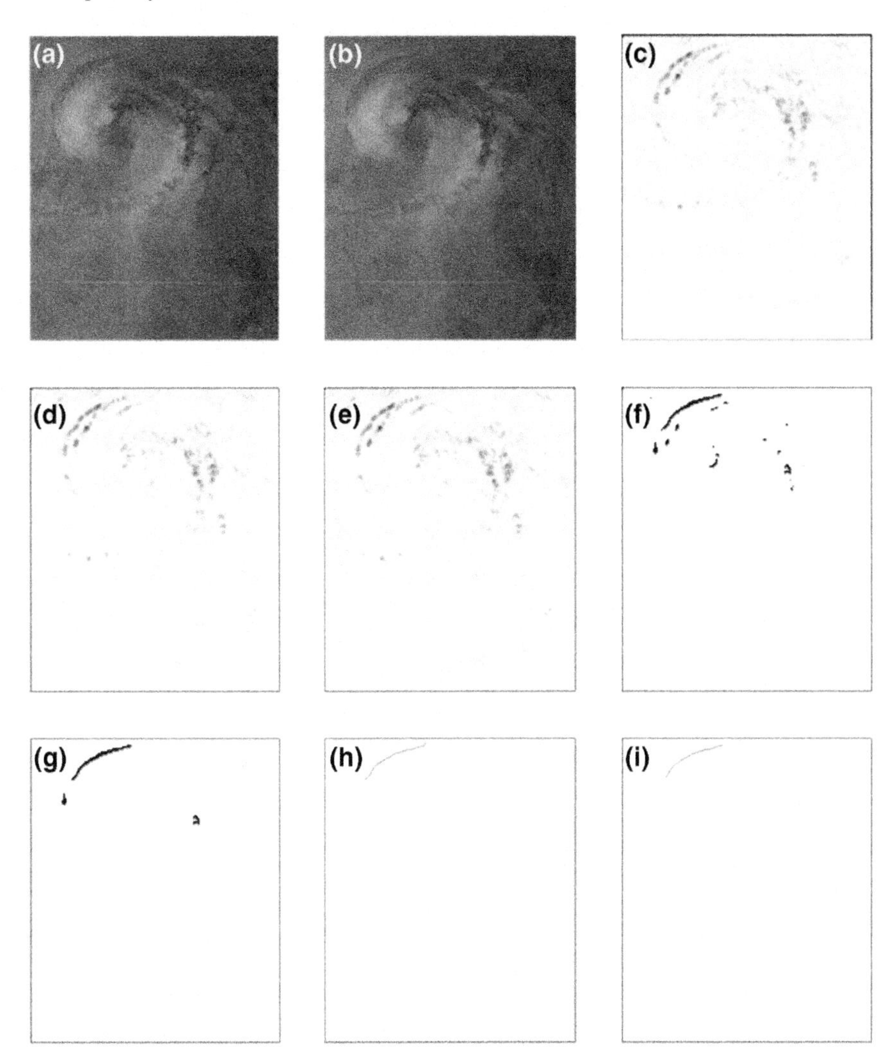

Fig. 8.16 **a** The SAR image of Tropical cyclone Franklin. **b** The denoised SAR image. **c** The gray value contrast feature map of the denoised SAR image. **d** The Gabor feature map. **e** The salient region map. **f** Remaining regions after selection filter criterion 1. **g** Remaining regions after selection filter criterion 2. **h** The skeleton lines of rain bands. **i** The smooth skeleton lines after pruning

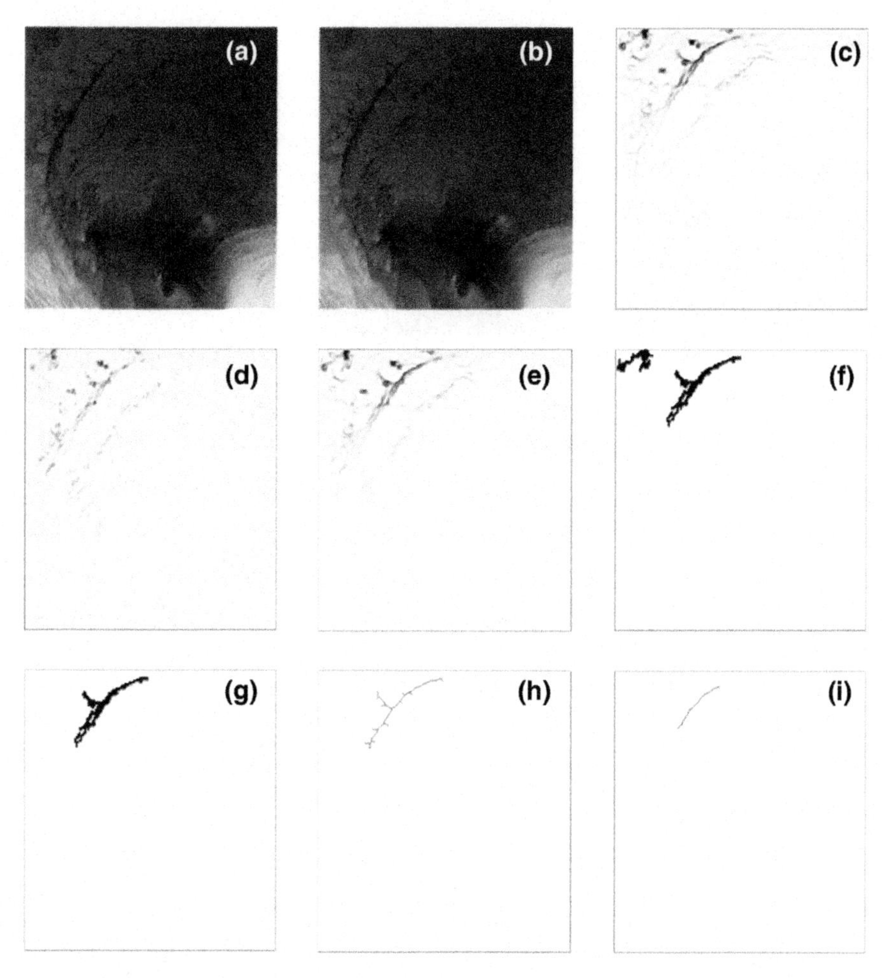

Fig. 8.17 Results of the SAR image of Tropical cyclone Bilis. Caption of each figure is the same as these in Fig. 8.16

more context change a patch has. Usually we use a threshold to remainder patches with a large standard deviation value. As shown in Figs. 8.16c, 8.17c, 8.18c, 8.19c, 8.20c and 8.21c, a gray value contrast feature map can indicate the position of rain bands. The parts with higher pixel values mainly contain rain bands.

Then we make use of orientation information by calculating the Gabor features. The Gabor features of a denoised tropical cyclone SAR image are calculated in four orientations (0°, 45°, 90°, 135°), and then the Gabor feature map is obtained using a weighted combination of four orientation maps. As shown in Fig. 8.13, experiments indicate that rain bands are more obvious in the 45° and 135° orientation angles. Therefore, to distinguish rain bands from other objects, an orientation map of 45° and an orientation map of 135° are given higher weights of 0.4, and an orientation

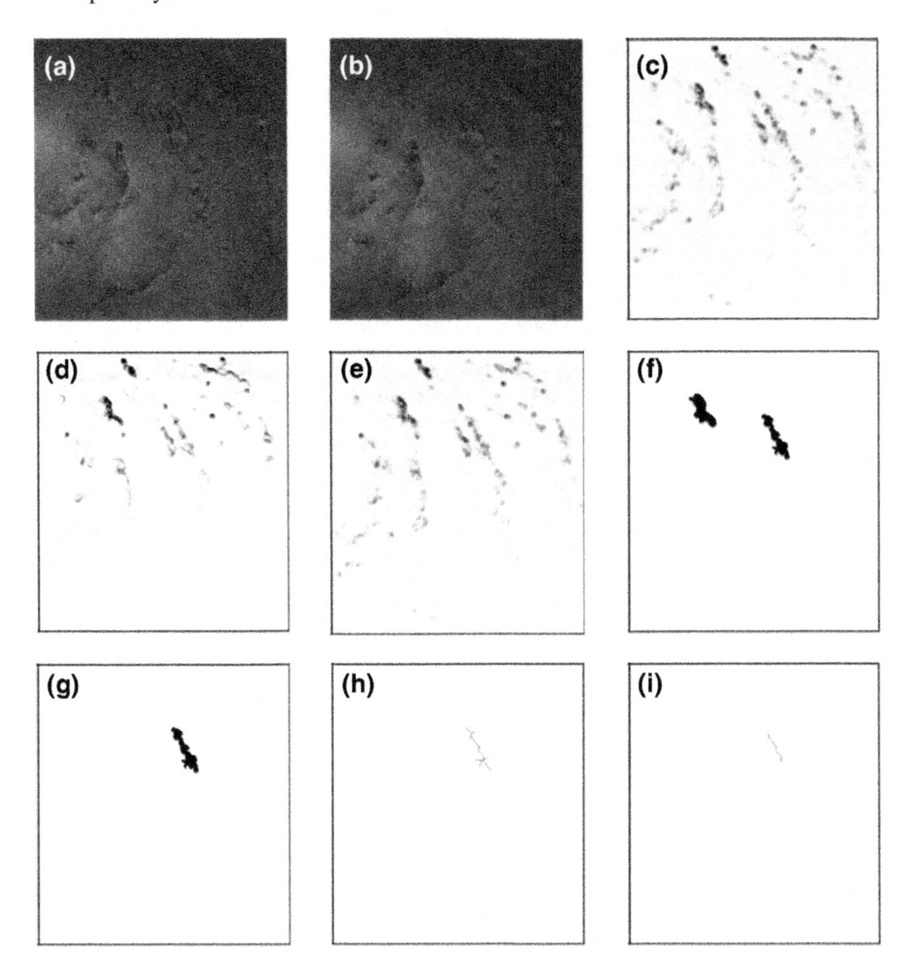

Fig. 8.18 Results of the SAR image of Tropical cyclone Karl. Caption of each figure is the same as these in Fig. 8.16

map of 0° and an orientation map of 90° are given lower weights of 0.1. As shown in Figs. 8.16d, 8.17d, 8.18d, 8.19d, 8.20d and 8.21d, the Gabor feature maps indicate the obvious rolling tendency and rain bands with higher pixel values (Figs. 8.14 and 8.15).

We get the final salient region map by a weighted combination of the gray value contrast feature map and the Gabor feature map. As shown in the Figs. 8.16e, 8.17e, 8.18e, 8.19e, 8.20e and 8.21e, rain bands are salient in the final salient region map. After that, we segment the final salient region maps into binary images. Regions with higher pixel values remain. For some small and unrelated regions in the binary image, we used the two filter criteria mentioned in Sect. 8.2.2 to select the rain bands regions. As shown in Figs. 8.16f, 8.17f, 8.18f, 8.19f, 8.20f and 8.21f, small and unrelated regions are deleted. The remaining region appears as long and thin lines,

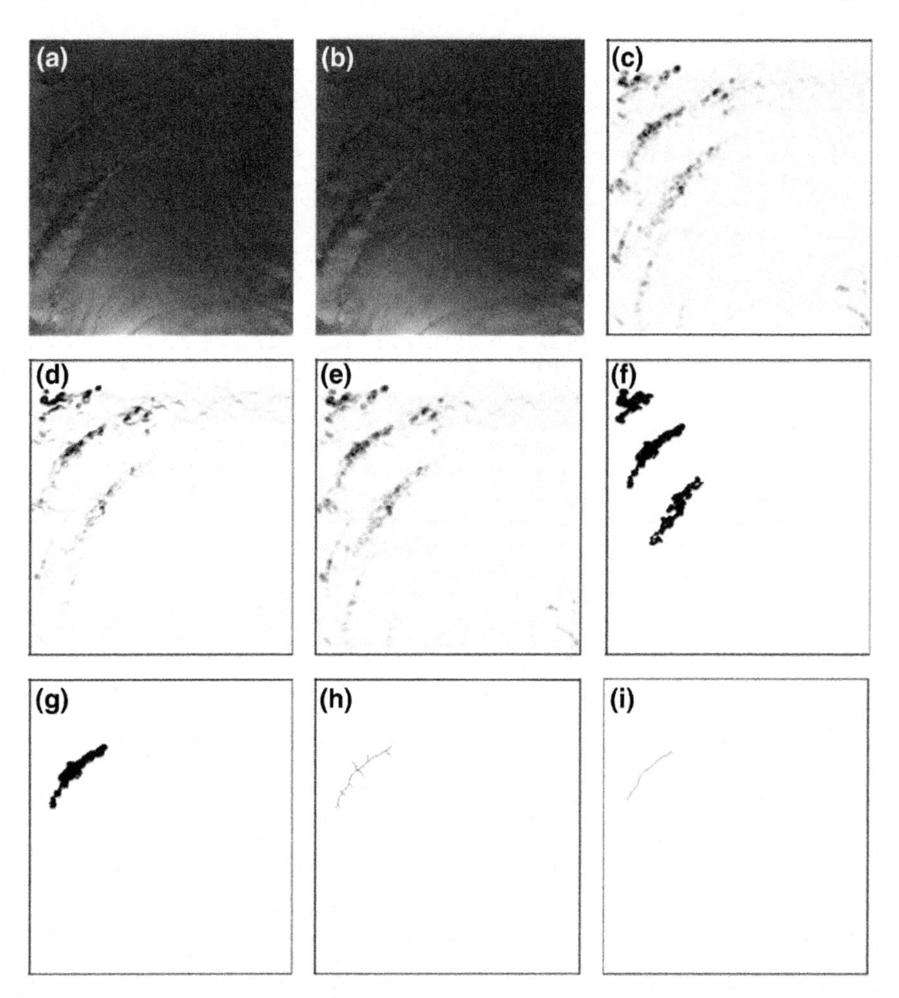

Fig. 8.19 Results of the SAR image of Tropical cyclone Nesat. Caption of each figure is the same as these in Fig. 8.16

which represent the characteristics of rain bands. However, there are still holes in some rain bands regions so we use the expansion operator to solve this problem. After extracting the skeleton lines, pruning is operated to avoid the influence of burrs. Expansion and pruning can be repeated to make the skeleton lines smoother as shown in Figs. 8.16a, 8.17a, 8.18a, 8.19a, 8.20a and 8.21a.

Finally, using the PSOA, we match each skeleton line with a given model. Here we set $c1 = c2 = 2$ and $w = 1$. The size of the particle is 20 and the number of iterations is 200. The center of a tropical cyclone may be out of the SAR image if there is no eye in the SAR image. Suppose the range of an image is mn, we expand the search range of PSOA to $3m \times 3n$ or $5m \times 5n$. The initial position and velocity of particles are given randomly. According to the pixel position set $(x_i, y_i)|(x_i, y_i) \in l_{ske}$ of a

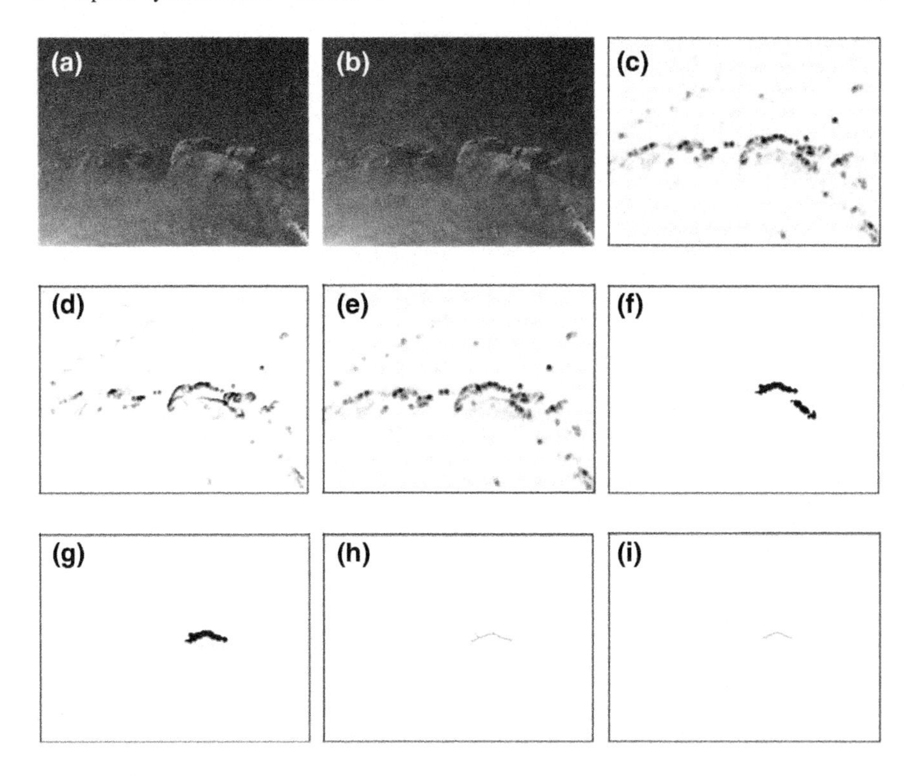

Fig. 8.20 Results of the SAR image of Tropical cyclone Etau. Caption of each figure is the same as these in Fig. 8.16

skeleton line using the procedures outlined in Sect. 8.2.4, we calculate the optimal solution of matching with the PSOA. If there are several skeleton lines in a binary image, we calculate the optimal solution of each matching skeleton line and then take an average of all the optima as the final tropical cyclone center. The center location results are shown in Fig. 8.22. We can see that the centers are within the images in Figs. 8.22a, b, but the centers are out of the images in the other four figures. We can not estimate whether the results are correct if there are no centers in the SAR images. In order to evaluate the accuracy of our center location results, we compare the estimated center with the NOAA Best Track Data sets in Table 8.3 and Fig. 8.23. The NOAA Best Track Data is an archive of global history tropical cyclones since 1842. The information is obtained from NOAAs program International Best Track Archive for Climate Stewardship (IBTrACS). The data contain the center position (usually it is the latitude and longitude) and the intensity (described by the maximum

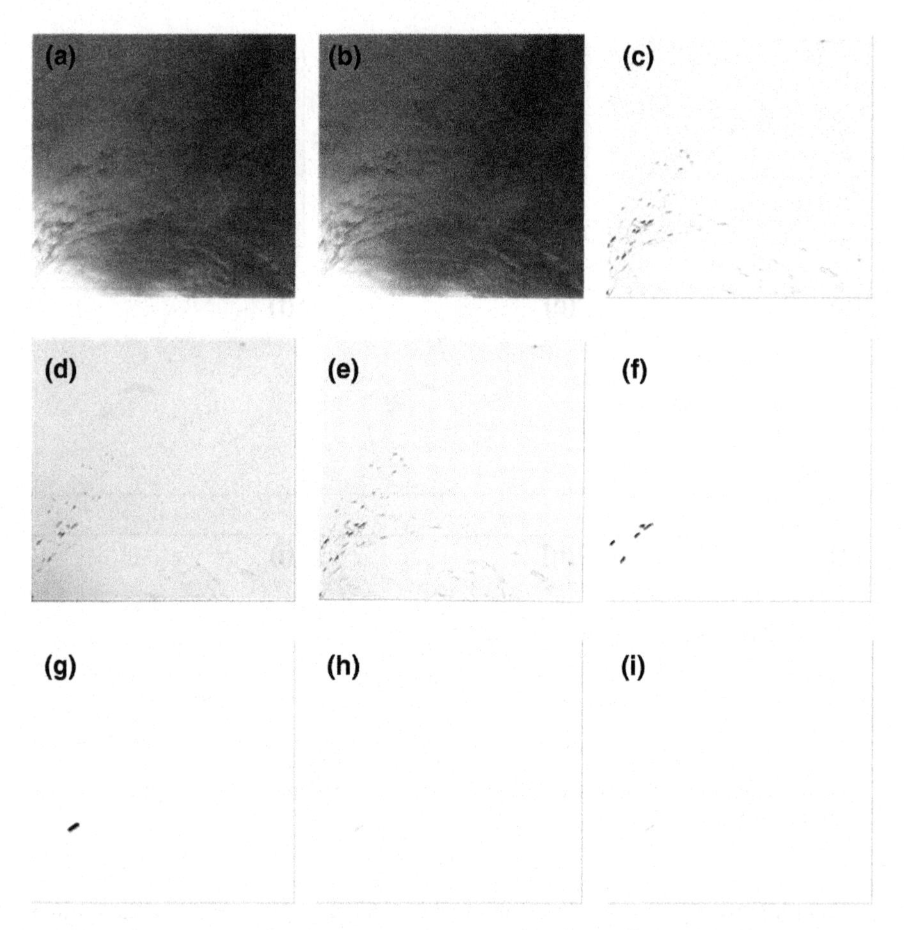

Fig. 8.21 Results of the SAR image of Tropical cyclone Karl. Caption of each figure is the same as these in Fig. 8.16

wind speed or the lowest central air pressure) of a tropical cyclone at a certain time and other information. The NOAA Best Track data is updated every six hours. As a result, the locations of the tropical cyclone centers at the SAR imaging times are interpolated from the two nearby Best Track data records. The estimated center position of a tropical cyclone should between the center positions at the two recorded times. Table 8.3 shows each estimated centers position and the two center positions on the two above recorded times in the best track data set. Using linear interpolation, we draw a straight line between the center positions before and after the time that

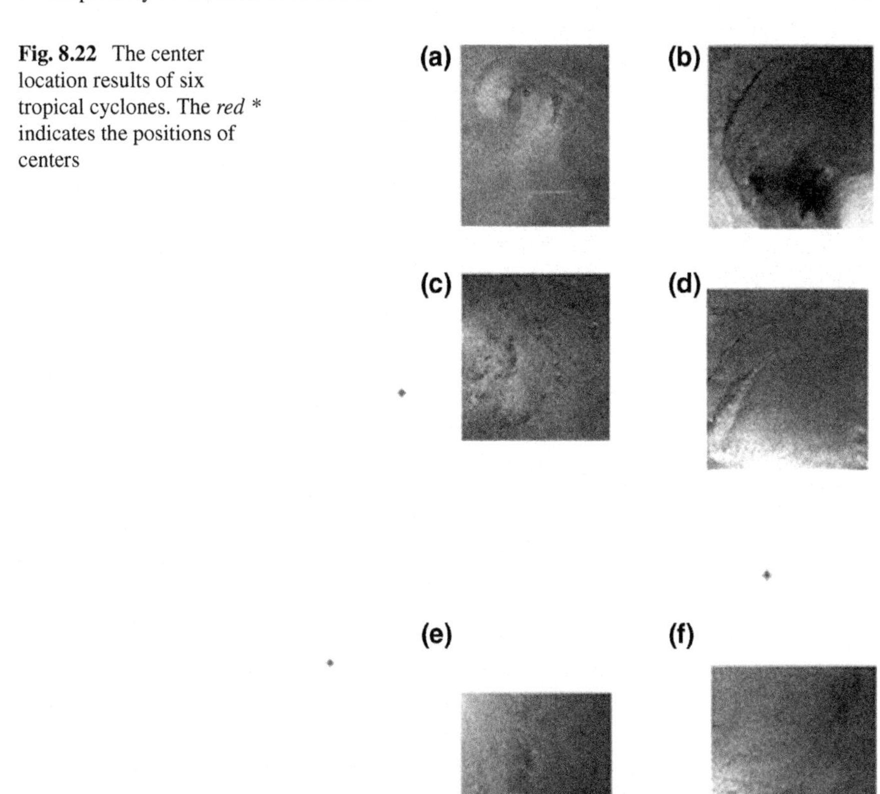

Fig. 8.22 The center location results of six tropical cyclones. The *red* * indicates the positions of centers

each SAR image was captured. Then we point out the estimated center position in the same figure. Theoretically the estimated center position should be on the line. We can see that the estimated centers of Tropical Cyclones Franklin and Bilis are almost on the straight lines, and those of Tropical Cyclone Karl, Nesat and Jova are close to the straight lines. Their longitude and latitude are respectively within the range of the center position recorded before the imaging time and the position recorded after the image capture time. However, the estimated center position of Tropical Cyclone

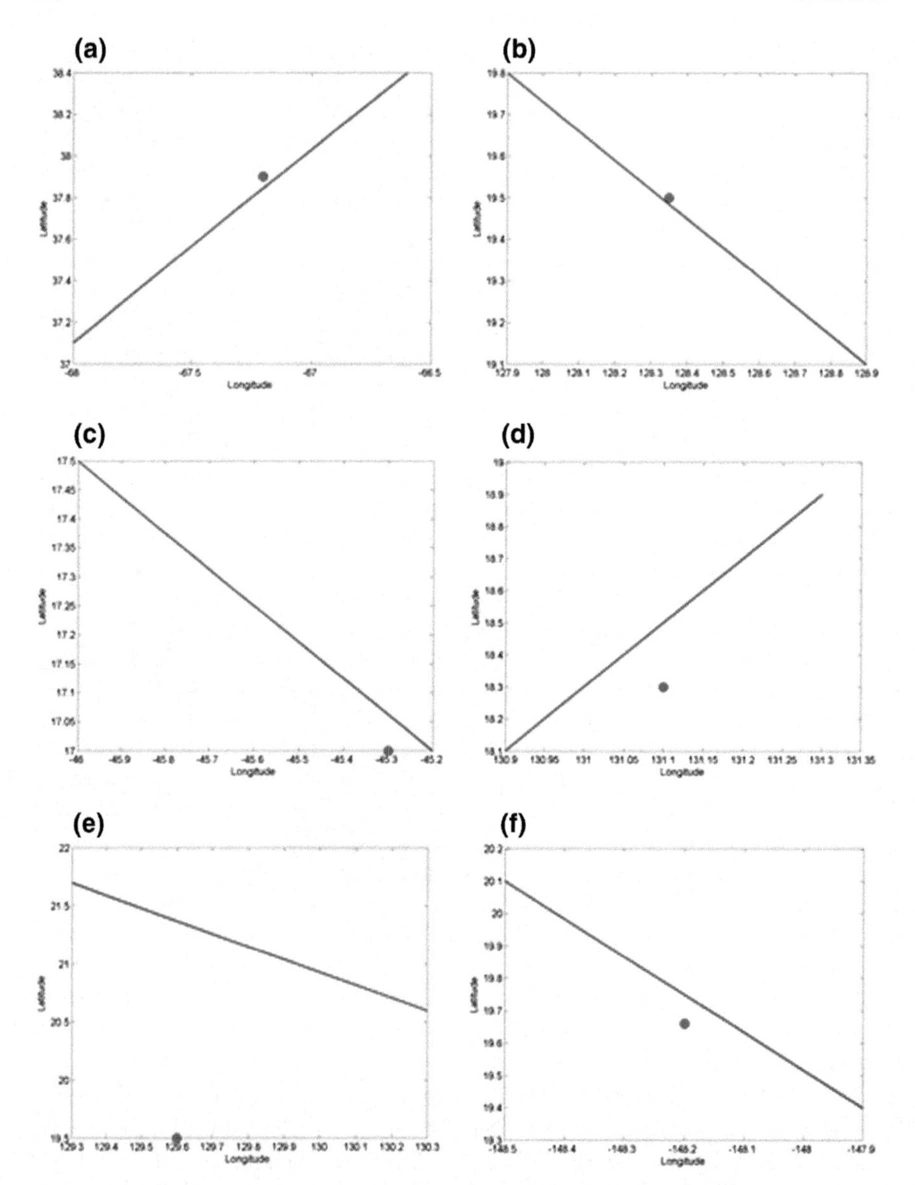

Fig. 8.23 The position of each estimated center and the straight line between the center positions in the best track data sets recorded before and after the time that the SAR images are captured. Theoretically the estimated center position should be on the straight line. **a** Tropical cyclone Franklin in Fig. 8.16. **b** Tropical cyclone Bilis in Fig. 8.17. **c** Tropical cyclone Karl in Fig. 8.18. **d** Tropical cyclone Nesat in Fig. 8.19. **e** Tropical cyclone Etau in Fig. 8.20. **f** Tropical cyclone Jova in Fig. 8.21

Table 8.3 Estimated centers with this method and centers from the Best Track Data sets records before and after the time that SAR images are captured on. Theoretically the estimated center of a cyclone is between the center positions on the two recorded time

Tropical cyclones	UTC time	Estimated center (Lat, Lon)	UTC time before	Best track center (Lat, Lon)	UTC time after	Best track center (Lat, Lon)
Franklin	2005.07.28 22:16:05	(37.9, −67.2)	2005.07.28 18:00:00	(37.1, −68.0)	2005.07.29 00:00:00	(38.4, −66.6)
Bilis	2006.07.11 09:34:35	(19.5, 128.4)	2006.07.11 06:00:00	(19.1, 128.9)	2006.07.11 12:00:00	(19.8, 127.9)
Karl	2004.09.20 08:56:44	(17.0, −45.3)	2004.09.20 06:00:00	(17.0, −45.2)	2004.09.20 12:00:00	(17.5, −46.0)
Nesat	2005.06.05 09:30:39	(18.3, 131.3)	2005.06.05 06:00:00	(18.1, 130.9)	2005.06.05 12:00:00	(18.9, 131.3)
Etau	2003.08.05 20:58:22	(19.5, 129.6)	2003.08.05 18:00:00	(20.5, 130.3)	2003.08.06 00:00:00	(21.5, 129.5)
Jova	2005.09.22 15:34:15	(19.7, −148.2)	2005.09.22 12:00:00	(19.4, −147.9)	2005.09.22 18:00:00	(20.1, −148.5)

Etau is far away from its straight line and its longitude and latitude are a little out of the range. These samples illustrate that our method is effective in most cases but it is not sufficiently good. A large number of experiments prove this. The inflow angle model is proposed based on the analysis of observation data. It is suitable for many tropical cyclones but not all tropical cyclones. In addition, the initialization of particles and the change strategy of particle positions of the PSOA will both affect the optimization results.

8.4 Conclusion

The study about center location of tropical cyclones in SAR images is presented in this chapter. Based on the analysis of the characteristics of tropical cyclones in SAR images, the grey level feature, the grey level contrast feature and the orientation feature are learned to extract tropical cyclone eyes or rain bands. Image processing methods and visual saliency are presented for the center location.

For tropical cyclones with their eyes in SAR images, an imaging processing chain is presented to extract tropical cyclone eyes from SAR images. First, filtering technique is use to reduce image speckle noise, and then image contrast is enhanced by applying top-hat transform and finally image segmentation is done based on morphology watershed algorithm. Three extraction results of tropical cyclone eyes from three ENVISAT images are shown as samples in Sect. 8.2.4. Results show that the combined method can effectively extract typhoon cyclone eye for most images. However, it is still hard to set up a general rule that can be applied to all SAR images.

The grey level is the only feature used in this method. Further work on utilizing texture, geometric features and semantic information should be carried.

Then the problem of the semi-automatic center location of tropical cyclones without eyes in SAR images is addressed. In order to get precise matching results with the assistance of spiral rain bands of tropical cyclones, a location method based on salient region detection and a PSOA for SAR images is presented. The work contains three main steps: (1) Propose a salient region detection method based on the contrast of gray values and the Gabor features. The final salient region map reflects the spiral structure of rain bands. Rain bands can be effectively extracted with this method. (2) Give two filter criteria to select rain bands that can be used for effective matching. (3) Transform the matching problem into an optimization problem after obtaining the skeleton line of spiral rain bands, and use the PSOA to search the optimum solution. Six location results of tropical cyclone centers from six RADARSAT-1 images are shown as samples in Sect. 8.3.5. Experiment results show that tropical cyclone centers can be precisely located.

This work is a meaningful attempt to estimate the tropical cyclone center with a single SAR image. It applies to cases that only contain partial structure of a tropical cyclone without eye. It also applies to cases that contain the whole structure of a tropical cyclone with fuzzy eye. It proposes a rain bands extraction method by taking advantage of the pixel gray level contrast between the tropical cyclone and the background and the orientation information in a SAR image. Experiments indicate that the method is effective in most cases. In practical applications it may be not so effective when there is an island in the tropical cyclone SAR image or the gray level of the tropical cyclone is close to that of its background. Then the continuous and smooth skeleton lines of rain bands are extracted by two filter criteria and morphologic operations. It is convenient for the following matching problem to use an inflow angle model. It can also use a geometric model such as the logarithmic spiral to match the skeleton lines to solve the center estimation problem. The two steps can be widely used for tropical cyclone analysis from the view of image processing. All SAR images used in this study are C-band horizontal polarization images. Although L-band (ALOS-1 and -2) and X-band SAR satellites (TerraSAR, Tandem-X, and Cosmo-Skymed) are currently in orbit, the number of images covering tropical cyclones is extremely rare. As a result, we do not have SAR images acquired in other bands to perform similar analyses. However, we believe that the technique developed here can be applied to SAR images acquired at different bands and polarizations.

There are some limitations that need improved in the future. Several parameters need adjusted in this work and this means the method will not be fully-automatic. Then an inflow angle model is proposed based on the analysis of near-surface inflow angles using the wind observation data of a number of tropical cyclones. This works well for some tropical cyclones but may not work well for all tropical cyclones. The universal applicability of a model should be considered.

This chapter focuses on estimating tropical cyclone centers with a single SAR image from the aspect of image processing. It would be interesting to solve this problem by combining other data information such as wind field, image sequence, etc. in future studies.

References

1. Zhang, Q.H., Q. Wei, and L.S. Chen. 2010. Impact of landfalling tropical cyclones in mainland China. *Science China Earth Sciences* 53 (10): 1559–1564.
2. Li, X., J.A. Zhang, X. Yang, W.G. Pichel, M. DeMaria, D. Long, and Z. Li. 2013. Tropical cyclone morphology from spaceborne synthetic aperture radar. *Bulletin of the American Meteorological Society* 94 (2): 215–230.
3. Li, X. 2015. The first Sentinel-1 SAR image of a typhoon. *Acta Oceanologica Sinica* 34 (1): 1–2.
4. Li, X., W. Pichel, M. He, S. Wu, K. Friedman, P. Clemente-Colon, and C. Zhao. 2002. Observation of hurricane-generated ocean swell refraction at the Gulf Stream North Wall with the RADARSAT-1 synthetic aperture radar. *IEEE Transactions on Geoscience and Remote Sensing* 40 (10): 2131–2142.
5. Lee, I., A. Shamsoddini, X. Li, J.C. Trinder, and Z. Li. 2016. Extracting hurricane eye morphology from spaceborne SAR images using morphological analysis. *ISPRS Journal of Photogrammetry and Remote Sensing* 117: 115–125.
6. Du, Y., and P.W. Vachon. 2003. Characterization of hurricane eyes in RADARSAT-1 images with wavelet analysis. *Canadian Journal of Remote Sensing* 29 (4): 491–498.
7. Liu, W.T. 2002. Progress in scatterometer application. *Journal of Oceanography* 58 (1): 121–136.
8. Wong, K.Y., and C.L. Yip. 2005. Tropical cyclone eye fix using genetic algorithm with temporal information. *Knowledge-Based Intelligent Information and Engineering Systems* 3681: 854–860.
9. Katsaros, K.B., P.W. Vachon, W.T. Liu, and P.G. Black. 2002. Microwave remote sensing of tropical cyclones from space. *Journal of Oceanography* 58 (1): 137–151.
10. Horstmann, J., D.R. Thompson, F. Monaldo, S. Iris, and H.C. Graber. 2005. Can synthetic aperture radars be used to estimate hurricane force winds? *Geophysical Research Letters* 32 (22).
11. Yang, X., X. Li, W.G. Pichel, and Z. Li. 2011. Comparison of ocean surface winds from ENVISAT ASAR, MetOp ASCAT scatterometer, buoy measurements, and NOGAPS model. *IEEE Transactions on Geoscience and Remote Sensing* 49 (12): 4743–4750.
12. Xu, F., X.F. Li, P. Wang, J. Yang, W.G. Pichel, and Y.Q. Jin. 2015. A backscattering model of rainfall over rough sea surface for synthetic aperture radar. *IEEE Transactions on Geoscience and Remote Sensing* 53 (6): 3042–3054.
13. Zheng, G., J.S. Yang, A.K. Liu, X.F. Li, W.G. Pichel, and S.Y. He. 2016. Comparison of typhoon centers from SAR and ir images and those from best track data sets. *IEEE Transactions on Geoscience and Remote Sensing* 54 (2): 1000–1012.
14. Wang, P., C.S. Guo, and Y.X. Luo. 2006. Local spiral curves simulating based on hough transformation and center auto-locating of developing typhoon. *Transactions of Tianjin University* 12 (2): 142–146.
15. Xu, J.W., P. Wang, and Y.Y. Xie. 2009. Image segmentation of typhoon spiral cloud bands based on support vector machine. *Machine Learning and Cybernetics, 2009 International Conference* 2: 1088–1093.
16. Wong, K.Y., C.L. Yip, and L.P. Wah. 2008. Automatic tropical cyclone eye fix using genetic algorithm. *Expert Systems with Applications* 34 (1): 643–656.

17. Oliver, C., and S. Quegan. 1998. *Understanding Synthetic Aperture Radar images with CDROM*. Boston: Artech House.
18. Lee, J.S. 1980. Digital image enhancement and noise filtering by use of local statistics. *IEEE Transactions on Pattern Analysis and Machine Intelligence* 2 (2): 165–168.
19. Kuan, D.T., A.A. Sawchuk, and T.C. Strand. 1985. Adaptive noise smoothing filter for images with signal dependent noise. *IEEE Transactions on Pattern Analysis and Machine Intelligence* 7 (2): 165–177.
20. Lopes, A., R. Touzi, and E. Nezry. 1990. Adaptive speckle filters and scene heterogeneity. *IEEE transactions on Geoscience and Remote Sensing* 28 (6): 992–1000.
21. Feng, H., B. Hou, and M. Gong. 2011. SAR image despeckling based on local homogeneous region segmentation by using pixel relativity measurement. *IEEE Transactions on Geoscience and Remote Sensing* 49 (7): 2724–2737.
22. Lee, J.S. 1983. Digital image smoothing and the sigma filter. *Computer Vision, Graphics and Image Processing* 24 (2): 255–269.
23. Park, J.M., W.J. Song, and W.A. Pearlman. 1999. Speckle filtering of sar images based on adaptive windowing. *Computer Vision, Graphics and Image Processing* 146 (4): 191–197.
24. Buades, A., B. Coll, and J. Morel. 2005. A non-local algorithm for image denoising. *Computer Vision and Pattern Recognition* 2: 60–65.
25. Deledalle, C.A., L. Denis, and F. Tupin. 2009. Iterative weighted maximum likelihood denoising with probabilistic patch-based weights. *IEEE Transactions on Image Processing* 18 (12): 2661–2672.
26. Polzehl, J., and V. Spokoiny. 2006. Propagation-Separation approach for local likelihood estimation. *Probability Theory and Related Fields* 135: 335–362. doi:10.1007/s00440-005-0464-1.
27. Kim, Y.T. 1997. Contrast enhancement using brightness preserving bi-histogram equalization. *IEEE Transactions on Consumer Electronics* 43 (1): 1–8.
28. Bai, X., and F. Zhou. 2010. Multi scale top-hat transform based algorithm for image enhancement. *Signal Processing (ICSP)* pp. 797–800.
29. Beucher, S., and C. Lantuejoul. 1997. Use of watersheds in contour detection. Rennes, France. In *Proc. Int. Workshop Image. Processing, Real-Time Edge and Motion Detection/Estimation*.
30. Beucher, S., and F. Meyer. 1993. *The morphological approach to segmentation: the watershed transformation*, pp. 433–482. New York.: Dekker
31. Roerdink, J.B.T.M., and A. Meijster. 2000. The watershed transform: definitions, algorithms and parallelization strategies. *Fundamenta Informaticae* 41 (1,2): 187–228.
32. Pikaz, A., and A. Averbuch. 1996. Digital image threshold based on topological stable-state. *Pattern Recognition* 29 (5): 829–843.
33. Haris, K., S.N. Efstratiadis, N. Maglaveras, and A. Katsaggelos. 1998. Hybrid image segmentation using watersheds and fast region merging. *IEEE Transactions on Image Processing* 7 (12): 1684–1699.
34. O'Callaghan, R.J., and D.R. Bull. 2005. Combined morphological spectral unsupervised image segmentation. *IEEE Transactions on Image Processing* 14 (1): 49–62.
35. Soille, P. 2002. *Morphological image analysis: principles and applications*. Berlin: Springer.
36. Wang, S., X.J. Zhang, L.C. Jiao, and X.R. Zhang. 2009. An improved watershed-based SAR image segmentation algorithm. In *Sixth International Symposium on Multispectral Image Processing and Pattern Recognition: International Society for ics and Photonics*.
37. Cave, K.R., and J.M. Wolfe. 1990. Modeling the role of parallel processing in visual search. *IEEE Transactions on Image Processing* 22 (2): 225–271.
38. Koch, C., and S. Ullman. 1990. Shifts in selective visual attention: towards the underlying neural circuitry. *Matters of Intelligence* 4: 115–141.
39. Mareschal, I., J.A. Henrie, and R.M. Shapley. 2002. A psychophysical correlate of contrast dependent changes in receptive field properties. *Vision Research* 42 (15): 1879–1887.
40. Itti, L., C. Koch, and E. Niebur. 2002. A model of saliency based visual attention for rapid scene analysis. *IEEE Transactions on Pattern Analysis and Machine Intelligence* 20 (11): 1254–1259.

41. Cheng, M., G. Zhang, N.J. Mitra, X. Huang, and S. Hu. 2011. Global contrast based salient region detection. *Computer Vision and Pattern Recognition* 37 (3): 409–416.
42. Duan, L., C. Wu, J. Miao, L. Qing, and Y. Fu. 2011. Visual saliency detection by spatially weighted dissimilarity. *Computer Vision and Pattern Recognition* pp. 473–480.
43. Liu, T., J. Sun, N.N. Zheng, and X.O. Tang. 2007. Learning to detect a salient object. *Computer Vision and Pattern Recognition* 33 (2): 1–8.
44. Kienzle, W., F.A. Wichmaan, B. Scholkopf, and M.O. Frane. 2007. A nonparametric approach to bottom up visual saliency. *Advances in Neural Information Processing Systems.*
45. Rosin, P.L. 2009. A simple method for detecting salient regions. *Pattern Recognition* 42 (11): 2363–2371.
46. Goferman, S., L. Zelnik-Manor, and A. Tal. 2012. Context-aware saliency detection. *IEEE Transactions on Pattern Analysis and Machine Intelligence* 34 (10): 1915–1926.
47. Daugman, J.G. 1988. Complete discrete 2d gabor transforms by neural networks for image analysis and compression. *IEEE Transactions on Acoustics, Speech, and Signal Processing* 36 (7): 1169–1179.
48. Lades, M., J.C. Vorbruggen, J. Buhmann, J. Lange, C. von der Malsburg, R.P. Wurtz, and W. Konen. 1993. Distortion invariant object recognition in the dynamic link architecture. *IEEE Transactions on Computers* 42 (3): 300–311.
49. Lee, J., R.M. Haralick, and L.G. Shapiro. 1987. Morphologic edge detection. *IEEE Journal of Robotics and Automation* 3 (2): 142–156.
50. Gonzalez, C.R., and R.E. Woods. 2002. *Digital Image Processing*, 2nd ed. Publishing House of Electronics Industry.
51. Zhang, J., and E. Uhlhorn. 2012. Hurricane sea-surface inflow angle and observation-based parametric model. *Monthly Weather Review* 140 (11): 3587–3605.
52. Eberhart, R., and J. Kennedy. 1995. A new imizer using particle swarm theory. In *Proceedings of the Sixth International Symposium on. IEEE*, pp. 39–43, Nagoya.
53. Eberhart, R.C., and Y. Shi. 2001. Particle swarm imization: developments, applications and resources. *Proceedings of the 2001 Congress on* 1: 81–86.
54. Lee, K., and M. El-Sharkawi. 2008. Fundamentals of particle swarm techniques. *Modern Heuristic imization Techniques: Theory and Applications to Power Systems*, 71–87.
55. Yoshida, H., K. Kawata, Y. Fukuyama, S. Takayama, and Y. Nakanishi. 2000. A particle swarm imization for reactive power and voltage control considering voltage security assessment. *IEEE Transactions on Power Systems* 15 (14): 1232–1239.
56. van den Bergh, F., and A. P. Engelbrecht. 2001. Training product unit networks using cooperative particle swarm imisers. In *Proceedings. IJCNN'01. International Joint Conference on, 2001*, vol. 1, pp. 126–131. Neural Networks.
57. Banal, S., S. Iris, and R. Saint-Jean. 2006. The Canadian Space Agency's Hurricane Watch Program: supporting research on wind field retrieval from RADARSAT-1 image. In *Proceedings Ocean SAR.*

Chapter 9
Observing Typhoons from Satellite-Derived Images

Gang Zheng, Jingsong Yang, Antony K. Liu, Xiaofeng Li, William G. Pichel, Shuangyan He and Shui Yu

Abstract This chapter compares the typhoon centers from the tropical cyclone best track (BT) datasets of three meteorological agencies and those from synthetic aperture radar (SAR) and infrared (IR) images. First, we carried out algorithm comparison, using two newly developed and one existing wavelet-based algorithms, which were used to extract typhoon eyes in six SAR images and two IR images. These case studies showed that the extracted eyes by the three algorithms are consistent with each other. The differences among them are relatively small. However, there is a systematic difference between these extracted centers and the typhoon centers from the three BT datasets, which are interpolated to the imaging times first. We then compared the typhoon centers determined from 25 SAR and 43 IR images with those from the three BT datasets to investigate the performance of the latter at the sea surface and at the cloud top, respectively. We found the typhoon centers from the three BT datasets are generally closer to the locations extracted from the SAR images showing sea surface imprints of the typhoons than those from the IR images showing

G. Zheng (✉) · J. Yang
State Key Laboratory of Satellite Ocean Environment Dynamics, Second Institute of Oceanography, State Oceanic Administration, Hangzhou 310012, China
e-mail: gang_zheng@outlook.com

A.K. Liu
NASA/GSFC, Greenbelt, MD 20771, USA
e-mail: tonyakliu@gmail.com

X. Li
GST, National Oceanic and Atmospheric Administration (NOAA)/NESDIS, College Park, MD, USA
e-mail: xiaofeng.li@noaa.gov

W.G. Pichel
NOAA/NESDIS/STAR, College Park, MD 20740, USA
e-mail: wpichel@verizon.net

S. He
Ocean College, Zhejiang University, Hangzhou 310058, China
e-mail: hesy103@gmail.com

S. Yu
Ocean College, Zhejiang University, Zhoushan 316021, China
e-mail: kerry_1993@163.com

X. Li (ed.), *Hurricane Monitoring With Spaceborne Synthetic Aperture Radar*, Springer Natural Hazards, DOI 10.1007/978-981-10-2893-9_9

cloud top structures of the typhoons. We also evaluate the effect caused by rain to the SAR wind field retrieval. By using RADARSAT-2 data, National Centers for Environmental Prediction (NCEP) reanalysis data and Tropical Rainfall Measuring Mission satellite (TRMM) precipitation radar rainfall data, rain-induced attenuation, raindrop volumetric scattering are calculated and the perturbation of the water surface is simulated by rainfall and incident angle. The performance of this model is further proved by one typhoon case.

9.1 Introduction

Typhoons are strong tropical cyclones developing in the northwestern part of the tropical and subtropical Pacific Ocean. They bring strong winds and heavy rains that could severely affect marine ecosystems as well as human lives near the coast. Therefore, monitoring, tracking and predicting typhoons are very important. With the development of satellite remote sensing, various visible (VIS), infrared (IR) and microwave spaceborne sensors are used for observing typhoons [1, 2]. Typhoons, usually accompanied by spiral clouds, can be observed by IR sensors, which receive radiation emitted or reflected by clouds. Different from IR sensors in principle, synthetic aperture radar (SAR) has cloud penetrating capability, so sea surface signatures of atmospheric phenomena can be observed even under cloudy condition. Within a typhoon eye, the winds are weaker than those in the surrounding area. Accordingly, the sea surface in the eye is calmer and appears as a darker area in SAR image because of lower backscattering. Vertical wind shear acts to tilt the typhoon eye, and consequently the eye locations at the sea surface and at cloud top are uncoordinated [3, 4]. The location of the eye near the sea surface may be more important than those at the cloud top in some cases, especially for ocean applications. Sometimes, a typhoon eye is partially covered by clouds; in this case, SAR could be a good monitoring tool because of its cloud penetrating capability. Figure 9.1 is an example showing an European Space Agency ENVISAT SAR image of Typhoon Sinlaku (a) and an IR image (b) from MTSAT (Multi-functional Transport Satellite) taken about three minutes earlier. Both images were acquired on 10 September 2008. Additionally, SAR has much higher spatial resolution than IR sensors and other microwave sensors (such as scatterometers and radiometers) onboard operational satellites. The finer features of a typhoon can be observed in SAR images (e.g., eye, rain bands, storm cells, etc.) [4–10]. Therefore, the application of SAR in typhoon monitoring has its advantages and drawn a lot of attention recently [4–10].

Wind can change the sea surface roughness and radar backscattering power through the effect on the sea surface waves. By using geophysical model function (GMF), the wind field can be obtained. Typhoons are often associated with a wide range of heavy rain, but the GMF function doesn't take rain into consideration. For a typhoon system, the inversion error in a wind vector even can reach 50% [11].

In the literatures, identifying tropical cyclone centers has been studied using VIS and IR cloud imageries [12–14], microwave radiometers [15], microwave

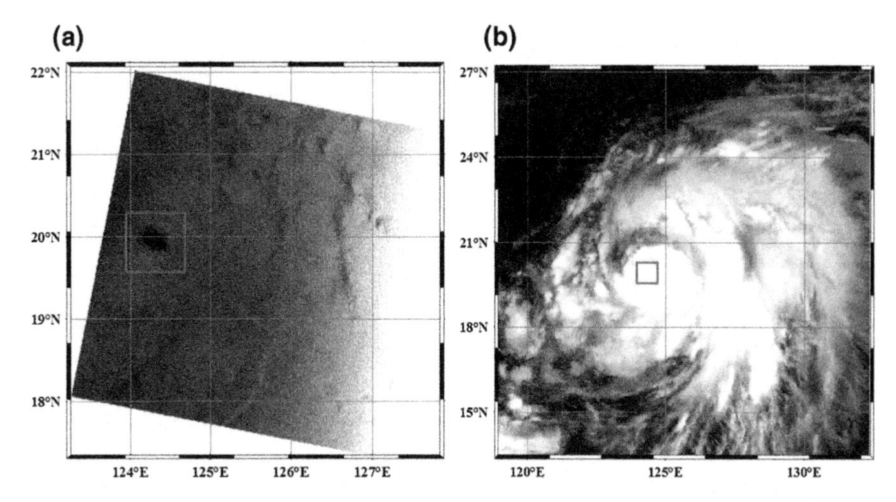

Fig. 9.1 **a** A SAR image of Typhoon Sinlaku. It was taken at 01:33 UTC 10 September 2008 by ENVISAT, and the eye can be clearly seen in the image, which is enclosed in a *red box*. **b** An IR image of Typhoon Sinlaku. It was taken at 01:30 UTC 10 September 2008 by MTSAT. The *red box* in this figure is the same to that in Fig. 9.1a, but the eye is invisible

scatterometers [16], Doppler radar systems [17], etc. Most studies tried to reduce human influences during identifying procedures. In this paper, typhoon centers are estimated based on SAR imagery. A common procedure to determine tropical cyclone centers in SAR images involves two steps: extract the area of the typhoon eye, and then calculate the geometric center of the area [4, 6, 7, 9]. Different from the eyes in IR images, the typhoon eyes in SAR images usually have fuzzy edges and speckle noises, making the area extraction difficult.

From view point of image segmentation and classification, wavelet transform, because of its multi-resolution quality, has been used to reduce the noise of extracting edges of typhoon eyes in SAR images [4, 6–9] and analyze other ocean phenomena [18, 19]. In this paper, two alternative algorithms are also applied for extracting typhoon eyes in SAR images. In the first algorithm, we apply different linear and non-linear smoothing operators to smooth SAR images and find typhoon-eye edges using contour lines. In the second algorithm, we extract typhoon-eye edges using several morphological image operators. Image smoothing is unnecessary in this algorithm. Compared with the first wavelet-based algorithm, there are no differential or difference operators in the two algorithms, which, as well known, are sensitive to noise. Although the two algorithms as well as the wavelet-based algorithm are used for SAR images, they are also applicable for IR images because typhoon eyes in IR images have much sharper edges. To determine typhoon centers in images, researchers may choose any algorithms they like. Thus, it is necessary to conduct an algorithm comparison in terms of extracting typhoon-center locations. For this purpose, all three algorithms are first tested in six SAR images and two MODIS (Moderate Resolution Imaging Spectroradiometer) IR images, and the results are compared with each other.

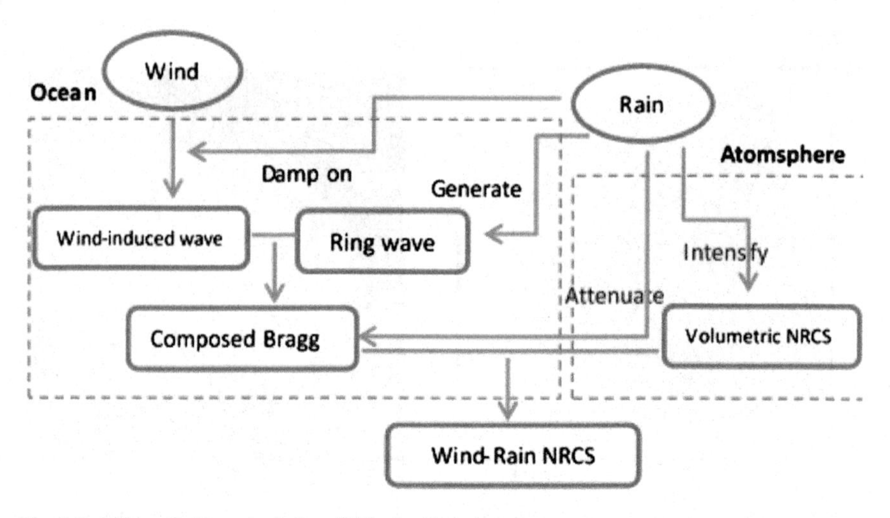

Fig. 9.2 Effect of rain and wind on SAR signals

In another aspect, in order to reduce the influence by rain in wind field retrieval, we firstly need to understand scattering mechanisms contributed by rainfall to the radar backscattering signals. Raindrops cause volumetric scattering and attenuation through wave propagating, after that, rain striking the water create disturbance including rings, stalks, and crowns [20]. These effects to the backscattering vary with the incident angle and polarization [21]. Figure 9.2 shows the effect of rain and wind on SAR signals. In this paper, based on C band scatterometer rain effect model and SAR rainfall backscattering model, rain-induced attenuation, raindrop volumetric scattering and rough sea surface scattering are calculated. Performance of these two models are further demonstrated by comparing RADARSAT-2 data with synchronous National Centers for Environmental Prediction (NCEP) reanalysis data and Tropical Rainfall Measuring Mission satellite (TRMM) precipitation radar rainfall data, and results prove that they both can improve the rain-induced inaccurate wind field retrieval.

The Chinese Meteorological Administration (CMA), the Regional Specialized Meteorological Center (RSMC) in Japan, and the Joint Typhoon Warning Center (JTWC) are three major agencies, which release tropical cyclone best track (BT) datasets for the western North Pacific basin. The BT datasets, including center locations of tropical cyclones [22], have been widely used for studying tropical cyclones and climate changes [23]. However, due to the differences in raw data inputs and analysis procedures among the agencies, the three BT datasets have discrepancies in terms of center locations and intensities as shown by [22–26]. Additionally, the typhoon eyes are tilted due to vertical wind shear [3, 4]. So this brings up a question: Are the center locations in the BT datasets close to the typhoon centers at the sea surface or those at the cloud top? The investigation of this question is quite necessary, and was made by remote sensing in this paper. For this purpose, the typhoon-center locations determined from 25 SAR images (capturing the typhoon eyes at the sea surfaces) and 43 MODIS IR images (capturing the typhoon eyes at the cloud top) are compared with those in the three BT datasets.

The rest of this paper is organized as follows. In Sects. 9.2 and 9.3, the two newly developed algorithms, as well as wavelet-based typhoon-eye extraction algorithm, are discussed. In Sect. 9.4, six SAR images and two IR images are used to test the three algorithms. In Sect. 9.5, the typhoon centers determined from 25 SAR images and 43 MODIS IR images are compared with the temporally interpolated centers from the three BT datasets. For simplicity, we call the former the determined centers and the later the interpolated centers. In Sect. 9.6, the effect by rain on SAR wind field retrieval is discussed. The perturbation of the water surface is simulated by rainfall and incident angle in Sect. 9.6. During the period of typhoon Megi, the rain effect to one of a SAR picture data are evaluated which also prove the good performance of this model. Section 9.8 summarizes the main findings.

9.2 Typhoon Eye Extraction Algorithms

The first algorithm we applied is an existing algorithm based on the two-dimensional (2D) Mexican hat continuous wavelet transform. The contour line of zero is selected as the edge after the SAR image is transformed [4, 8]. This transform can be considered as the combination of the Gaussian smoothing operator and the Laplacian operator. There are many ways to combine different smoothing operators with the Laplacian operator to extract typhoon eyes, and the Laplacian operator can itself have other forms (e.g., finite-difference form). Therefore, a class of similar algorithms can be developed actually.

The second algorithm involves two steps. In Step 1, the SAR image is smoothed to reduce noise. In Step 2, a contour line with a proper value is selected as the edge of the typhoon eye. Determination of the contour value will be discussed in Sect. 9.3. The following 2D linear smoothing operator is adopted in Step 1.

$$S(r', a) = \iint s(r)\phi\left(\frac{r-r'}{a}\right)dr \Big/ \iint \phi\left(\frac{r}{a}\right)dr \tag{9.1}$$

where $\phi\left(\frac{r-r'}{a}\right)$ is weighting window of the smoothing (such as rectangular window or Gaussian window), and its size is determined by a. $s(r)$ and $S(r', a)$ respectively represent the original image and the image smoothed by the window with size a. r and r' are position vectors in the original and smoothed images, respectively. This image smoothing process can also be done more efficiently in frequency domain through the Fast Fourier Transform. Like the previous algorithm, other smoothing operators can also be adopted in this algorithm, such as the median smoothing operator [27]. The impact of selecting a smoothing operator will be investigated in Sect. 9.4.

In the third algorithm, we first select a proper threshold value for a SAR image. The determination of the threshold value will be discussed in Sect. 9.3. Pixels with the gray level values smaller than the threshold value are assigned to one, and the other pixels are assigned to zero. Then, the largest connected component of the pixels with one is considered as the initial eye area. This component is treated with the dilation operator [28] and the erosion operator [28] to reduce the noises. The treated component is considered as the eye area, and its edge can be found by traversing the pixels and judging whether the pixels have connected neighborhood pixels with zero values. Unlike the previous two algorithms that must include smoothing operators, this algorithm based on morphological image operators can preserve finer features of edge in the SAR image without using any smoothing operators.

9.3 Contour Value and Threshold Value Selection by Gray Level Analysis

The contour value and the threshold value need to be set for selecting proper contour line and image binaryzation in the two newly developed algorithms, respectively. The two values can be determined from the following gray level analysis.

For example, Fig. 9.3a shows a RADARSAT-1 SAR image with the typhoon eye as a dark feature in the middle of the image. We divide the SAR image into three boxes. The biggest blue box includes the eye area, the transitional area and the surrounding background area. The smaller green box covers the eye area and the transitional area. Thus, the area between the blue and the green boxes belongs to the surrounding background area. The smallest red box covers part of the eye area. The SAR image is first smoothed to reduce noise. Then, the probability density distribution of gray

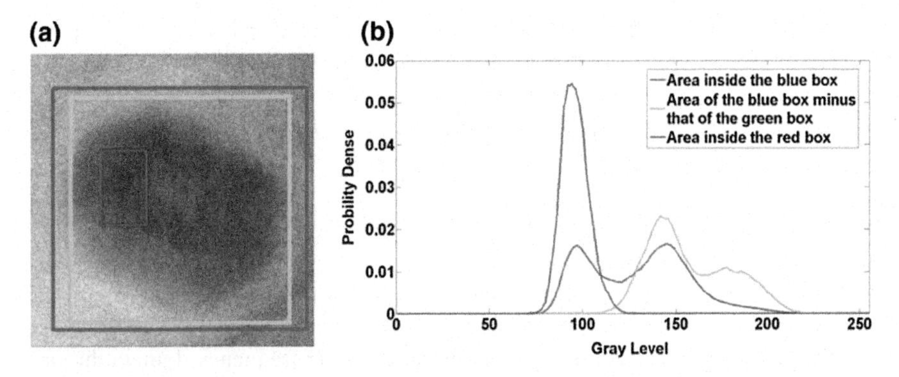

Fig. 9.3 The *gray level* analysis of the SAR image of Typhoon Kirogi. It was taken at 21:04 UTC 15 October 2005 by RADARSAT-1. **a** The SAR image of the typhoon eye and its surrounding area are divided by three boxes. **b** The probability density distributions of the *gray level* values in the *blue box* (*blue curve*), in the area between the *blue box* and the *green box* (*green curve*), and in the *red box* (*red curve*)

level values in the blue box is calculated and given in Fig. 9.3b (blue curve). For comparison, the probability density distributions of gray level values in the area between the blue box and the green box and in the area of the red box are also calculated and given in Fig. 9.3b as green and red curves, respectively.

The surrounding background area and eye area are considered to be homogeneous, so the green and red curves can be considered as the probability density distributions of gray level values in the areas. It can be seen that the two peaks of the blue curve approximately occurred at the peak of the green curve and of the red curve. Thus, the two peaks of the blue curve are respectively considered to correspond to the surrounding background area and the eye area, and the trough is considered to correspond to the transitional area.

In the two algorithms, edges of typhoon eye are extracted in terms of contour line. (The edge extracted by the algorithm with morphological image operators is also a contour line actually, and the contour value is the threshold value in image binaryzation.) The edge is the fastest varying area in the transitional area. Therefore, if we try to find a contour line as an edge, the gray level values on its two sides should vary fastest. In other words, the probability of the gray level values in the transitional area falling in the neighborhood of the contour value is the smallest. Thus, the minimum value in the trough of the blue curve is used as the contour value and the threshold value in the two algorithms. Compared to the wavelet-based algorithm, an advantage of the two new algorithms is that there are no differential or difference operators, which, as well known, are sensitive to noise. In the wavelet-based algorithm, the contour value usually needs to be artificially tuned from zero to a proper value to fit the observations in practice [8], and this is largely based on users' experiences.

To objectively determine a typhoon center in a SAR image, various algorithms may be used; the selection of one algorithm over the others may affect the result. Therefore, it is necessary to evaluate the impact of algorithm selection. Next, we conduct case studies to answer this question.

9.4 Comparison of the Algorithms

Two ENVISAT (wide swath mode) and four RADARSAT-1 (scanSAR wide mode) C-band SAR images are used for investigating the performance of the three algorithms. All the images are geometrically corrected. For clarity, the typhoons in the six SAR images are listed in Table 9.1. The typhoon eyes in these images are extracted by the three algorithms using different smoothing operators. There are total of five approaches listed in Table 9.2.

The six cases are given in Fig. 9.4, and the edges extracted by the five approaches are delineated and marked in different colors in Table 9.2. Overall, the edges extracted by the five approaches are in good agreement. The best case is Case 4. Since the eye edge is relatively sharper in the original image in Case 4, the extracted edges by different approaches almost overlap. In the other five cases, the edges extracted by

Table 9.1 The typhoons in the six SAR images

Typhoon name	UTC time hh:mm, dd/mm/yyyy	SAR source	Pixel size (m)
Guchol	08:18, 22/08/2005	RADARSAT-1	50
Khanun	01:47, 11/09/2005	ENVISAT	75
Kirogi	21:05, 15/10/2005	RADARSAT-1	50
Yagi	20:19, 21/09/2006	RADARSAT-1	50
Man-Yi	09:33, 13/07/2007	RADARSAT-1	50
Sinlaku	01:33, 10/09/2008	ENVISAT	75

Table 9.2 Typhoon eye extraction approaches

No.	Approach	Marking color
1	Algorithm based on the 2D Mexican hat continuous wavelet transform	Magenta
2	Algorithm only with the 2D Gaussian smoothing operator	Red
3	Algorithm only with the 2D rectangular smoothing operator	Green
4	Algorithm only with the median smoothing operator	Blue
5	Algorithm based on the morphological image operators	Yellow

approaches No. 2–5 show excellent agreement, but approach No. 1 performs slightly differently.

Next, we calculate the geometric centers of typhoon eyes and evaluate the location differences among different approaches. Definition of the geometric center of typhoon eye in the SAR image is given by:

$$P_c = \sum_{P_n \in A} P_n / N \tag{9.2}$$

where A and N respectively represent the set of the pixels inside the edge and the number of these pixels. P_n is the position vector to pixel n. The determined centers from the extracted eyes are marked in Fig. 9.4 in five different colors noted in Table 9.2. It can be seen that the extracted typhoon-eye centers by different approaches are very close to each other. For comparison, the typhoon centers reported in the three BT datasets by CMA, RSMC and JTWC are also used. The reported centers are temporally interpolated to the SAR imaging times. The determined centers from the SAR images and the centers based on the best tracks of by CMA (blue line), RSMC (green line) and JTWC (black line) are all shown in Fig. 9.5. The dots are the determined centers (marked in five different colors given in Table 9.2), the diamonds are the reported centers, and the pentagrams are the temporally interpolated centers. From Fig. 9.5, we can see that the typhoon-eye centers from SAR and those from BT datasets have obvious location differences. The distances between them are from 4.9 to 23.5 km.

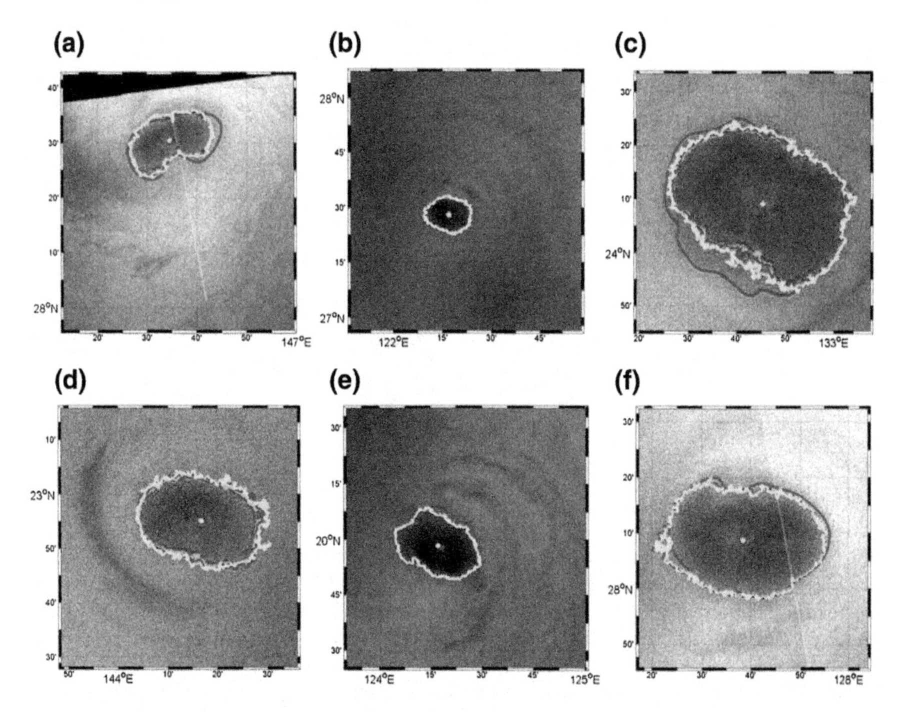

Fig. 9.4 Six SAR images with extracted typhoon eyes and centers. *Magenta, red, green, blue*, and *yellow* are respectively used to mark the results of the algorithm based on the 2D Mexican hat continuous wavelet transform, the algorithm only with the 2D Gaussian smoothing operator, the algorithm only with the 2D rectangular smoothing operator, the algorithm only with the median smoothing operator, and the algorithm based on the morphological image operators. **a** Case 1: Typhoon Guchol; **b** Case 2: Typhoon Khanun; **c** Case 3: Typhoon Kirogi; **d** Case 4: Typhoon Yagi; **e** Case 5: Typhoon Man-Yi; and **f** Case 6: Typhoon Sinlaku

The following is an intuitive reason for these obvious location differences. Typhoon eye is a tilted structure due to vertical wind shear [3, 4]. The eye in a SAR image is at the bottom of the tilted structure because of the cloud penetrating capability of SAR. Consequently, the center determined from SAR image is the center of the typhoon eye at the sea surface. However, the typhoon centers reported in the BT datasets are the synthesized typhoon centers from post-event analysis. Figure 9.5 also shows that the choice of algorithm will affect the determination of typhoon-eye center in SAR image. However, the location differences among the determined centers are much smaller than the location differences between the so-determined centers and the interpolated centers. The maximum location difference among the determined centers in Case 3, Typhoon Kirogi, is the largest in the six cases, which is 3.7 km. However, the location differences between the determined centers and the three interpolated centers from the BT datasets in Case 3 are obviously larger, ranging from 7.7 to 11.8 km. The maximum location differences among the determined centers in the other five cases are within 2.2 km. The maximum location differences

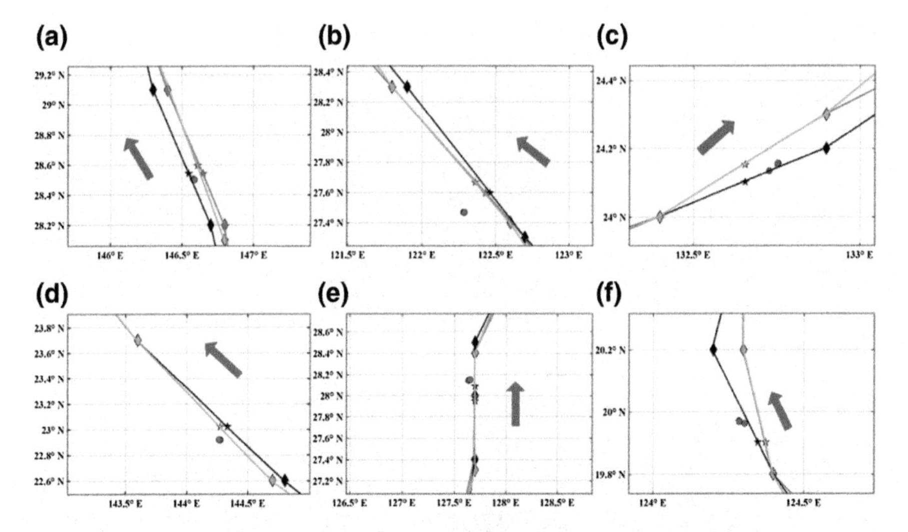

Fig. 9.5 The *blue, green* and *black lines* are the best tracks of CMA, RSMC and JTWC, respectively. The diamonds are the reported centers by the CMA, RSMC and JTWC, and the pentagrams are the temporally interpolated centers at the SAR imaging times. The dots are the determined typhoon centers by the approaches from the SAR images, and their colors are given in Table 9.2. The *red arrows* are the general directions of typhoon movements. **a** Case 1: Typhoon Guchol; **b** Case 2: Typhoon Khanun; **c** Case 3: Typhoon Kirogi; **d** Case 4: Typhoon Yagi; **e** Case 5: Typhoon Man-Yi; and **f** Case 6: Typhoon Sinlaku

among the determined centers of the eyes extracted by approaches No. 2–5 are within 0.5 km in all six cases. The mean distance between the determined centers and the interpolated centers in the six cases is 13.3 km, but the mean value of the maximum location differences among the determined centers is only 1.8 km. Therefore, the impact of algorithm selection is much smaller, and the centers determined from SAR images can be used to investigate the near-sea-surface performance of the typhoon-center locations in the three BT datasets.

Typhoon eyes in IR images usually have much sharper edges than those in SAR images. Thus, the algorithms are also applicable for extracting edges of typhoon eyes in IR images. Two MODIS IR images are used for testing (listed in Table 9.3). These images have a pixel size of about 1 km, and are also geometrically corrected. The extracted edges and determined centers by the approaches are given in Fig. 9.6. It can be seen that they are in good agreement with each other. The determined centers and the temporally interpolated centers from the BT datasets are shown in Fig. 9.7. Like the cases of using SAR images, the location differences among the determined centers are much smaller than those between the so-determined centers and the interpolated centers.

The above comparison shows that the center locations from the BT datasets are obviously different from both the center locations from the SAR and IR images. Comparing these location differences, the effect caused by extraction algorithm selection

Table 9.3 The Typhoons in the two MODIS IR images

Typhoon name	UTC time hh:mm, dd/mm/yyyy
Melor	01:55, 05/10/2009
Roke	01:30, 20/09/2011

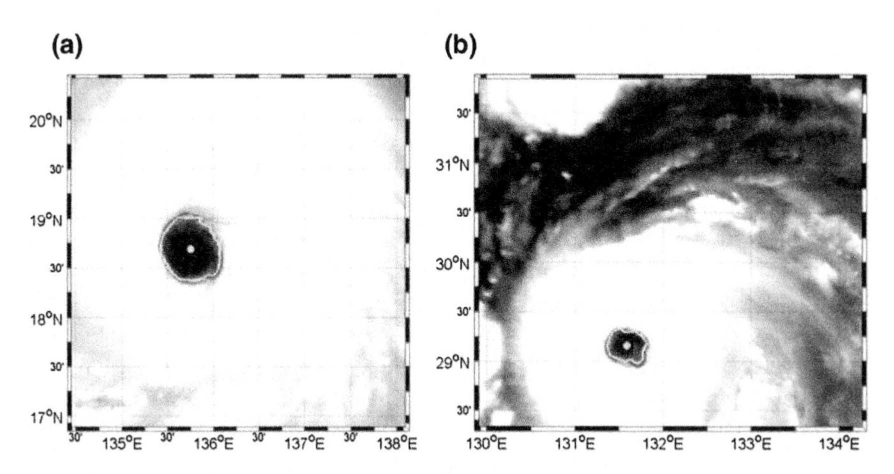

Fig. 9.6 Two MODIS IR images with extracted typhoon eyes and centers. *Magenta, red, green, blue*, and *yellow* are respectively used to mark the results of the algorithm based on the 2D Mexican hat continuous wavelet transform, the algorithm only with the 2D Gaussian smoothing operator, the algorithm only with the 2D rectangular smoothing operator, the algorithm only with the median smoothing operator, and the algorithm based on the morphological image operators. **a** Case 1: Typhoon Melor; and **b** Case 2: Typhoon Roke

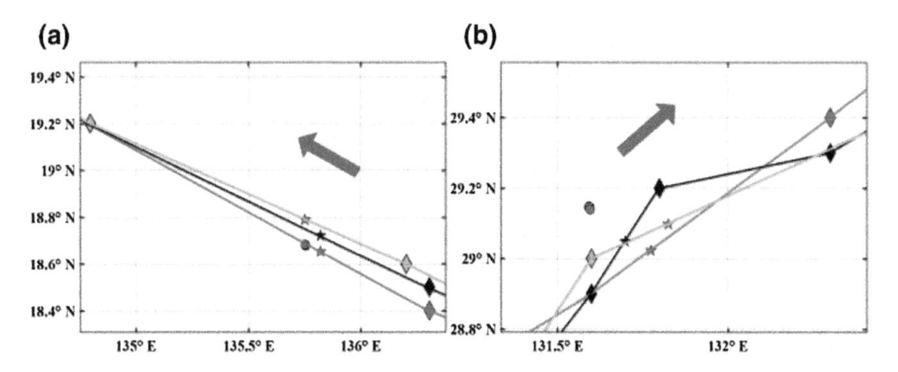

Fig. 9.7 The *blue, green* and *black lines* are the best tracks of CMA, RSMC, and JTWC, respectively. The diamonds are the reported centers by CMA, RSMC, and JTWC, and the pentagrams are the temporally interpolated centers at the MODIS IR imaging times. The dots are the determined typhoon centers by the approaches from the MODIS IR images, and their colors are given in Table 9.2. The *red arrows* are the general directions of typhoon movements. **a** Case 1: Typhoon Melor, and **b** Case 2: Typhoon Roke

is not significant. Thus, the centers from SAR and IR images can be used to investigate the performance of the centers from the BT datasets at the sea surface and at the cloud top, respectively.

9.5 Statistics of Typhoon Center Differences Between Image-Based and BT Dataset-Based

In order to investigate the near-sea-surface performance of the typhoon-center locations in the BT datasets, 25 SAR images (one ENVISAT SAR image, global mode; six ENVISAT SAR images, wide swath mode; 18 RADARSAT-1 SAR images, scanSAR wide mode) with typhoon eyes are collected. The typhoon eyes are extracted by the algorithm with morphological image operators, and the eye centers are determined using Eq. (9.2). They are numbered and given in Table 9.4 along with typhoon names and imaging times. The spatial distribution of the typhoon centers is shown in Fig. 9.8, approximately between 15°N and 30°N. The typhoon-center locations in the three BT datasets are temporally interpolated to the imaging times. They are listed in Table 9.4 with the corresponding BT directions (represented by the angle from north to the BT direction counterclockwise). BT direction is the direction from a reported center to the next reported center. The distances (denoted as $D_{\mathrm{SAR-BT}}$) between the SAR-derived centers and the interpolated centers from the three BT datasets are calculated and also listed in Table 9.4. Figure 9.9 shows the histograms of $D_{\mathrm{SAR-BT}}$. Most values of $D_{\mathrm{SAR-BT}}$ are within 30 km. The mean values and standard deviations of $D_{\mathrm{SAR-BT}}$ are given in Table 9.5. The mean values for the three BT datasets are close, which are from 13.7 to 15.6 km. The standard deviations are from 6.6 to 10.0 km. The standard deviation for the RSMC is a little bigger than those for CMA and JTWC, by about 3.5 and 2.7 km, respectively.

To investigate whether the typhoon-center location discrepancy between the BT datasets and the 25 SAR extracted values are related to geographical direction or BT direction, we examine the distributions of the location discrepancy in two coordinate systems given in Table 9.6. The geographical direction is fixed for the typhoons, which is the eastward direction here. The BT direction is derived from the three BT datasets of CMA, RSMC and JTWC, which represents general direction of typhoon movement. Different from the geographical direction, the BT direction is variable. In other words, the BT directions are different for different typhoons and different BT datasets.

In the first coordinate system, the SAR-derived centers are used as the origin, and the relative positions of the interpolated centers from the three BT datasets are represented by dots in Fig. 9.10a. For clarity, we call the relative positions in this coordinate system "SAR-BT discrepancy". Obviously, the distributions of SAR-BT discrepancy in the radial direction should be close to the distributions of $D_{\mathrm{SAR-BT}}$ in Fig. 9.9. As shown in Fig. 9.9, most values of $D_{\mathrm{SAR-BT}}$ are within 30 km. Correspondingly, most dots in Fig. 9.10a are inside the circle with the radius of 0.3°.

Table 9.4 The typhoon centers determined from the SAR images and those temporally interpolated from the three BT datasets are listed in this table as well as the distances between them. Other information is also given, including typhoon name, date, imaging time, SAR name, and BT direction (represented by counterclockwise angle form north to BT direction)

No.	Name	UTC time		SAR			CMA				RSMC				JTWC			
		dd/mm/yyyy	hh:mm	SAR	Lat	Lon	Lat	Lon	Dist.	Dir.	Lat	Lon	Dist.	Dir.	Lat	Lon	Dist.	Dir.
					(°N)	(°E)	(°N)	(°E)	(km)	(°)	(°N)	(°E)	(km)	(°)	(°N)	(°E)	(km)	(°)
1	Krovanh	23/08/2003	22:15	R1	18.432	115.244	18.595	115.163	19.9	52	18.824	114.963	52.5	56	18.524	115.193	11.5	59
2	Aere	25/08/2004	1:53	Env.	25.494	121.028	25.5	121.086	5.89	90	25.431	121.117	11.3	84	25.463	121.254	23	80
3	Songda	06/09/2004	13:24	Env.	30.264	127.467	30.257	127.387	7.77	324	30.087	127.393	20.9	333	30.157	127.387	14.2	324
4	Meari	28/09/2004	9:26	R1	29.509	127.338	29.53	127.559	21.5	324	29.473	127.459	12.3	321	29.488	127.701	35.3	330
5	Nock-Ten	23/10/2004	21:19	R1	19.675	126.588	19.687	126.759	18	68	19.732	126.714	14.7	69	19.687	126.759	18	68
6	Nesat	06/06/2005	9:03	R1	21.644	133.648	21.556	133.603	10.7	330	21.605	133.603	6.26	326	21.556	133.705	11.4	319
7	Matsa	05/08/2005	9:55	R1	27.186	122.389	27.156	122.514	12.8	52	27.156	122.514	12.8	52	27.186	122.549	15.8	48
8	Guchol	22/08/2005	8:18	R1	28.508	146.576	28.545	146.647	8.09	24	28.545	146.547	4.94	24	28.598	146.609	10.5	21
9	Mawar	22/08/2005	20:39	R1	25.099	138.117	24.954	138.023	18.7	27	24.954	138.023	18.7	27	24.954	137.979	21.3	32
10	Talim	30/08/2005	1:25	Env.	21.43	129.148	21.471	129.417	28.2	76	21.471	129.417	28.2	76	21.494	129.24	11.9	70
11	Khanun	11/09/2005	1:47	Env.	27.468	122.283	27.597	122.432	20.6	42	27.597	122.462	22.8	39	27.668	122.362	23.5	42
12	Kirogi	14/10/2005	9:12	R1	22.904	131.338	22.953	131.194	15.8	63	22.9	131.194	14.8	90	22.953	131.147	20.4	45
13	Kirogi	15/10/2005	21:05	R1	24.153	132.755	24.154	132.657	9.97	301	24.103	132.657	11.4	292	24.154	132.657	9.97	301
14	Ewiniar	03/07/2006	20:54	R1	15.003	133.175	14.991	133.258	8.99	40	14.991	133.258	8.99	40	15.091	133.209	10.4	45
15	Saomai	10/08/2006	10:02	R1	27.095	120.395	27.135	120.392	4.45	81	27.135	120.392	4.45	81	27.135	120.459	7.75	80
16	Yagi	21/09/2006	20:19	R1	22.918	144.279	23.024	144.338	13.1	47	23.024	144.338	13.1	47	23.024	144.277	11.7	45
17	Man-Yi	11/07/2007	21:14	R1	20.526	129.58	20.492	129.569	3.9	36	20.492	129.569	3.9	36	20.446	129.515	11.1	37
18	Man-Yi	13/07/2007	9:33	R1	28.144	127.64	27.992	127.7	17.8	360	28.092	127.7	8.22	360	27.951	127.7	22.2	360
19	Usagi	01/08/2007	20:59	R1	29.769	133.162	29.844	133.354	20.3	31	29.894	133.205	14.5	37	29.794	133.354	18.8	29

(continued)

Table 9.4 (continued)

No.	Name	UTC time		SAR			CMA				RSMC				JTWC			
		dd/mm/yyyy	hh:mm	SAR	Lat	Lon	Lat	Lon	Dist.	Dir.	Lat	Lon	Dist.	Dir.	Lat	Lon	Dist.	Dir.
					(°N)	(°E)	(°N)	(°E)	(km)	(°)	(°N)	(°E)	(km)	(°)	(°N)	(°E)	(km)	(°)
20	Fitow	31/08/2007	19:44	R1	27.839	152.936	27.786	152.87	8.69	69	27.786	152.941	5.82	72	27.758	152.841	12.9	77
21	Krosa	04/10/2007	21:35	R1	20.327	125.424	20.319	125.483	6.19	74	20.319	125.502	8.18	68	20.319	125.402	2.45	68
22	Kajiki	20/10/2007	20:26	R1	26.846	142.801	26.769	142.829	8.97	317	26.81	142.748	6.67	321	26.769	142.769	9.1	315
23	Sinlaku	10/09/2008	1:33	Env.	19.971	124.285	19.903	124.374	12	14	19.903	124.348	10	27	19.903	124.374	12	14
24	Megi	17/10/2010	1:27	Env.	18.674	127.039	18.652	127.187	15.8	99	18.652	127.187	15.8	99	18.652	127.187	15.8	99
25	Nalgae	30/09/2011	13:54	Env.	17.262	125.221	17.474	125.194	23.5	104	17.379	125.263	13.6	112	17.505	125.094	30	101

Table 9.5 Statistics of the comparisons between the determined typhoon centers from the SAR or MODIS IR images and the temporally interpolated typhoon centers from the three BT datasets

DSAR-BT or DMODIS-BT				Statistics of discrepancy					
				Discrepancy in coordinate system 1			Discrepancy in coordinate system 2		
				(Relative positions of the centers from the BT datasets with respective to those from the SAR or MODIS IR images)			(Relative positions of the centers from the SAR or MODIS IR images with respective to those from the BT datasets)		
				Bias		Std. dev. (°)	Bias		Std. dev. (°)
Ref.	BT	Mean value (km)	Std. dev. (km)	Lat	Lon		Left	Front	
				(°N)	(°E)		(°**L**)	(°F)	
SAR	CMA	13.6699	6.5511	0.0079	0.0453	0.1396	0.0473	0.0391	0.1394
	RSMC	13.7987	10.042	0.0086	0.0263	0.1597	0.0375	0.0331	0.1569
	JTWC	15.6414	7.3456	0.0036	0.0347	0.165	0.049	0.0253	0.166
MODIS IR	CMA	15.3559	7.8062	0.02	−0.0086	0.1639	0.0036	−0.0128	0.1657
	RSMC	16.2453	9.9421	0.0112	−0.0338	0.1795	−0.0105	−0.0243	0.1827
	JTWC	16.7801	8.6393	0.018	−0.0274	0.1779	−0.0087	−0.0364	0.1792

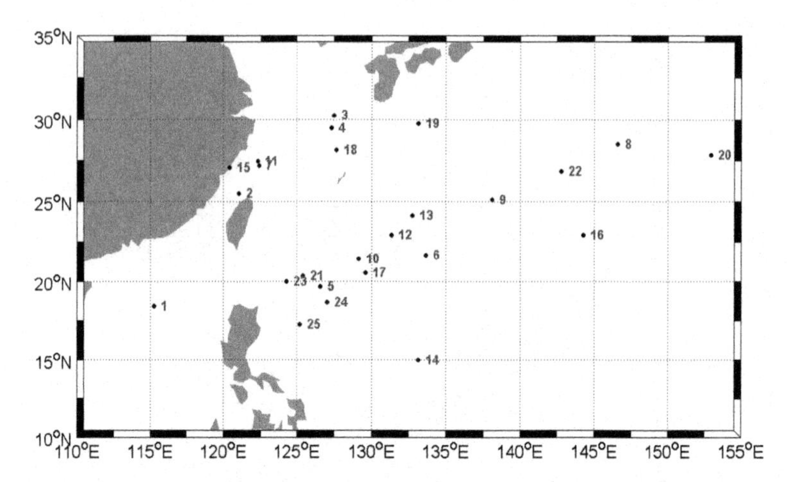

Fig. 9.8 Spatial distribution of the typhoon centers determined from the 25 SAR images for investigating the performance of typhoon-center locations in the three BT datasets at the sea surface

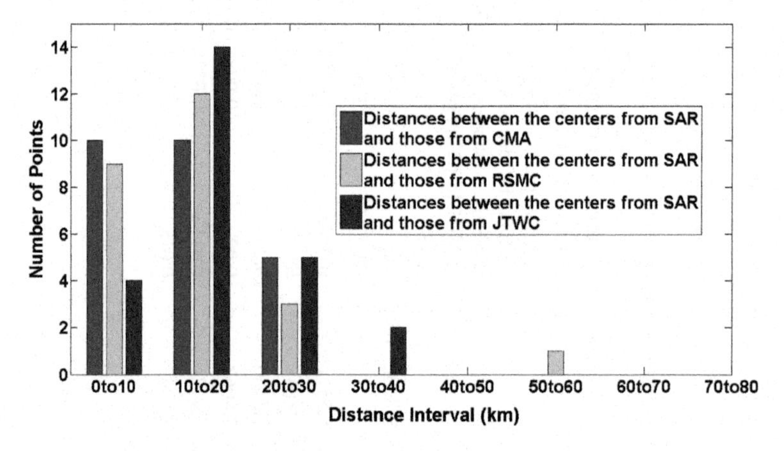

Fig. 9.9 Histograms of the distances (D_{SAR-BT}) between the determined centers from the SAR images and the interpolated centers from the three BT datasets

Table 9.6 Two coordinate systems

No.	Origin	Axes	Dots
1	The SAR- or MODIS-derived centers	Eastward and northward directions are used as the horizontal and vertical axes, respectively	Positions of the interpolated centers from the BT datasets relative to the SAR- or MODIS- derived centers
2	The interpolated centers from the three BT datasets	BT direction is used as the horizontal axis, and the vertical axis is 90° counterclockwise from BT direction	Positions of the SAR- or MODIS- derived centers relative to the interpolated centers from the BT datasets

However, the number of dots on the right side of the vertical axis (north) is obviously more than those on the left side. We divide the range of polar angle (the counterclockwise angle from north) into eight 45° intervals. The number of the dots in each interval is plotted in Fig. 9.11a. We can see that all three BT datasets have a consistent dominant lobe in the circumferential distributions of SAR-BT discrepancy, which points to east. The dots in the interval from 247.5° to 292.5°, which is near-east, are obviously more than those in the other intervals. Then, the mean position (denoted as P_{mean}) and standard deviation (denoted as σ) of the dots are calculated using

$$P_{mean} = \sum_{n=1}^{N} P_n / N \tag{9.3}$$

and

$$\sigma = \sqrt{\frac{\sum_{n=1}^{N} \| P_n - P_{mean} \|^2}{N - 1}} \tag{9.4}$$

where P_n is the position vector to the nth dot, and N is the total number of the dots. As the dots represent the SAR-BT discrepancy between the SAR-derived centers and the interpolated centers, P_{mean} represents the bias between them actually. σ represents the deviation of the dots to P_{mean}. The results are given in Table 9.5. The biases for the three BT datasets are consistent and near-east relative to the origin (i.e., the SAR-derived centers). This is consistent with the dominant lobes shown in Fig. 9.11a. The standard deviations are close, which range from 0.14° to 0.17°.

The axes of first coordinate system are fixed, and typhoon moving direction is not considered. The second coordinate system is used for investigate the relationship between typhoon-center location discrepancy and typhoon moving direction. In the second coordinate system, the interpolated centers from the three BT datasets are used as the origin. The horizontal axis is the BT direction, and the vertical axis is the direction that is 90° counterclockwise from the BT direction. Different form the axes (which are the fixed geographical directions, i.e., the eastward and northward directions) in the first coordinate system, the axes in this coordinate system vary according to the BT direction. The relative positions of the SAR-derived centers to the interpolated centers in this coordinate system are shown in Fig. 9.10b. For clarity, we call these relative positions "BT-SAR discrepancy". The radial distributions of the dots in Fig. 9.10b are the same to those in Fig. 9.10a because distance is a scalar. The numbers of the dots in the eight polar angle intervals are given in Fig. 9.11b, and all three BT datasets show similar inhomogeneity in the circumferential distributions of the dots. Figure 9.11b shows that the dots in Fig. 9.9b above the line for 135° and 315° (relative to the BT direction) are remarkably more than those below the line for all three BT datasets. The statistics of BT-SAR discrepancy are calculated using Eqs. (9.3) and (9.4) and given in Table 9.5. The biases are basically consistent, and

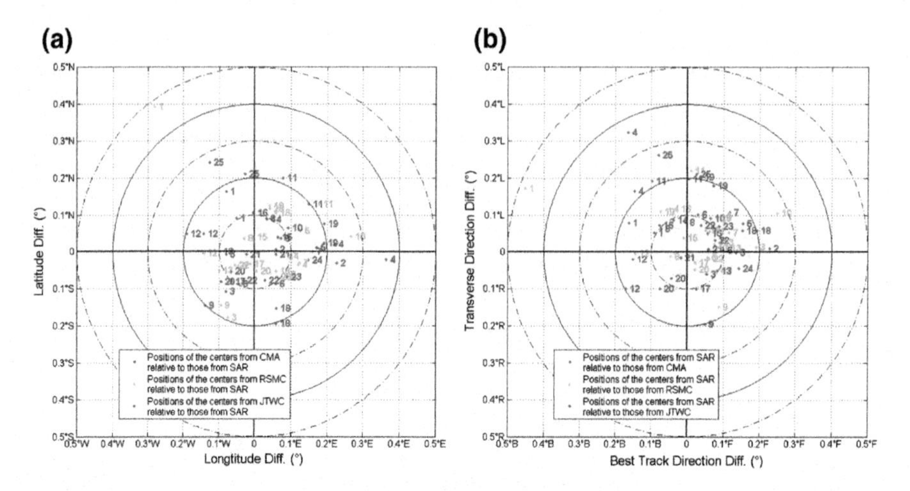

Fig. 9.10 **a** Positions of the interpolated centers relative to the centers determined from the SAR images in the first coordinate system (listed in Table 9.6), where the determined centers are used as the origins, and the eastward and northward directions are used as the horizontal and vertical axes. **b** Positions of the centers determined from the SAR images relative to the interpolated centers in the second coordinate system (also listed in Table 9.6). The origin is the interpolated centers. The horizontal axis is the BT direction, and the vertical axis is the direction that is 90° counterclockwise from the BT direction

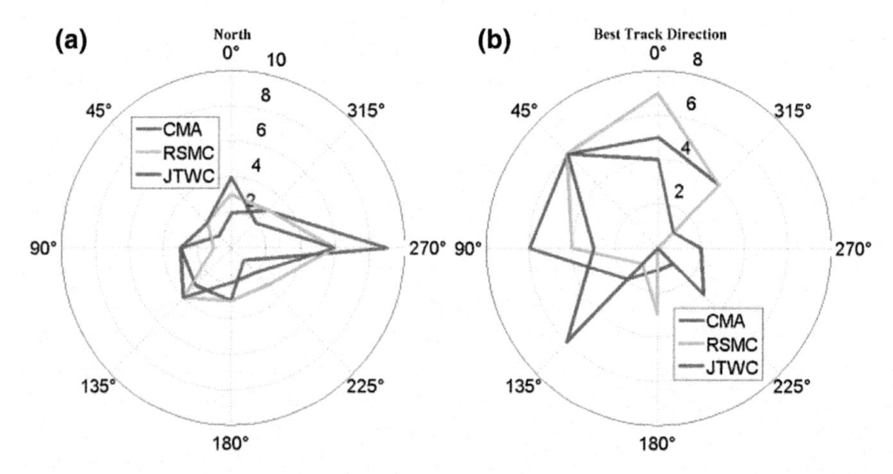

Fig. 9.11 Numbers of the dots in the eight polar angle intervals (**a**) in Fig. 9.10a in the first coordinate system, and (**b**) in Fig. 9.10b in the second coordinate system. The range of polar angle is divided into eight intervals, and the centers of the intervals are 0°, 45°, 90°, 135°, 180°, 225°, 270° and 315°

located in the directions of 50.4°, 48.6° and 62.7° from the BT directions counter-clockwise for CMA, JMA and JTWC, respectively.

Forty-three MODIS IR images with typhoon eyes are collected for investigating the performance of the typhoon-center locations in the BT datasets at the cloud top.

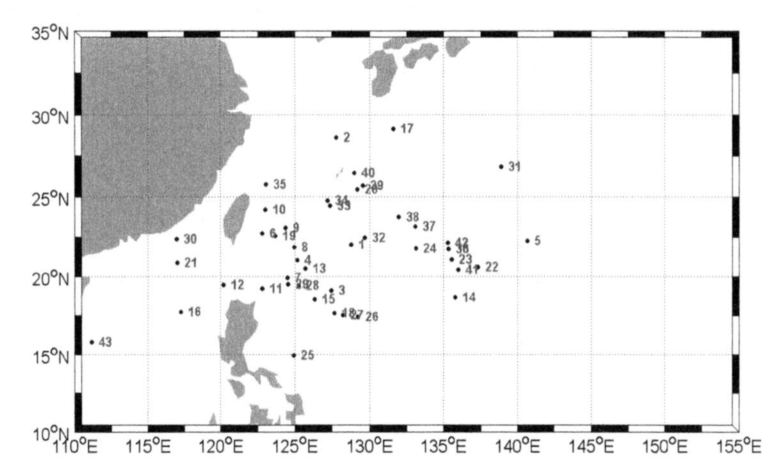

Fig. 9.12 Spatial distribution of the typhoon centers determined from the 43 MODIS IR images for investigating the performance of typhoon-center locations in the three BT datasets at the cloud top

The eye centers are determined from the MODIS IR images by the same procedure, and the spatial distribution of them is given in Fig. 9.12. Like the SAR cases, the centers are located approximately between 15°N and 30°N. The same comparison is made between the MODIS-derived centers and the temporally interpolated centers from the three BT datasets. The locations of the MODIS-derived centers and the interpolated centers are listed in Table 9.7. The histograms of the distances between them (denoted as $D_{\mathrm{MODIS-BT}}$) are given in Fig. 9.13. Similar to the distribution of $D_{\mathrm{SAR-BT}}$ in Fig. 9.9, most values of $D_{\mathrm{MODIS-BT}}$ are also within 30 km. The statistics of $D_{\mathrm{MODIS-BT}}$ are given in Table 9.5. The mean values and standard deviations for the three BT datasets are close, which are from 15.4 to 16.8 km and from 7.8 to 9.9 km, respectively.

Then, the relative positions between the MODIS-derived centers and the interpolated centers from the three BT datasets are investigated using the two coordinate systems (listed in Table 9.6), which are based on geographical direction and BT direction, respectively. For clarity, the relative positions in the first (geographical direction based) and second (BT direction based) coordinate systems are called "MODIS-BT discrepancy" and "BT-MODIS discrepancy", respectively. They are represented by the dots in Fig. 9.14a, b, respectively. Most of the dots are inside the circle with the radius of 0.3° in both coordinate systems. However, in the second coordinate system, the dots in the forth quadrant (i.e., from 270° to 360°) are obviously fewer than those in the other three quadrants (see Fig. 9.14b). Thus, in order to clearly show the circumferential distributions of the dots, the numbers of the dots in the eight polar angle intervals are plotted in Fig. 9.15. The circumferential distributions show similar patterns for all three BT datasets, especially for CMA and RSMC in the second coordinate system. Unlike the SAR cases (see Fig. 9.11a), no obvious dominant lobe is found in the first coordinate system in Fig. 9.15a. However, Fig. 9.15b shows the

Table 9.7 The typhoon centers determined from the MODIS IR images and those temporally interpolated from the three BT datasets are listed in this table as well as the distances between them. Other information is also given, including typhoon name, date, imaging time, and BT direction (represented by counterclockwise angle form north to BT direction)

No.	Name	UTC time		MODIS IR		CMA				RSMC				JTWC			
		dd/mm/yyyy	hh:mm	Lat	Lon	Lat	Lon	Dist.	Dir.	Lat	Lon	Dist.	Dir.	Lat	Lon	Dist.	Dir.
				(°N)	(°E)	(°N)	(°E)	(km)	(°)	(°N)	(°E)	(km)	(°)	(°N)	(°E)	(km)	(°)
1	Man-Yi	12/07/2007	5:00	22.019	128.766	22	128.7	7.15	27	22	128.7	7.15	27	22.183	128.6	25	25
2	Man-Yi	13/07/2007	13:35	28.635	127.785	28.664	127.832	5.6	333	28.711	127.806	8.7	333	28.738	127.906	16.4	336
3	Krosa	04/10/2007	4:35	19.103	127.429	19.058	127.542	12.8	45	19.035	127.542	14	41	19.058	127.565	15.2	49
4	Krosa	05/10/2007	5:20	21.01	125.174	20.933	125.022	17.9	18	20.933	125.033	16.9	27	20.933	125.022	17.9	18
5	Kajiki	20/10/2007	1:30	22.254	140.67	22.4	140.675	16.2	5	22.325	140.675	7.9	4	22.4	140.6	17.7	360
6	Fung-Wong	27/07/2008	13:55	22.733	122.804	22.796	122.876	10.2	67	22.764	122.844	5.38	76	22.76	122.844	5.11	58
7	Sinlaku	10/09/2008	2:30	19.915	124.515	19.967	124.358	17.4	14	19.967	124.317	21.5	27	19.967	124.358	17.4	14
8	Sinlaku	11/09/2008	14:10	21.867	124.958	22.044	124.764	28	14	22.072	124.8	27.9	360	22.008	124.764	25.4	18
9	Sinlaku	12/09/2008	2:20	23.083	124.357	23.033	124.306	7.63	40	23.111	124.344	3.37	27	23.033	124.406	7.41	40
10	Sinlaku	13/09/2008	3:00	24.22	123.022	24.2	122.65	37.9	56	24.2	122.6	43	63	24.2	122.7	32.8	45
11	Hagupit	22/09/2008	2:55	19.244	122.806	19.146	122.768	11.5	77	19.143	122.768	11.8	69	19.146	122.768	11.5	77
12	Hagupit	22/09/2008	13:50	19.473	120.2	19.622	120.281	18.6	77	19.653	120.35	25.4	74	19.592	120.35	20.6	81
13	Jangmi	27/09/2008	4:45	20.493	125.688	20.471	125.788	10.7	39	20.471	125.788	10.7	39	20.392	125.788	15.3	42
14	Melor	05/10/2009	1:55	18.679	135.755	18.656	135.821	7.43	62	18.724	135.821	8.57	65	18.792	135.753	12.5	67
15	Megi	17/10/2010	4:55	18.531	126.304	18.536	126.435	13.8	99	18.536	126.435	13.8	99	18.536	126.417	11.9	99
16	Megi	20/10/2010	5:25	17.733	117.224	17.742	117.21	1.76	9	17.751	117.29	7.32	349	17.661	117.21	8.13	14
17	Roke	20/09/2011	1:30	29.146	131.592	29.025	131.775	22.3	306	29.05	131.7	14.9	326	29.1	131.825	23.2	294
18	Guchol	17/06/2012	2:30	17.675	127.634	17.717	127.475	17.4	17	17.658	127.475	16.9	15	17.758	127.533	14	20
19	Tembin	23/08/2012	3:00	22.562	123.681	22.55	123.45	23.8	72	22.55	123.35	34.1	79	22.65	123.35	35.4	79
20	Bolaven	26/08/2012	1:50	25.468	129.204	25.414	129.386	19.3	45	25.483	129.317	11.5	45	25.483	129.317	11.5	45
21	Tembin	26/08/2012	3:30	20.853	117.012	20.883	116.858	16.3	207	20.925	116.7	33.4	180	20.942	116.742	29.8	135

(continued)

Table 9.7 (continued)

No.	Name	UTC time		MODIS IR		CMA				RSMC				JTWC			
		dd/mm/yyyy	hh:mm	Lat	Lon	Lat	Lon	Dist.	Dir.	Lat	Lon	Dist.	Dir.	Lat	Lon	Dist.	Dir.
				(°N)	(°E)	(°N)	(°E)	(km)	(°)	(°N)	(°E)	(km)	(°)	(°N)	(°E)	(km)	(°)
22	Soulik	09/07/2013	16:40	20.581	137.293	20.581	137.322	6.5	73	20.611	137.322	4.47	68	20.533	137.4	12.3	72
23	Soulik	10/07/2013	2:05	21.069	135.544	21.069	135.453	21	68	21.204	135.453	17.7	73	21.204	135.383	22.4	76
24	Soulik	10/07/2013	13:00	21.772	133.124	21.772	133.3	18.9	85	21.817	133.3	18.9	85	21.833	133.283	17.9	81
25	Utor	11/08/2013	5:10	14.972	124.921	14.972	124.95	5.97	61	15.017	124.953	5.97	61	14.917	124.953	7.04	61
26	Usagi	18/09/2013	14:00	17.428	129.219	17.428	129.3	19.9	96	17.233	129.267	22.2	101	17.267	129.333	21.6	95
27	Usagi	19/09/2013	5:20	17.538	128.243	17.538	128.03	24.5	31	17.433	128.033	25.1	27	17.533	128.011	24.6	9
28	Usagi	20/09/2013	2:55	19.409	125.238	19.409	125.06	20.7	66	19.494	125.063	20.7	66	19.492	125.014	25.2	59
29	Usagi	20/09/2013	4:25	19.494	124.558	19.494	124.84	31.3	66	19.594	124.838	31.3	66	19.642	124.764	27.1	59
30	Usagi	22/09/2013	5:50	22.381	116.942	22.381	116.83	11.5	61	22.383	116.831	11.5	61	22.383	116.831	11.5	61
31	Pabuk	24/09/2013	13:25	26.848	138.885	26.848	138.81	10.7	24	26.889	138.853	5.57	14	26.86	138.958	7.36	29
32	Fitow	04/10/2013	4:35	22.448	129.676	22.448	129.65	15.2	22	22.558	129.571	16.4	27	22.582	129.724	15.7	11
33	Fitow	05/10/2013	2:10	24.432	127.374	24.432	127.45	11.2	67	24.472	127.383	4.57	72	24.472	127.383	4.57	72
34	Fitow	05/10/2013	5:20	24.766	127.198	24.766	127.08	16.5	67	24.656	126.833	38.9	72	24.578	127.067	24.8	72
35	Fitow	06/10/2013	4:25	25.789	123.041	25.789	123.21	28.4	63	25.894	123.064	11.9	68	25.921	123.232	24	59
36	Danas	06/10/2013	4:25	21.745	135.321	21.745	135.4	12.7	56	21.789	135.396	9.15	62	21.789	135.343	5.4	58
37	Danas	06/10/2013	13:50	23.138	133.086	23.138	132.95	14.3	56	23.067	132.981	13.4	55	23.097	132.85	24.6	54
38	Danas	06/10/2013	16:35	23.737	131.957	23.737	132.13	17.3	56	23.617	132.201	28.3	55	23.693	132.025	8.54	54
39	Danas	07/10/2013	2:00	25.704	129.539	25.704	129.73	19.9	45	25.667	129.533	4.21	55	25.733	129.567	4.24	45
40	Danas	07/10/2013	5:10	26.466	128.979	26.466	128.99	6.82	45	26.478	128.839	14.1	32	26.419	128.881	11.1	45
41	Wipha	13/10/2013	16:40	20.421	135.975	20.421	136.01	3.82	32	20.344	136.089	14.6	30	20.344	136.011	9.26	36
42	Wipha	14/10/2013	2:05	22.114	135.294	22.114	135.29	4.17	25	22.217	135.292	11.4	27	22.151	135.292	4.17	25
43	Nari	14/10/2013	3:45	15.817	111.138	15.817	111	15.2	63	15.75	110.925	24	56	15.763	110.825	34.1	81

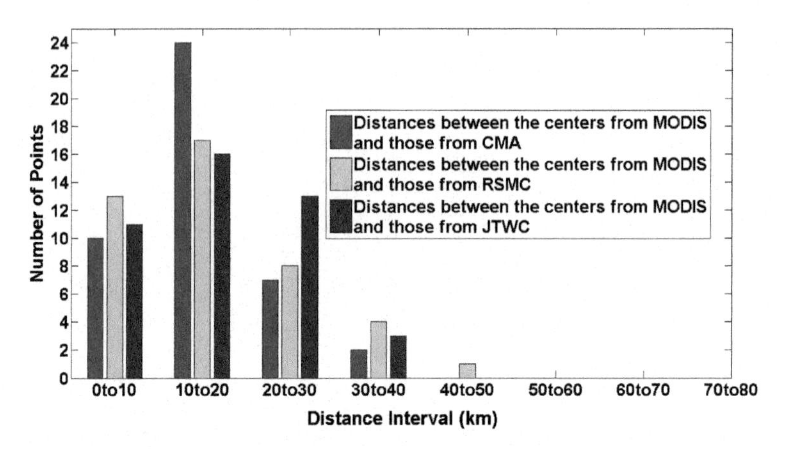

Fig. 9.13 Histograms of the distances ($D_{\mathrm{MODIS-BT}}$) between the determined centers from the MODIS IR images and the interpolated centers from the three BT datasets

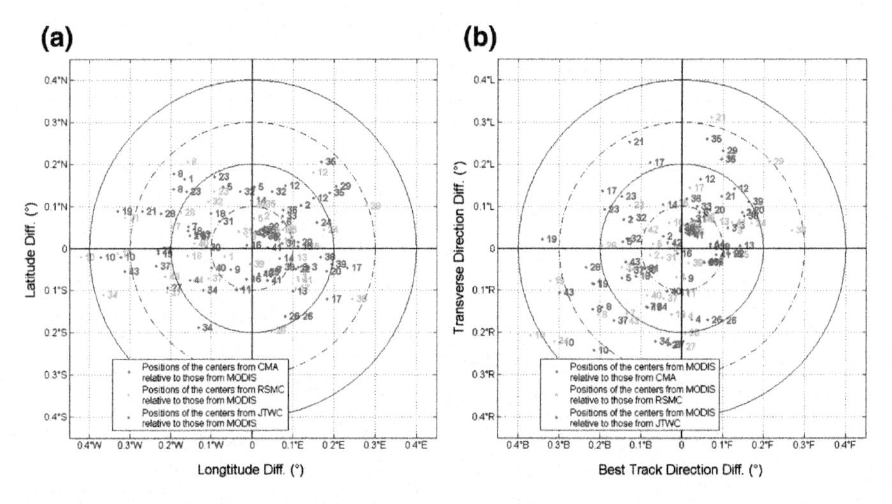

Fig. 9.14 **a** Positions of the interpolated centers relative to the centers determined from the MODIS IR images in the first coordinate system (listed in Table 9.6), where the determined centers are used as the origins, and the eastward and northward directions are used as the horizontal and vertical axes. **b** Positions of the centers determined from the MODIS IR images relative to the interpolated centers in the second coordinate system (also listed in Table 9.6). The origin is the interpolated centers. The horizontal axis is the BT direction, and the vertical axis is the direction that is 90° counterclockwise from the BT direction

dots in the interval of 292.5° to 337.5° (relative to the BT direction) are fewer than those in the other intervals, especially for CMA and RSMC, in the second coordinate system. The statistics of MODIS-BT and BT-MODIS discrepancies are given in Table 9.5. The biases and standard deviations are close.

Table 9.5 shows that the statistics of the three BT datasets are close. The CMA BT dataset is the best in comparison with both SAR- and MODIS-derived centers. The

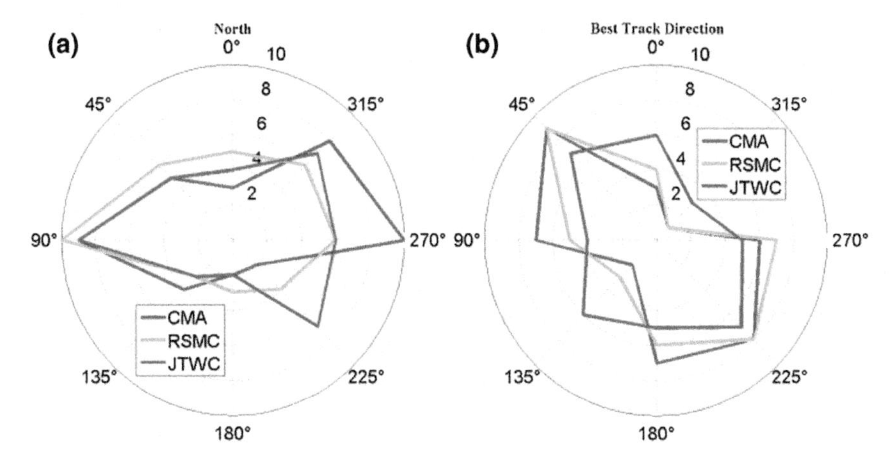

Fig. 9.15 15 Numbers of the dots in the eight polar angle intervals (**a**) in Fig. 9.14a in the first coordinate system, and (**b**) in Fig. 9.14b in the second coordinate system. The range of polar angle is divided into eight intervals, and the centers of these intervals are 0°, 45°, 90°, 135°, 180°, 225°, 270° and 315°

mean values and standard deviations of $D_{\mathrm{MODIS-BT}}$ are larger than those of $D_{\mathrm{SAR-BT}}$, except for the standard deviations for the RSMC. Note that in the first SAR case, Typhoon Krovanh in 2003, the value of $D_{\mathrm{SAR-BT}}$ is larger than 50 km for the RSMC, but those for CMA and JTWC are small, with 20 and 12 km, respectively. This case causes the standard deviation of $D_{\mathrm{SAR-BT}}$ to be a litter bigger for the RSMC. Thus, overall, the typhoon centers from the BT datasets are closer to the SAR-derived centers at the sea surface than the MODIS-derived centers at the cloud top.

9.6 Rain Effect Model

SAR plays an important role in wind field retrieval, because it has the advantage of high spatial resolution and can work in all-weather condition. However, typhoon always accompany with heavy rain which may alter the wind-induced backscatter signature, therefore influence the accuracy of wind speed retrieval. The influence of heavy rainfall on radar signals including three factors: rain-induced attenuation in atmosphere, raindrop volumetric scattering and rough sea surface scattering. Through quantitative analysis on these aspects, a backscatter model can be expressed as [29]:

$$\sigma_0 = \sigma_{wind}^0 \alpha_{atm} + \sigma_{eff} \tag{9.5}$$

where σ_0 is the total normalized radar-backscattering cross section (NRCS) received by radar, σ_{wind}^0 is the wind-induced backscatter, α_{atm} is the rain-induced atmospheric attenuation, and σ_{eff} is the summing of attenuated surface perturbation and the atmospheric scattering.

9.6.1 Rain-Induced Attenuation

Under the condition of rainfall, the signal in the process of transmission will be affected by the influence of precipitation particles. On the one hand, signal projected to the raindrops, energy will be absorbed by rain, on the other hand, energy will be scattered by the rain particles, which also caused the SAR signal attenuation. To simplify the calculation, the raindrops can approximate as Rayleigh particles, rain-induced attenuation can be expressed as [30]:

$$\alpha_{vv}^2 = \exp\left(-2H\sec\theta\frac{4n_0\pi}{k}\mathrm{Im}\langle S_{vv}^f\rangle\right) \tag{9.6}$$

where subscript vv denotes VV polarization, n_0 denotes number of raindrops per unit volume, k denotes wavenumber, ε denotes complex relative permittivity, H denotes rainfall layer depth, θ is incident angle, $\mathrm{Im}\langle S_{vv}^f\rangle$ is the parameter which related to rain features and rain rate.

9.6.2 Raindrop Volumetric Backscattering

During signal transmission, signals may go through the raindrops, so the energy will return back to the scattering direction which lead to the creation of the volumetric backscattering. Raindrops are regarded as Rayleigh spherical particles. The raindrops volumetric scattering and rainfall penetration attenuation can be expressed as [30]:

$$V_{vv}^2 = n_0 k^4 \left|\frac{\varepsilon - 1}{\varepsilon + 2}\right| \frac{H}{\tan\theta} \langle t_v^2\rangle \tag{9.7}$$

where $\langle t_v^2\rangle$ is the parameter which related to rain features and rain rate.

The relationship between rain rate, incident angle and rain-induced attenuation, raindrop volumetric backscattering is shown in Fig. 9.16. We can see that with the increase of rain rate in the same incident angle, attenuation coefficient decreases, raindrop volumetric backscattering increase. With the increase of incident angle in same rain rate, the variation of attenuation coefficient is very small, the raindrop volumetric backscattering is slightly decreased.

The influence of rainfall on the sea surface is very complex, so it is difficult to accurately judge the relationship between sea surface disturbance and rain rate. To explore the sea surface disturbance caused by rain, we used RADARSAT-2 SAR data,

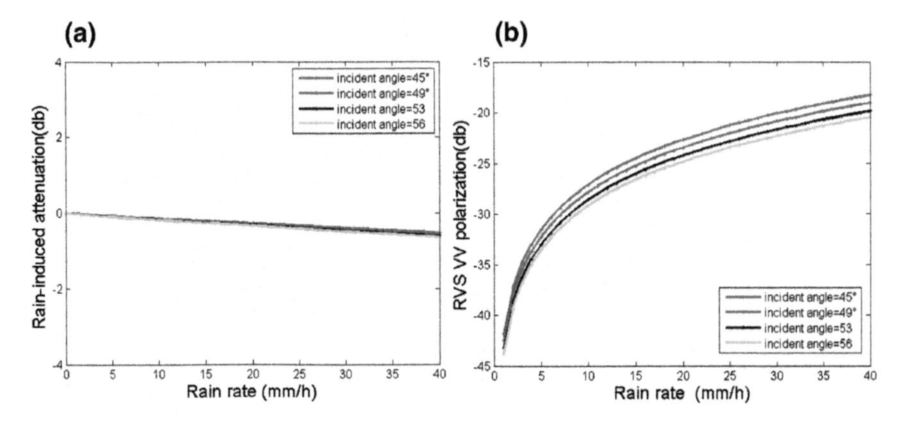

Fig. 9.16 **a** Relationship between rain-induced attenuation and rain rate or incident angle; **b** Relationship between rain volumetric backscattering and rain rate or incident angle

TRMM precipitation data and NCEP reanalysis data to fit the relationship between rainfall and sea surface disturbance according to Eq. (9.5).

RADARSAT-2 is a C-band space-borne SAR which was launched on 14 December 2007. Its image is acquired in dual-polarization imaging mode; the incidence angles in the near and far range are 19.4° and 49.4°, respectively. The pixel spacing is 50 m, and the swath is about 500 × 500 km.

Precipitation data used here is TRMM precipitation radar 3B42 data. The TRMM precipitation radar is a Ku-band pulsed radar operating at 13.8 GHz in a horizontal polarization. The 3B42 rain rate is 3-hourly average centered at the middle of each 3-h period, and is gridded in 0.25° × 0.25°.

9.7 Approach and Case Study

In order to study the rain effect to the sea surface in different incident angle, we divided one obvious rain region into three parts according to the rain rate. The rainfall of these three parts respectively are around 20, 10, and 15 mm/h. The location of these three area is showed in σ_0 and rain rate diagram in Fig. 9.17a, b.

In order to get reasonable wind filed, we linearly interpolated the wind speeds along the radical direction to substitute the wind with a deviation of 6 m/s from average wind speeds. Figure 9.18a–f show the wind speed distribution before and after correction. We took the final wind speeds as the standard winds which were not affected by rain.

Rain-induced attenuation and raindrop volumetric backscattering can calculate by theoretical equation. With the Eq. (9.5), we fit the rainfall data with sea surface disturbance backscattering to get the quantitative relationship between them.

Using a linear function to fit the data, as shown in Eq. (9.8)

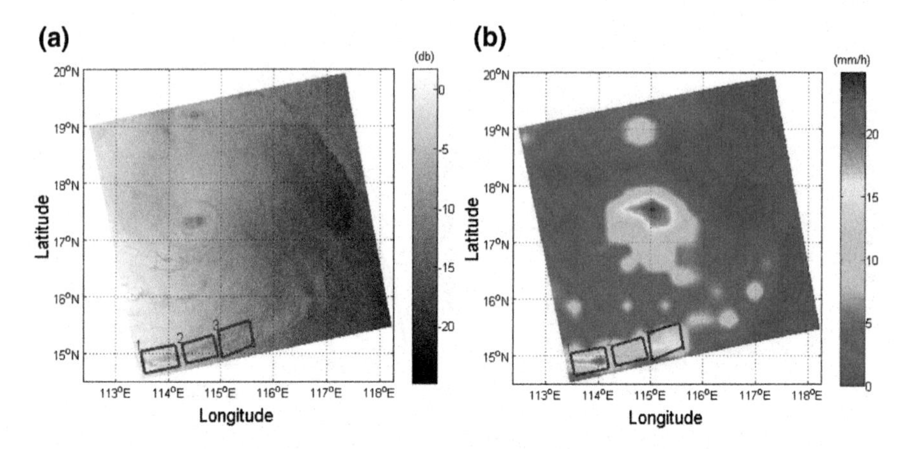

Fig. 9.17 Location of study areas are showed in σ_0 (**a**) and rain rate (**b**)

$$\sigma_{rain} = \sum_{n=0}^{N} P_n(\theta) R^n \tag{9.8}$$

where P is fitting coefficient. N is fitting polynomial order, in our research, N is 1. The result and corresponding fitting coefficient is shown in Fig. 9.19 and Table 9.8.

To validate the proposed model under the condition of rainfall. We choose one ENVISAT ASAT data which acquired in October 17 in 2010 in typhoon Megi period. The pixel spacing is 75 m. With the comparison to the rain rate data and wind speeds data observed by Step Frequency Microwave Radiometer (SFMR), we can verify the performance of our model. Figure 9.20 shows the SFMR path (blue solid line) and China Meteorological Administration(CMA) typhoon best track (red solid line) in Typhoon Megi ENVISAT ASAR image.

For each selected set of wind speeds and rain rates measured by SFMR, the mapping procedure consists of following steps: (1) according to the best track move the SAR images to collocate the SAR center; (2) determine whether the SFMR data set is within the mapped SAR image; (3) interpolated SAR data within one hour; (4) because of the limitation of CMOD5, we choose data that wind speeds is smaller than 30 m/s. Figure 9.21 is wind speeds of SFMR and its corresponding SAR normalized radar cross section (NRCS).

By using SFMR rainfall and wind speeds data, we calculated the rain attenuation and raindrop volumetric backscattering. According to the fitting coefficient mentioned above, we calculated the sea surface backscattering caused by rain. Considering all these factors together, we can get a total NRCS. Comparing SAR observed NRCS with our calculation, we can see the performance of our model. Figure 9.22a, c, e are comparison before using our model, Fig. 9.22b, d, f are comparison between our model and observations. Table 9.9 shows the results.

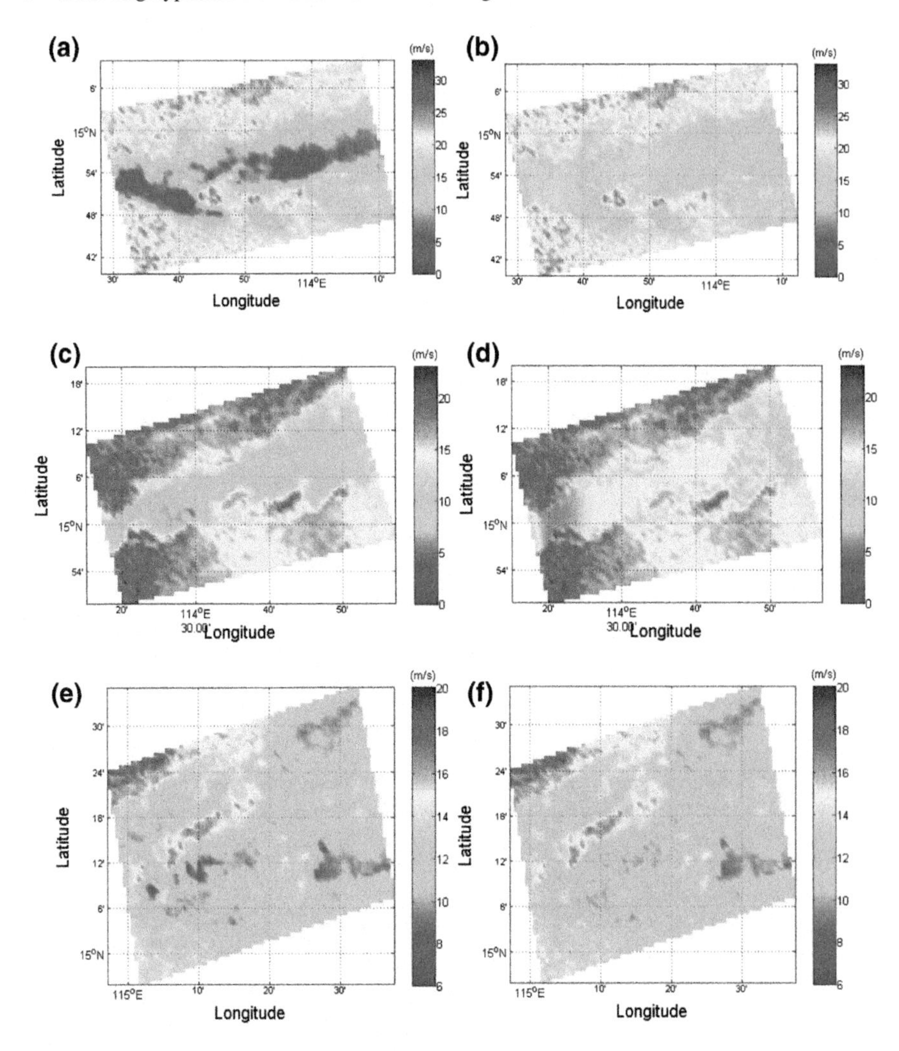

Fig. 9.18 Regional 1, 2 and 3 wind speed distribution (**a**), (**c**), (**e**) before correction and (**b**), (**d**), (**f**) after correction

Table 9.8 Fitting coeefficient

Incident angle (°)	p_0	p_1
20.20° ~ 25.16°	0.001871	−0.4709
25.80° ~ 29.63°	−0.01284	0.03075
30.45° ~ 35.09°	−0.0003267	−0.0409

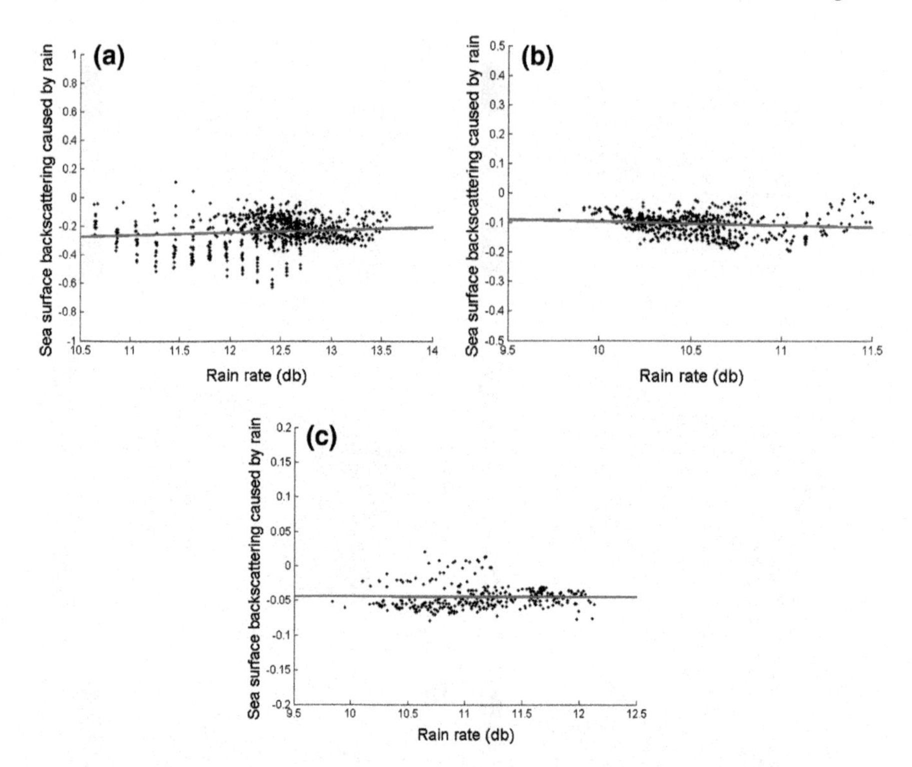

Fig. 9.19 Relationship between rain rate and sea surface backscattering caused by rain in different incident angle **a** incident angle is 20.20° ~ 25.16° **b** incident angle is 25.80° ~ 29.63° **c** incident angle is 30.45° ~ 35.09°

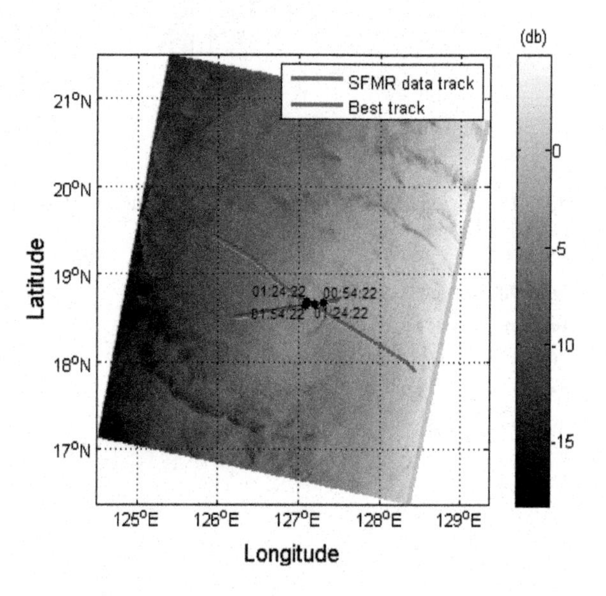

Fig. 9.20 Typhoon best track and SFMR path in 2010 10.17 01:24:22 Megi ENVISAT ASAR data

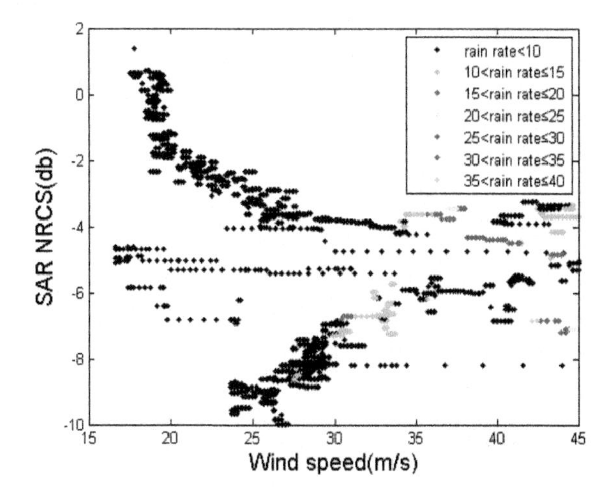

Fig. 9.21 SFMR rain and wind speeds data and its corresponding SAR NRCS

Table 9.9 NRCS comparison results between SAR observations and SFMR calculation data

Incident angle	Number N		Bias		RMS	
	Before	After	Before	After	Before	After
20.20° ~ 25.16°	74	74	0.33977	0.33691	0.3486	0.33737
25.80° ~ 29.63°	34	34	0.15705	0.07533	0.1765	0.07867
30.45° ~ 35.09°	52	52	0.13247	0.04701	0.13296	0.04703

The results revealed that in three incident angle range, our model has good correction effectiveness. Bias and root mean square decrease after applying our model.

9.8 Summary

SAR has high spatial resolution, and can capture footprints of typhoon eyes at the sea surface. IR sensors can provide typhoon eyes at the cloud top. This study compared the typhoon centers determined from SAR and IR images with those from the BT datasets of the three operational agencies, namely, CMA, RSMC and JTWC. For this objective, two alternative algorithms as well as the existing wavelet-based algorithm were used and compared for extracting typhoon eyes in SAR and IR images. The investigation showed that the location differences among the determined centers from the images due to different algorithms are much smaller than those between the so-determined centers and the centers (interpolated to the imaging times) from the three BT datasets.

The typhoon centers interpolated from the three BT datasets and those determined from 25 SAR images and 43 MODIS IR images were compared in detail for investigating the performance of the center locations in the three BT datasets at the sea

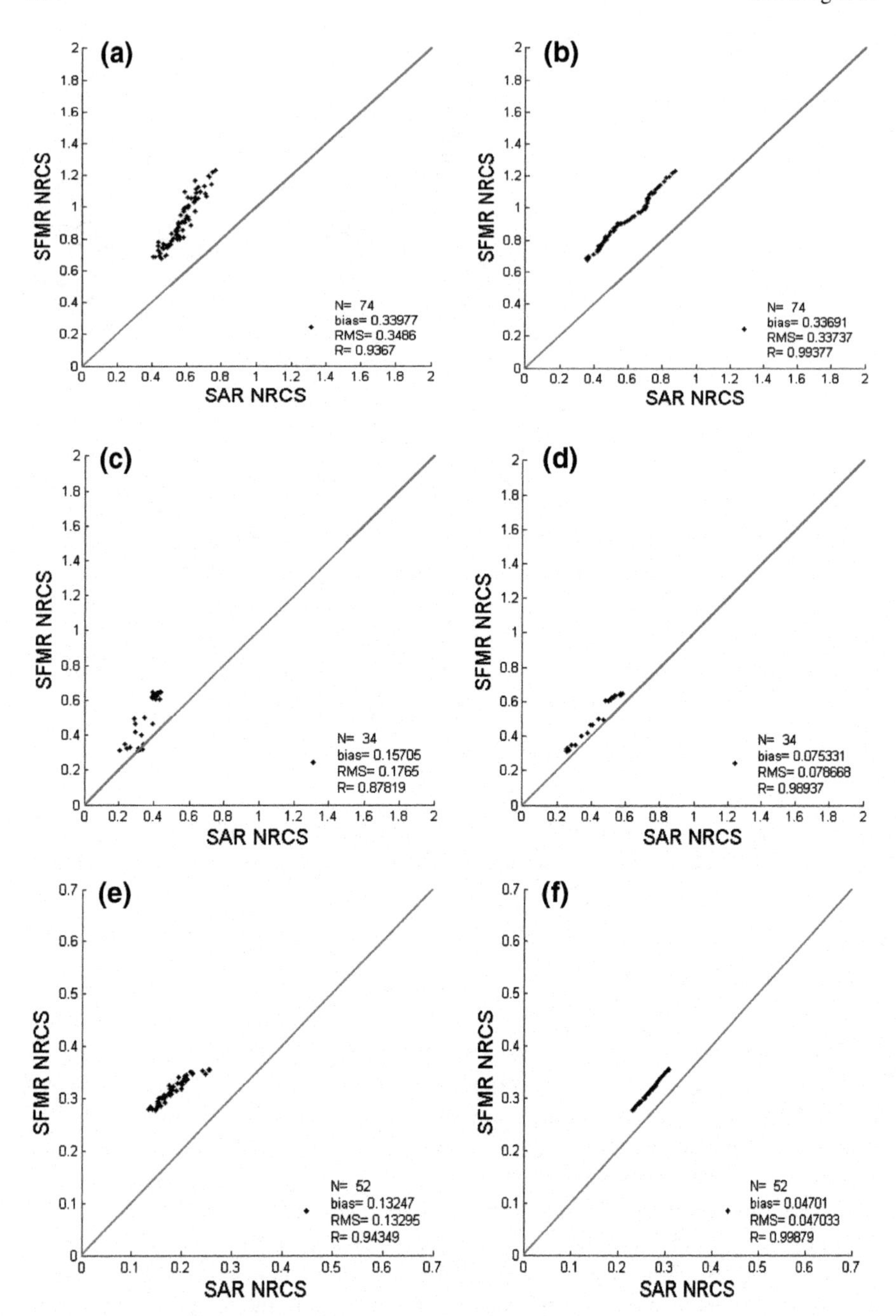

Fig. 9.22 NRCS comparison between SAR observations and SFMR calculation NRCS before **a**, **c**, **e** and after **b**, **d**, **f** using our model

surface and at the cloud top, respectively. Overall, the statistics of the location differences between the interpolated and determined centers are close for all the three BT datasets. We found the typhoon centers from the BT datasets are generally closer to these from SAR at the sea surface than those from MODIS IR at the cloud top. The directional (circumferential) distributions of the relative positions between the interpolated and determined centers were also investigated. The distributions showed similar patterns for all three BT datasets, although they were generated by different meteorological agencies. Similar dominant lobes were found in the patterns for all three BT datasets in comparison with SAR.

This paper also provided a method to evaluate the influence of rain on SAR wind field retrieval and proposed a model to calculate this effect. This model took rain attenuation, raindrop backscattering and sea surface disturbance caused by rain into consideration. A case study using an ENVISAT ASAR image with collocated SFMR wind and rain data shows that using our rain effect model is better than only considering wind effect. It also revealed that the understanding of rain effects still needs further improvement.

References

1. Velden, C., and J. Hawkins. 2002. The increasing role of weather satellites in tropical cyclone analysis and forecasting. *Proceedings of the Fifth WMO International Workshop on Tropical Cyclones (IWTC-V)*. Cairns, Queensland, Australia.
2. Velden, C., J. Hawkins, R. Edson, D. Herndon, R. Zehr, and J. Kossin. 2010. Satellite observations of tropical cyclones. *Global Perspectives of Tropical Cyclones*, 201–226.
3. Elsner, J.B., J.P. Kossin, and T.H. Jagger. 2008. The increasing intensity of the strongest tropical cyclones. *Nature* 455 (7209): 92–95.
4. Cheng, Y.H., S.J. Huang, A.K. Liu, C.R. Ho, and N.J. Kuo. 2012. Observation of typhoon eyes on the sea surface using multi-sensors. *Remote Sensing of Environment* 123: 434–442.
5. Friedman, K.S., and X. Li. 2000. Monitoring hurricanes over the ocean with wide swath SAR. *Johns Hopkins APL Technical Digest* 21 (1): 80–85.
6. Du, Y., and P.W. Vachon. 2003. Characterization of hurricane eyes in RADARSAT-1 images with wavelet analysis. *Canadian Journal of Remote Sensing* 29 (4): 491–498.
7. Li, X., J.A. Zhang, X. Yang, W.G. Pichel, M. DeMaria, D. Long, and Z. Li. 2013. Tropical cyclone morphology from spaceborne synthetic aperture radar. *Bulletin of the American Meteorological Society* 94 (2): 215–230.
8. Pan, Y., A.K. Liu, S. He, J. Yang, and M.X. He. 2013. Comparison of typhoon locations over ocean surface observed by various satellite sensors. *Remote Sensing* 5 (7): 3172–3189.
9. Reppucci, A., S. Lehner, J. Schulz-Stellenfleth, and S. Brusch. 2010. Tropical cyclone intensity estimated from wide-swath SAR images. *IEEE Transactions on Geoscience and Remote Sensing* 48 (4): 1639–1649.
10. Jin, S., S. Wang, and X. Li. 2014. Typhoon eye extraction with an automatic SAR image segmentation method. *International Journal of Remote Sensing* 35 (11–12): 3978–3993.
11. Zou, J.H., M.S. Lin, D.L. Pan, Z.H. Chen, and L. Yang. 2009. Applications of QuikSCAT in typhoon observation and tracking. *Journal of Remote Sensing* 13 (5): 840–846.
12. Piñeros, M.F., E.A. Ritchie, and J.S. Tyo. 2008. Objective measures of tropical cyclone structure and intensity change from remotely sensed infrared image data. *IEEE Transactions on Geoscience and Remote sensing* 46 (11): 3574–3580.

13. Chaurasia, S., C. Kishtawal, and P. Pal. 2010. An objective method of cyclone centre determination from geostationary satellite observations. *International Journal of Remote Sensing* 31 (9): 2429–2440.
14. Jaiswal, N., and C.M. Kishtawal. 2011. Automatic determination of center of tropical cyclone in satellite-generated ir images. *IEEE Geoscience and Remote Sensing Letters* 8 (3): 460–463.
15. Wimmers, A.J., and C.S. Velden. 2010. Objectively determining the rotational center of tropical cyclones in passive microwave satellite imagery. *Journal of Applied Meteorology and Climatology* 49 (9): 2013–2034.
16. Zhang, D., Y. Zhang, T. Hu, B. Xie, and J. Xu. 2014. A comparison of HY-2 and QuikSCAT vector wind products for tropical cyclone track and intensity development monitoring. *IEEE Geoscience and Remote Sensing Letters* 11 (8): 1365–1369.
17. Chang, P.L., and B. Jong-Dao. 2009. Jou, and J. Zhang. An algorithm for tracking eyes of tropical cyclones. *Weather and Forecasting* 24 (1): 245–261.
18. Liu, A.K., S.Y. Wu, and Y. Zhao. 2003. Wavelet analysis of satellite images in ocean applications. *Frontiers of Remote Sensing Information Processing* 7: 141–162.
19. Wu, S., and A. Liu. 2003. Towards an automated ocean feature detection, extraction and classification scheme for SAR imagery. *International Journal of Remote Sensing* 24 (5): 935–951.
20. Bliven, L., P. Sobieski, and C. Craeye. 1997. Rain generated ring-waves: measurements and modelling for remote sensing. *International Journal of Remote Sensing* 18 (1): 221–228.
21. Nie, C., and D.G. Long. 2007. A C-band wind/rain backscatter model. *IEEE Transactions on Geoscience and Remote Sensing* 45 (3): 621–631.
22. Ying, M., W. Zhang, H. Yu, X. Lu, J. Feng, Y. Fan, Y. Zhu, and D. Chen. 2014. An overview of the China Meteorological Administration tropical cyclone database. *Journal of Atmospheric and Oceanic Technology* 31 (2): 287–301.
23. Schreck III, C.J., K.R. Knapp, and J.P. Kossin. 2014. The impact of best track discrepancies on global tropical cyclone climatologies using IBTrACS. *Monthly Weather Review* 142 (10): 3881–3899.
24. Song, J.J., Y. Wang, and L. Wu. 2010. Trend discrepancies among three best track data sets of western North Pacific tropical cyclones. *Journal of Geophysical Research: Atmospheres* 115 (D12).
25. Lei, X. 2001. The precision analysis of the best positioning on WNP TC. *Journal of Tropical Meteorology* 17: 65–70.
26. Lander, M.A., and C. Guard. 2006. The urgent need for a re-analysis of western North Pacific tropical cyclones. *Preprints, 27th Conference on Hurricanes and Tropical Meteorology*, vol. 5, American Meteorological Society B, Monterey, CA.
27. Chin, R.T., and C.L. Yeh. 1983. Quantitative evaluation of some edge-preserving noise-smoothing techniques. *Computer Vision, Graphics, and Image Processing* 23 (1): 67–91.
28. Gonzalez, R.C., R.E. Woods, and S.L. Eddins. 2008. Digital Image Processing Using MATLAB. Publishing House of Electronics Industry.
29. Nie, C., and D.G. Long. 2008. A C-band scatterometer simultaneous wind/rain retrieval method. *IEEE Transactions on Geoscience and Remote Sensing* 46 (11): 3618–3631.
30. Xu, F., X. Li, P. Wang, J. Yang, W.G. Pichel, and Y.Q. Jin. 2015. A backscattering model of rainfall over rough sea surface for synthetic aperture radar. *IEEE Transactions on Geoscience and Remote Sensing* 53 (6): 3042–3054.

Chapter 10
Coupled Nature of Hurricane Wind and Wave Properties for Ocean Remote Sensing of Hurricane Wind Speed

Paul A. Hwang, Xiaofeng Li and Biao Zhang

Abstract Wind measurement using microwave radar suffers decreased or loss of sensitivity of the return signal in high winds. The 2D wavenumber spectra collected by airborne scanning radar altimeter in hurricane hunter missions are used to investigate the fetch- and duration-limited nature of wave growth inside hurricanes. Despite the much more complex wind forcing conditions, the dimensionless growth curves obtained with the wind wave triplets (reference wind velocity, significant wave height and dominant wave period) inside hurricanes except near the eye region are comparable to the reference similarity counterparts constructed with the wind wave triplets collected in field experiments conducted under ideal quasi-steady fetch-limited conditions. Making use of this property, the hurricane wind speed is retrievable from the dominant wave parameter (significant wave height or dominant wave period) using the fetch- or duration-limited wave growth functions. An algorithm based on such consideration is developed and applied to two hurricanes of different strengths (category 2 and 4). The retrieved wind speeds are in good agreement with the reference wind speeds from hurricane hunter measurements, and there is no indication of saturation problem in the wind retrieval using the dominant wave parameters.

P.A. Hwang (✉)
Remote Sensing Division, Naval Research Laboratory, Washington, DC, USA
e-mail: paul.hwang@nrl.navy.mil

X. Li
GST, National Oceanic and Atmospheric Administration (NOAA)/NESDIS,
College Park, MD, USA
e-mail: xiaofeng.li@noaa.gov

B. Zhang
School of Marine Sciences, Nanjing University of Information Science and Technology,
and Jiangsu Research Center for Ocean Survey and Technology, Nanjing, Jiangsu, China
e-mail: zhangbiao@nuist.edu.cn

X. Li (ed.), *Hurricane Monitoring With Spaceborne Synthetic Aperture Radar*, Springer Natural Hazards, DOI 10.1007/978-981-10-2893-9_10

10.1 Introduction

Airborne and satellite “remote sensing of surface winds” is a misnomer. What really have been measured by various microwave “wind” sensors are ocean surface waves. This is why satellite sensors produce global ocean synoptic surface vector winds but not land surface winds although humans are land dwellers and it would be useful to provide the detailed wind fields over the land habitat.

The microwave wind (wave) sensing techniques can be roughly divided into three broad categories according to the length scales of waves sensed by the instrument: (a) short scale (Bragg resonance roughness), (b) intermediate scale (tilting facets), and (c) long scale (energetic dominant waves). Category (a) includes active scatterometer, synthetic aperture radar (SAR) and passive radiometer. Category (b) includes radar altimeter and Global Navigation Satellite Systems reflectometry (GNSS-R). Category (c) includes radar altimeter, SAR, and potentially GNSS-R.

Scatterometer and radiometer are the primary spaceborne instruments providing the global ocean surface vector wind measurements. At the present, the spatial resolution is about 12.5 km. For highly variable wind fields such as those in coastal areas, mountain gap winds and hurricanes, SAR is frequently employed for its high spatial resolution, which reaches sub-km scale and represents a powerful means for resolving delicate features such as the hurricane eye structure [1] and sharp horizontal gradient in the wind field [2].

One limiting factor of the SAR or scatterometer wind sensing is the signal saturation in high winds. Several detailed airborne measurements of Ku- and C-band (13.5 and 5.3 GHz) ocean surface backscattering in hurricane hunter flights provide ample evidence of decreased sensitivity toward high winds, as well as saturation and dampening of co-polarized (VV and HH for vertical transmit vertical receive and horizontal transmit horizontal receive) normalized radar cross section (NRCS) over a wide range of incidence angle, azimuth angle and wind speed [3, 4].

Early analyses of RADARSAT-2 C-band full-polarization measurements reveal two unusual properties of the cross-polarized backscattering (VH or HV for vertical transmit horizontal receive or vice versa): (a) the wind speed sensitivity increases toward higher wind speeds, and (b) no evidence of wind speed saturation in the available datasets up to about 26 m/s [5–7]. As more cross-polarization data are reported [8, 9], VH signal saturation at wind speeds as low as 32 m/s is noticed [10, 11]. Meissner et al. (2014) [12] report the analysis result of L-band VV, HH and VH NRCS from the Aquarius satellite. The highest wind speed in the data set is close to 35 m/s. The signal saturation problem is also evident for all polarizations.

This chapter addresses two important aspects of the signal saturation issue: (a) its underlying cause from the point of view of surface wave properties, and (b) how to overcome the signal saturation problem. In Sect. 10.2, we investigate the first subject and examine several published reports on microwave radar backscattering from the ocean surface in high winds. The connection between the NRCS and ocean surface roughness is discussed in Sect. 10.3. The properties of the short and intermediate scale waves are mainly determined by the ocean surface wind stress, which is connected to the surface wind speed by a drag coefficient. Recent measurements have shown a nonmonotonic behavior of the ocean surface drag coefficient as a function of wind speed. The main cause of the NRCS saturation may be attributed to the nonmonotonic surface drag coefficient in high winds (Sect. 10.4).

In contrast to the short and intermediate scale ocean surface roughness, for which the generation by wind stress is the main driving mechanism, the dominant wave parameters (significant wave height and spectral peak wave period) depend on wind speed monotonically. This is because the nonlinear wave-wave interaction plays an important role in the dynamics of the energetic wave components near the spectral peak that contribute the lion's share to the dominant wave parameters (Sect. 10.5).

In Sect. 10.6, we present a hurricane wind retrieval algorithm using dominant wave parameters. The full set of the wind-wave triplets (U_{10}, H_s, T_p) can be calculated with the fetch- or duration-limited growth function knowing only one of three variables and accompanied with the fetch or duration information [13]. The wind retrieval algorithm is developed from the fetch-limited wave growth functions based on about one quarter of the total dataset of wind and wave measurements collected inside Hurricane Bonnie (1998) [13, 14]. Application of the algorithm to the full dataset serves as verification. The fetch-law wind retrieval algorithm is then applied to two SAR images of two hurricanes. Compared to the wind retrieved with the geophysical model function (GMF), the result shows good agreement in the azimuthal segment within about 30° of the radar pointing direction. Section 10.7 describes an update of the fetch and duration model for the hurricane wind field and Sect. 10.8 is a summary.

The materials presented in this book chapter are mainly based on the presentations given in two recent American Meteorological Society conferences: 32nd Conf. Hurricanes and Tropical Meteorol., San Juan, Puerto Rico [15], and 20th Air-Sea Interaction Conference, Madison, Wisconsin [16].

10.2 Saturation of Microwave Backscattering

As discussed in the Introduction, many field measurements of airborne NRCS have shown signal saturation in high winds [3, 4]. Figure 10.1 plots the Ku- and C-band VV and HH NRCS (σ_0) in hurricane conditions reported by Fernandez et al. (2006) [4]. These are digitized from their Figs. 10.5, 10.6, 10.7 and 10.8, showing the NRCS at 4 similar incidence angles each (between 29° and 50°) for the two frequency bands and two polarizations. For the Ku band, the nonmonotonic wind speed dependency for both polarizations is quite prominent over a wide range of incidence angles. The nonmonotonic trend for the C band is somewhat milder than the Ku band but the loss of wind sensitivity toward high winds is obvious.

Meissner et al. (2014) [12] report L band (1.4 GHz) VV, HH, and VH NRCS data from the Aquarius satellite with wind speed coverage from 0 to 35 m/s and 3 incidence angles (29.4°, 38.4° and 46.3°). The results are plotted in Fig. 10.1c. The wind speed saturation behavior is very similar for the three polarization states.

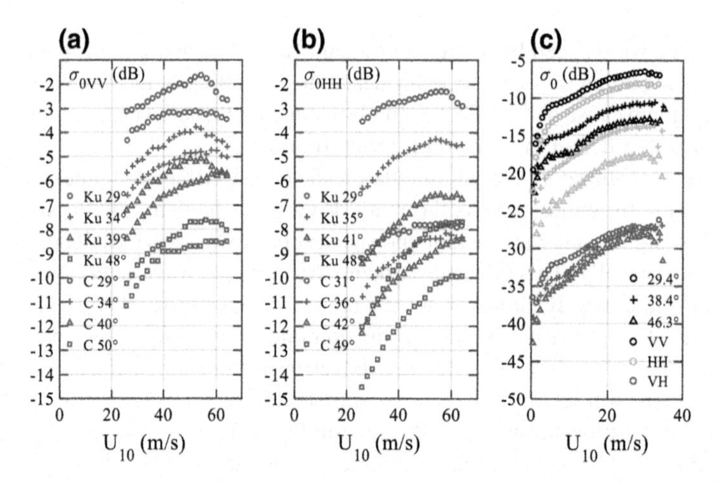

Fig. 10.1 **a** VV, and **b** HH polarization Ku- and C-band radar backscattering dependence on wind speed from airborne measurements in hurricanes, and **c** L-band VV, HH and VH radar backscattering dependence on wind speed from spaceborne measurements; **a**, **b** are digitized from Figs. 10.1, 10.2, 10.3, 10.4, 10.5, 10.6, 10.7 and 10.8 of Fernandez et al. (2006) [4], **c** is digitized from Fig. 10.1 of Meissner et al. (2014) [12]

SAR has been used as a high spatial resolution scatterometer for studying hurricanes and mountain gap winds in coastal regions [2, 5, 7–11, 17–21]. Figure 10.2 shows several examples of spaceborne RADARSAT-2 dual-polarization (VV and VH) NRCS measurements plotted against reference wind speed U_{10} [11]. The wind sources are ocean wind wave buoys, stepped frequency microwave radiometer (SFMR) and H*Wind product for the left column [9], exclusively SFMR measurements in several hurricane hunter missions for the middle column [8], and European Center for Mid-range Weather Forecast (ECMWF) numerical model output for the right column [8]. The top row shows VV and the bottom row shows VH. The trend of signal saturation is detectable in both VV and VH. The possible exception of the VH saturation in the ECMWF dataset (Fig. 10.2f) may be caused by the much coarser spatial resolution of the numerical model (the SAR resolution is degraded to match the numerical model resolution for this dataset). In general, the critical wind speed (incidence angle dependent) at which NRCS starts to saturate is lower in VV than in VH. Also, the range of incidence angles with signal saturation problem is broader in VV than in VH.

10.3 Surface Roughness Dependence on Wind Stress

Understanding the NRCS dependence on short scale waves is one of the main motivations of studying short scale waves by oceanographers. Here we examine the causes of NRCS signal saturation through the connection of the NRCS and the surface

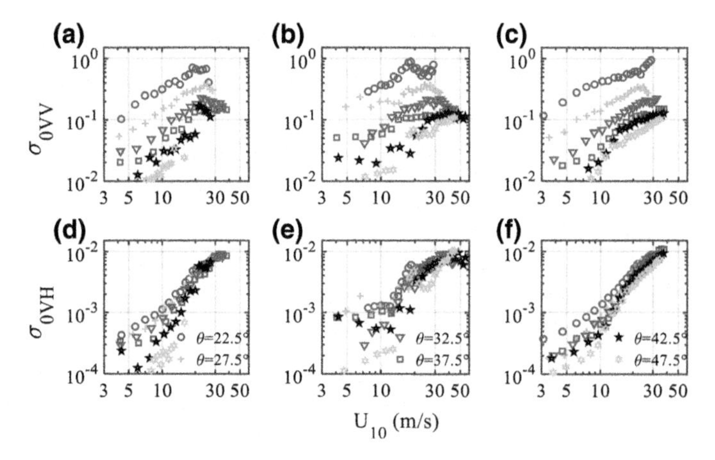

Fig. 10.2 RADARSAT-2 C-band radar backscattering dependence on wind speed: (*top row* **a–c**) VV polarization, (*bottom row* **d–f**) VH polarization; sources of wind speeds in the *left column* (**a, d**) are combined buoy, SFMR and H*Wind dataset; SFMR for the *middle column* (**b, e**); and ECMWF numerical output in the *right column* (**c, f**). Further description of these data is given in Hwang et al. (2015) [11]

roughness, which is contributed by the ocean surface waves for radar backscattering from the ocean surface. For the vertical polarization (VV), Bragg resonance is the dominant contributor of backscattering in moderate incidence angles. The VV NRCS σ_{0vv} can be expressed symbolically as [22, 23]:

$$\sigma_{0vv}(f, \theta, U_{10}) = G_{vv}(f, \theta, U_{10}) B(k_B, U_{10}) \tag{10.1}$$

where G_{vv} is the scattering coefficient, B is the dimensionless surface roughness spectrum, and k is wavenumber and subscript B indicates the Bragg resonance surface roughness component. The 1D dimensionless spectrum $B(k)$ is related to the 1D surface waves spectrum $S(k)$ by $B(k) = k^3 S(k)$. The modification of the relative permittivity from air entrained by wave breaking impacts the Fresnel reflection coefficient so the scattering coefficient is a function of f, θ and U_{10}. The net impact of the entrained air is to decrease reflection and increase transmission; thus the effects of increasing roughness and air entrainment from increasing wind are additive in enhancing passive microwave emission but counter each other for active microwave scattering. The detail is discussed by Hwang (2012) [24] and Hwang and Fois (2015) [22].

Here we focus on the ocean surface roughness. A brief review of field measurements of short waves and more extensive discussions of the connection between NRCS and surface roughness is given by Hwang et al. (2013) [25]. The efforts of surface roughness measurements in the ocean have led to the discovery of a similarity relationship expressing the dimensionless roughness spectral component $B(k)$ as a function of dimensionless wind forcing factor given by the ratio u_*/c, where u_*

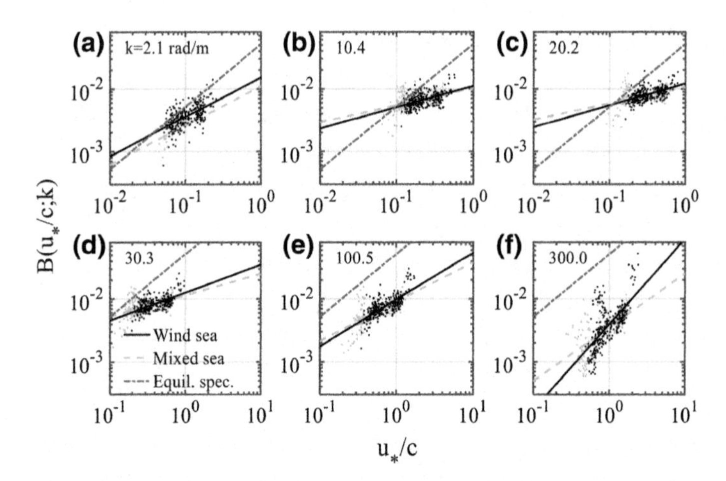

Fig. 10.3 Similarity relationship of short and intermediate scale ocean surface wave spectral components: **a** $k = 2.1$ rad/m, **b** $k = 10.4$ rad/m, **c** $k = 20.2$ rad/m, **d** $k = 30.3$ rad/m, **e** $k = 100.5$ rad/m, and **f** $k = 300.0$ rad/m. *Dark-colored dots* are wind seas and *light-colored dots* are mixed seas; the corresponding fitted curves are shown with line segments of the same colors. The *dashed-dotted line* segment (*magenta*) is the equilibrium spectrum. Further description of these data is given in Hwang and Wang (2004) [26]

is wind friction velocity and c is the phase speed of the surface roughness spectral component [26]:

$$B\left(\frac{u_*}{c}; k\right) = A(k)\left(\frac{u_*}{c}\right)^{a(k)} \tag{10.2}$$

The proportionality constant A and exponent a of the power-law function vary with the wavenumber. Figure 10.3 shows examples of scatter plots illustrating the similarity relationship for several wave components with k ranging from 2.1 to 300 rad/m for the wind-sea and mixed-sea data groups. The least squared fitting curve for each group is shown with a line segment of the same color as the data. Also shown in the figure is the equilibrium spectrum [27, 28]: $B_e(u_*/c; k) = 5.2 \times 10^{-2}(u_*/c)$ with a dashed-dotted line for each spectral component. Increasing deviation from the equilibrium spectral function toward the shorter wave component is clearly shown.

Because k and c are connected by the dispersion relationship: $c^2 = g/k + \tau k$, the roughness spectral similarity relationship is used as a parameterization function of the H roughness spectrum model [25, 29], i.e., $B(u_*/c; k) = B(k; U_{10})$. The spectral parameterization function is initially built on in-situ measurements of $A(k)$ and $a(k)$ using fast-response wave gauges mounted on a free-drifting and wave-following instrument platform. These results were obtained from several years field campaigns conducted in the Gulf of Mexico [26]. Figure 10.4a, b plot the $A(k)$ and $a(k)$ derived from the wind-sea and mixed-sea field data groups with solid and dashed curves, respectively.

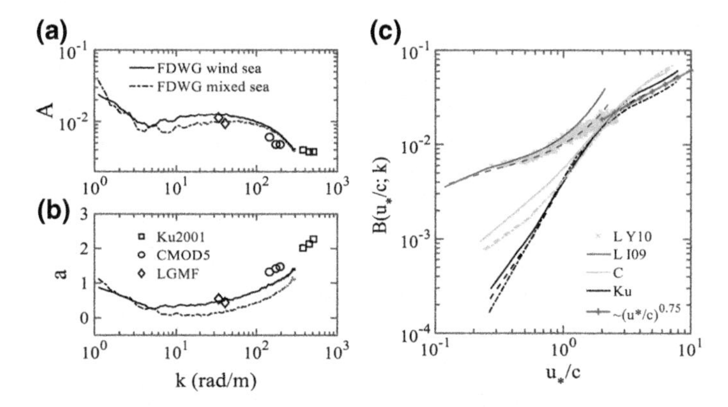

Fig. 10.4 **a** Proportionality coefficient, and **b** exponent of the power-law similarity relationship of short and intermediate scale ocean surface wave spectral components; **c** surface roughness spectral components inverted from Ku-, C- and L-band GMFs using the radar spectrometer analysis and expressed in the similarity relation function (10.2). Further description of these data is given in Hwang and Wang (2004) [26] and Hwang et al. (2013) [25]

For the vertical polarization radar backscattering, Bragg resonance plays a critical role (10.1), thus microwave radar can be treated as a spectrometer of the ocean surface roughness. Using the Ku-, C- and L-bands GMFs, Hwang et al. (2013) [25] retrieve the Bragg resonance surface roughness spectral components of the three microwave frequency bands (Fig. 10.4c). Expressed in the similarity function form (10.2), for low to moderate wind conditions ($u_*/c <\sim 3$) the results of microwave spectrometer analysis (shown with diamonds, circles and squares for Ku-, C- and L-bands in Fig. 10.4a, b) are in good agreement with in-situ free-drifting wave gauge (FDWG) data. For higher wind conditions ($u_*/c >\sim 3$), the wind speed exponent (a) becomes almost identical for all wave components (Fig. 10.4c).

The GMFs are derived from global radar data with wind speed coverage ranging from calm to hurricane wind conditions. Their analysis results compensate for the limited coverage of the in-situ data in both geographical locations and environmental conditions. Combining in-situ FDWG data and GMF radar spectrometer analysis, Hwang and Fois (2015) [22] show that NRCS computations using the H spectrum are in agreement with Ku-, C-, and L-band VV GMFs to within about +3 and −2 dB for wind speeds less than 60 m/s and incidence angles between 20° and 50°.

The results outlined in this subsection represent a critical link between microwave scattering and the ocean surface roughness. The link is particularly useful for examining the wind speed dependency of both NRCS and ocean surface roughness.

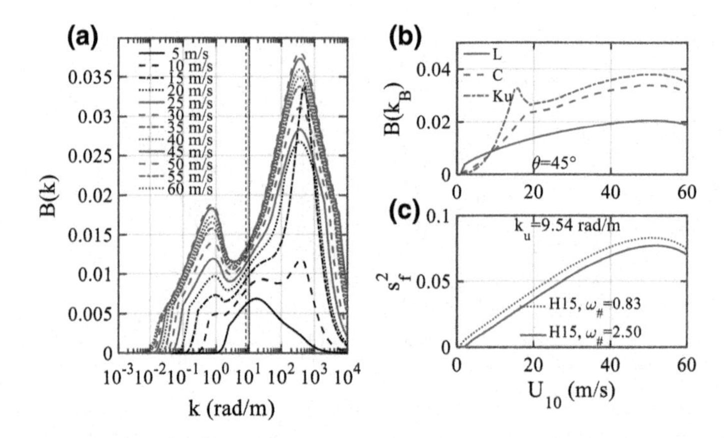

Fig. 10.5 **a** Dimensionless H spectra at $U_{10} = 5, 10, \ldots 60$ m/s, wave age is 0.4 ($\omega_{\#} = 2.5$); **b** examples of the Bragg resonance spectral components at 45° incidence angle for L-, C- and Ku-band frequencies; and **c** low-pass integrated tilting mean square slopes with upper cutoff wavenumber $k_u = 9.54$ rad/m (L band) for mature ($\omega_{\#} = 0.83$) and young ($\omega_{\#} = 2.5$) seas

10.4 Surface Roughness Saturation

Figure 10.5a shows the 1D roughness spectra for wind speeds from 5 to 60 m/s in 5 m/s steps and $\omega_{\#} = U_{10}/c_p = \omega_p U_{10}/g = 2.5$, where c and $\omega = 2\pi/T$ are the phase speed and angular frequency of a wave component, subscript p indicates the spectral peak component, T is wave period and g is gravitational acceleration. The dimensionless spectrum $B(k) = k^3 S(k)$ is presented here since the cubic k weighting emphasizes the short and intermediate scale components (large k). The relatively young wave age ($c_p/U_{10} = 1/\omega_{\#} = 0.4$) is selected for the representative conditions observed inside the broad region of the hurricane coverage except near the eye region [13] based on analyzing the 60 wave spectra collected by airborne scanning radar altimeter (SRA) reported by Wright et al. (2001) [14], further discussion of the analysis of hurricane waves is given in Sect. 10.6.

Figure 10.5b plots the wind speed dependency of the Bragg resonance spectral components $B(k_B)$ at Ku, C and L bands calculated for the 45° incidence angle. There are two notable regions where the short waves show apparent non-monotonic dependence on U_{10}: (i) between U_{10} = 15 and 20 m/s in the neighborhood of the Ku-band Bragg resonant wave components (around k = 400 rad/m), and (ii) for $U_{10} >$ 50 m/s over a broad wavenumber range. The former [(i)] is caused by the change of wind speed exponent a of the similarity Eq. (10.2) in the region $u_*/c >\sim 3$ as illustrated in Fig. 10.4c. This is an artifact resulted from approximating the gradual variation of the exponent in the neighborhood of $u_*/c = 3$ by two linear segments employed by Hwang et al. (2013) [25]. A more sophisticated representation of the u_*/c transition in Fig. 10.4c for the roughness spectrum parameterization should be able to remove this artifact.

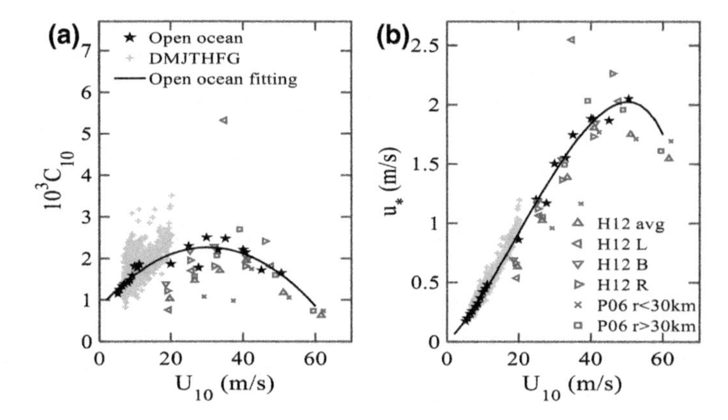

Fig. 10.6 **a** Ocean surface drag coefficient $C_{10}(U_{10})$, and **b** wind friction velocity $u_*(U_{10})$. Description of field datasets ("Open ocean" and "DMJTHFG") has been given in Hwang et al. (2013) [25]. The *black line* is fitted curve of the open ocean data. Additional measurements in hurricanes (P06: Powell (2006) [31], and H12: Holthuijsen et al. (2012) [30]) are shown with different symbols, see text for further description

The latter [(ii)] is a consequence of the non-monotonic behavior of the ocean surface wind drag coefficient and surface wind stress (proportional to the square of the wind friction velocity u_*) as a function of wind speed (Fig. 10.6). An extensive discussion of the drag coefficient and its effect on NRCS computation has been presented by Hwang et al. (2013) [25]. In Fig. 10.6, additional drag coefficient data collected inside hurricanes [30, 31] are added (labelled P06 and H12, respectively). Of special interest are the two groups of P06 data collected inside and outside the 30-km circle from the hurricane center. The drag coefficients for the inside group are considerably lower than those of the outside group. This likely reflects the swell effect reducing the surface wind stress [32–35]. In a recent analysis of the 2D wavenumber spectra collected by airborne SRA in hurricane hunter missions reported by Wright et al. (2001) [14], the characteristic wave ages are distinctly different inside and outside the 30-km circle from the hurricane center. The wave conditions inside the circle are clearly contaminated by swell (mixed seas) and those outside the circle are relatively young wind seas (Hwang 2016) [13].

The hurricane data labelled H12 in Fig. 10.6 include those based on the average of a large number (1452) of wind profiles as well as subgroups sorted into left, right and rear (back) sectors with respect to the hurricane heading. The solid line is the fitted curve using the data marked "Open ocean" as discussed in Hwang (2011) [36] and Hwang et al. (2013) [25]

$$C_{10} = 10^{-5}(-0.16U_{10}^2 + 9.67U_{10} + 80.58) \tag{10.3}$$

This drag coefficient formula captures the feature of saturation and dampening of the wind stress in high winds and is used in the H spectrum computation shown in Fig. 10.5. According to these drag coefficient observations, the wind stresses, or correspondingly wind friction velocities, above and below 50 m/s may have the same value and therefore produce similar ocean surface roughness. For example, as shown in Fig. 10.5a the roughness spectrum in the short and intermediate length scales at 60 m/s (Category-4 hurricane wind speed) is the same as that at 38 m/s (Category-1 hurricane). This may result in serious underprediction of the hurricane intensity from SAR imagery.

The tilting slopes of intermediate scale waves represent the critical surface roughness property important to microwave altimeter and GNSS-R remote sensing applications. An example of the tilting slopes integrating to upper cutoff wavenumber $ku = 9.54$ rad/m (corresponding to the tilting scale of GNSS-R L-band frequency) is shown in Fig. 10.5c for mature ($\omega_\# = 0.83$) and young ($\omega_\# = 2.50$) seas. Saturation of tilting slopes is also expected as a consequence of the non-monotonic wind stress dependence on wind speed.

It is emphasized here that the saturation of short and intermediate scale waves occurs in other wave spectral models using wind stress (with a non-monotonic dependence on wind speed) as the driving force [28, 37, 38], e.g., see the NRCS computations using the E [38] and the H spectra presented in Hwang and Fois (2015) [22].

10.5 Dominant Wave Properties

Although short and intermediate scale waves contribute significantly to the ocean surface roughness, their role in the dominant wave properties such as the significant wave height is negligible because the surface displacement spectrum decreases as a power-law function of wavenumber. For the azimuthally integrated 1D spectrum $S(k)$, the exponent of the power law is about -2.5, thus the spectral density levels at wavenumbers 2, 4, 8 and 16 times of the spectral peak wavenumber are 0.18, 0.031, 0.0055 and 0.00098 times of the spectral peak value. The contribution to the significant wave height thus decreases rapidly toward high wavenumber.

The most important factor in the similarity relationship of the surface wave spectrum in the energetic region near the spectral peak is $\omega_\# = U_{10}/c_p$ [39–43]. Consequently, for the same wave growth condition, i.e., same $\omega_\#$, the spectral peak downshifts monotonically with wind speed; the spectral similarity thus leads to the monotonic relationship between U_{10} and the dominant wave properties H_s and T_p.

Figure 10.7a shows the same set of wave spectra as those of Fig. 10.5a ($U_{10} = 5$, 10, 15, ... 60 m/s, and $\omega_\# = U_{10}/c_p = 2.5$) but the displacement spectra $S(k)$ are presented to highlight the monotonic downshifting of the energetic portion of the spectrum. For any other inverse wave age $\omega_\#$, the downshift of the spectral peak follows the dispersion relationship $k_p = g/c_p^2 = g\omega_\#^{-2}U_{10}^{-2}$, therefore also with a monotonic dependency between k_p and U_{10}.

The significant wave height integrated from the wave spectrum (labelled H15) is shown in Fig. 10.7b for $\omega_\# = 0.83$ and 2.5, illustrating the monotonic increase with wind speed; the result based on the Pierson and Moskowitz (1964) [44] fully developed spectral model (labelled PM) is also shown for comparison with the mature case. Note that for very high wind speeds (usually in hurricanes with limited fetch and finite duration over a region), the conditions for wind-wave full development rarely occur, the theoretical ~80 m wave height at 60 m/s is unlikely to happen (fortunately).

10.6 Coupled Nature of Hurricane Wind and Wave Properties and Remote Sensing Application

An extensive literature exists showing that the surface waves generated by hurricane winds can be described by the fetch-limited wave growth functions established with data collected under steady wind forcing conditions, e.g., see detailed analyses of buoy recordings within 8 times the hurricane maximum wind radius described in Young (1998, 2006) [45, 46] and Hu and Chen (2011) [47]. The robust wave growth similarity relation is also applicable to the wave fields produced by unsteady wind fields such as the rapidly increasing or decreasing mountain gap winds [48–51]. In a recent study, Hwang (2016) [13] uses the data of wind-wave triplets obtained by an SRA inside Hurricane Bonnie (1998) [14] to derive an empirical formula for the effective wind fetch and effective wind duration in the three major sectors of a hurricane: right, left and back [52]. Figure 10.8 summarizes the wind-wave similarity relations expressed as (a) fetch-limited growth, (b) duration-limited growth, and (c) a wave age similarity function. For reference, the dashed and solid curves are the first- and second-order polynomial curves fitting through the reference dataset: BHDDB,

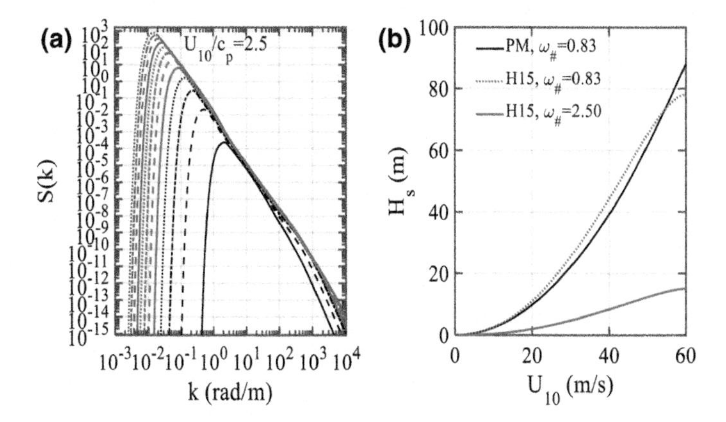

Fig. 10.7 **a** Displacement H spectra at $U_{10} = 5, 10, \ldots$ 60 m/s, wave age is 0.4 ($\omega_\# = 2.5$), the line style is the same as that of Fig. 10.5a; and **b** the significant wave height based on the H spectra for mature ($\omega_\# = 0.83$) and young ($\omega_\# = 2.5$) seas. For comparison of the mature sea condition, the result based on the fully developed PM spectrum [44] is also shown

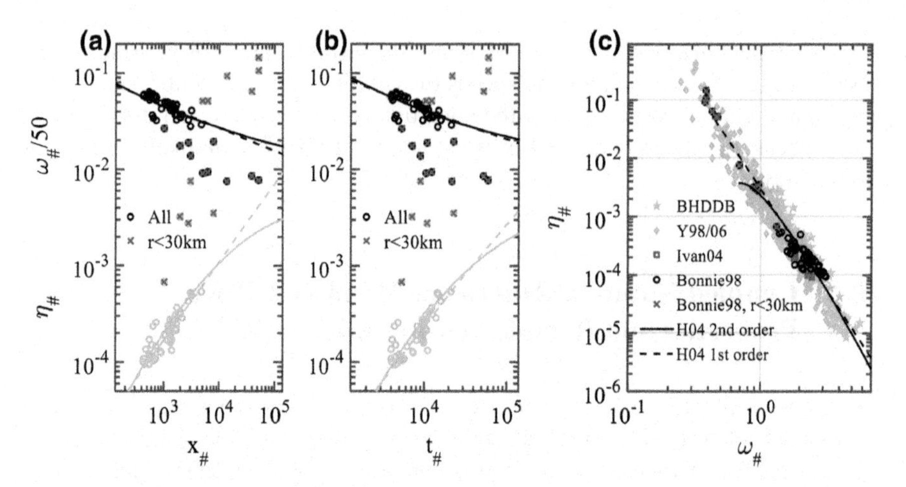

Fig. 10.8 Fetch- and duration-limited nature of wave development inside hurricanes: **a** $\omega_\#(x_\#)$ and $\eta_\#(x_\#)$, **b** $\omega_\#(t_\#)$ and $\eta_\#(t_\#)$, and **c** $\eta_\#(\omega_\#)$. Data displayed in **a**, **b** are from the 60 SRA spectra collected in Bonnie (1998) [14], the ones inside the 30 km circle from the hurricane center (marked with an x) are contaminated by swell. In **c** other hurricane data: SRA measurements from Ivan (2004) [52] and directional buoy data reported by Young (1998, 2006) [45, 46], and non-hurricane data (BHDDB) are also superimposed. The *solid* and *dashed curves* are the 2nd and 1st order fitting curves of the BHDDB data. Further descriptions of these data are given in Hwang and Wang (2004) [53] and Hwang (2016) [13]

which is an assembly of five field experiments under quasi-steady wind forcing and neutral stability conditions discussed in Hwang and Wang (2004) [26].

In the wave age range of hurricane waves that can be classified as wind seas (wave age $1/\omega_\#$ greater than about 1.2), the data can be fitted by the first- and second-order curves equally well; the first-order equations are much simpler to work with:

$$\begin{aligned} \eta_\# &= 6.17 \times 10^{-7} x_\#^{0.81};\ \omega_\# = 11.86 x_\#^{-0.24} \\ \eta_\# &= 1.27 \times 10^{-8} t_\#^{1.06};\ \omega_\# = 36.92 t_\#^{-0.31} \\ \eta_\# &= 2.94 \times 10^{-3} \omega_\#^{-3.42} \end{aligned} \tag{10.4}$$

The dimensionless fetch and duration are given by $x_\# = x_f g/U_{10}^2$, $t_\# = t_d g/U_{10}$ and the dimensionless variance is $\eta_\# = \eta_{rms}^2 g^2/U_{10}^4$, where $\eta_{rms} = H_s/4$. As illustrated in Fig. 10.8, the simultaneous measurements of wind speed, wave height and wave period obtained in hurricane hunter missions can be described equally well by three sets of similarity relations of Eq. (10.4) describing the fetch-, and duration-limited growth or wave age dependency of the wind wave energy.

The full set of the wind-wave triplets (U_{10}, H_s, T_p) can be calculated with the fetch- or duration-limited growth function knowing only one of the three variables and accompanied with the fetch or duration information. For example, expressing the fetch-limited equations in dimensional variables:

$$\frac{\eta_{rms}^2 g^2}{U_{10}^4} = 6.19 \times 10^{-7} \left(\frac{x_f g}{U_{10}^2}\right)^{0.81}$$
$$\frac{\omega_p U_{10}}{g} = 11.86 \left(\frac{x_f g}{U_{10}^2}\right)^{-0.24} \tag{10.5}$$

which provides two equations for the four unknowns ($\eta_{rms} = H_s/4$, $\omega_p = 2\pi/T_p$, U_{10} and x_f; $g = 9.8\text{m/s}^2$ is a constant), so the wind-wave triplets can be solved with any one of the three variables together with fetch x_f. Similarly, the duration-limited functions can be written out in the same fashion, and the wind-wave triplets can be solved with any one of the three variables together with duration t_d. For example, from the fetch-limited growth function, the wind speed can be calculated with the H_s and $x_{\eta x}$ input by solving the first equality in Eq. (10.5)

$$U_{10}(H_s, x) = 397.46 H_s^{0.841} x_f^{-0.341} \tag{10.6}$$

The application of the concept to the SRA dataset shows very encouraging results [13]. The agreement between the fetch- or duration-function derived wave parameters from wind speed or wind speed from wave parameters are in very good agreement with the reference SRA wave measurements and HRD wind speed, except for the region near the hurricane eye. The regression statistics (based on the data outside the 30 km circle) of the bias, slope of linear fitting curve, root mean squares (rms) difference and correlation coefficient, respectively B, s, D and R^2, are listed in Table 10.1. The correlation coefficient is greater than 0.88 for H_s and T_p from U_{10}; 0.85 for U_{10} from H_s and T_p using the fetch function, and 0.60 and 0.65 using the duration function. The quality of U_{10} retrieval using H_s is considerably better than that of using T_p (correlation coefficient of 0.85 vs. 0.60–0.64; rms difference of 2.8–3.1 m/s vs. 4.5–5.2 m/s).

Making use of these results, here we make an attempt to retrieve hurricane wind speed using the SAR-derived dominant wave properties. Hwang (2016) [13] calculated the fetch $x_{\eta x}$ and duration $t_{\omega x}$ for the cyclonic hurricane wind field (analogous to a race track) by solving Eq. (10.5) with the 60 SRA wind-wave triplets reported in Wright et al. (2001) [14]. The fetch is fitted with a linear function of the distance from the hurricane center r (both length parameters are in km in the next equation) for each of in three hurricane sectors: right, left and back respectively for the azimuth angles 0–135°, 135–225° and 225–360° referenced to the hurricane heading [the angle increases counterclockwise (CCW)]

$$x_{\eta x} = \begin{cases} -0.26r + 259.79, & \textit{right} \\ 1.25r + 58.25, & \textit{left} \\ 0.71r + 30.02, & \textit{back} \end{cases} \tag{10.7}$$

Table 10.1 Regression statistics (bias, slope of linear fitting curve, rms difference and correlation coefficient, B, s, D and R^2, respectively) of wind and wave parameters retrieved from one of the three variables in the triplets (U_{10}, H_s, T_p), combining with the empirical design fetch and duration Eq. (10.7) and the wind wave growth Eq. (10.5). Reproduced from Hwang (2016) [13]

	B (m/s)	s	D (m/s)	R^2
$T_p(U_{10}, x_{\omega x})$	−0.11	0.99	0.75	0.90
$H_s(U_{10}, x_{\eta x})$	−0.03	1.00	0.74	0.92
$T_p(U_{10}, t_{\omega t})$	−0.14	0.99	0.85	0.88
$H_s(U_{10}, x_{\eta t})$	−0.07	1.00	0.83	0.91
$U_{10}(T_p, x_{\omega x})$	0.98	1.03	5.24	0.60
$U_{10}(H_s, x_{\eta x})$	0.27	1.01	3.07	0.85
$U_{10}(T_p, x_{\omega t})$	0.88	1.02	4.49	0.64
$U_{10}(H_s, x_{\eta t})$	0.42	1.01	2.83	0.85

Subscript ηx in Eq. (10.7) indicates that the fetch is derived from fetch-limited growth function governing the wave variance, and correspondingly the significant wave height [the first equality of Eq. (10.5)]. This algorithm is established with the published 60 SRA spectra collected in Bonnie (1998) on 24–25 August (Wright et al. 2001) [14]. The full set of measurements during the mission contains 233 spectra. Here, we apply the algorithm to the full dataset for verification. The results are shown in Fig. 10.9. Panel (a) shows the scatter plot of $U_{10}(H_s, x_{\eta x})$ versus the Hurricane Research Division (HRD) reference: U_{10}(HRD). The statistics of [B (m/s), s, D (m/s), R^2] for those measurements with r $\geq$ 45 km are [−1.51, 0.96, 3.59, 0.82]. The ratio $R_U = U_{10}(H_s, x_{\eta x})/U_{10}(HRD)$ is shown in panels (b)–(d) as functions of r, ϕ and $\omega_{\#}$, respectively. Except in the region near the hurricane eye, the ratio mostly stays within 10% of the reference, that is, R_U is mostly within 1 ± 0.1, and there is no indication of signal saturation problem in high winds as that encountered in scatterometer wind retrieval; the maximum wind speed in the dataset is 46 m/s. The region with poor results occurs for r < 45 km, whereas the radius of hurricane coverage is about 250 km; so U_{10} derived using the fetch growth function is in good agreement with the HRD U_{10} over more than 95% of the hurricane coverage area. These results demonstrate the usefulness of the algorithm of hurricane wind retrieval using the dominant wave parameters.

For application to the spaceborne SAR measurements, two RADARSAT (C-band) images (one each for Bonnie at 23:20:11 UTC 25 Aug 1998 and Ivan at 09:06:19 UTC 06 Sep 2004) are used for the case study. The SAR-derived H_s uses the algorithm of Monaldo and Lyzenga (1986) [54], which has been implemented at NOAA/NESDIS for operational application. For comparison with the fetch function retrieved U_{10}, the reference wind speed is based on the CMOD5 GMF applied to the RADARSAT NRCS.

Figure 10.10 presents the result of wind retrieval using the fetch growth function. Limiting the data to within $\pm$ 30° of the radar look direction and the distance between 50 and 300 km from the hurricane center, the statistics of [B (m/s), s, D (m/s), R^2]

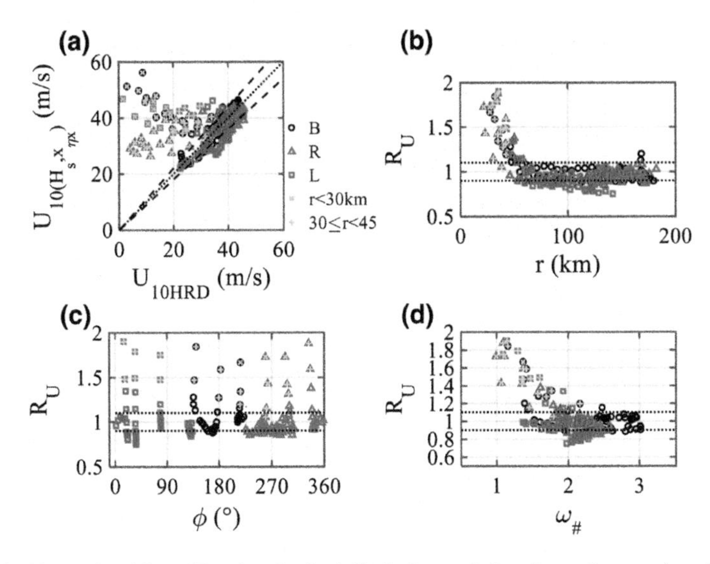

Fig. 10.9 U_{10} retrieval from H_s using the fetch-limited growth function: **a** Scatter plot of retrieved U_{10} versus HRD reference; **b** R_U as a function of r; **c** R_U as a function of ϕ; **d** R_U as a function of $\omega_\#$

are [1.13, 1.06, 2.99, 0.53] for the Bonnie image (Fig. 10.9a), and [1.01, 1.05, 2.50, 0.69] for the Ivan image (Fig. 10.9b).

10.7 Update of Hurricane Wind Field Fetch and Duration Model

Since the fetch- and duration-limited nature of wave growth inside hurricanes was first presented with the case study of Bonnie (1998) [13], the effort continues to search for a more general fetch and duration model of hurricane wind field suitable for scaling to a wider range of hurricane conditions.

The wind and wave measurements from the hurricane hunter missions provide the necessary data to calculate the effective fetches and durations for the locations where the wind and wave data are acquired [13, 55]. Explicitly, the fetch and duration are derived by rearranging the variables in the fetch- and duration-limited wave growth functions, i.e., the first two sets of equations in Eq. (10.4)

$$x_{\eta x} = 4.24 \times 10^7 U_{10}^{-2.93} H_s^{2.47}, \; x_{\omega x} = 2.29 \times 10^4 U_{10}^{-2.22} T_p^{4.22} \tag{10.8}$$

$$t_{\eta t} = 1.75 \times 10^4 U_{10}^{-2.77} H_s^{3.77}, \; t_{\omega t} = 4.81 \times 10^4 U_{10}^{-2.22} T_p^{3.22} \tag{10.9}$$

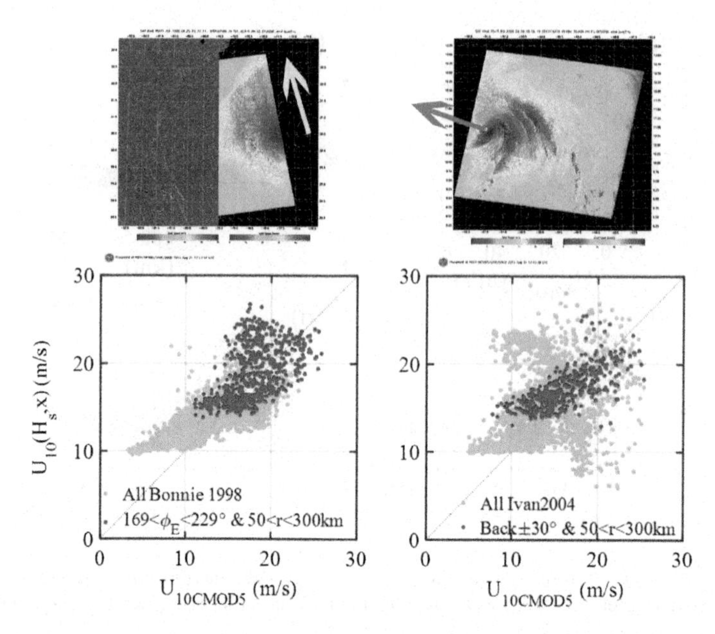

Fig. 10.10 Comparison of wind speeds retrieved by the fetch-limited growth function and the CMOD5 GMF. **a** Bonnie (1998); **b** Ivan (2004). Shown on top of each panel is the corresponding wind field derived from the RADARSAT image, the superimposed arrow shows the hurricane heading with the root of the arrow at the hurricane center

The analyses of Hwang (2016) [13] and Hwang and Walsh (2016) [55] show that x_f and t_d can be represented by linear functions of the radial distance r from the hurricane center:

$$x_{\eta x} = s_{\eta x}(\phi)r + I_{\eta x}(\phi), x_{\omega x} = s_{\omega x}(\phi)r + I_{\omega x}(\phi) \tag{10.10}$$

$$t_{\eta t} = s_{\eta t}(\phi)r + I_{\eta t}(\phi), t_{\omega t} = s_{\omega t}(\phi)r + I_{\omega t}(\phi) \tag{10.11}$$

where ϕ is the azimuth angle referenced to the hurricane heading, positive counterclockwise (CCW). The intercepts I and slopes s of the linear functions in Eqs. (10.10) and (10.11) are obtained from processing the four hurricane scenes. The fitting parameters of slope and intercept are expressed in Fourier series:

$$q = a_0 + 2\sum_{n=1}^{N}(a_{n,q}\cos n\phi + b_{n,q}\sin n\phi) \tag{10.12}$$

where q can be $s_{\eta x}$, $I_{\eta x}$, $s_{\omega x}$, $I_{\omega x}$, $s_{\eta t}$, $I_{\eta t}$, $s_{\omega t}$ or $I_{\omega t}$.

The harmonics a_{nq} and b_{nq} display a systematic quasi-linear variation with the radius of maximal wind speed r_m.

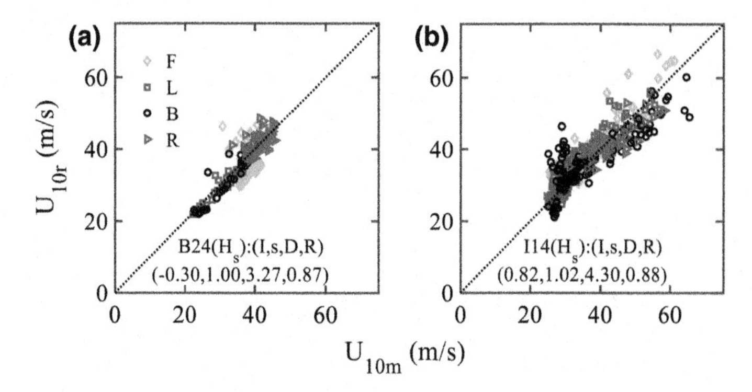

Fig. 10.11 Comparison of U_{10} retrieval from H_s using the fetch-limited growth function with the new fetch model presented in Sect. 10.7 and the HRD reference. Results from two hurricane hunter missions are presented: **a** Bonnie (1998); **b** Ivan (2004)

$$Y = p_{1Y} r_m + p_{2Y} \tag{10.13}$$

where Y represents $a_{n,q}$ and $b_{n,q}$ in Eq. (10.12). The fitting coefficients p_{1Y} and p_{2Y} are listed in Table 10.2 (the headers s_ex, I_ex, s_ox, I_ox, s_et, I_et, s_ot and I_ot represent $s_{\eta x}$, $I_{\eta x}$, $s_{\omega x}$, $I_{\omega x}$, $s_{\eta t}$, $I_{\eta t}$, $s_{\omega t}$ and $I_{\omega t}$).

Figure 10.11 shows the comparison of the wind speed retrieved using the new fetch model and the hurricane hunter measurements for Hurricane Bonnie (1998) and Ivan (2004). The highest wind speed in the dataset is 44.4 m/s for the former and 61.6 m/s for the latter. To minimize swell contamination, only those data between 45 and 200 km from the hurricane center are used. The statistics of the intercept and slope of linear regression, rms difference and correlation coefficient (I, s, D, R) are (−0.30 m/s, 1.00, 3.27 m/s, 0.87) for Bonnie (1998) and (0.82 m/s, 1.02, 4.30 m/s, 0.88) for Ivan (2004).

10.8 Summary

Field measurements and radar spectrometer analysis of ocean surface roughness indicate that the growth of short and intermediate scale roughness spectral components is controlled by the surface wind stress, which is proportional to the wind friction velocity squared and relates to wind speed squared by a drag coefficient. Recent wind profile measurements inside hurricanes show that the drag coefficient dependence on wind speed is nonmonotonic (Fig. 10.6). Consequently, u_* reaches a maximum at U_{10} about 50 m/s and then decreases as U_{10} increases further. As a result, the high wind ocean surface roughness conditions above and below 50 m/s may not be distinguishable. For example, based on the C_{10} formula established on the dropsonde data inside

Table 10.2 Linear fitting coefficients of fetch and duration harmonics as a function of radius of maximum winds $Y = p_{1Y} r_m + p_{2Y}$ Eq. (10.13), where Y represents $a_{n,q}$ and $b_{n,q}$ in Eq. (10.12); see text for further explanation

q	s_ex	I_ex	s_ox	I_ox	s_et	I_et	s_ot	I_ot
a0,q								
p1	−1.40E−02	1.95E+00	−3.04E−02	3.22E+00	−7.88E−04	1.07E−01	−1.47E−03	1.60E−01
p2	1.57E+00	−3.54E+01	3.27E+00	−1.42E+02	9.04E−02	−2.42E+00	1.58E−01	−6.73E+00
a1,q								
p1	−1.02E−02	1.26E+00	−1.34E−02	1.09E+00	−5.17E−04	5.76E−02	−6.11E−04	4.81E−02
p2	5.10E−01	−4.77E+01	9.85E−01	−5.22E+01	2.39E−02	−2.23E+00	3.99E−02	−2.18E+00
b1,q								
p1	5.21E−03	−8.23E−01	−1.27E−02	1.16E+00	1.54E−04	−2.81E−02	−5.75E−04	5.30E−02
p2	−6.87E−02	3.02E+01	1.51E+00	−1.02E+02	5.25E−03	8.52E−01	6.90E−02	−4.50E+00
a2,q								
p1	5.55E−03	−4.11E−01	1.23E−02	−1.10E+00	3.22E−04	−2.31E−02	6.14E−04	−5.28E−02
p2	−4.79E−01	2.25E+01	−1.01E+00	6.93E+01	−2.63E−02	1.27E+00	−4.91E−02	3.24E+00
b2,q								
p1	−3.32E−03	−1.52E−01	−2.00E−02	1.51E+00	−2.88E−04	5.96E−03	−9.14E−04	6.68E−02
p2	2.11E−01	8.86E+00	1.26E+00	−9.80E+01	1.58E−02	−2.16E−01	5.43E−02	−4.14E+00
a3,q								
p1	6.44E−03	−5.32E−01	1.45E−02	−9.21E−01	3.19E−04	−2.61E−02	6.23E−04	−3.91E−02
p2	−3.29E−01	2.85E+01	−8.47E−01	5.45E+01	−1.52E−02	1.31E+00	−3.41E−02	2.18E+00
b3,q								
p1	1.91E−03	−2.28E−01	−5.70E−03	1.01E+00	1.09E−04	−9.53E−03	−1.37E−04	3.62E−02
p2	−2.25E−01	2.11E+01	1.80E−01	−4.71E+01	−1.06E−02	8.72E−01	3.06E−03	−1.68E+00

hurricanes as shown in Fig. 10.6, the surface roughness spectrum (Fig. 10.5a) in the short and intermediate length scales important to microwave remote sensing at 60 m/s (Category-4 hurricane) is essentially the same as that at about 38 m/s (Category-1 hurricane). Wind retrieval methods relying on the microwave signatures reflecting the short and intermediate scale roughness properties, such as scatterometers and altimeters, may have difficulty separating the two wind speed conditions.

To avoid the ambiguity, we seek methods of hurricane wind retrieval using the dominant wave information (significant wave height and spectral peak wave period). Because nonlinear wave-wave interaction plays an important role in the evolution near the energetic spectral peak region, the continuous downshift of the spectral peak component in increasing wind results in a monotonic relationship between wind speed and significant wave height or dominant wave period.

Making use of the wind wave analyses showing that the wave fields inside hurricanes are primarily wind seas except in a small area near the hurricane center [13], an algorithm is developed for hurricane wind speed retrieval using the dominant wave information. The database for the algorithm development is the 60 SRA wave spectra collected in Category-2 Hurricane Bonnie (1998) [14]. The full set of the wind-wave triplets (U_{10}, H_s, T_p) can be calculated with the fetch- or duration-limited growth functions knowing only one of three variables and accompanied with the fetch or duration information. Examination of additional simultaneous wind and wave data from additional hurricane hunter mission further confirms the robustness of the coupled nature of wind and waves inside hurricanes consistent with those observed in the fetch- and duration-limited wave growth functions under more ideal steady wind forcing conditions [15, 16, 55].

The results are used to develop a wind retrieval algorithm to obtain hurricane wind speed using the SAR-derived H_s. Applying the algorithm to two SAR images of hurricanes, the fetch-law and GMF derived wind speeds are in good agreement in an azimuthal sector (about $\pm 30°$ wide) in the radar look direction.

References

1. Li, X., J.A. Zhang, X. Yang, W.G. Pichel, M. DeMaria, D. Long, and Z. Li. 2013. Tropical cyclone morphology from spaceborne synthetic aperture radar. *Bulletin of the American Meteorological Society* 94 (2): 215–230.
2. Monaldo, F. 2000. The Alaska SAR demonstration and near-real-time synthetic aperture radar winds. *Johns Hopkins APL Technical Digest* 21 (1): 75–79.
3. Donnelly, W.J., J.R. Carswell, R.E. McIntosh, P.S. Chang, J. Wilkerson, F. Marks, and P.G. Black. 1999. Revised ocean backscatter models at C and Ku band under high-wind conditions. *Journal of Geophysical Research* 104 (C5): 11485–11497.
4. Fernandez, D.E., J. Carswell, S. Frasier, P. Chang, P. Black, and F. Marks. 2006. Dual-polarized C-and Ku-band ocean backscatter response to hurricane-force winds. *Journal of Geophysical Research: Oceans* 111 (C8).
5. Hwang, P.A., B. Zhang, J.V. Toporkov, and W. Perrie. 2010. Comparison of composite Bragg theory and quad-polarization radar backscatter from RADARSAT-2: With applications to wave breaking and high wind retrieval. *Journal of Geophysical Research: Oceans* 115 (C8).

6. Vachon, P.W., and J. Wolfe. 2011. C-band cross-polarization wind speed retrieval. *IEEE Geoscience and Remote Sensing Letters* 8 (3): 456–459.
7. Zhang, B., W. Perrie, and Y. He. 2011. Wind speed retrieval from RADARSAT-2 quad-polarization images using a new polarization ratio model. *Journal of Geophysical Research: Oceans* 116 (C8).
8. Van Zadelhoff, G.J., A. Stoffelen, P. Vachon, J. Wolfe, J. Horstmann, and M. Belmonte Rivas. 2014. Retrieving hurricane wind speeds using cross-polarization C-band measurements. *Atmospheric Measurement Techniques* 7 (2): 437–449.
9. Zhang, B., W. Perrie, J.A. Zhang, E.W. Uhlhorn, and Y. He. 2014. High-resolution hurricane vector winds from C-band dual-polarization SAR observations. *Journal of Atmospheric and Oceanic Technology* 31 (2): 272–286.
10. Hwang, P.A., W. Perrie, and B. Zhang. 2014. Cross-polarization radar backscattering from the ocean surface and its dependence on wind velocity. *IEEE Geoscience and Remote Sensing Letters* 11 (12): 2188–2192.
11. Hwang, P.A., A. Stoffelen, G.J. Zadelhoff, W. Perrie, B. Zhang, H. Li, and H. Shen. 2015. Cross-polarization geophysical model function for C-band radar backscattering from the ocean surface and wind speed retrieval. *Journal of Geophysical Research: Oceans* 120 (2): 893–909.
12. Meissner, T., F.J. Wentz, and L. Ricciardulli. 2014. The emission and scattering of L-band microwave radiation from rough ocean surfaces and wind speed measurements from the Aquarius sensor. *Journal of Geophysical Research: Oceans* 119 (9): 6499–6522.
13. Hwang, P.A. 2016. Fetch-and duration-limited nature of surface wave growth inside tropical cyclones: With applications to air-sea exchange and remote sensing. *Journal of Physical Oceanography* 46 (1): 41–56.
14. Wright, C.W., E.J. Walsh, D. Vandemark, W.B. Krabill, A.W. Garcia, S.H. Houston, M.D. Powell, P.G. Black, and F.D. Marks. 2001. Hurricane directional wave spectrum spatial variation in the open ocean. *Journal of Physical Oceanography* 31 (8): 2472–2488.
15. Hwang, P., X. Li, B. Zhang, and E.J. Walsh. 2016. Fetch- and duration-limited nature of surface wave growth inside tropical cyclones and microwave remote sensing of hurricane wind speed using dominant wave parameters. In *AMS Hurricane Conference*.
16. Hwang, P.A., and Y. Fan. 2016. Estimating air-sea energy and momentum exchanges inside tropical cyclones using the fetch-limited wave growth properties. In *AMS Air-Sea Interaction Conference*.
17. Chunchuzov, I., P.W. Vachon, and X. Li. 2000. Analysis and modeling of atmospheric gravity waves observed in RADARSAT SAR images. *Remote Sensing of Environment* 74 (3): 343–361.
18. Li, X., C. Dong, P. Clemente-Colón, W.G. Pichel, and K.S. Friedman. 2004. Synthetic aperture radar observation of the sea surface imprints of upstream atmospheric solitons generated by flow impeded by an island. *Journal of Geophysical Research: Oceans*, 109 (C2).
19. Monaldo, F.M., C.R. Jackson, W.G. Pichel, and X. Li. 2015. A Weather Eye on Coastal Winds. *Eos Transactions American Geophysical Union* 96 (17): 16–19.
20. Monaldo, F.M., D.R. Thompson, R.C. Beal, W.G. Pichel, and P. Clemente-Colón. 2001. Comparison of SAR-derived wind speed with model predictions and ocean buoy measurements. *IEEE Transactions on Geoscience and Remote Sensing* 39 (12): 2587–2600.
21. Zhang, B., and W. Perrie. 2012. Cross-polarized synthetic aperture radar: A new potential measurement technique for hurricanes. *Bulletin of the American Meteorological Society* 93 (4): 531.
22. Hwang, P.A., and F. Fois. 2015. Surface roughness and breaking wave properties retrieved from polarimetric microwave radar backscattering. *Journal of Geophysical Research: Oceans* 120 (5): 3640–3657.
23. Valenzuela, G.R. 1978. Theories for the interaction of electromagnetic and oceanic waves-A review. *Boundary-Layer Meteorology* 13 (1–4): 61–85.
24. Hwang, P.A. 2012. Foam and roughness effects on passive microwave remote sensing of the ocean. *IEEE Transactions on Geoscience and Remote Sensing* 50 (8): 2978–2985.
25. Hwang, P.A., D.M. Burrage, D.W. Wang, and J.C. Wesson. 2013. Ocean surface roughness spectrum in high wind condition for microwave backscatter and emission computations. *Journal of Atmospheric and Oceanic Technology* 30 (9): 2168–2188.

26. Hwang, P.A., and D.W. Wang. 2004. An empirical investigation of source term balance of small scale surface waves. *Geophysical Research Letters* 31 (15).
27. Hwang, P.A., D.W. Wang, E.J. Walsh, W.B. Krabill, and R.N. Swift. 2000. Airborne measurements of the wavenumber spectra of ocean surface waves. Part I: Spectral slope and dimensionless spectral coefficient. *Journal of Physical Oceanography* 30 (11): 2753–2767.
28. Phillips, O.M. 1985. Spectral and statistical properties of the equilibrium range in wind-generated gravity waves. *Journal of Fluid Mechanics* 156 (1): 501–531.
29. Hwang, P.A. 2006. Duration-and fetch-limited growth functions of wind-generated waves parameterized with three different scaling wind velocities. *Journal of Geophysical Research: Oceans* 111 (C2).
30. Holthuijsen, L.H., M.D. Powell, and J.D. Pietrzak. 2012. Wind and waves in extreme hurricanes. *Journal of Geophysical Research: Oceans* 117 (C9).
31. Powell, M., and I. Ginis. 2006. Drag coefficient distribution and wind speed dependence in tropical cyclones. *Final Report to the National Oceanic and Atmospheric Administration (NOAA) Joint Hurricane Testbed (JHT) Program.*
32. García-Nava, H., F.J. Ocampo-Torres, P. Hwang, and P. Osuna. 2012. Reduction of wind stress due to swell at high wind conditions. *Journal of Geophysical Research: Oceans* 117 (C11).
33. Hwang, P.A., H. García-Nava, and F.J. Ocampo-Torres. 2011a. Dimensionally consistent similarity relation of ocean surface friction coefficient in mixed seas. *Journal of Physical Oceanography* 41 (6): 1227–1238.
34. Pan, J., D.W. Wang, and P.A. Hwang. 2005. A study of wave effects on wind stress over the ocean in a fetch-limited case. *Journal of Geophysical Research: Oceans* 110 (C2).
35. Potter, H. 2015. Swell and the drag coefficient. *Ocean Dynamics* 65 (3): 375–384.
36. Hwang, P.A. 2011. A note on the ocean surface roughness spectrum. *Journal of Atmospheric and Oceanic Technology* 28 (3): 436–443.
37. Donelan, M.A., and W.J. Pierson. 1987. Radar scattering and equilibrium ranges in wind-generated waves with application to scatterometry. *Journal of Geophysical Research: Oceans* 92 (C5): 4971–5029.
38. Elfouhaily, T., B. Chapron, K. Katsaros, and D. Vandemark. 1997. A unified directional spectrum for long and short wind-driven waves. *Journal of Geophysical Research: Oceans* 102 (C7): 15781–15796.
39. Donelan, M.A., J. Hamilton, and W. Hui. 1985. Directional spectra of wind-generated waves. *Philosophical Transactions of the Royal Society of London A: Mathematical, Physical and Engineering Sciences* 315 (1534): 509–562.
40. Hasselmann, K., T. Barnett, E. Bouws, H. Carlson, D. Cartwright, K. Enke, J. Ewing, H. Gienapp, D. Hasselmann, P. Kruseman, et al. 1973. Measurements of wind-wave growth and swell decay during the Joint North Sea Wave Project (JONSWAP). Technical report, Deutsches Hydrographisches Institut.
41. Janssen, P. 2004. *The interaction of ocean waves and wind*. Cambridge: Cambridge University Press.
42. Komen, G.J., L. Cavaleri, M. Donelan, K. Hasselmann, S. Hasselmann, and P. Janssen. 1994. *Dynamics and modelling of ocean waves*. Cambridge: Cambridge University Press.
43. Young, I.R. 1999. *Wind generated ocean waves*, vol. 2. Amsterdam: Elsevier.
44. Pierson, W.J., and L. Moskowitz. 1964. A proposed spectral form for fully developed wind seas based on the similarity theory of SA Kitaigorodskii. *Journal of Geophysical Research* 69 (24): 5181–5190.
45. Young, I. 1998. Observations of the spectra of hurricane generated waves. *Ocean Engineering* 25 (4): 261–276.
46. Young, I.R. 2006. Directional spectra of hurricane wind waves. *Journal of Geophysical Research: Oceans* 111 (C8).
47. Hu, K., and Q. Chen. 2011. Directional spectra of hurricane-generated waves in the Gulf of Mexico. *Geophysical Research Letters* 38 (19).
48. García-Nava, H., F. Ocampo-Torres, P. Osuna, and M. Donelan. 2009. Wind stress in the presence of swell under moderate to strong wind conditions. *Journal of Geophysical Research: Oceans* 114 (C12).

49. Hwang, P.A., H. García-Nava, and F.J. Ocampo-Torres. 2011. Observations of wind wave development in mixed seas and unsteady wind forcing. *Journal of Physical Oceanography* 41 (12): 2343–2362.
50. Ocampo-Torres, F., H. García-Nava, R. Durazo, P. Osuna, G.D. Méndez, and H.C. Graber. 2011. The INTOA Experiment: A study of ocean-atmosphere interactions under moderate to strong offshore winds and opposing swell conditions in the Gulf of Tehuantepec, Mexico. *Boundary-Layer Meteorology* 138 (3): 433–451.
51. Romero, L., and W.K. Melville. 2010. Airborne observations of fetch-limited waves in the Gulf of Tehuantepec. *Journal of Physical Oceanography* 40 (3): 441–465.
52. Black, P.G., E.A. D'Asaro, W.M. Drennan, J.R. French, et al. 2007. Air-sea exchange in hurricanes: Synthesis of observations from the coupled boundary layer air-sea transfer experiment. *Bulletin of the American Meteorological Society* 88 (3): 357–374.
53. Hwang, P.A., and D.W. Wang. 2004. Field measurements of duration-limited growth of wind-generated ocean surface waves at young stage of development. *Journal of Physical Oceanography* 34 (10): 2316–2326.
54. Monaldo, F.M., and D.R. Lyzenga. 1986. On the estimation of wave slope-and height-variance spectra from SAR imagery. *IEEE Transactions on Geoscience and Remote Sensing* 4: 543–551.
55. Hwang, P.A., and E.J. Walsh. 2016. Azimuthal and radial variation of wind-generated surface waves inside tropical cyclones. *Journal of Physical Oceanography* 46 (9): 2605–2621.

Chapter 11
Hurricane Winds Retrieval from C Band Co-pol SAR

Hui Shen, Will Perrie and Yijun He

Abstract Since the beginning of modern radar technology, the co-polarization (co-pol) mode has been predominantly used for marine navigation, target detection, ocean and atmospheric monitoring etc. In remote sensing of ocean surface winds, co-pol is operationally used in satellite scatterometer missions and some of SAR winds demonstration projects. Compared to cross-polarization modes, co-pol radar measurements have much stronger backscattered intensity and signal to noise ratio; thus, cross-pol measurements could sense atmospheric and oceanic processes which could affect small scale waves on the ocean surface. Co-pol measurements also have stronger sensitivity to wind direction than cross-pol, which made it suitable for monitoring ocean surface wind direction as well as wind speed. However, due to the different radar configuration of cross-pol SAR and the complicated dynamical processes that develop under hurricane conditions, wind monitoring under hurricane forces has formidable challenges. These include the demands for wind direction a priori, saturation of radar backscattered signals, speed ambiguity with decreased radar NRCS response under high wind speeds in low incidence angles. This chapter will summarize some recent progress in these fields. The demands for combining the advantages of both co-pol and cross-pol measurements and appropriate data assimilation methodology to ingest SAR winds into marine and hurricane numerical weather forecasting will also be discussed.

H. Shen (✉) · Y. He
Nanjing University of Informatics and Technology, Nanjing, Jiangsu, China
e-mail: Hui.Shen@dfo-mpo.gc.ca

Y. He
e-mail: yjhe@nuist.edu.cn

H. Shen
Bedford Institute of Oceanography, Dartmouth, NS, Canada

W. Perrie
Fisheries and Oceans Canada, Bedford Institute of Oceanography,
Dartmouth, NS, Canada
e-mail: William.Perrie@dfo-mpo.gc.ca

X. Li (ed.), *Hurricane Monitoring With Spaceborne Synthetic Aperture Radar*, Springer Natural Hazards, DOI 10.1007/978-981-10-2893-9_11

11.1 Introduction

From the very first spaceborne Synthetic Aperture Radar - SEASAT, storm signatures have been observed, originally from the unique structure of storm related rain footprints [1, 2]. These unique mesoscale features accompanying hurricanes have also been observed by following on SAR missions, such as rain band [3], eye morphology [4], vortex [5]. A big effort on quantitative information retrieval for ocean surface wind from SAR has been conducted in sea surface roughness, thus, after several decades of development, especially benefitted from large datasets of same microwave frequency band scatterometer that were collected, the wind algorithms of SAR wind retrieval from co-polarization mode (co-pol hereafter) are reaching operational applications for conventional sea conditions [6]. Based on the continuing achievements of wind speed retrievals under conventional wind conditions, present research efforts now ambitiously include efforts to test the potential capability of algorithms for high wind speed retrievals from SAR, such as for conditions that develop in hurricanes. These efforts are motivated by the desperate needs for increased numbers of wind observations during hurricanes, specifically over the ocean, before the storms make landfall. For conventional wind conditions, it is comparatively straightforward to apply the SAR wind algorithm to SAR hurricane images, where wind direction information is known as a priori, and wind speed is obtained through a geophysical model function. However, challenges remain even here. Firstly, the wind direction needs to be known a priori so that the wind speed can be retrieved, because the SAR radar observation technique is different compared to scatterometer methodology. Secondly, the accuracy of the geophysical model function determines the validity of the retrieved wind speed. However, current GMFs are all based on moderate wind conditions. Thus applicability of GMFs to high wind speed needs to be evaluated. Thirdly, under high wind conditions, different physical processes become dominant, such as wave breaking induced sea spray and foam. How do these processes affect SAR observations and what are the consequences for SAR wind retrieval? This chapter tries to summarize some progress on these topics.

11.2 Hurricane Wind Vector Retrieval Based on Vortex Structure

Generally, two main methods are used for SAR near-surface (usually 10 m above sea level) wind speed retrieval from co-pol SAR [7, 8]. One method is the SAR wind direction algorithm (SWDA); the other is the SAR wind algorithm (SWA). In the SWDA method, wind direction must be known a priori, and is usually inferred from wind-aligned features visible in the SAR images, numerical weather prediction (NWP) winds, in-situ measurements (e.g. buoys), or remotely sensed measurements (e.g. scatterometers). Wind speed is retrieved from a geophysical model function (GMF), which constraints of the relationship among the ocean wind speed, the normalized radar cross section (NRCS), the radar relative wind direction and the local incidence angle. In the SWA method, the near surface wind speed is estimated from the degree of azimuth cut-off of the SAR image spectrum [9]. This method requires models to describe the relationship between the spectral width of the azimuth spectrum, the ocean wave spectrum and the wind speed [10]. The SWA method requires high spatial resolution; however, the most ScanSAR mode images do not meet the high resolution required by the SWA method [11]. For the SWDA method to work, the challenge is to get precise wind directions to input into the GMF. As previously noted, directions inferred from wind-aligned features (e.g. wind streaks and roll vortices [12]) found in SAR images are often applied in studies [13]. However, not all images have visible wind-aligned features, and directions from wind-streaks/atmosphere boundary rolls may be biased relative to the real wind [14]. Moreover, other secondary atmospheric flow phenomena exist that at times mimic wind-aligned features but in reality these are not wind-aligned (e.g. gravity waves [10]). For extreme weather conditions, waves generated from the early stages of the hurricane life cycle may dominate the local waves, which can greatly bias the retrieved wind directions. Additional contamination effects include wave breaking, whitecaps and foam formed by locally strong hurricane winds [15].

Besides SWDA and SAR wind algorithm methods, He et al. (2005) [7] developed a gradient method for SAR wind retrieval. This method is based on the assumption that the wind field is quasi-uniform, allowing retrieval of wind vectors from two-neighboring sub-image blocks with slightly different incidence angles. Perrie et al. (2006) [16] extended the application of the GM formulation to the hurricane case. However, external wind direction information is still needed to find the best solution for the cost function needed to define the parameterization. Shen et al. (2006) [17, 18] modified the GM formulation, making it possible to retrieve wind pairs (speed and direction) from SAR images without external information, in certain circumstances. However, an apparent bias still exists for hurricane directions in studies of both Perrie et al. (2006) [16] and Shen et al. (2006) [17, 18].

Different from other dynamical processes, hurricanes are large, nearly circular storms that form and intensify over warm ocean waters. They are characterized by an extremely low surface air pressure center with a steep pressure gradient from the center outward, producing a series of closed concentric isobars. It is shown that this unique structure can be adapted to hurricane wind vector retrieval from SAR imagery [17, 18]. Both wind speed and direction can be retrieved directly from the NRCSs and wind direction ambiguity can be removed.

11.2.1 Methodology

Mature hurricanes typically exhibit a quasi-axisymmetric vortex in hydrostatic and rotational balance [19]. Previous studies reveal the main patterns of the hurricanes. Inside the radius of maximum wind, the core is nearly in solid-body rotation. Wind speed near the hurricane eye is almost uniform [20], and the hurricane structure is

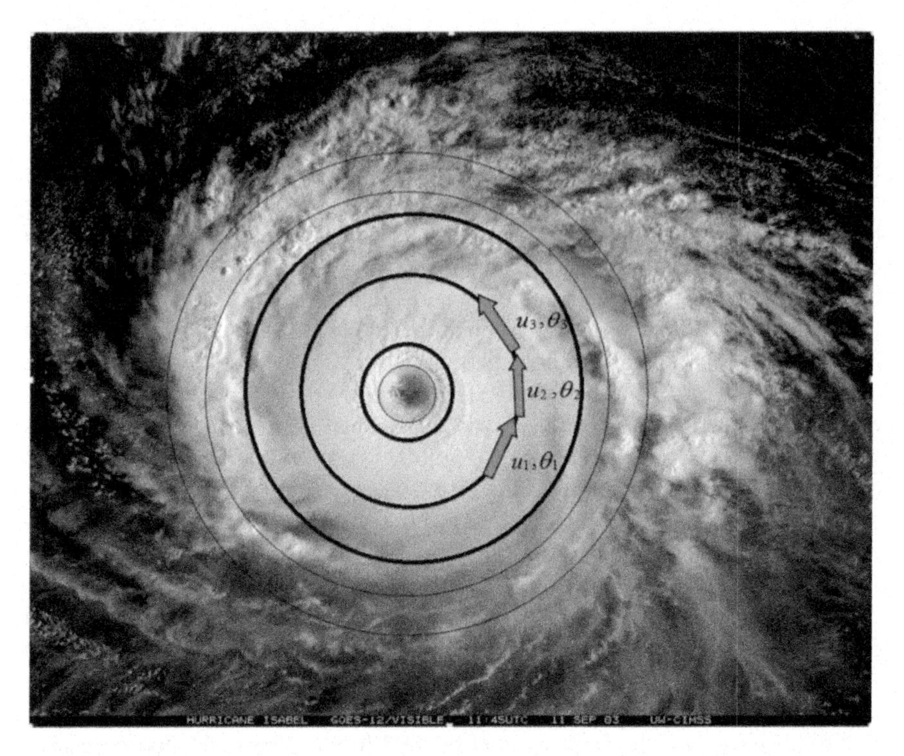

Fig. 11.1 Concentric wind patterns overlaid visible imagery of hurricane Isabel in Sep.11, 2003

circular in shape. Therefore, wind speed along any concentric circle centered on the hurricane eye, especially those near the eye, is quasi-constant and wind direction is approximately tangential and circularly varying. Thus, it is reasonable to assume wind speeds in three neighboring sub-image blocks of a given concentric circle are constant and wind directions are uniformly varying (Fig. 11.1). The relationship can be formulated as Eq. (11.1).

$$\begin{aligned} u_1 &= u_2 = u_3 \\ \theta_1 &= \theta_2 - \Delta\theta \\ \theta_2 &= \theta_3 - \Delta\theta \end{aligned} \tag{11.1}$$

where $(u_i, \theta_i)i = 1, 2, 3$ are wind speed and wind direction of the i − th sub-image block along a specific radial circle, and $\Delta\theta$ is the wind direction difference of i − th block relative to the $(i - 1)$ − th sub-image block.

Wind speed and direction are related to NRCS by the GMF, whose general form is,

$$\sigma_0 = a(\phi)u^{\gamma(\phi)}(1 + b(u, \phi)\cos\theta + c(u, \phi)\cos 2\theta) \tag{11.2}$$

where σ_0 represents NRCS, u is wind speed, ϕ is the nadir incidence angle, θ represents the relative angle between radar look direction and the wind direction, and a, b, c, γ are empirical parameters. The CMOD5 GMF [21] is presently chosen for hurricane wind analysis because of its ability to be more reliable [13] in high winds compared to previous versions of this formulation, such as CMOD4 [22], or CMOD-IFR2 [23], which have obvious tendencies to underestimate winds in high wind conditions. It should be noted that these GMFs are mainly designed for low to moderate wind speeds, and only contains limited number of data for high winds. Based on the concentric structure of hurricane wind pattern, polar coordinates are more appropriate than Cartesian coordinates. A detailed introduction on how to design the grid for a given desired resolution can be found in studies by Shen et al. (2006) [17, 18].

11.2.2 Wind Retrieval Method

For each sub-image block, wind speed and direction, NRCS and radar incidence angle are determined by CMOD5. Equations for three neighboring blocks are,

$$\sigma_{0i} = \mathrm{cmod5}(u_i, \theta_i, inc_i) \quad i = 1, 2, 3 \tag{11.3}$$

Replacing u_1, u_3 and θ_1, θ_3 in Eq. (11.3) by (11.1), the set of Eq. (11.3) is reduced to three functions with three variables of u_2, θ_2 and $\Delta\theta$.

A cost function based on a least mean square algorithm is defined to solve for the wind direction and wind speed for each sub-image blocks.

$$J = \sum_{i=1}^{3} (\sigma_{0i} - \sigma_{ci})^2 \tag{11.4}$$

where σ_{0i} is observed from the SAR data and σ_{ci} is the calculated NRCS in the $i-$th sub-image block.

If hurricane vortices are perfectly symmetric and if CMOD5 could exactly represent the relationship between the NRCS and wind speed and direction for the associated sea state conditions, wind vectors would be the sole solution of Eq. (11.4). However, in actual hurricanes, surface friction adds asymmetric effects to the hurricane vortices, which lead to noisy solutions. Under such circumstances, several solutions may merge in the domain defined by the truncation parameter of the cost function (ε, for example 1e-5). Moreover, the solution with the smallest truncation parameter is not always the best solution. Therefore, a reference parameter needs to be introduced to get the optimal solution, which is usually taken from external sources, such as *in situ* wind directions used in He et al. (2005) [7] and NWP wind directions applied by Perrie et al. (2006) [24]. We established a new method to

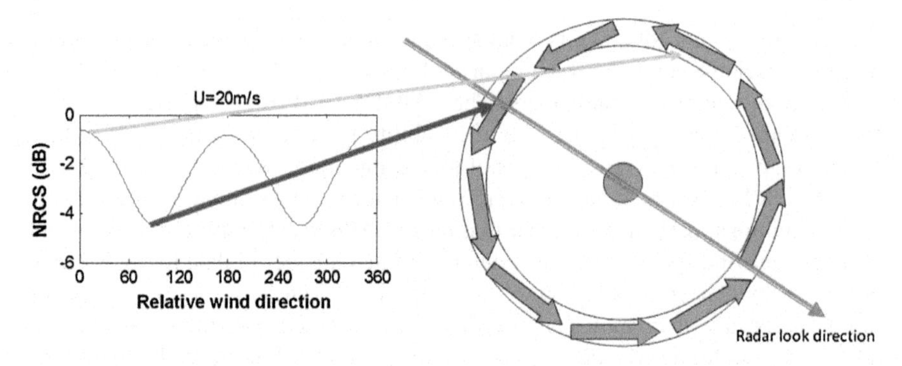

Fig. 11.2 Methodology to obtain reference wind speed for each concentric cycle centered on hurricane eye. Reference wind speed is taken as the average of wind solutions for minimum and maximum NRCS in the cycle

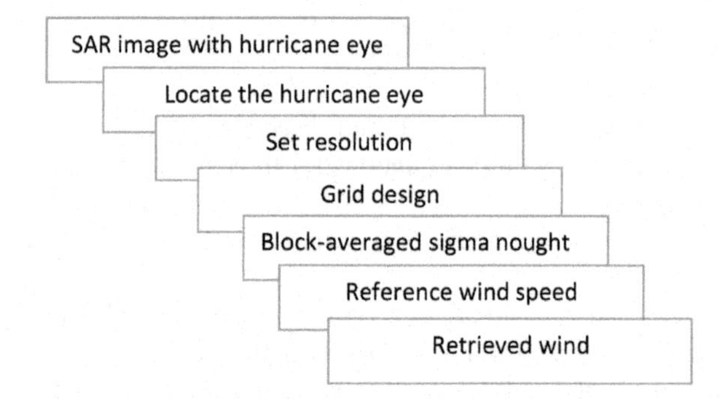

Fig. 11.3 Flow chart of wind retrieval from SAR based on circle method (CM)

determine the reference parameter, without external information [17, 18]. A schematic figure describing the concept of the reference wind speed is shown in Fig. 11.2.

This reference wind is used to find the optimal wind solution when multi-solutions exists for Eq. (11.4). Since the reference wind speed is not directly taken as the retrieved wind speed and the cost Eq. (11.4) is solved by a least-squares method, departures from the quasi-axisymmetric assumption for actual hurricanes are acceptable. As can be seen from Fig. 11.2a, some NRCS values can be linked to as many as 4 wind directions for a given wind speed. Therefore, the retrieved wind direction based on Fig. 11.3 can have multiple solutions. To remove the ambiguity of the retrieved wind direction, additional data are needed, such as island wake features visible in SAR images, and NWP estimates [10]. For hurricanes, where quasi-circular rotational wind structures are centered on their eyes, the wind direction near the ocean surface are cyclonic with an inflow angle relative to the tangential angle, which we approximate as 20° [24]. This inflow angle can be used as a reference to remove directional ambiguity. Of the four possible directional solutions to the CMOD5 GMF, the angle closest to the inward spiral angle, is chosen as the final retrieved wind direction.

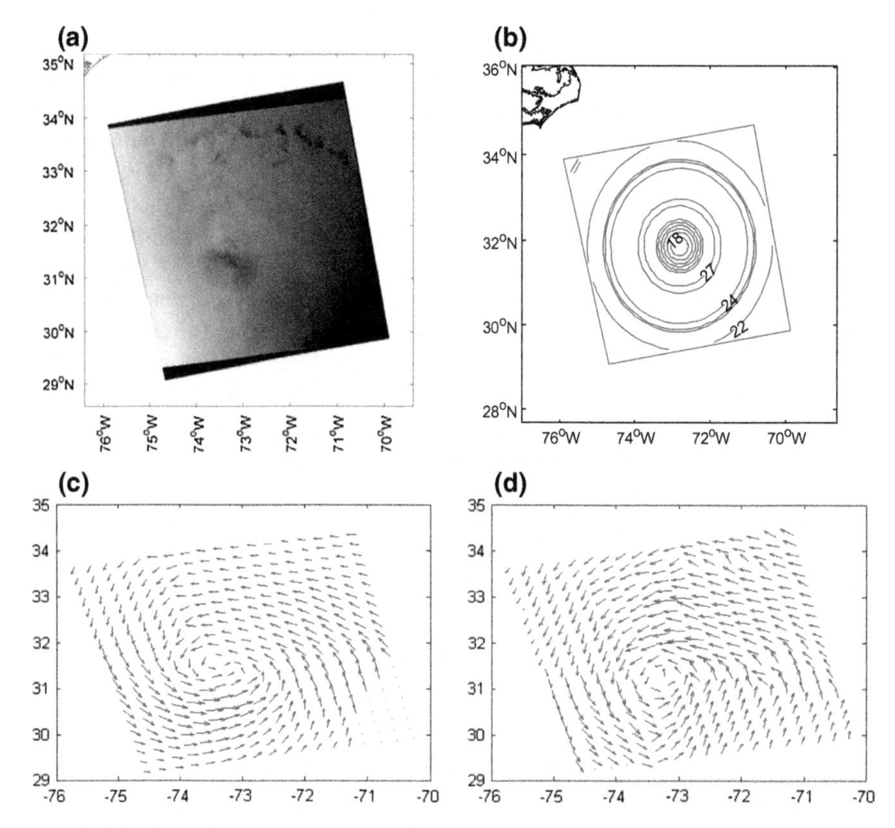

Fig. 11.4 Surface wind vectors retrieved from RADARSAT-1 SAR image of Hurricane Isabel (2003). **a** RADARSAT-1 NRCS **b** reference wind speed **c** scatterometer wind field from QuikSCAT **d** SAR wind based on circle method. **a**, **c**, **d** are from Shen et al. (2006) [17, 18]

Figure 11.4 presents an example of wind vector retrieval for Hurricane Isabel from a co-pol HH RADARSAT-1 image. Since CMOD5 is originally designed for VV polarization, a polarization ratio (e.g. Thompson 1998 [25]) is required to convert the HH NRCS to VV polarization.

11.3 Evaluation of Geophysical Model Functions for Application in High Wind Speed

In laboratory experiments, Donelan et al. (2004) [26] found that under low to moderate wind conditions, the NRCS also increases correspondingly, up to a limit, which depends on the incident angle as wind speed increases. For even stronger winds, the NRCS then starts to decrease. These same characteristics were obtained in airborne scatterometer measurements obtained from hurricanes [27]. For high winds

[25–60 m/s], the NRCS experiences dampening in both VV polarized and HH polarized data under certain radar incident angles. The effect is more severe for VV polarization. These different characteristics for NRCS under high wind imply that it is important to evaluate the validity of different GMFs for their application to high winds.

In this section, we will study four different GMFs which were intended to be useful for high wind studies, which includes: CMOD4HW [28], CMOD5 [21], HWGMF_VV and HWGMF_HH [27]. CMOD4HW is an update of CMOD4 tuned to high winds. CMOD5 was designed using high wind speed measurements. HWGMF (both HH and VV polarized GMFs) was developed with in-situ and aircraft measurements of high winds in actual hurricanes. Before HWGMF_HH was proposed, HH polarized SAR measurements were analyzed by a VV polarized GMF with a hybrid polarization ratio (PR) model (e.g. Thompson et al. 1998 [25]), used in wind retrievals from RADARSAT-1 SAR images. The difference in so-called "alpha" parameters in these parameterizations can lead to different wind results. A detailed comparison has been given by He et al. (2007) [29]. Here, GMFs are studied for their applications in both HH and VV polarized radar measurements. The hybrid polarization ratio model is applied for VV polarized GMFs in HH polarized applications, and vice versa, enabling us to test these polarization models for both VV and HH polarized measurements.

11.3.1 NRCS Dependence on High Wind Speed in Various GMFs

Figure 11.5 shows the relationship between wind speed and NRCS for winds from 25 to 80 m/s, for both HH and VV polarizations. Obvious differences exist among these GMFs. CMOD4HW generally has higher NRCS values compared with the other GMFs under the same situations. CMOD5 is intermediate between HWGMF_V and HWGMF_HH with the PR model (1) [25] applied when necessary. HWGMF_HH has the smallest NRCS. In the wind speed domain of 25–80 m/s, all GMFs exhibit the wind speed ambiguity problem, except CMOD4HW. The uppercase letter 'S' in Fig. 11.5 shows the location where the NRCS saturates, after which the radar backscattered signals start to decrease with increasing winds. Compared with CMOD5, the dampening of HWGMF_HH and HWGMF_VV occurs at higher winds, especially for winds blowing toward the radar (Fig. 11.1). Therefore, the wind speed ambiguity problem starts to occur in less severe hurricanes for CMOD5, compared to the HWGMF models. For CMOD4HW, no wind speed ambiguity problem is evident.

$$\sigma_0^H = \sigma_0^V \frac{(1+\alpha\tan^2\theta)^2}{(1+2\tan^2\theta)^2} \tag{11.5}$$

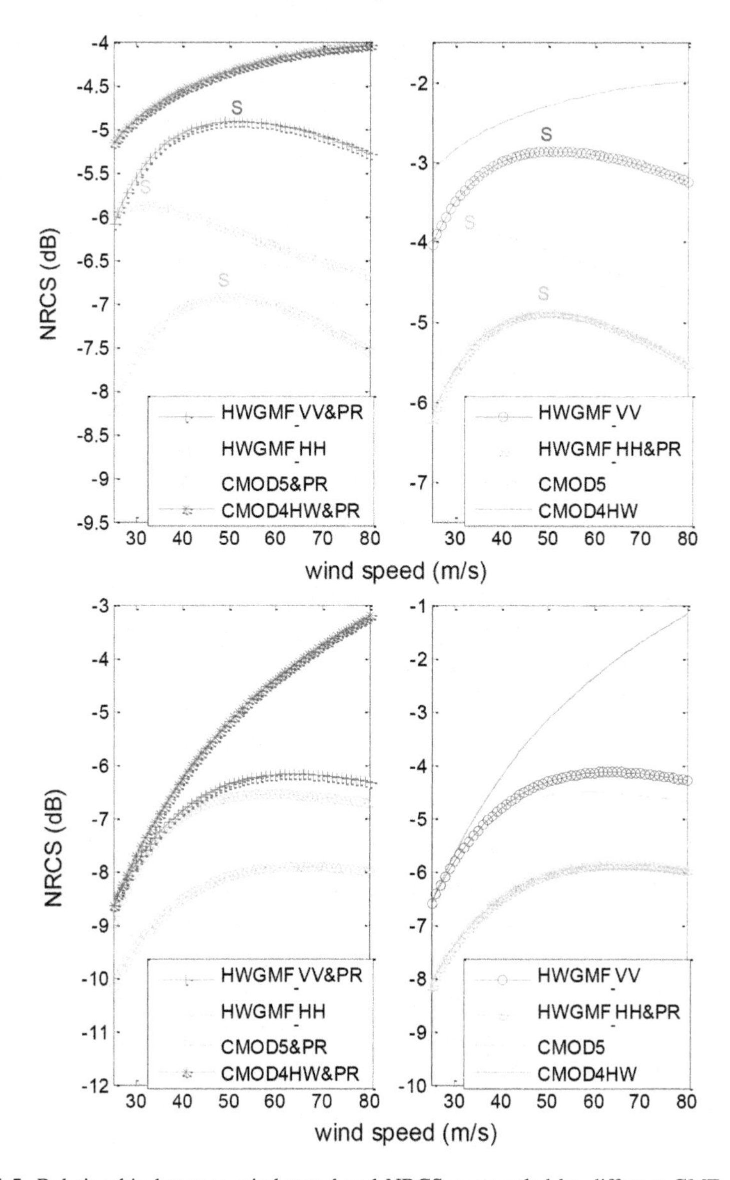

Fig. 11.5 Relationship between wind speed and NRCS as revealed by different GMFs. *Upper panels* use wind direction = 0°, and incident angle = 31°, whereas *lower panels* use wind direction = 90°, and incident angle = 31°. *Left panels* are HH polarization and *right panels* are VV polarization. S shows the saturation speed of the corresponding GMFs (from Shen et al. 2009 [30]). Used with the permission of the Canadian Aeronautics and Space Institute

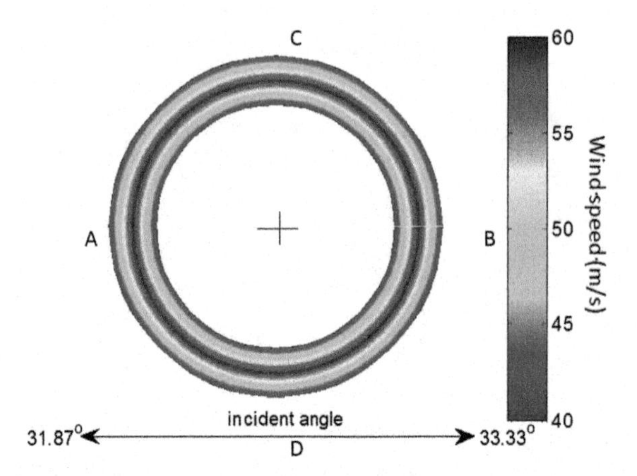

Fig. 11.6 Hypothetical symmetric hurricane wind speed field near the eyewall with related radar incident angles. Hurricane eye is indicated by +. (from Shen et al. 2009 [30]) Used with the permission of the Canadian Aeronautics and Space Institute

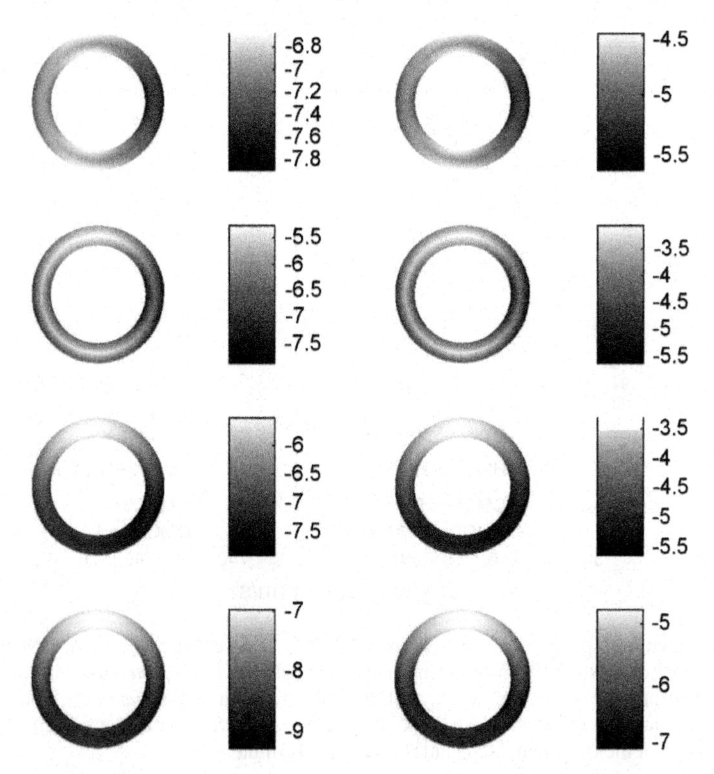

Fig. 11.7 NRCS (dB) for wind speed in Fig. 11.1 as generated by different GMFs. *Left panels* are HH polarization and *right panels* are VV polarization (PR was applied when necessary). From *top to bottom*, models are CMOD5, COMD4HW, HWGMF_VV, and HWGMF_HH (from Shen et al. 2009 [30])

where θ is the incidence angle, parameter α is zero for Bragg scattering theory, and $\alpha = 1$ for Kirchhoff scattering theory. The parameter α is subject to ongoing research, here we take $\alpha = 1$ following Vachon et al. (2000) [10].

The differences among these GMFs lead to different radar backscatter results for the same wind. Tropical hurricanes have surface wind velocity fields that have approximately circular geometry. This result often occurs even when these storms propagate to extratropical latitudes, retaining tropical characteristics. To take advantage of the circular geometry of the wind field, we consider the maximum winds near the eyewall, where the hurricane wind speed field is essentially circular. For a typical hurricane wind structure (shown in Fig. 11.6), with the position of hurricane eye indicated and quasi-constant wind speeds along concentric circles, the associated simulated radar backscattered signals are presented in Fig. 11.7. The high wind circle (60 m/s) represents the position of the hurricane eyewall. The incident angles are shown in Fig. 11.6. For simplicity, the wind direction is taken as the tangential angle of the circle. In Fig. 11.7, all the simulated radar backscattered signals show the dependence of radar incident angles and wind directions, at different directions. Among the four models, the HWGMFs show much stronger asymmetry of the radar backscattered signals for winds blowing to the radar, compared with winds blowing away from the radar. Basically, high wind speeds lead to high NRCS values, giving the CMOD4HW simulation results. However, when the wind speed ambiguity takes effect, high wind conditions may lead to low wind speed results (Fig. 11.7).

11.3.2 Concept of Speed Ambiguity for High Wind Retrieval from SAR

To demonstrate how the different speed dependence of NRCS under high winds can affect inverse procedure of wind retrieval from SAR. Wind speed solutions are sought from each corresponding simulated NRCS introduced in Sect. 11.2.1. As can be expected, for GMFs that saturate under high winds, two solutions for each individual pair of NRCS values and wind directions exist. The left panels of Fig. 11.8 give the smaller wind speed solutions for each of the GMF retrievals and the right panels are the associated larger solutions. All these solutions are different from the original 'real' wind fields in Fig. 11.6 except for the results of CMOD4HW where no wind speed ambiguity problem exists. For the other three GMFs, wind speed ambiguity makes neither the smaller solutions nor the larger acceptable as representatives of the real winds. Therefore, for the retrieval of high winds from radar measurements, wind speed ambiguities are an important problem.

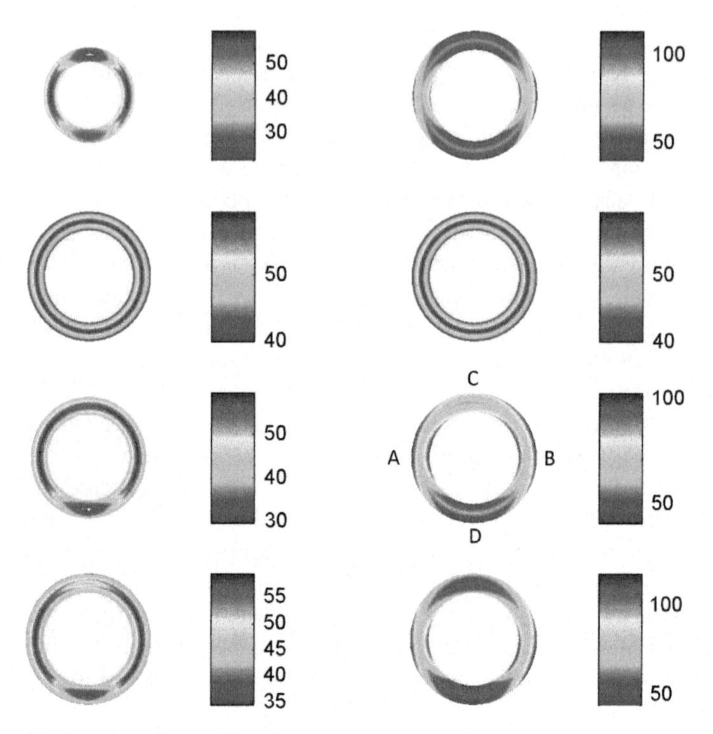

Fig. 11.8 Ambiguous wind speeds are shown, with 'smaller solutions' in *left panels* and 'larger solutions' in *right panels* (unit: m/s). These plots are retrieved from Fig. 11.3 (*right panels*) by the corresponding VV polarized GMF. From *top to bottom*, models are CMOD5, COMD4HW, HWGMF_VV, and HWGMF_HH plus application of PR. (from Shen et al. 2009, [30]) Used with the permission of the Canadian Aeronautics and Space Institute

11.4 Speed Ambiguity of Hurricane Wind Retrieval from SAR

In addition to the wind direction ambiguity problem one may commonly encounter when trying to deduce wind direction from SAR images, there is another ambiguity in SAR wind speed retrievals, as introduced in previous section. Ancillary studies show that the drag coefficient on the ocean surface decreases under high wind speed conditions (>33 m/s) [31, 32], which is a mirror for the changes of NRCS under strong winds [13]. The physical mechanism for such saturation might be due to the high wind induced wave breaking which contributes to increased sea spray and foam coverage over the sea surface. Consequently, with a priori wind directional information, two wind speed solutions exist for these ambiguity-affected radar backscattered signals. Thus, an additional under-determination for wind retrieval from SAR images is created by the ambiguity problem in wind speed retrievals.

11.4.1 Wind Speed Ambiguity Characteristics

Figure 11.9 shows the relationship of the NRCS with wind speed (at 10 m height), radar-relative wind direction and radar incident angle as suggested by CMOD5. For low-to-moderate winds, the NRCS increases with increasing wind speed for all incident angles and all radar-relative wind directions. However, as wind speed increases further, the NRCS reaches a maximum and then begins to decrease, for the near range of radar incident angles employed by space borne SARs (Fig. 11.9a, b). These wind speeds (0–80 m/s) are within the range of values observed in extreme hurricanes in nature. For very large incident angles, such as 45° (upper limit for ENVISAT ASAR measurements), wind speed ambiguity does not exist in CMOD5 (Fig. 11.9c). However, Fig. 11.9a, b show that for a large domain of incident angles of operational SAR sensors, wind speed ambiguity does exist. It therefore cannot be ignored. For the speed ambiguity affected NRCSs, there are always two possible wind speed solutions. As the radar-relative wind direction becomes increasingly close to the radar range direction, the two wind ambiguous wind solutions of a given NRCS become closer. In these cases, it is more difficult to remove the ambiguity (by experience), because both ambiguous wind speeds are realistic possible solutions. Similarly, the ambiguity problem becomes more difficult to resolve when the radar

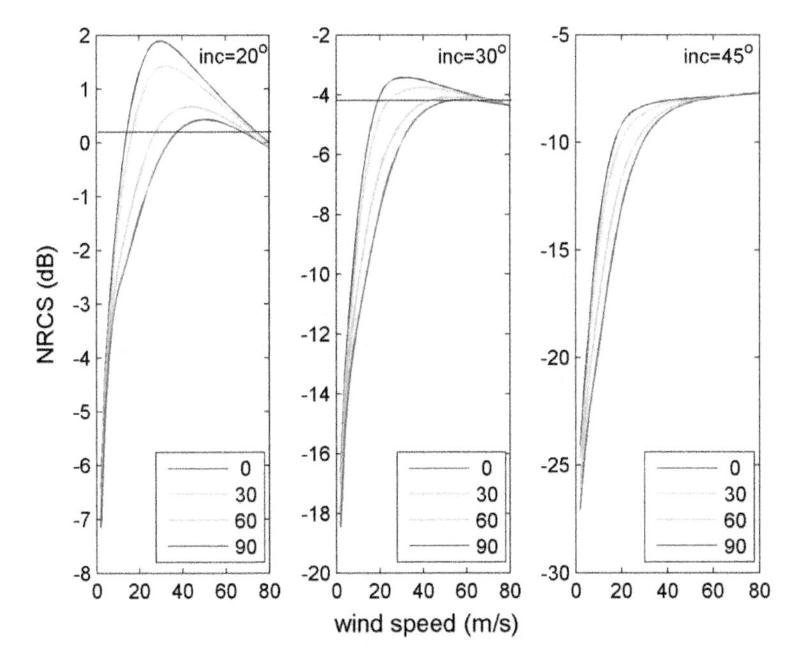

Fig. 11.9 Relationship of NRCS and wind speed revealed by CMOD5 at three incident angles and four radar-relative wind directions (as labeled in the *upper left corner* and *lower left corner* of each panel). *Solid lines* in **a** and **b** illustrate the wind speed ambiguity problem in wind retrieval, where two wind speeds are possible for a given NRCS (from Shen et al. 2009 [33])

incident angle becomes small (Fig. 11.9a, b). In fact, in the near range of SAR images, the ambiguity problem also exists in low-to-moderate wind conditions. However, the ambiguous wind speed solution is much higher than the corresponding real wind, in some cases exceeding the limit of all observed wind in nature. In such cases, the wind speed ambiguity is typically resolved by taking the lesser of each pair of possible wind solutions. However, for known moderate-to-strong wind speeds (e.g. hurricanes), such a choice may be incorrect.

Tropical cyclones and hurricanes can be very strong. According to the National Hurricane Centers (NHC) analysis, the highest ocean surface wind speed so far reached over 90 m/s in hurricane Patricia (2015). For such extreme conditions, the wind speed ambiguity problem may affect almost all of the incident angles of the SAR measurements. In cases where a hurricane is captured in the SAR image and the hurricane eye lies on the near range of the radar image, neglecting the ambiguity problem leads to severe bias in the retrieved wind speed. One example is the VV polarized ENVISAT ASAR image of Hurricane Rita acquired at 03:44 UTC on September 22, 2005.

11.4.2 Wind Speed Ambiguity Removal Method

Figure 11.10a shows the variation of U_{10} with radius for a typical hurricane, for winds in the range 0–80 m/s, as simulated by the Holland wind model (Holland 1980 [34]). To retrieve the corresponding wind speeds from radar images, an ambiguity removal method needs to account for the typical hurricane U_{10} profile. Figure 11.10b shows two possible wind solution profiles (denoted min-U and max-U in Fig. 11.2b) corresponding to upper and lower solutions for NRCS profiles of the wind profile in Fig. 11.10a, as deduced by CMOD5.

The smaller of the two wind solutions in Fig. 11.10b, min-U, increases with increasing radial distance from the hurricane eye. After reaching its first maximum at about 28 km, min-U decreases in magnitude, reaching a minimum at about 40 km from the storm centre. Thereafter, a second maximum is reached, at about 73 km radius, after which min-U decreases with further distance from the storm centre. The variation of the larger of the ambiguous wind solutions in Fig. 11.10b, max-U, is almost the reverse of the variation of min-U, having a local minima at 28 km and at about 73 km, corresponding to the locations of the maxima in the min-U profile. The threshold radii, which determine whether min-U or max-U should be taken as the final wind retrieval result, lie at intersections of min-U and max-U ($r1$ and $r2$ in Fig. 11.10a).

Therefore, for a typical hurricane wind structure where wind speed increases in the radial direction from the hurricane eye outward, reaches a maximum and decreases thereafter (Fig. 11.10a), the wind speed ambiguity problem can be removed. Firstly, all the possible wind speed solutions are retrieved based on NRCS, incident angle, and radar-relative wind direction. Secondly, the radial positions of the first two min-U maxima along the radial direction are obtained, which are also the locations of

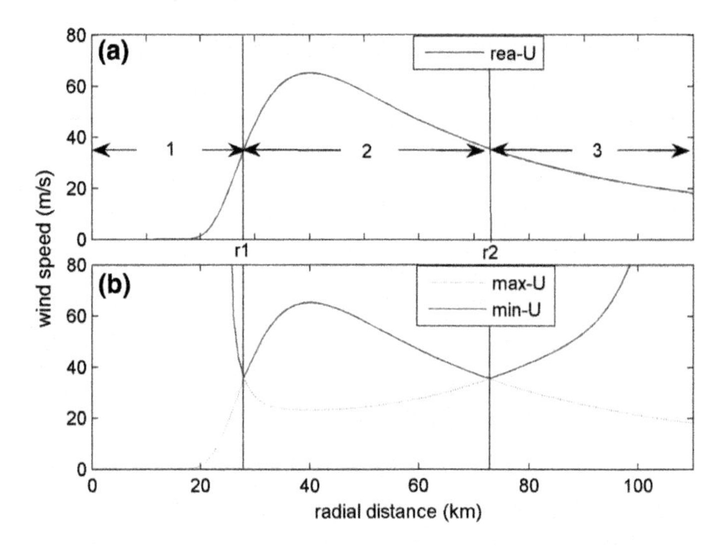

Fig. 11.10 **a** A typical hurricane wind profile simulated by the Holland model (1980) with incident angle of 35° and radar-relative wind direction of 0°; and **b** possible wind solutions deduced by CMOD5. The *vertical solid lines* in **a** show the positions ($r1$, $r2$) of the threshold wind. The numbers 1, 2, and 3 represent three domains divided by the threshold (from Shen et al. 2009 [33])

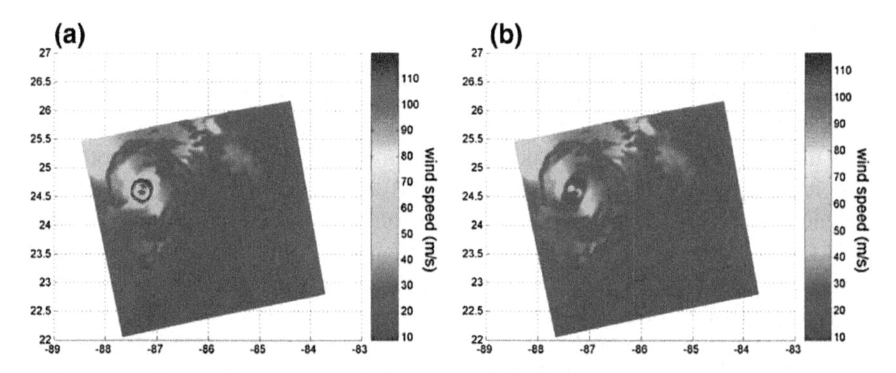

Fig. 11.11 Retrieved wind speed from the ENVISAT ASAR image of Hurricane Rita using CMOD5 and taking the tangential direction with a 20° inflow angle as the wind direction, using: **a** the new ambiguity removal method introduced; and **b** the minimum wind solution for each pair of possible ambiguous solutions (from Shen et al. 2009 [33])

the first two max-U minima. Finally, we identify the actual U_{10} as min-U, from the hurricane eye to the first threshold radius $r1$ (domain 1 in Fig. 11.10a), where min-U reaches its first maximum; between the two threshold radii (domain 2 in Fig. 11.10a), we identify U_{10} with max-U; for radii greater than the second threshold radius $r2$ (domain 3 in Fig. 11.10a), we identify U_{10} with min-U.

11.4.3 Strong Wind Speed Retrieval from SAR

Figure 11.11a, b show the retrieved wind speeds from the actual ENVISAT ASAR image of Hurricane Rita (03:44 UTC, September 22, 2005) with spatial resolution of 1 km, taking the tangential direction of the hurricane eye with an inflow angle of 20° as the wind direction (Powell et al. 1996). The contributions to differences between the SAR retrieved winds and HRD winds, which include the hurricane eyewall, are possibly due to radiometric accuracy, rain effect, error in CMOD5, and errors in NHC winds. However, these errors are additional sources of error that compound the wind speed ambiguity problem. The latter is present because of the fact that the NRCS is dampened in strong wind speed conditions and at specific incident angles.

Obvious improvements in the retrieved wind results are obtained when our wind speed ambiguity removal method is applied (Fig. 11.11a): the eyewall is now clearly exhibited. Additional possible errors occur in the retrieved eyewall winds due to heavy radiometric attenuation from rain [3, 35] or errors in the applied GMF. However, for the retrieved position of the hurricane eyewall, our results are more consistent with the HRD wind analysis than results obtained by simply taking the minimum of each pair of possible wind solutions [33]. More examples of speed ambiguity problem can be found in Shen et al. (2014, 2016) [36, 37].

11.5 Perspectives

Ocean surface wind vector data from co-pol SAR under hurricane conditions provided important potential support to high wind monitoring over the ocean. In turn these data can promote improved accuracy of hurricane forecast and enhanced information for decision making. To enhance this effort, suitable data fusion or assimilation methodology need to be developed to accommodate the unique characteristics of SAR winds, for example, very high resolution and irregular temporal interval. A variational data assimilation method has been applied for Hurricanes Gustav (2002) and Isabel (2003) to produce enhanced marine wind estimates [38]; the newly constructed mid-latitude wind fields represent an improvement compared to background wind field estimates or QSCAT/NCEP blended winds. The impact of the new winds on simulations of waves and upper ocean currents has also been studied.

Unlike scatterometer, which continuously measures the Earth, SAR works on a programming procedure, whereby targeted acquisitions need to be programed beforehand. Therefore, the general application of SAR involves a specialized pre-order procedure which may be constrained by budget factors. For hurricane application perspective, a professional output need to be constructed, such as the 'Hurricane Watch' program [39]. Future satellite missions will provide increased coverage and revisit periodicities of SAR measurement, which could provide more valuable data for hurricane wind speed retrievals.

It is worth noting that although co-pol SAR suffers from the speed ambiguity problem that hinders its application for very strong hurricane winds, cross-pol SAR has demonstrated increased dependency of NRCS on wind speed under high wind conditions, with monotonically proportional relationship between wind speed, and less dependency on wind direction and radar incidence angle, which makes cross-pol SAR suitable for hurricane wind speed retrieval [36, 37, 40]. Meanwhile, co-pol SAR measurements are still of great value for hurricane monitoring, for detailed footprints of hurricane-related air-sea interaction processes, which is less visible in cross-pol SAR images due to different imaging mechanisms.

References

1. Fu, L.L., and B. Holt. 1982. *Seasat views oceans and sea ice with synthetic aperture radar*. Pasadena: JPL Publication.
2. Atlas, D. 1994. Origin of storm footprints on the sea seen by synthetic aperture radar. *Science* 266 (25): 1364–1366.
3. Katsaros, K.B., P.W. Vachon, W.T. Liu, and P.G. Black. 2002. Microwave remote sensing of tropical cyclones from space. *Journal of Oceanography* 58 (1): 137–151.
4. Li, X., J.A. Zhang, X. Yang, W.G. Pichel, M. DeMaria, D. Long, and Z. Li. 2013. Tropical cyclone morphology from spaceborne synthetic aperture radar. *Bulletin of the American Meteorological Society* 94 (2): 215–230.
5. Sikora, T., G. Young, R. Beal, F. Monaldo, and P. Vachon. 2006. Applications of synthetic aperture radar in marine meteorology. *Atmosphere Ocean Interactions* 2: 83–105.
6. Monaldo, F.M., C.R. Jackson, W.G. Pichel, and X. Li. 2015. A Weather Eye on Coastal Winds. *Eos Transactions American Geophysical Union* 96 (17): 16–19.
7. He, Y., W. Perrie, Q. Zou, and P.W. Vachon. 2005. A new wind vector algorithm for C-band SAR. *IEEE Transactions on Geoscience and Remote Sensing* 43 (7): 1453–1458.
8. Monaldo, F., and V. Kerbaol, et al. 2004. The SAR measurement of ocean surface winds: an overview for the 2nd workshop on coastal and marine applications of SAR. In *Proceedings of the 2nd workshop on* SAR *coastal and marine applications*, pp. 8–12.
9. Kerbaol, V., B. Chapron, and P.W. Vachon. 1998. Analysis of ERS-1/2 synthetic aperture radar wave mode imagettes. *Journal of Geophysical Research: Oceans* 103 (C4): 7833–7846.
10. Vachon, P.W., and F.W. Dobson. 2000. Wind retrieval from RADARSAT SAR images: Selection of a suitable C-band HH polarization wind retrieval model. *Canadian Journal of Remote Sensing* 26 (4): 306–313.
11. Horstmann, J., W. Koch, S. Lehner, and R. Tonboe. 2000. Wind retrieval over the ocean using synthetic aperture radar with C-band HH polarization. *IEEE Transactions on Geoscience and Remote Sensing* 38 (5): 2122–2131.
12. Young, G.S., D.A. Kristovich, M.R. Hjelmfelt, and R.C. Foster. 2002. Rolls, streets, waves, and more: a review. *Bulletin of the American Meteorological Society* 83 (7): 997–1001.
13. Horstmann, J., D. Thompson, F. Monaldo, S. Iris, and H. Graber. 2005. Can synthetic aperture radars be used to estimate hurricane force winds. *Geophysical Research Letters* 32 (22).
14. Thompson, D.R., and R.C. Beal. 2000. Mapping high-resolution wind fields using synthetic aperture radar. *Johns Hopkins APL Technical Digest* 21 (1): 58–67.
15. Zhang, W., W. Perrie, and W. Li. 2006. Impacts of waves and sea spray on midlatitude storm structure and intensity. *Monthly Weather Review* 134 (9): 2418–2442.
16. Perrie, W., Y. He, and H. Shen. 2006. On determination of wind vectors for C-band SAR for high wind speeds. In *Proceedings of the SEASAR Workshop*, pp. 23–26.
17. Shen, H., W. Perrie, and Y. He. 2006. A new hurricane wind retrieval algorithm for SAR images. *Geophysical Research Letters* 33 (21).
18. Shen, H., W. Perrie, and Y. He. 2006. Progress in determination of wind vectors from SAR images. In *2006 IEEE International Symposium on Geoscience and Remote Sensing*, pp. 2228–2231. IEEE.

19. Emanuel, K.A. 1991. The theory of hurricanes. *Annual Review of Fluid Mechanics* 23 (1): 179–196.
20. Yueh, S.H., B.W. Stiles, W.Y. Tsai, H. Hu, and W.T. Liu. 2001. Quikscat geophysical model function for tropical cyclones and application to Hurricane Floyd. *IEEE Transactions on Geoscience and Remote Sensing* 39 (12): 2601–2612.
21. Hersbach, H., A. Stoffelen, and S. De Haan. 2007. An improved C-band scatterometer ocean geophysical model function: CMOD5. *Journal of Geophysical Research: Oceans* 112 (C3).
22. Stoffelen, A., and D. Anderson. 1997. Scatterometer data interpretation: Estimation and validation of the transfer function CMOD4. *Journal of Geophysical Research: Oceans* 102 (C3): 5767–5780.
23. Quilfen, Y., B. Chapron, T. Elfouhaily, K. Katsaros, and J. Tournadre. 1998. Observation of tropical cyclones by high-resolution scatterometry. *Journal of Geophysical Research: Oceans* 103 (C4): 7767–7786.
24. Powell, M.D., S.H. Houston, and T.A. Reinhold. 1996. Hurricane Andrew's landfall in south Florida. Part I: Standardizing measurements for documentation of surface wind fields. *Weather and Forecasting* 11 (3): 304–328.
25. Thompson, D., T. Elfouhaily, and B. Chapron. 1998. Polarization ratio for microwave backscattering from ocean surface at low to moderate incidence angles. In *IEEE International Proceedings on Geoscience and Remote Sensing Symposium.*
26. Donelan, M., B. Haus, N. Reul, W. Plant, M. Stiassnie, H. Graber, O. Brown, and E. Saltzman. 2004. On the limiting aerodynamic roughness of the ocean in very strong winds. *Geophysical Research Letters* 31 (18).
27. Fernandez, D.E., J. Carswell, S. Frasier, P. Chang, P. Black, and F. Marks. 2006. Dual-polarized C-and Ku-band ocean backscatter response to hurricane-force winds. *Journal of Geophysical Research: Oceans*, 111 (C8).
28. Donnelly, W.J., J.R. Carswell, R.E. McIntosh, P.S. Chang, J. Wilkerson, F. Marks, and P.G. Black. 1999. Revised ocean backscatter models at C and Ku band under high-wind conditions. *Journal of Geophysical Research* 104 (C5): 11485–11497.
29. He, Y., H. Shen, J. Guo, and W. Perrie. 2007. A comparison of models for retrieving high wind speeds. In *2007 IEEE International Geoscience and Remote Sensing Symposium*, pp. 2527–2530. IEEE.
30. Shen, H., W. Perrie, and Y. He. 2009. On SAR wind speed ambiguities and related geophysical model functions. *Canadian Journal of Remote Sensing* 35 (3): 310–319.
31. Powell, M.D., P.J. Vickery, and T.A. Reinhold. 2003. Reduced drag coefficient for high wind speeds in tropical cyclones. *Nature* 422 (6929): 279–283.
32. Jarosz, E., D.A. Mitchell, D.W. Wang, and W.J. Teague. 2007. Bottom-up determination of air-sea momentum exchange under a major tropical cyclone. *Science* 315 (5819): 1707–1709.
33. Shen, H., Y. He, and W. Perrie. 2009. Speed ambiguity in hurricane wind retrieval from SAR imagery. *International Journal of Remote Sensing* 30 (11): 2827–2836.
34. Holland, G.J. 1980. An analytic model of the wind and pressure profiles in hurricanes. *Monthly Weather Review* 108 (8): 1212–1218.
35. Zhang, G., X. Li, W. Perrie, B. Zhang, and L. Wang. 2016. Rain effects on the hurricane observations over the ocean by C-band Synthetic Aperture Radar. *Journal of Geophysical Research: Oceans* 121 (1): 14–26.
36. Shen, H., W. Perrie, Y. He, and G. Liu. 2014. Wind speed retrieval from VH dual-polarization RADARSAT-2 SAR images. *IEEE Transactions on Geoscience and Remote Sensing* 52 (9): 5820–5826.
37. Shen, H., W. Perrie, and Y. He. 2016. Evaluation of hurricane wind speed retrieval from cross-dual-pol SAR. *International Journal of Remote Sensing* 37 (3): 599–614.
38. Perrie, W., W. Zhang, M. Bourassa, H. Shen, and P.W. Vachon. 2008. Impact of satellite winds on marine wind simulations. *Weather and Forecasting* 23 (2): 290–303.
39. Michaux, M.P. 2005. CSA hurricane watch 2005. In *Canadian Space Agency.*
40. Dagestad, K.F., J. Horstmann, A. Mouche, W. Perrie, H. Shen, B. Zhang, X. Li, F. Monaldo, W. Pichel, S. Lehner, et al. 2012. Wind retrieval from synthetic aperture radar-an overview. In *4th SAR Oceanography Workshop (SEASAR 2012).*

Chapter 12
Sea-Level Pressure Retrieval from SAR Images of Tropical Cyclones

Ralph Foster

Abstract Validation and calibration of wind vector retrievals from synthetic aperture radar images of tropical cyclones remains a serious challenge. The basic wind vector measurements in tropical cyclones come from the approximately ten to twenty drop sonde profiles that are obtained during reconnaissance flights. In the highly turbulent tropical cyclone boundary layer, any given drop sonde profile represents a single realization of the virtual ensemble that must be averaged to estimate a mean surface wind vector. This is the quantity of interest because geophysical model functions are calibrated in terms of mean surface wind speeds. Even if a mean surface wind vector can be estimated from a single sonde profile, it can only be compared to the nearest wind retrieval. Furthermore, in the high wind regime geophysical model functions can be extremely sensitive to small errors in either backscatter measurements or assumed wind direction. Drop sondes do provide very accurate atmospheric pressure data, which has the beneficial property of being a scalar mean flow quantity. A method for calculating surface pressure fields from synthetic aperture radar images of tropical cyclones is presented. These fields are very accurate, with an RMS error of about 3 mb. Importantly, the surface pressure field represents an integral of the full wind vector field. Hence, comparing the pair-wise (between drop sonde splash locations) aircraft- and satellite-derived pressure differences provides a means by which the overall quality of the wind vector retrievals can be assessed. Finally, wind vector fields derived from the surface pressure fields can provide significantly improved estimates in regions of the image where objective quality flags reject the raw winds.

12.1 Introduction

There are two primary difficulties involved in retrieving surface wind vectors from SAR images of tropical cyclones (TCs) over the ocean. The first is that the co-polarization geophysical model functions (GMFs), which relate the normalized radar cross section (NRCS or backscatter) to the surface wind vector, are generally

R. Foster (✉)
Applied Physics Laboratory, University of Washington, Seattle, WA 98105-6698, USA
e-mail: ralph@apl.washington.edu

X. Li (ed.), *Hurricane Monitoring With Spaceborne Synthetic Aperture Radar*, Springer Natural Hazards, DOI 10.1007/978-981-10-2893-9_12

less reliable when the wind speed reaches hurricane force ($\sim$33 m/s). This problem is worst when the high winds occur in the low incidence angle portion of the image. A second issue is the calibration and validation of high wind retrievals using the limited set of in situ wind vector measurements in hurricane conditions. The standard methodology is restricted to analyzing a relatively small number of collocated measurements. Furthermore, extracting a mean surface wind from single measurements in turbulent flow is problematic. In this chapter we discuss how retrievals of sea-level pressure (SLP) from the SAR images can mitigate these issues.

It may at first seem strange that TC SLP patterns can be determined from high-resolution measurements of NRCS, but the physical basis is straightforward and the computation is inexpensive. The wind-induced stress field (which is directly related to the surface wind field) that is impressed on the sea surface is the net result of the downward flux of momentum from the TC interior flow above the atmospheric boundary layer. Because the pressure gradient force dominates the momentum budget, the imprint of the surface stress field can be used to estimate the surface pressure gradient field through the use of a diagnostic Tropical Cyclone Boundary Layer (TCBL) model. Given a field of surface pressure gradient vectors derived from the SAR NRCS and TCBL model, the corresponding zero-mean SLP pattern that best explains the surface pressure gradient field can be calculated using least-squares.

If pressure observations are available, the average difference between them and the zero-mean SLP field is the optimal estimate of what is effectively the integration constant resulting from converting pressure gradients into a pressure field. It is important to emphasize that, even without using ancillary data to set the absolute value of the pressure field, the bulk pressure difference between any two points in the SAR-derived pressure field is the optimal estimate of that pressure difference based on the entire field of SAR wind vectors. The premise of using pressure data for SAR wind Cal/Val is based on this fact. Furthermore, the derived SLP patterns may be used as inputs to the TCBL model to re-derive an SLP-filtered surface wind product. This product is a scene-wide surface wind retrieval from the input NRCS that ensures consistency (within the accuracy of the TCBL model) between the wind vectors and the pressure fields. Also, the dynamic constraints implicit in the TCBL model allow for SLP retrieval even when many of the highest wind retrievals fail.

In what follows we focus on C-band ($\sim$5.5 cm wavelength) co-polarization retrievals since these are presently most commonly available imagery. However, the methodology is broadly applicable to remotely-sensed ocean vector winds from a variety of sources, e.g. cross-polarization, X-band, etc. SAR images. The requirement is relatively dense observations and reasonable accuracy at least up to hurricane force winds. The two primary limitations are the applicability of the TCBL model and the ability to retrieve the highest wind speeds. (The method has been tested up to Category 3 storms.) These are both areas of active research.

12.2 High Wind Ambiguity and Wind Retrieval Confidence

The atmospheric surface layer (very roughly this is the lowest 10% of the boundary layer depth) is characterized by large speed shear and negligible turning shear. Consequently, remotely sensed ocean surface wind vectors are defined at a standard reference height of 10 m above the mean sea surface. The magnitude, U_{10}^N, is defined as the virtual wind speed in a neutrally-stratified surface layer that would produce the same surface stress, as the actual wind in the actual stratification (and ocean surface current) conditions. The surface wind stress magnitude is $\tau = \rho u_*^2 = \rho C_D^N U_{10}^{N^2}$, the air density is ρ, and, the 10 m neutral drag coefficient is C_D^N. (The effective stratification in TC conditions is close to neutral.) The neutral drag coefficient has been well-established up to near hurricane force winds [1]. At higher wind speeds, it is believed that it either holds constant or decreases with increasing wind speed [2, 3]. The wind-induced surface stress vector is in essentially the same direction as the surface wind vector.

SAR NRCS is effectively a measurement of the ocean surface roughness in the Bragg scattering regime. This cm-scale roughness is largely due to the capillary and small gravity waves on the sea surface (at least up to 30 m/s), whose generation and growth are directly related to the surface stress. The basis for vector wind retrieval from SAR and scatterometers is that the NRCS depends on the magnitude of the surface stress, the radar beam incidence angle and the relative azimuth angle between the radar beam and the surface stress vector.

Over the ocean in the range $\sim 3 < U_{10}^N < \sim 30$ m/s, and for any given wind direction, the co-polarization NRCS increases with decreasing incidence angle ($\sim 45° > \theta > \sim 20°$), which is the range of incidence angles used in SAR. For a fixed incidence angle and wind direction, the NRCS increases with U_{10}^N up to $\sim$30 m/s. For a fixed θ and U_{10}^N, the NRCS is highest for the radar beam looking into the wind (relative azimuth angle, $\Psi = 0°$) and lowest looking across the wind ($\Psi = 90°$). There is a secondary maximum for NRCS looking downwind ($\Psi = 180°$). GMFs, such as CMOD5N [4], describe the average radar backscatter from the sea surface in terms of the three parameters (U_{10}^N, θ, Ψ).

Scatterometers, which arguably produce the best global ocean vector wind data set, obtain approximately simultaneous looks at the same patch of the sea surface from multiple viewing geometries (at least fore and aft looks). The radar look angles are known, so it is possible to use a maximum likelihood method to estimate likely wind speed and direction solutions that best explain the separately measured NRCS on a pixel-by-pixel basis. The highest-ranked ranked solutions (up to four) are median filtered against forecast or analyzed surface winds to improve the maximum likelihood solution. As a consequence of achieving much higher spatial resolution than scatterometers (O(0.5–1 km) versus O(12–25 km)), SAR only measures a single NRCS measurement at each patch of the sea surface. Hence, the wind direction must be estimated by other means. Koch (2004) [5], Horstmann and Koch (2005) [6] and Wackerman et al. (2006) [7] describe methods for obtaining the wind direction from the signature of wind streaks in the NRCS. These streaks are induced by coherent

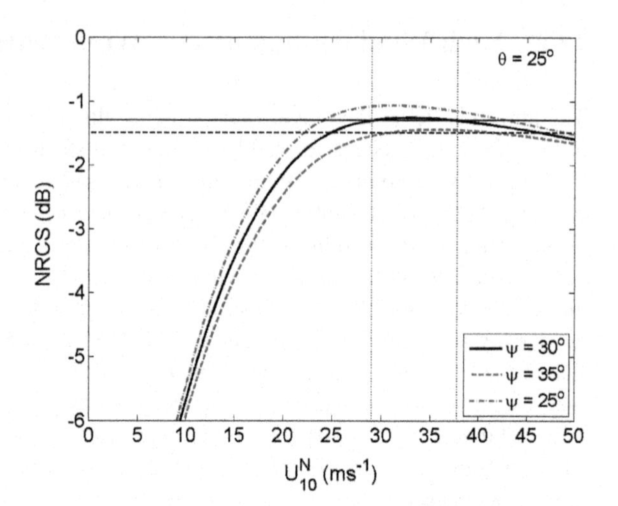

Fig. 12.1 Dependence of C-band NRCS on the surface wind speed for an incidence angle of $\theta = 25°$ and relative azimuth angle of $\Psi = 30°$ from the CMOD5N GMF

structures that form in the TCBL [8–14]. In what follows, the wind directions were obtained using a blend of the Horstmann, Koch and Wackerman wind directions and the CMOD5N GMF, which is currently the standard GMF for VV C-band images. For images obtained in HH polarization, we use the Thompson and Beal (2000) [15] polarization ratio with the parameter $\alpha = 0.8$. An important part of the initial wind vector processing is removal of image processing artifacts such as seams between any individual sub-images in the case of ScanSAR imagery [16] and scallop filtering [17] We refer to the pixel-by-pixel wind vector retrievals as the RAW winds.

When the TC eye is sampled in the high incidence part of the SAR image, the wind speed retrievals can have ambiguous high and low wind speed solutions and high sensitivity to small errors in either measured NRCS or wind direction. This is because the NRCS begins to decrease at very high wind speeds and this effect is pronounced in the much brighter low incidence portion of the SAR image (Fig. 12.1). The black line shows the CMOD5N-predicted NRCS as a function of U_{10}^{N} for a low incidence angle $\theta = 25°$ and for a relative azimuth angle $\Psi = 30°$. The maximum NRCS (-1.26 dB) occurs at 33 m/s. Assuming the SAR measures, for example, a correct $NRCS = -1.3$ dB, CMOD5N predicts a wind speed of either 29.1 or 37.8 m/s. If the SAR measures an NRCS that is 0.2 dB too high, no wind speed is found. If the measured SAR NRCS is 0.2 dB too low, the CMOD5N wind speeds are 24.8 and 46.6 m/s. Similarly, a 5° error in wind direction can lead to either no solution or the ambiguity pair 24.1 and 42.9 m/s. Horstmann et al. (2013) [16] summarized these GMF sensitivities into a set of wind retrieval quality flags, which make it convenient to objectively filter the SAR wind retrievals and include just those wind vectors for which the confidence is highest.

12.3 SLP Retrieval

The basic SLP retrieval methodology has been adapted from a similar methodology that was developed for scatterometer wind vectors in non-TC conditions [18–21]. The important difference between the generic scatterometer and SAR TC applications is that the fundamental boundary layer dynamics are quite different, which necessitates the development of a new boundary layer model. Aside from the TCBL model discussed below, the SLP retrieval method is the same as for the scatterometer. Figure 12.2 illustrates the basic steps. First the SAR image is converted into a RAW wind vector image with the attendant wind vector quality flags. The two components of each wind vector provide two input data points per pixel. However, because SLP is a scalar, the system is over-determined. Thus, we can make the best use of the wind vector quality flags to exclude (mask) any input wind vectors for which we have low confidence. For each wind vector that passes the quality control, we use the TCBL model to calculate the pressure gradient vector that is consistent with the retrieved surface stress. Least-squares optimization is used to calculate the best-fit, zero-mean SLP pattern that is consistent with the input pressure gradients. Note that the SLP field is calculated even over the data voids where the input wind vectors were rejected. If pressure observations are available, the SLP pattern is normalized by the mean difference between them and the SLP pattern.

With or without normalization, a new set of pressure gradients are calculated from the SLP pattern, which are then used as inputs to the TCBL model to derive a set of SLP-filtered surface wind vectors. In this way, we have found an overall best (within the accuracy limits of the TCBL model) scene-wide wind vector retrieval from the original SAR image. Importantly, SLP-filtered wind vectors are found for the SAR pixels whose wind vectors were rejected by the retrieval confidence flags.

12.4 TCBL Model

The primary characteristic of atmospheric boundary layer flow is that turbulent fluxes (the ensemble variances and covariances of the fluctuating velocities, temperature and humidity) are leading order contributions. The classic analysis arises from adding a small perturbation to an assumed mean flow, taking an ensemble average and performing a scale analysis to determine which terms dominate for the given conditions. The turbulent momentum fluxes (Reynolds stresses) arise from the nonlinear advection terms in the total momentum equations. For standard atmospheric boundary layers, in which the relevant horizontal length scale is much larger than the relevant vertical length scale, the nonlinear terms that include only the mean flow are found to be relatively small and can be dropped. However, the TCBL mean flow has strong radial shear and is highly curved (often nearly circular). Consequently, scale analysis shows that the equivalent nonlinear mean flow terms are of leading order and must be retained (e.g. [12, 22]).

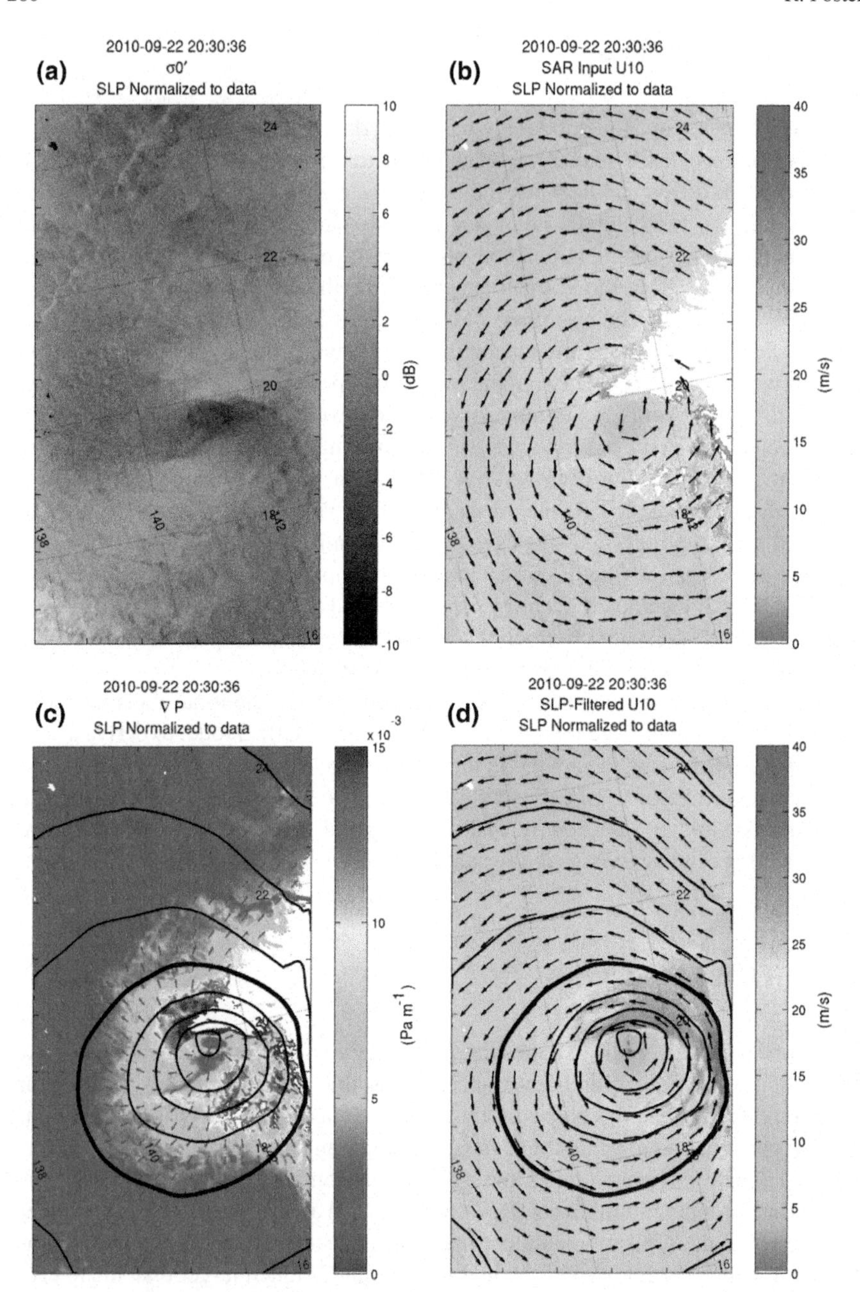

Fig. 12.2 Basic steps in retrieving sea-level pressure from a SAR image of a tropical cyclone. The example shows Typhoon Malakas, 22 Sep 2010, 20:30 UTC. **a** SAR backscatter (with mean incidence angle NRCS decrease with incidence angle subtracted). **b** RAW surface wind speeds with low confidence winds masked. Wind directions are shown every 40 km. **c** Surface pressure gradient vectors calculated from RAW winds with SLP contours superposed. **d** SLP-filtered surface wind vectors calculated from the SLP pattern shown in **c**

Mean flow solutions for a turbulent boundary layer may be found if the Reynolds stresses can be parameterized in terms of the mean flow variables. The scatterometer SLP retrieval PBL model is an analytic two-layer model that matches a surface layer similarity model to standard parameterizations in the outer PBL [18–21, 23]. The solution matching provides a simple turbulence closure based on the neutral drag coefficient. Outside of the tropics negligible PBL-top momentum entrainment is assumed. In the tropics, scale analysis shows that momentum entrainment at the top of the PBL is of leading order [24] and we employ a simple entrainment model in the outer layer [18]. In either case, the nonlinear mean flow terms are not included. Both the inner surface layer and the outer PBL models include the effects of stratification and baroclinic shear. When necessary, the contribution of boundary layer organized large eddies are included. We have demonstrated that the SLP retrievals are very accurate and that they correctly resolve sharper frontal zones, stronger gradients and deeper lows than manual forecasts [19, 25].

Foster (2009) [22] describes a flexible diagnostic similarity boundary layer model that is scaled for use in the TCBL. The inner layer follows Monin–Obukhov similarity and the outer layer is Ekman-like, but the dynamics are scaled for the strong radial shear and flow curvature. The key aspects of the model are: a neutral drag coefficient that matches the COARE model [1] up to hurricane force winds and is constant for higher wind speeds; dynamic shallowing of the TCBL depth as the storm center is approached; nonlinear mean flow accelerations; vertical advection of horizontal turbulent momentum flux; and, it allows a wide range of turbulence closures that smoothly match into the surface layer model. At present, this level of complexity is too computationally intensive for SLP retrieval applications. As a compromise, we adapted the scatterometer PBL model to include all of the above except for the vertical advection terms.

In the mid-latitudes, the inverse of the Coriolis parameter, f, sets an adjustment timescale and the geostrophic wind speed sets a characteristic velocity scale. The Coriolis parameter can be combined with a characteristic value of the eddy viscosity, K, to define the so-called Ekman frictional depth, $\delta_E = \sqrt{2K/f}$ as the characteristic vertical length scale. We seek analogs for these scalings that are valid for the TCBL.

The dominant nonlinear mean flow contribution in the TCBL is related to the centrifugal force and it nearly balances the strong radial pressure gradients. This fact is used to define the gradient wind speed, which is used as the characteristic velocity scale in the TCBL: $\frac{1}{\rho}\frac{\partial P}{\partial n} = \frac{V_g^2}{R} + fV_g$, in which P is the pressure, n is the direction normal to the isobars, and, R is the local radius of curvature in the flow. Curved flow is inertially stable only if the angular momentum increases with radius, so that radial displacements from an equilibrium mean state induce restoring forces. Foster (2009) [22] followed Eliassen and Lystads (1977) [26] assumption that the inverse of the inertial stability of the TC vortex, $I^2 = (f + \frac{2V_g}{r})(f + \frac{V_g}{r} + \frac{\partial V_g}{\partial r})$, establishes an adjustment timescale that is equivalent to the inverse of the Coriolis parameter. The inertial stability is a measure of the stiffness of the vortex to radial flow and may be derived from Rayleighs criterion [Chandrasekhar, 1961]. Here $V_g(r)$ is the radial profile of the gradient wind speed as a function of distance from the TC center. Within the TC core, $I \gg f$, while in the limit of large radius, $I \to f$. So, we define the characteristic vertical length scale as $\delta = \sqrt{2K/I}$, which results in a much shallower boundary layer near the TC center that asymptotes smoothly toward the ambient vertical length scale, δ_E, far from the storm center. To calculate the gradient wind, we found that the radius of curvature could be parameterized as $R = 0.65r$. These quantities are found iteratively as part of the inner and outer TCBL matching. The effects of the nonlinear mean flow accelerations are parameterized by iteratively applying a storm-relative gradient wind correction to the pressure gradient calculation. (The propagation speed of the tropical cyclone is subtracted from SAR wind vectors in the TC inner core before the correction is applied.) It is possible to extract the storm propagation vector from the SAR wind vectors through a limited domain kinematic flow partition [27], although external sources may also be used.

12.5 Sea-Level Pressure Validation

Demonstrating the accuracy of the SLP retrievals requires SAR image acquisitions that are close in time to in situ data collection, which mostly come from aircraft reconnaissance missions. The relevant data provided by the aircraft are drop sondes, Stepped Frequency Microwave Radiometer (SFMR) surface wind speeds [28], and, the flight level meteorological measurements that can be used to estimate the surface pressure along the flight track. It is important to compensate for the fact that SAR TC images are essentially snapshots in time whereas aircraft missions require about 6 hours for a complete storm survey. During the flight, the storm will propagate and may even change intensity. Thus, one of the major challenges in validating SLP retrievals is an accurate storm track that can be used to map the observations into the storm-relative coordinate system at the time of the SAR overpass.

We focus on two primary data sets. The first is the RadarSAT-1 Hurricane Watch image set provided under an Announcement of Opportunity by the Canadian Space Agency and NOAA. We use four matchups between the Hurricane watch images and reconnaissance flights. These are: Hurricanes: Lili (30 Sep, 2002); Katrina (27 Aug, 2005); Helene (20 Sep, 2006) and Ike (13 Sep, 2008). The second data set came from the Office of Naval Research-sponsored Impact of Typhoons on the Pacific (ITOP) field program in 2010 [29]. One of the major goals of ITOP was to provide in situ aircraft observations for calibration and validation of SAR retrievals of surface winds, waves and sea-level pressure. ITOP provided one very close matchup with a SAR image of Typhoon Malakas.

The SAR image of Typhoon Malakas was acquired on September 22, 2010 at 20:30 UTC during ITOP (Fig. 12.2a). An Air Force C-130J aircraft surveyed the region of the storm that was captured in the SAR image for around five hours starting 15 min before the image was acquired. During the flight, 28 drop sondes successfully splashed at locations that, accounting for storm motion, landed within the SAR scene.

An enlargement of the SLP retrieved in the inner core is shown in Fig. 12.3a. The storm track-adjusted drop sonde splash locations are shown as filled boxes using the

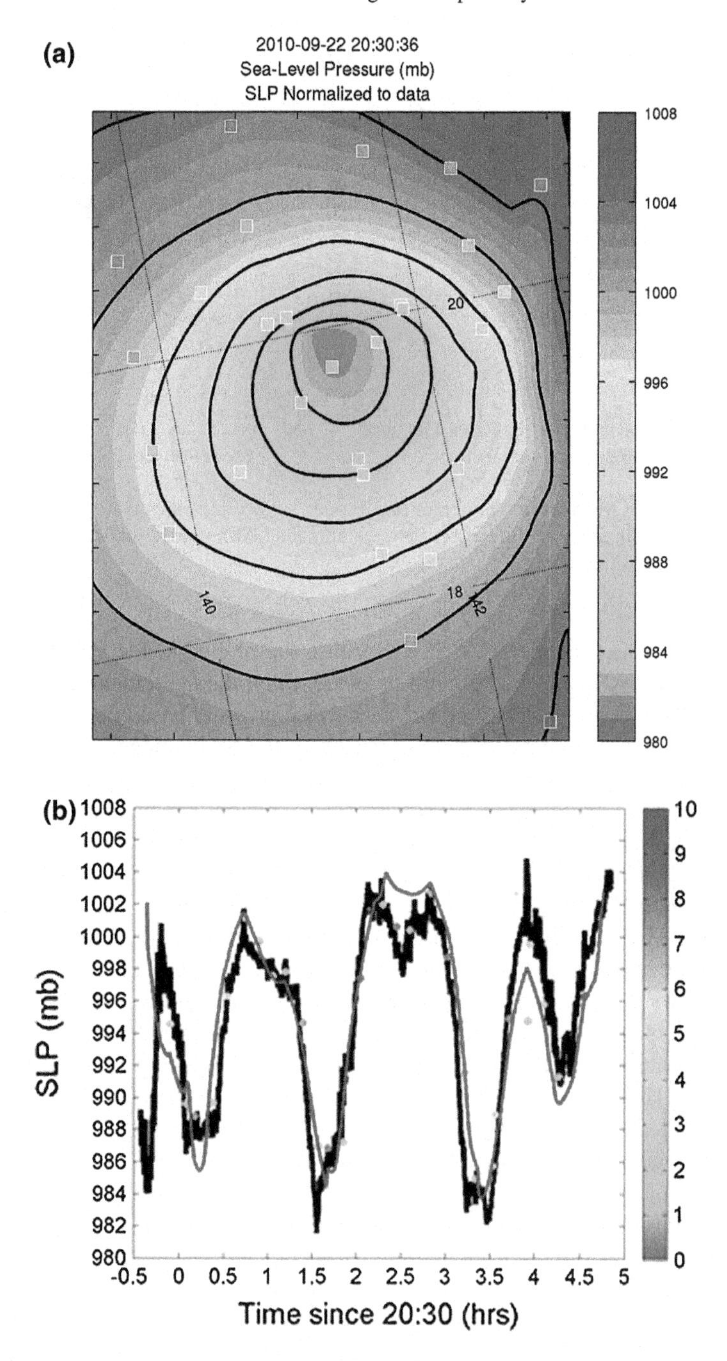

Fig. 12.3 Surface pressures for Typhoon Malakas. **a** Comparison of SAR-derived SLP with drop sonde surface pressures with splash locations adjusted to account for storm propagation. **b** *Black line* Surface pressure estimated from aircraft flight level meteorological data; *Red Line* Surface pressure derived from SAR image

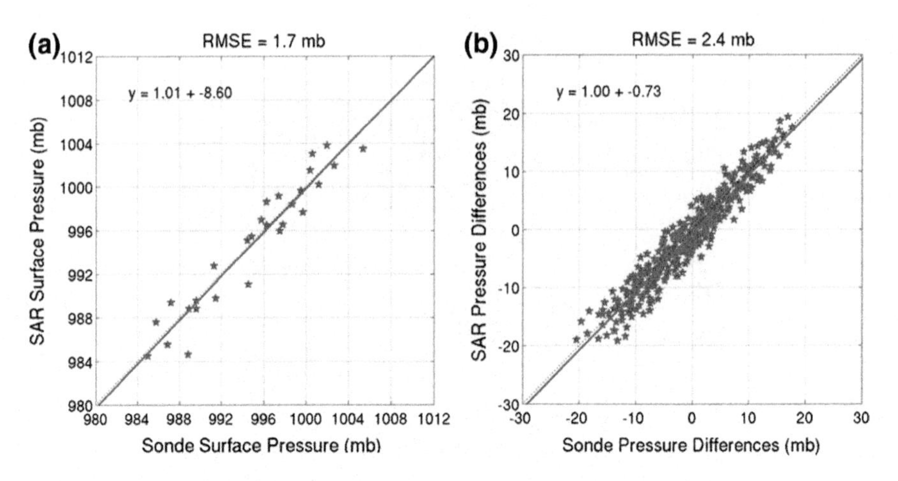

Fig. 12.4 Typhoon Malakas. **a** Scatterplot of 28 sonde surface pressures and SAR-derived SLP. **b** Scatterplot of bulk pressure differences between all pairs (378) of drop sondes compared to the SAR-derived SLP differences

same color map as for the SAR SLP. Because the plotted sonde splash pressures blend into the SAR SLP background, is evident that the agreement is quite close. Figure 12.3b compares the estimate of the surface pressure based on the flight-level meteorological instruments to that derived from the SAR image. The two curves track each other quite closely. The drop sonde surface pressures are also shown; but it is important to recognize that the sonde splash locations can be many km from the aircraft track because they drift with the prevailing wind during their descent.

Figure 12.4a is a scatterplot of the observed and retrieved SLP. Overall, the RMS between the observations and satellite-derived SLP is 1.7 mb. Even though we can validate the SLP retrievals with point wise pressure data, we prefer to use pairs of pressure observations to validate both the retrieved SLP pattern and the wind vector retrievals over a large number of wind vector cells. For N drop sondes, we have $N(N-1)/2$ pairs of surface pressure observations defining bulk pressure gradients that can be compared to the bulk pressure gradients determined from the SAR image (Fig. 12.4b). Here the RMS is less than 2.5 mb. This is the most stringent test of the SLP retrieval, and, it is also a sensitive test of the wind vector retrievals scene-wide. Systematic biases in the SAR surface wind vectors in either speed or direction would induce systematic biases in the surface pressure. High bias speeds introduce a local region of high biased pressure gradient magnitudes. Similarly, wind directional biases will introduce biases in the local pressure gradient directions. Either of these would introduce biases in the shape of the overall pressure surface that would be evident in the bulk pressure difference scatter plots. Good agreement between derived and measured pressure differences demonstrates that we have captured the shape of the surface pressure pattern and that the surface wind vectors are of overall good quality. Figure 12.5 shows similar SAR versus dropsonde scatterplots for five SAR scenes. Overall the RMS differences are approximately 3 mb.

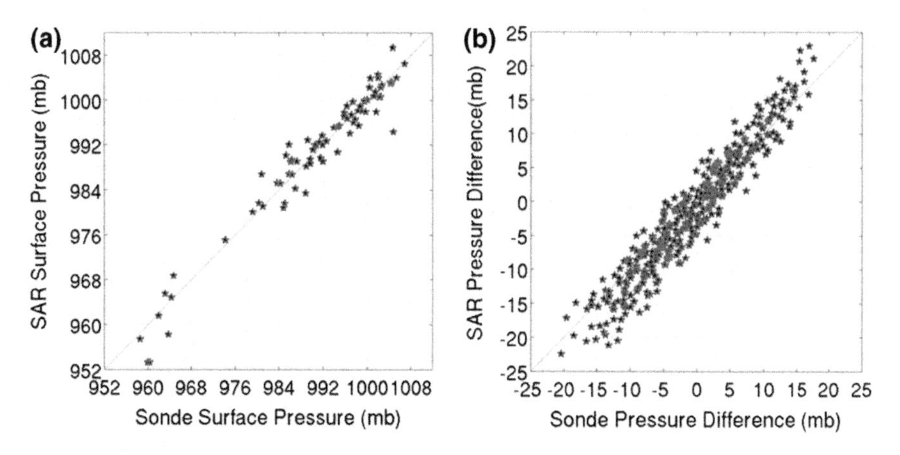

Fig. 12.5 As in Fig. 12.4, but summarizing the results for five SAR scenes

12.6 SLP-Filtered Surface Wind Retrievals

The classic evaluation of SAR wind vector retrievals requires point-wise comparisons of SAR surface wind speeds and directions with the best estimates of the mean surface wind from the drop sondes. Because the TCBL flow is highly turbulent, any individual drop sonde wind profile represents a single member of the ensemble of wind profiles that would be necessary to define the mean wind profile at that location. This means that simple interpolation or extrapolation of the drop sonde profiles will not provide useable mean surface winds. In order to estimate the surface wind, we fit a Monin–Obukov surface layer similarity solution to the near-surface wind, temperature and humidity profiles treating each data point on a profile as an independent measurement. A second set of surface wind speed data is obtained from the SFMR measurements, which are obtained along the flight track.

Figure 12.6a shows the C-130J time-adjusted flight track shown as SFMR wind speeds and drop sonde surface wind vectors superposed on the CMOD5N-derived (RAW) surface wind vectors, using the same SAR wind speed color mapping. Whenever the wind speeds agree, the observed speeds fade into the background. The drop sonde wind directions are shown in magenta, and they agree well with the SAR-derived directions, except for a drop sonde in the eye where wind speeds are low and direction is poorly defined. The most noticeable feature of this plot is the large masked region of the SAR image to the north-east of the eye for which the wind vector retrievals did not pass the SLP retrieval input quality control. Figure 12.6b is the same as for Fig. 12.6a, except now we show the SLP-filtered surface winds. Evidently SLP-filtered winds, i.e. the scene-wide wind vector retrieval, also agree well with the observations even in the region where the input winds did not pass quality control.

Figure 12.7a, b shows scatterplots of observed and SAR-derived surface wind speeds and directions for both the CMOD5N and SLP-filtered wind vectors. The

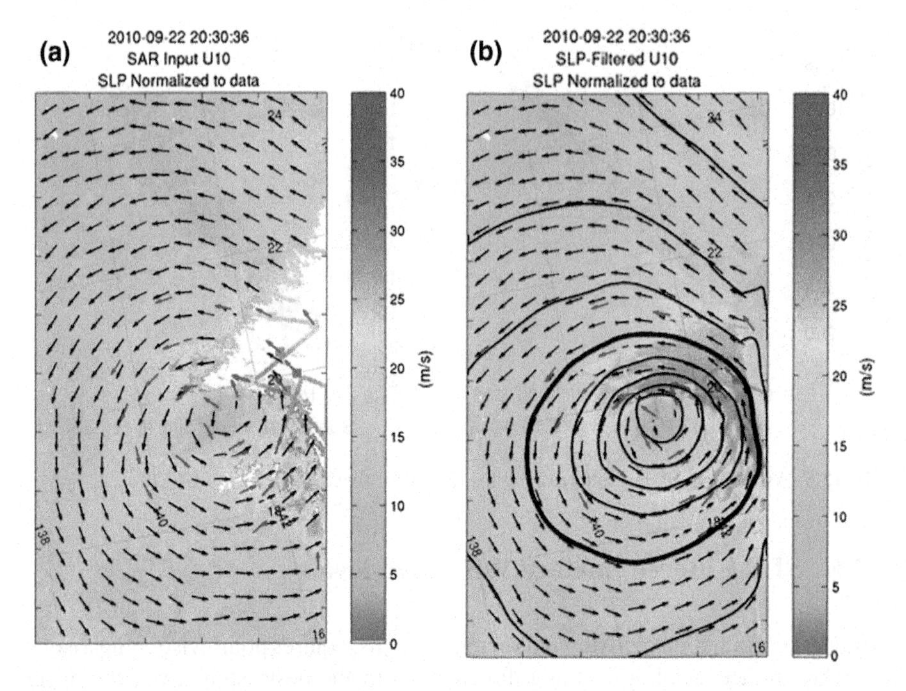

Fig. 12.6 Typhoon Malakas. Surface wind vectors from SAR with SFMR speeds and drop sonde wind vectors. **a** RAW wind vectors. **b** SLP-filtered wind vectors

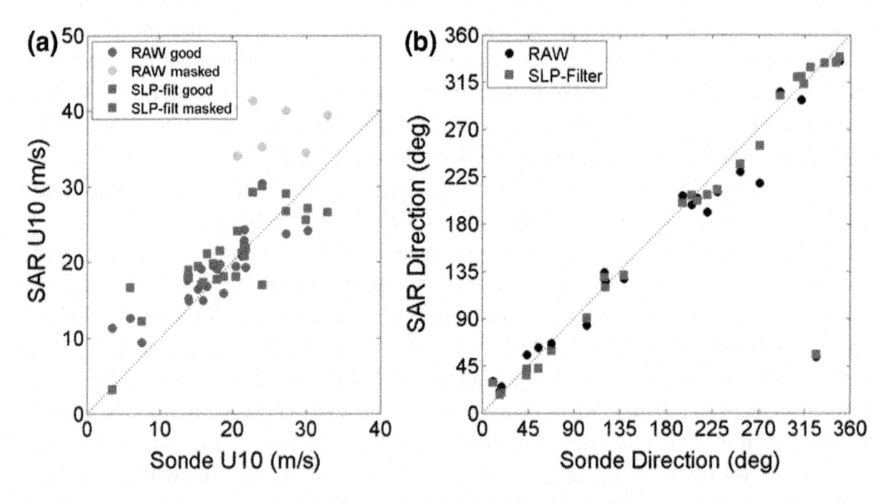

Fig. 12.7 Typhoon Malakas. **a** Scatterplot of drop sonde surface wind speeds compared to SAR surface wind speeds. *Blue dots* RAW winds in unmasked region; *Cyan dots* RAW winds in masked region; *Red squares* SLP-filtered in unmasked region; Magenta squares: SLP-filtered in masked regions. **b** Scatterplot of wind directions. *Black* RAW; *Red* SLP-filtered

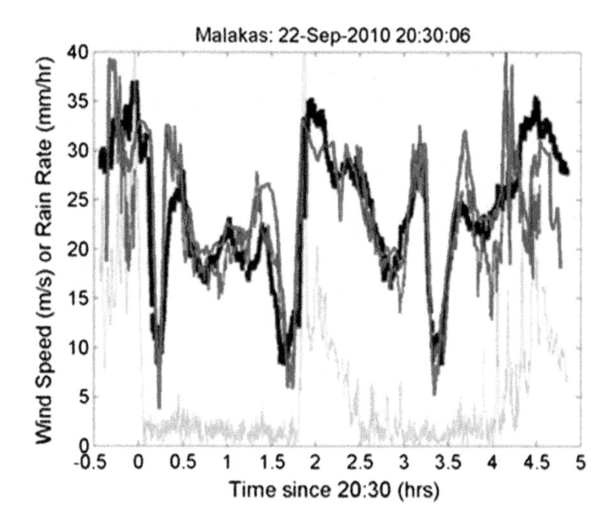

Fig. 12.8 Typhoon Malakas. Comparisons of SFMR wind speeds (*black*) with RAW winds in unmasked region (*blue*) and SLP-filtered speeds (*red*). The SFMR-estimated rainrate is shown in *cyan*

scatterplot further separates the six observations obtained in the masked region of the SAR image from the 22 where the wind retrievals are of high confidence. We find that the RMS for the RAW and SLP-filtered speeds in the high confidence regions are approximately the same (3.3 m/s). If we consider all of drop sondes, including the masked region, the RAW RMS increases to 5.6 m/s. This reinforces the importance of the wind retrieval confidence flags in SAR TC applications. In contrast, the SLP-filtered wind speeds have an overall RMS of 3.9 m/s. We conclude that there is skill in the SLP-filtered wind vector retrievals in the masked regions of the SAR image. There is little difference between the SLP-filtered and RAW wind directions and the RMS differences versus the drop sondes are small and approximately the same. However, the high wind sensitivity of CMOD5N in the low incidence angle regime to small changes in wind direction shown in Fig. 12.1 is important. We have found that re-running CMOD5N using the directions determined by the SLP-filtering procedure usually shrinks the region of low confidence wind speed retrievals.

Figure 12.8 compares the RAW and SLP-filtered surface wind speeds to those estimated by the SFMR instrument for the Malakas image. There are differences in the effective time averaging interval over which the mean speeds are estimated in these separate remotely-sensed measurements. The details of effective time averaging estimates are explored in Harper et al. (2010) [30]. In the course of many comparisons between SFMR and SAR, we determined an empirical correction factor of 1.15, which is somewhat less that the factor of 1.2 used to convert 10-min averages to 1-min averages. However, because the SFMR model function has been subject to revision and recalibration, our empirical factor may not work for all SAR and SFMR comparisons.

Overall, both the RAW and SLP-filtered wind speeds track the SFMR wind speeds quite closely. In high confidence regions the RMS between SFMR and either the RAW or SLP-filtered wind speeds is 3.4 m/s. The RAW winds in the low confidence region are not included in the plot. Including both the masked and unmasked regions, the SLP-filtered RMS is 4.2 m/s and the RAW winds RMS is 7.8 m/s. Careful examination of the plot shows that the peak winds in the SLP-filtered product are lower than the RAW winds, which is consistent with the inherent smoothing of the SLP retrieval process.

Very similar results are found when we examine five SAR scenes. The overall RMS in the unmasked regions are 5.7 and 5.5 m/s respectively for RAW and SLP-filtered wind speeds. Combining the masked and unmasked regions, we find RMS of 7.4 and 5.4 m/s, respectively. From this we conclude that in regions where confidence in the wind retrievals is high, the SLP-filtered winds are comparable in quality, but smoother than the RAW winds. It is unlikely that they will resolve the maximum wind speed at the radius of maximum winds. SLP retrieval allows us to retrieve wind vectors in the low confidence region of the image although these winds should be used with care.

12.7 Summary

In this chapter we describe a general method by which sea-level pressure patterns can be derived from SAR images of tropic cyclones with an RMS accuracy on the order of 3 mb (although the sample size is fairly small). The key to this algorithm is the development of a diagnostic tropical cyclone boundary layer model that can be used to estimate realistic pressure gradient vectors from the high quality and high resolution surface wind vectors that can be retrieved from SAR images. It is important to note important caveats. First, the input wind vectors must be generated with care. In the high wind regime, the speed retrievals can be highly sensitive to wind direction errors; consequently, estimates of wind direction from the image itself (rather than external data sources) should be employed. This further necessitates careful pre-processing of the SAR image to remove image artifacts such as scalloping noise and seams between sub-images in ScanSAR data so that the wind direction algorithms can be applied. Secondly, we have shown that wind retrieval quality flags are needed for quality control on the input wind vectors. (In fact, based on our comparisons with in situ wind observations, winds that fail the SLP wind vector input quality control are likely of no practical use.) The SLP pattern can be estimated over the masked regions of the image which allows us to derive SLP-filtered surface wind vectors for the entire image. In contrast to the RAW wind vectors in the masked part of the image, the SLP-filtered winds have skill because the retrieval methodology requires them to be dynamically consistent (within the accuracy of the TCBL model) with the wind vectors retrieved in the high confidence regions of the SAR image. However, these wind vectors are less accurate than those retrieved using convention means in the high confidence region of the SAR image. Further development of this technique focuses primarily on improving the TCBL model. Ideally a moving vortex version of Foster (2009) [22] should replace the current model.

References

1. Edson, J.B., V. Jampana, R.A. Weller, S.P. Bigorre, A.J. Plueddemann, C.W. Fairall, S.D. Miller, L. Mahrt, D. Vickers, and H. Hersbach. 2013. On the exchange of momentum over the open ocean. *Journal of Physical Oceanography* 43 (8): 1589–1610.
2. Powell, M.D., P.J. Vickery, and T.A. Reinhold. 2003. Reduced drag coefficient for high wind speeds in tropical cyclones. *Nature* 422 (6929): 279–283.
3. Donelan, M.A., B.K. Haus, N. Reul, W.J. Plant, M. Stiassnie, H.C. Graber, O.B. Brown, and E.S. Saltzman. 2004. On the limiting aerodynamic roughness of the ocean in very strong winds. *Geophysical Research Letters*, 31(18).
4. Hersbach, H. 2008. CMOD5.N: A C-band geophysical model function for equivalent neutral wind, technical memorendam. *Technical Memorendam European Centre for Medium-Range Weather Forecasting* 554.
5. Koch, W. 2004. Directional analysis of SAR images aiming at wind direction. *IEEE Transactions on Geoscience and Remote Sensing* 42 (4): 702–710.
6. Horstmann, J., and W. Koch. 2005. Measurement of sea surface winds using synthetic aperture radars. *IEEE Journal of Oceanic Engineering* 30 (3): 508–515.
7. Wackerman, C., W.G. Pichel, X. Li, and P. Clemente-Colon. 2006. Estimation of surface winds from SAR using a projection algorithm.
8. Gerling, T.W. 1986. Structure of the surface wind field from the Seasat SAR. *Journal of Geophysical Research: Oceans* 91 (C2): 2308–2320.
9. Wurman, J., and J. Winslow. 2003. Intense sub-kilometer-scale boundary layer rolls observed in Hurricane Fran. *Science* 280 (5363): 555–557.
10. Morrison, I., S. Businger, F. Marks, P. Dodge, and J.A. Businger. 2005. An observational case for the prevalence of roll vortices in the hurricane boundary layer. *Journal of the Atmospheric Sciences* 62 (8): 2662–2673.
11. Foster, R.C. 1997. Structure and energetic of optimal Ekman layer perturbations. *Journal of Fluid Mechanics* 333: 97–123.
12. Foster, R.C. 2005. Why rolls are prevalent in the hurricane boundary layer. *Journal of the Atmospheric Sciences* 62 (8): 2647–2661.
13. Foster, R.C. 2013. Signature of large aspect ratio roll vortices in SAR images of tropical cyclones. *Oceanography* 26 (2): 58–67.
14. Lorsolo, S., J.L. Schroeder, P. Dodge, and J.F. Marks. 2008. Observational study of hurricane boundary layer small-scale coherent structures. *Monthly Weather Review* 136 (8): 2871–2893.
15. Thompson, D.R., and R.C. Beal. 2000. Mapping high-resolution wind fields using synthetic aperture radar. *Johns Hopkins APL Technical Digest* 21 (1): 58–67.
16. Horstmann, J., C. Wackerman, S. Falchetti, and S. Maresca. 2013. Tropical cyclone winds retrieved from synthetic aperture radars. *Oceanography* 26: 46–57.
17. Romeiser, R., J. Horstmann, M.J. Caruso, and H.C. Graber. 2013. A descalloping post-processor for ScanSAR images of ocean scenes. *IEEE Transactions on Geoscience and Remote Sensing* 51 (6): 3259–3272.
18. Patoux, J., R.C. Foster, and R.A. Brown. 2003. Global pressure fields from scatterometer winds. *Journal of Applied Meteorology* 42 (6): 813–826.
19. Patoux, J., R.C. Foster, and R.A. Brown. 2008. An evaluation of scatterometer-derived oceanic surface pressure fields. *Journal of Applied Meteorology and Climatology* 47 (3): 835–852.
20. Patoux, J., R. Foster, and R. Brown. 2010. A method for including mesoscale and synoptic-scale information in scatterometer wind retrievals. *Journal of Geophysical Research: Atmospheres*, 115(D11).
21. Patoux, J., and R. Foster. 2012. Cross-validation of scatterometer measurements via sea-level pressure retrieval. *IEEE Transactions on Geoscience and Remote Sensing* 50 (7): 2507–2517.
22. Foster, R.C. 2009. Boundary-layer similarity under an axisymmetric, gradient wind vortex. *Boundary-Layer Meteorology* 131 (3): 321–344.
23. Brown, R.A., and W.T. Liu. 1982. An operational large-scale marine planetary boundary layer model. *Journal of Applied Meteorology* 21 (3): 261–269.

24. Stevens, B., J. Duan, J.C. McWilliams, M. Mnnich, and J.D. Neelin. 2002. Entrainment, Rayleigh friction, and boundary layer winds over the tropical Pacific. *Journal of Climate* 15 (1): 30–44.
25. Ahn, J.M.V., J.M. Sienkiewicz, and J. Patoux. 2006. A comparison of sea level pressure analyses derived from QuikSCAT winds to manual surface analyses produced in the NOAA Ocean Prediction Center. In *21st Conference on Weather Analysis and Forecasting/17th Conference on Numerical Weather Prediction.*
26. Elliassen, A., and M. Lystad. 1977. On the Ekman layer in a circular vortex: a numerical and theoretical study. *Geophysica Norvegica* 31: 1–16.
27. Bishop, C. 1996. Domain-independent attribution. Part 1: Reconstructing the wind from estimates of vorticity and divergence using free space Greens functions. *Journal of the Atmospheric Sciences* 53 (2): 241–252.
28. Uhlhorn, and P.G. Black. 2003. Verification of remotely sensed sea surface winds in hurricanes. *Journal of Atmospheric and Oceanic Technology*, 20 (1):99–116.
29. D'Asaro, P. Black, L. Centurioni, Y.T. Chang, S. Chen, R. Foster, H. Graber, P. Harr, V. Hormann, R.C. Lien, I.I. Lin, T. Sanford, T.Y. Tang, and C.C. Wu. 2014. Impact of typhoons on the ocean in the Pacific: ITOP. *Bulletin of the American Meteorological Society* 95 (9): 1405–1418.
30. Harper, B.A., J.D. Kepert, and J.D. Ginger. 2010. Guidelines for converting between various wind averaging periods in tropical cyclone conditions. In *World Meteorological Organization.*

Chapter 13
Electromagnetic Scattering of Rainfall and Tropical Cyclones over Ocean

Feng Xu and Xiaofeng Li

Abstract This chapter introduces a physics-based radiative transfer model to capture the scattering behavior of rainfall over a rough sea surface. Raindrops are modeled as Rayleigh scattering nonspherical particles, while the rain-induced rough surface is described by the Log-Gaussian ring-wave spectrum. The model is validated against both empirical models and measurements. A case study of collocated Envisat ASAR data and NEXRAD rain data is presented. To showcase the capability of the developed scattering model in studying cyclones, we use regional WRF weather model to simulate the Hurricane Hermine occurred in North America at September 2016. Finally, numerical analyses suggest that rain-related scattering becomes significant as compared to wind-related scattering when the frequency is above C-band, while the raindrop volumetric scattering becomes significant above X-band.

13.1 Introduction

Monitoring of tropical cyclones is critical for disaster prevention and relief, and understanding of its birth, migration and death. Weather satellites working at infrared and optical bands are routinely used for constant monitoring of tropical cyclones. Occasional in situ observations with airborne sensors are very useful for studying the mechanism of such events. Due to the rapid advancement of spaceborne synthetic aperture radar (SAR) programs over the past two decades, high-resolution SAR has become a new useful tool for monitoring of rain events and tropical cyclones. Microwave SAR with its unique high-resolution, all-weather, and day-night imaging capability has been attracting additional attention from the research community. However, many interesting rain-related phenomena revealed by SAR images are

F. Xu (✉)
Key Lab for Information Science of Electromagnetic Waves (MoE),
Fudan University, Shanghai 200433, China
e-mail: fengxu@fudan.edu.cn

X. Li
GST, National Oceanic and Atmospheric Administration (NOAA)/NESDIS,
College Park, MD, USA
e-mail: Xiaofeng.Li@noaa.gov

X. Li (ed.), *Hurricane Monitoring With Spaceborne Synthetic Aperture Radar*, Springer Natural Hazards, DOI 10.1007/978-981-10-2893-9_13

still not fully understood due to poor theoretical modeling of the rain-wind-ocean interactions and their electromagnetic scattering behaviors.

With SAR images, localized rainfall cells and tropical cyclones can now be observed in a never-before level of detail. Many interesting and mysterious phenomena have been reported and qualitatively analyzed during the course of development of SAR technology and its applications [1–4]. Earlier studies focused on higher frequency microwave bands such as Ku-band [e.g. [5, 6]], as raindrops are usually believed to be transparent to lower frequency bands such as C/L-band. Extensive work, both experiments and analyses, have been carried out towards a better understanding of rains effect on Ku-band scatterometry [e.g. [7–9]]. However, the general consensus now is that rain-induced surface roughness has observable impact on the total scattering power even in lower frequency -band SAR images.

Some empirical models have been developed using remote sensing and weather prediction matchup datasets [e.g. [10–13]], leading to proposed rain/wind retrieval methods [e.g. [14, 15]]. However, empirical models are highly dependent on the selection of fitting data points, susceptible to data error/noise and may have overfitting issues. Derived from specific datasets, empirical models may not be generally applicable in different settings, e.g. incident angle, frequency bands, polarization, and environment conditions.

Theoretically, Tournadre and Quilfen (2003) [16] proposed a radiative transfer model (RTM) of rain for Ku-band scatterometers which takes into account raindrop volumetric scattering and attenuation but not rain perturbation to the sea surface, which would become dominant at lower frequency bands. Contreras and Plant (2006) [17] proposed a sea surface wave model incorporating two types of rain perturbation, namely, ring-wave and turbulence damping. However, the model does not include rainfall scattering and attenuation and it is tuned for scatterometers.

A general physics-based scattering model capable of incorporating different types of rain factors would be the key to better understanding and interpreting various phenomena observed by SAR. First, we need to understand various scattering mechanisms contributed by rainfall to the radar backscattering signals. In general, besides the volumetric scattering from raindrops and the attenuation of wave propagating through rainfall, sea surface roughness alteration caused by raindrops hitting the sea surface and its associated effects are other major contributors to the overall measured scattering. It is very difficult to model the complicated air-sea interface when coupled with dynamic raindrop-induced splashes and turbulences. The major factors through which rainfall may affect the sea surface are generally believed to be [3, 18–21]

(a) Raindrops striking the water surface create ring-waves, stalks and crowns;
(b) Raindrops generate turbulence in the upper water layer which attenuates the short gravity wave spectrum;
(c) Downspread airflow associated with a rain event may alter the original wind field and then further alter the sea surface spectrum.

Wetzel (1990) [18] first studied the waterdrops splash via experiments assisted with high-speed photography and found that stalks, crowns and ring-waves are the three major scatters. The ring-waves are the most-studied factor among these three. Empirical log-Gaussian ring-wave spectra were derived via rain experiments [19, 22]. Corresponding scattering models were focused on the ring-waves [e.g. [23]] as it is generally believed that stalks and crowns would only become dominant at low grazing angles [18]. Factors (b) and (c) have not been well studied and are not completely understood. Although Contreras and Plant (2006) [17] proposed a modified version of the Kudryavtsev short wave spectrum [24] incorporating the rain damping effect, it contains parameters tuned together with a scattering model using specific datasets and has not been separately validated in terms of the spectrum itself. A recent study by [25] suggests immaturities in existing short-wave spectra including the one by Kudryavtsev et al. (1999). To avoid introducing uncertainty into our model analyses, we will restrict our efforts to the well-studied ring-wave scattering mechanism. Nevertheless, previous analyses suggest that the damping effect by turbulence only becomes observable at frequencies lower than C-band or, if at C-band, under small off-nadir incident angles ($<30°$) [e.g. [3, 12] which specifies a validity condition for us to neglect the turbulence damping. The downspread airflow contributes to SAR indirectly through alteration of the wind field [1]. It has been shown that such alteration can be detected in scatterometer-retrieved wind fields [26]. Modeling of such processes is left for oceanic/metrological physicists, while our model could take the altered wind field as input and simulate the desired effect in the future.

This study attempts to establish a general scattering model which consists of the radiative transfer model of a rainfall layer and the rough sea surface model of a linearly combined wind-driven gravity-wave spectrum and the rain-induced ring-wave spectrum [27]. Raindrops are modeled as Rayleigh spherical particles while the empirical Fung-Lee spectra [28] and the empirical log-Gaussian spectra [22] are used to describe the wind-driven and rain-impacted sea surface, respectively. The major objective is to make the initial attempt towards a comprehensive scattering model of the complex system of rain, wind and sea surface and lay the groundwork for further refining and modeling of each individual scattering mechanism.

13.2 Parameterized Radiative Transfer Model

13.2.1 Volume Scattering

We assume a locally uniform distribution of rain over a rough sea surface with a cloud height of H. A pixel cell in the range dimension of the SAR image corresponds to the summation of the scattered powers from all scatterers in the strip perpendicular to radar incidence direction. According to the SAR mapping and projection principle [29], the scattering power of the corresponding cell in the SAR image can be written as a sum of all volume scatterers and the corresponding area of surface scatterer. For an infinitesimal volume along the strip, its scattering contribution can be written as [29]:

$$dM_{vol} = n_0 dRdxdy\exp(-x\tan\theta\kappa^{+}) \cdot \mathrm{P} \cdot \exp(-x\tan\theta\kappa^{-}) \qquad (13.1)$$

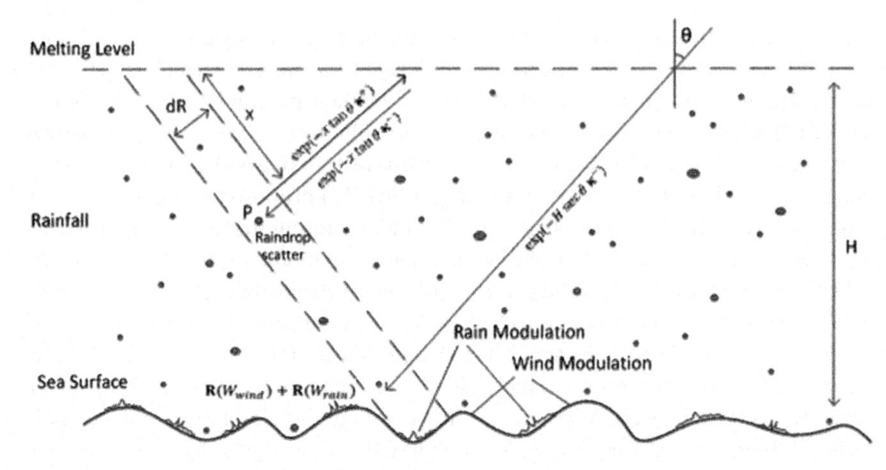

Fig. 13.1 Mapping and projection of SAR imaging of rainfall over a sea surface, including raindrop volumetric scattering, rain-modulated rough surface scattering and wind-modulated rough surface scattering

where R and y corresponds to slant-range and azimuth dimension, respectively, and together with the 3rd dimension x, form the Cartesian system (R, y, x). Therefore, the product dRdxdy denotes the differential volume. n_0 denotes the number of raindrop particles per unit volume; P and κ denotes the phase matrix and extinction matrix of an individual raindrop particle, respectively. The superscript of κ denotes upward and downward propagation direction, respectively. θ is the incidence angle. x denotes the slant distance of the scatterer to the melting level (labeled in Fig. 13.1).

Note that Mueller matrix (M_{vol}), phase matrix (P) and extinction matrix (κ) are 4 by 4 real matrices (in bold) which describe the fully polarimetric scattering characteristics.

From matrix algebra, we have

$$\exp(-x\tan\theta\kappa^{+}) = \mathrm{E}^{+} \cdot \mathrm{D}(-x\tan\theta\beta^{+}) \cdot \mathrm{E}^{+^{-1}} \tag{13.2}$$

where β^{+} denotes eigenvalues of κ^{+}; E^{+} denotes the eigenvector matrix of κ^{+}; $\mathrm{D}(-x\tan\theta\beta^{+})$ denotes the diagonal matrix with diagonal elements $\exp(-x\tan\theta\beta^{+})$. This equation also holds for downward direction. Assuming

$$\mathrm{P}' = \mathrm{E}^{+^{-1}} \cdot \mathrm{P} \cdot \mathrm{E}^{-} \tag{13.3}$$

the scattering matrix can be rewritten as

$$d\mathrm{M}_{vol} = n_0 dRdxdy\mathrm{E}^{+} \cdot \left\{\exp[-x\tan\theta(\beta_i^{+} + \beta_j^{-})]P'_{ij}\right\}_{4\times4} \cdot \mathrm{E}^{-^{-1}} \tag{13.4}$$

Hence, the integration over the entire strip can be evaluated as:

$$dM_{vol} = An_0 E^+ \cdot \int_0^{\frac{H}{\sin\theta}} dx \left\{ \exp[-x\tan\theta(\beta_i^+ + \beta_j^-)] P'_{ij} \right\}_{4\times4} \cdot E^{-^{-1}} \tag{13.5}$$

where A is the cross section area equivalent to one pixel. It yields

$$dM_{vol} = An_0 E^+ \cdot \left\{ \frac{1 - \exp[-H\sec\theta(\beta_i^+ + \beta_j^-)]}{\tan\theta(\beta_i^+ + \beta_j^-)} P'_{ij} \right\}_{4\times4} \cdot E^{-^{-1}} \tag{13.6}$$

Interestingly, given that $\beta_i^{\pm}, \beta_j^{\pm} \ll 1$, which is the case for raindrops in microwave frequencies (e.g. $\beta_{i,j}^{\pm} \approx 0.0001$ for 50 mm/hr rain at Ku-band according to Eq. (13.18)), the above expression can be approximated as

$$M_{vol} = n_0 A \frac{H}{\sin\theta} P \tag{13.7}$$

which indicates that the volumetric scattering from raindrops only depends on the individual raindrop scattering coefficient and rainfall layer depth, but not the attenuation coefficients.

For scatterometers, the received power is an integral along the incident direction. Hence, Eq. (13.6) should be written as

$$M_{vol} = n_0 A E^+ \cdot \int_0^{\frac{H}{\cos\theta}} dx \left\{ \exp[-x(\beta_i^+ + \beta_j^-)] P'_{ij} \right\}_{4\times4} \cdot E^{--1} = n_0 A \frac{H}{\cos\theta} P \tag{13.8}$$

Equation (13.8) differs from SAR expression, Eq. (13.7), only in the factor $\tan\theta$.

13.2.2 Surface Scattering

According to the mapping and projection principle, the corresponding small area of the underlying sea surface contributes to the SAR image as:

$$M_{surf} = A\sec\theta\exp(-H\sec\theta\kappa^+) \cdot R \cdot \exp(-H\sec\theta\kappa^-) \tag{13.9}$$

where R is the rough surface Mueller matrix of a unit area.

The nonlinear wind-rain interaction phenomenon, i.e. the 3rd and 4th factors mentioned in Sect. 13.1, is beyond the scope of this study. We assume that the rough sea surface is modulated by a linear sum of two major forces, i.e. wind and rain. The surface height correlation function can be written as the sum of wind-modulated correlation and rain-modulated correlation. The same linearity applies to the spectral density function, i.e. the Fourier transform of the correlation function.

Following from the integral equation method (IEM) [30] for a randomly rough surface, the Mueller matrix is a linear function of the spectral density function, W, of the rough surface, i.e.

$$\mathrm{R}(W_{wind} + W_{rain}) = \mathrm{R}(W_{wind}) + \mathrm{R}(W_{rain}) \tag{13.10}$$

Note that multiple scatterings between raindrop and sea surface are not taken into account as these terms are generally smaller than the direct scattering terms [31].

13.3 Components of the Radiative Transfer Model

13.3.1 Raindrop Approximated as Rayleigh Particle

Most of the recent SAR systems are in the frequency range between L-band and X-band. As most raindrop size is less than 3 mm, we can assume the Rayleigh particle approximation for raindrop scatterers. This assumption greatly simplifies the expression of the raindrop scattering matrix. Note that this assumption would break-down for frequencies above Ku-band, where the Rayleigh approximation condition would be violated and raindrop multiple scatterings would also become significant [32, 33].

The backscattering matrix of a vertically-oriented spheroid under Rayleigh approximation can be written as (cf. Eq. (13.26) in [31]):

$$\begin{aligned} S_{vv} &= \frac{2}{3k}[T_1 + (T_0 - T_1)\sin^2\theta] \\ S_{vh} &= S_{hv} = 0 \\ S_{hh} &= -\frac{3}{2k}T_1 \end{aligned} \tag{13.11}$$

where

$$\begin{aligned} T_0 &= t_0 + jt_0^2 \\ T_1 &= t_1 + jt_1^2 \\ t_0 &= \frac{k^{3v_0}}{6\pi}\frac{\varepsilon_r - 1}{1 + (\varepsilon_r - 1)g_c} \\ t_0 &= \frac{k^{3v_0}}{6\pi}\frac{\varepsilon_r - 1}{1 + (\varepsilon_r - 1)g_a} \end{aligned} \tag{13.12}$$

Wheras k denotes wavenumber, $j = \sqrt{-1}$, denotes incident angle, v_0 denotes raindrop volume, ε_r denotes complex relative permittivity, g_a,g_c are the geometric parameters of a Rayleigh scattering particle (for definitions, see Eq. (13.24) of [31]) which are determined by the shape of the raindrop, cf. Eq. (13.24). S_{pq} denotes the scattering matrix element with q-polarization incidence and p-polarization scattering.

For natural raindrops, we have water permittivity $\varepsilon_r \gg 1$ which yields (via Eq. (13.24)) the general ranges of $g_a \in [\frac{1}{5}, \frac{1}{3}]$, $g_c \in [\frac{1}{3}, \frac{1}{2}]$. Hence we obtain the following approximation:

$$\begin{aligned} t_0 &\approx \frac{1}{g_c}\frac{k^3 v_0}{6\pi}\frac{\varepsilon_r - 1}{\varepsilon_r + 2} \\ t_1 &\approx \frac{1}{g_a}\frac{k^3 v_0}{6\pi}\frac{\varepsilon_r - 1}{\varepsilon_r + 2} \end{aligned} \tag{13.13}$$

For Rayleigh scattering, $t_0, t_1 \ll 1$. Therefore, the quadratic terms of t_0, t_1 can be ignored when calculating the backscattering matrix, i.e.

$$\begin{aligned} S_{vv} &= k^2\frac{\varepsilon_r - 1}{\varepsilon_r + 2}\tau_v \\ S_{vh} &= S_{hv} = 0 \\ S_{hh} &= -k^2\frac{\varepsilon_r - 1}{\varepsilon_r + 2}\tau_h \end{aligned} \tag{13.14}$$

This yields the phase matrix [31]:

$$P = k^4\left|\frac{\varepsilon - 1}{\varepsilon + 2}\right|^2 \begin{bmatrix} \langle \tau_v'^2 \rangle & 0 & 0 & 0 \\ 0 & \langle \tau_h^2 \rangle & 0 & 0 \\ 0 & 0 & -\langle \tau_v \tau_h \rangle & 0 \\ 0 & 0 & 0 & -\langle \tau_v \tau_h \rangle \end{bmatrix} \tag{13.15}$$

where the bracket denotes ensemble average, and where τ_v and $\tau_v'^2$ are:

$$\begin{aligned} \tau_v &= \frac{a^3}{3}\left[\frac{1}{g_a} + \left(\frac{1}{g_c} - \frac{1}{g_a}\right)\sin^2\theta\right] \\ \tau_h &= \frac{a^3}{3g_a} \\ \tau_v'^2 &= \frac{a^6}{9}\left[\frac{1}{g_a^2} + \left(\frac{1}{g_c^2} - \frac{1}{g_a^2}\right)\sin^2\theta\right] \end{aligned} \tag{13.16}$$

where a denotes the equivalent raindrop radius.

However, the quadratic terms have to be kept when evaluating the extinction matrix which is determined by the imaginary part of the forward scattering matrix. The forward scattering matrix is:

$$\begin{aligned} \mathrm{Im}\left\langle S_{vv}^f \right\rangle &\approx 3k^2\frac{\mathrm{Im}(\varepsilon_r)}{|\varepsilon + 2|^2}\langle \tau_v \rangle + \frac{2k^5}{3}\left|\frac{\varepsilon_r - 1}{\varepsilon_r + 2}\right|^2\left\langle \tau_v'^2 \right\rangle \\ \mathrm{Im}\left\langle S_{hh}^f \right\rangle &\approx 3k^2\frac{\mathrm{Im}(\varepsilon_r)}{|\varepsilon_r + 2|^2}\langle \tau_h \rangle + \frac{2k^5}{3}\left|\frac{\varepsilon_r - 1}{\varepsilon_r + 2}\right|^2\langle \tau_h^2 \rangle \end{aligned} \tag{13.17}$$

where the superscript f denotes the forward scattering direction.

Under the condition that forward scattering $S_{vh}^f = S_{hv}^f = 0$, the eigenvectors/eigenvalues of the extinction matrix can be directly derived as [34]:

$$\mathrm{E}^+ = \mathrm{E}^- = \mathrm{E} = \begin{bmatrix} 1 & 0 & 0 & 0 \\ 0 & 1 & 0 & 0 \\ 0 & 0 & 1 & 1 \\ 0 & 0 & j & -j \end{bmatrix}$$

$$[\beta_i^+] = [\beta_i^-] = [\beta_i] = \frac{2n_0\pi}{k}\begin{bmatrix} 2\mathrm{Im}\left\langle S_{vv}^f \right\rangle \\ 2\mathrm{Im}\left\langle S_{hh}^f \right\rangle \\ \mathrm{Im}\left\langle S_{vv}^f \right\rangle + \mathrm{Im}\left\langle S_{hh}^f \right\rangle \\ \mathrm{Im}\left\langle S_{vv}^f \right\rangle + \mathrm{Im}\left\langle S_{hh}^f \right\rangle \end{bmatrix} \tag{13.18}$$

Substituting into Eq. (13.7), it yields:

$$\mathrm{M}_{vol} = n_0 A k^4 \left|\frac{\varepsilon - 1}{\varepsilon + 2}\right|^2 \frac{H}{\sin\theta} \begin{bmatrix} \langle \tau_v^{'2} \rangle & 0 & 0 & 0 \\ 0 & \langle \tau_h^2 \rangle & 0 & 0 \\ 0 & 0 & -\langle \tau_v \tau_h \rangle & 0 \\ 0 & 0 & 0 & -\langle \tau_v \tau_h \rangle \end{bmatrix} \tag{13.19}$$

Substituting into Eq. (13.8), the surface scattering is written as

$$\mathrm{M}_{surf} = A\sec\theta \mathrm{E} \cdot \mathrm{D}(-H\sec\theta\beta) \cdot \mathrm{E}^{-1} \cdot \mathrm{R} \cdot \mathrm{E} \cdot \mathrm{D}(-H\sec\theta\beta) \cdot \mathrm{E}^{-1} \tag{13.20}$$

Given that $R_{vh} = R_{hv} = 0$, i.e. cross-polarization backscattering from the surface is zero (which is true for the IEM rough surface model [30] employed here), the above expression can be further simplified as:

$$M_{surf} = A\sec\theta R' \tag{13.21}$$

with the attenuated surface scattering expressed as:

$$\begin{aligned} R'_{vv} &= R_{vv}\exp(-2H\sec\theta\frac{2n_0\pi}{k}\mathrm{Im}\langle S_{vv}^f \rangle) \\ R'_{hh} &= R_{hh}\exp(-2H\sec\theta\frac{2n_0\pi}{k}\mathrm{Im}\langle S_{vv}^f \rangle) \end{aligned} \tag{13.22}$$

In summary, the overall normalized radar cross section (NRCS) which appears in the SAR image is:

$$\begin{aligned}\sigma_{pp}^2 &= 4\pi V_{pp}^2 + 4\pi R_{pp}^2 \alpha_{pp}^2 \\ V_{pp}^2 &= n_0 k^4 \left|\frac{\varepsilon - 1}{\varepsilon + 2}\right|^2 \frac{H}{\tan\theta} \langle \tau_p^2 \rangle \\ \alpha_{pp}^2 &= \exp\left(-2H\sec\theta \frac{4n_0\pi}{k} \mathrm{Im}\langle S_{pp}^f \rangle\right)\end{aligned} \tag{13.23}$$

where V_{pp}^2 denotes raindrop volumetric scattering, R_{pp}^2 denotes sea surface scattering, α_{pp}^2 denotes rainfall penetration attenuation applied to sea surface scattering, the subscript pp denotes *vv* or *hh* for co-polarization. Note that the correlation term $\langle \sigma_{vv}\sigma_{hh}^* \rangle$ can also be calculated in the same manner following Eqs. (13.19) and (13.20), which is useful for dual/quad-polarization data.

13.3.2 Raindrop Size and Shape Distribution

In order to evaluate Eq. (13.23), the ensemble averages of $\tau_v, \tau_h, \tau_v'^2$ found in Eq. (13.16) need to be evaluated for all sizes of natural raindrops. For a Rayleigh spheroid, the shape parameter g_c is determined by the raindrop axial ratio r as (cf. Eq. (13.26) in [31]):

$$g_c = \frac{r^2}{r^2 - 1}\left[1 - \frac{1}{\sqrt{r^2 - 1}} \mathrm{atan}\sqrt{r^2 - 1}\right], g_a = \frac{1 - g_c}{2} \tag{13.24}$$

A polynomial expression of the axis ratio is given by the Chuang–Beard model [35]

$$r = 1.01668 - 0.98055a - 2.52686a^2 + 3.75061a^3 + 1.68692a^4 \tag{13.25}$$

where the equivalent radius a is in cm. The shapes of raindrops at different sizes governed by this model are shown in Fig. 13.2.

Substituting Eq. (13.25) into (13.24), we can plot $1/g_c$, $1/g_a$, $1/g_c^2$ and $1/g_a^2$ as a function of raindrop radius in Fig. 13.3. To simplify the ensemble average evaluation, the following linear approximations are employed:

$$\frac{1}{g_c} \approx 3 - \frac{a}{4l_0'}, \frac{1}{g_a} \approx 3 + \frac{a}{4l_0}, \frac{1}{g_c^2} \approx 9 - \frac{3a}{2l_0'}, \frac{1}{g_a^2} \approx 9 + \frac{3a}{2l_0} \tag{13.26}$$

where l_0 denotes unit length, which is used here to avoid confusion of units. The corresponding linear fitting lines are plotted in Fig. 13.3 for comparison. Hence, the ensemble averages are rewritten as

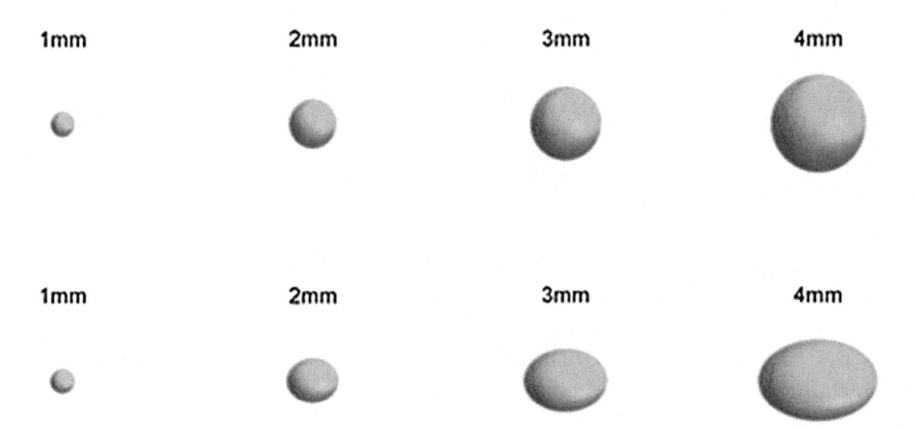

Fig. 13.2 Rendering of the Chuang–Beard raindrop (*bottom row*) versus the equivalent-volume spherical raindrop (*top row*) at different sizes

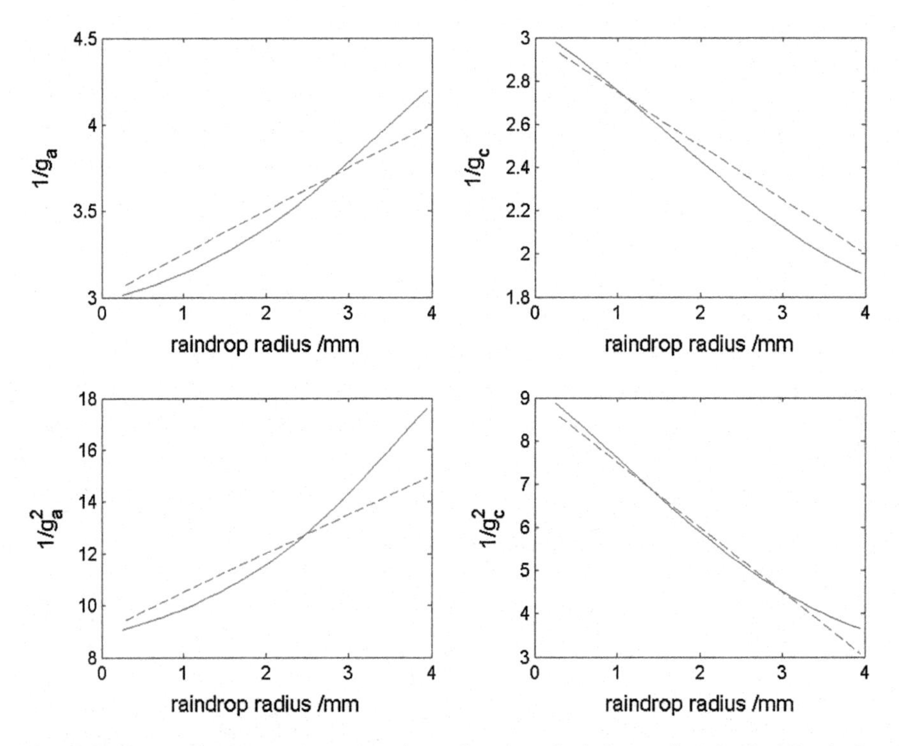

Fig. 13.3 Shape-related parameters plotted as a function of raindrop radius. *Dashed line* is the approximated linear fit employed in this study, cf. Eq. (13.26)

$$
\begin{aligned}
\langle \tau_v \rangle &= \langle a^3 \rangle + \frac{\cos 2\theta}{12} \langle a^4 \rangle l_0^{-1} \\
\langle \tau_h \rangle &= \langle a^3 \rangle + \frac{1}{12} \langle a^4 \rangle l_0^{-1} \\
\langle \tau_v^2 \rangle \approx \langle \tau_v'^2 \rangle &= \langle a^6 \rangle + \frac{\cos 2\theta}{6} \langle a^7 \rangle l_0^{-1} \\
\langle \tau_h^2 \rangle &= \langle a^6 \rangle + \frac{1}{6} \langle a^7 \rangle l_0^{-1} \\
\langle \tau_v \tau_h \rangle &= \langle a^6 \rangle + \frac{\cos^2\theta}{6} \langle a^7 \rangle l_0^{-1}
\end{aligned}
\tag{13.27}
$$

To evaluate the ensemble averages on a, consider the Marshall–Palmer (M-P) distribution of natural raindrop size [34]:

$$N(a) = 1.6 \times 10^4 \exp(-8.2aR^{-0.21}) \tag{13.28}$$

where the radius of raindrops a is in mm; R denotes rain rate (mm/h); N denotes raindrop density ($\text{m}^{-3}\text{mm}^{-1}$). The number of raindrops per unit volume can be calculated as:

$$n_0 = \int_0^\infty N(a)da \tag{13.29}$$

The probability distribution function of raindrop size is now estimated as

$$p(a) = \frac{N(a)}{n_0} \tag{13.30}$$

and therefore, the ensemble averages can be derived as:

$$\langle a^m \rangle = \int_0^\infty p(a) a^m da \tag{13.31}$$

These entities can be explicitly derived as [27]:

$$
\begin{aligned}
n_0 &= 1950R^{0.21} (\text{m}^{-3}) \\
\langle a^3 \rangle &= 1.09 \times 10^{-11} R^{0.63} (\text{m}^3) \\
\langle a^4 \rangle l_0^{-1} &= 5.31 \times 10^{-12} R^{0.84} (\text{m}^3) \\
\langle a^6 \rangle &= 2.37 \times 10^{-21} R^{1.26} (\text{m}^6) \\
\langle a^7 \rangle l_0^{-1} &= 2.02 \times 10^{-21} R^{1.47} (\text{m}^6)
\end{aligned}
\tag{13.32}
$$

where R is in mm/hr.

13.3.3 Rain-Induced Ring-Wave Spectrum

Ring wave is the dominant scatterer at about 30° incident angle, while stalks become dominant at grazing angles [21]. Since most SAR observations are taken at 15°–50° incidences (e.g. the incident angle of the ENVISAT ASAR image used later is within 16°–42°), we focus on the scattering from ring waves. A Log-Gaussian type function is used to approximate the spectral density function of the rain roughened surface [19, 22]. Lemaire et al. (2002) [22] conducted polydisperse rain experiments with three dropsizes and proposed an empirical ring-wave spectrum as:

$$W_{rain}(R, f) = \sum_{i=1}^{3} \exp\left[\frac{R_i}{68}\left(1 - \frac{R_i}{420}\right) - \frac{R}{68}\left(1 - \frac{R}{420}\right)\right] W_i(R_i, f) \quad (13.33)$$

where R_i and W_i denotes rain rate (mm/h) and individual spectra of the i-th dropsize, respectively.

$$W_i(R, f) = \begin{cases} (e_i R^{f_i}) \exp\left[-\pi \left(\dfrac{ln\frac{f}{a_i + b_i R}}{\dfrac{c_i + d_i R}{a_i + b_i R}}\right)^2\right] & f \le f_i^{up} \\ (e_i R^{f_i}) \exp\left[-\pi \left(\dfrac{ln\frac{f_i^{up}}{a_i + b_i R}}{\dfrac{c_i + d_i R}{a_i + b_i R}}\right)^2\right] \left(\dfrac{f}{f_i^{up}}\right)^{-3,3} & f > f_i^{up} \end{cases} \quad (13.34)$$

where the empirical coefficients and the corresponding drop diameters are given in Table 13.1.

Clearly, the spectrum consists of upper and lower frequency portions separated by frequency f_i^{up}.

Note that rain rate of an individual dropsize is defined as

$$R_i = R \frac{n_i D_i^3}{\sum_{i=1}^{3} n_i D_i^3} \quad (13.35)$$

Table 13.1 Empirical coefficients of ring-wave spectra given by Lemaire et al. (2002) [22]

i	Diameter (mm)	a_i	b_i	c_i	d_i	e_i	f_i	f_i^{up} (Hz)
1	4.2	4.24	−0.006	2.59	0.003	2.49e-3	0.481	9
2	2.8	4.73	−0.001	3.46	0.009	3.29e-4	0.563	11
3	2.3	5.07	0.019	4.31	0.027	0.45e-4	0.854	12

where n_i denotes relative rain density; D_i denotes drop diameter. Intuitively, one would consider deriving rain-rate-dependent n_i from the M-P dropsize distribution of natural rain. However, our experience indicates that the empirical fixed selection of $n_1 = 15$, $n_2 = 250$, $n_3 = 835$ used by the original author [22] produces better results when compared to measurements.

For rough surface scattering calculation, the frequency domain spectrum needs to be further converted to the wavenumber domain. Following Bliven et al. (1997) [19], we have

$$\begin{aligned} W_{rain}(R,k) &= \frac{\partial f}{\partial k} W_{rain}(R,f) \\ f &= \frac{1}{2\pi}\sqrt{gk + \frac{\tau}{\rho}k^3} \end{aligned} \tag{13.36}$$

where k is in cm^{-1}; gravity g =10 m/s^2; sea water density ρ =1.025 g/cm^3; sea water surface tension τ =74 dyn/cm. Note that Eq. (13.31) represents a 1D uniform spectrum with units of m^{-3}. A factor of $1/((2\pi k))$ should be included to calculate a

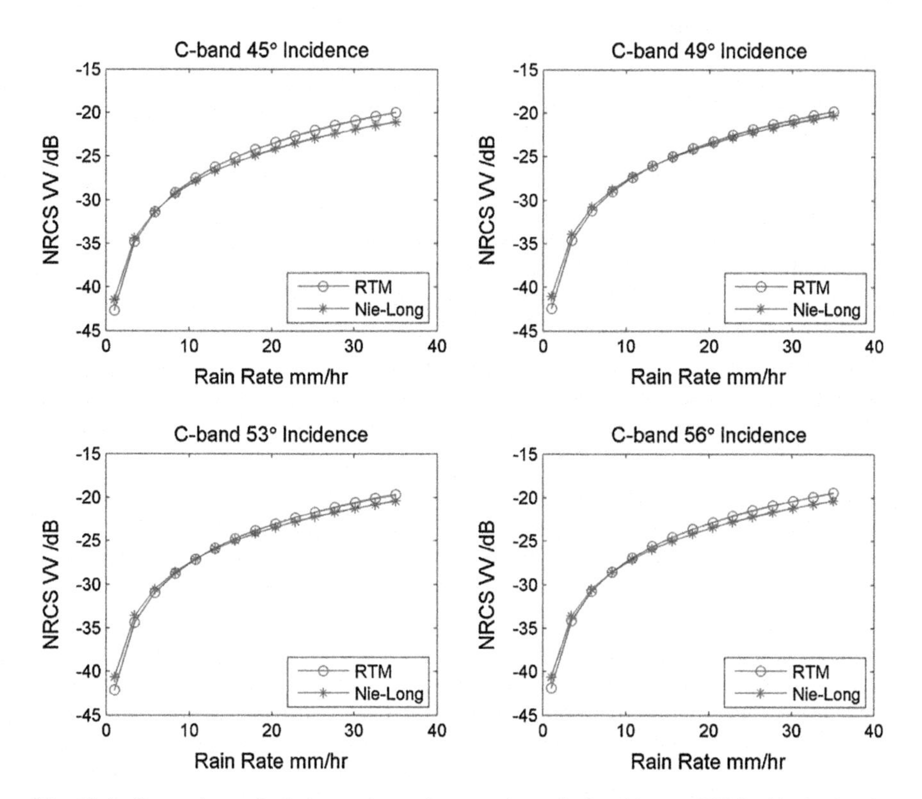

Fig. 13.4 Comparison of raindrop volumetric scattering calculated by our RTM with that by the Nie-Long empirical model under different rain conditions and incident angles at C-band

2D spectrum with units of m^{-4}. In summary, the rain-induced ring-wave spectra can be calculated given any rain rate.

13.3.4 Wind Driven Sea Surface Spectrum

The Fung-Lee spectrum [28] is used with the IEM model [30] to model wind-driven sea surface scattering. It is selected here because it was the original spectrum used to tune the IEM ocean scattering model by Chen and Fung (1992) [36]. As shown later in Fig. 13.7, the combination of the IEM and Fung-Lee spectrum has reasonable accuracy in predicting wind-driven sea surface scattering. However, it has to be pointed out that advanced spectra proposed by various authors [e.g. [37, 38] are believed to describe the actual sea roughness better. Incorporating better spectra into the IEM will be studied in future.

13.4 Validation

13.4.1 Comparison to Empirical Models and Measurements

The proposed RT Model is validated by comparing with known empirical models and measurement data.

In Fig. 13.4, the volumetric scattering from raindrops are calculated and compared with results from the empirical model given by Nie and Long (2007) [12] at C-band VV polarization. The results match well at different incidence angles under different rain rates. Note that the rainfall height is chosen as a uniform 4 km. According to [39], most rain types have the melting layer located at around 5 km. Considering that the effect of raindrop density decreases as altitude increases, we choose a uniform 4 Km as an approximation of the real scenario. Note that the raindrop volumetric scattering is linearly proportional to rainfall height (see Eq. (13.10)).

In Fig. 13.5, we compare the total attenuation as the radar wave propagates through the rainfall. The Nie-Long empirical model predicts ~1 dB higher attenuation than our RT model. Such a level of difference is considered within the margin of error of both models.

Subsequently in Fig. 13.6, the results of rain-induced surface scattering modeled by the IEM and the Log-Gaussian ring-wave spectrum are compared to the empirical model of [11] at Ku-band and the Nie-Long model at C-band. The comparison indicates a very good match between IEM/LogGaussian and Draper-Long, and a difference up to 4 dB between IEM/LogGaussian and Nie-Long at the low rain rate end.

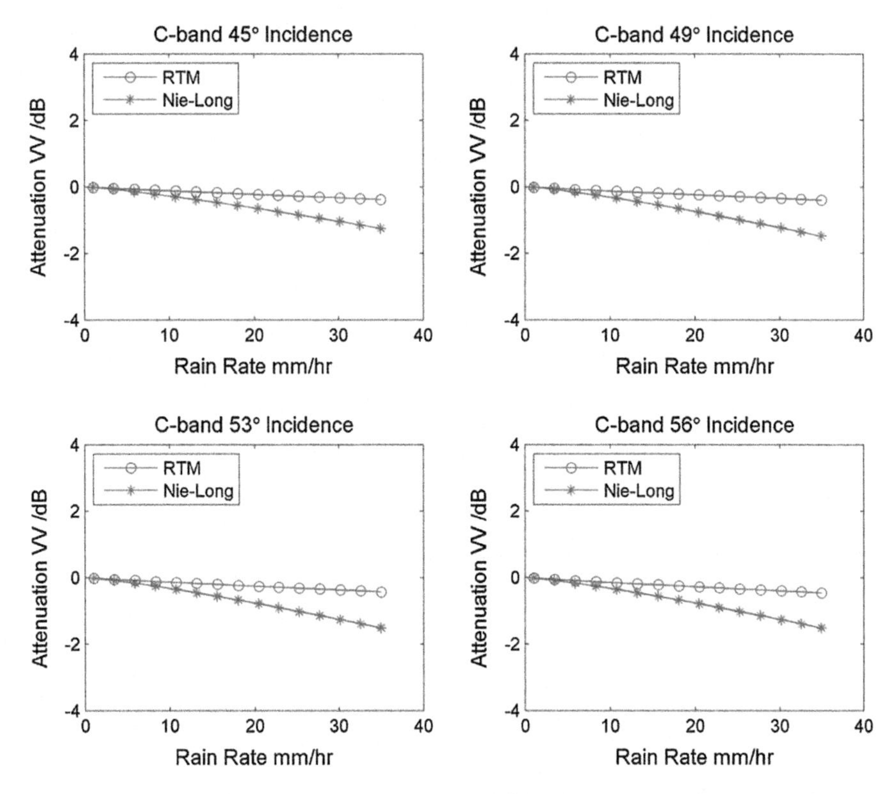

Fig. 13.5 Comparison of total attenuation through rainfall calculated by our RTM with that by the Nie-Long empirical model under different rain conditions and incident angles at C-band

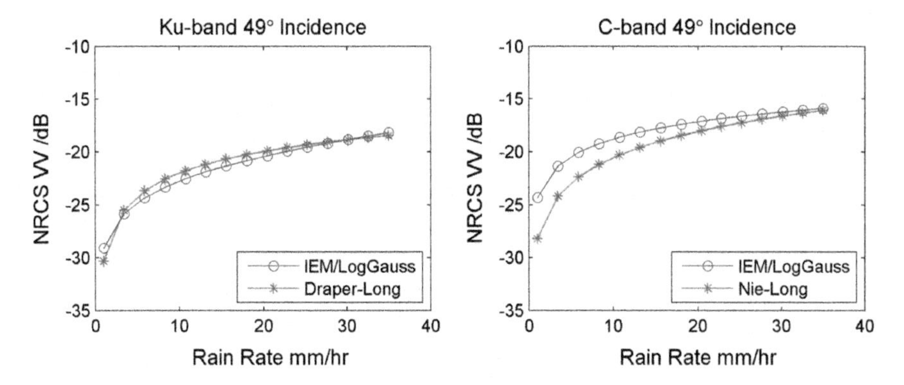

Fig. 13.6 Rain-induced sea surface ring-wave backscattering calculated by our model versus that by the Draper-Long and Nie-Long empirical models at Ku-band (*Left*) and C-band (*Right*), respectively. The IEM rough surface scattering model is used along with the Log-Gaussian ring-wave spectrum proposed by Lemaire et al. (2002) [22]

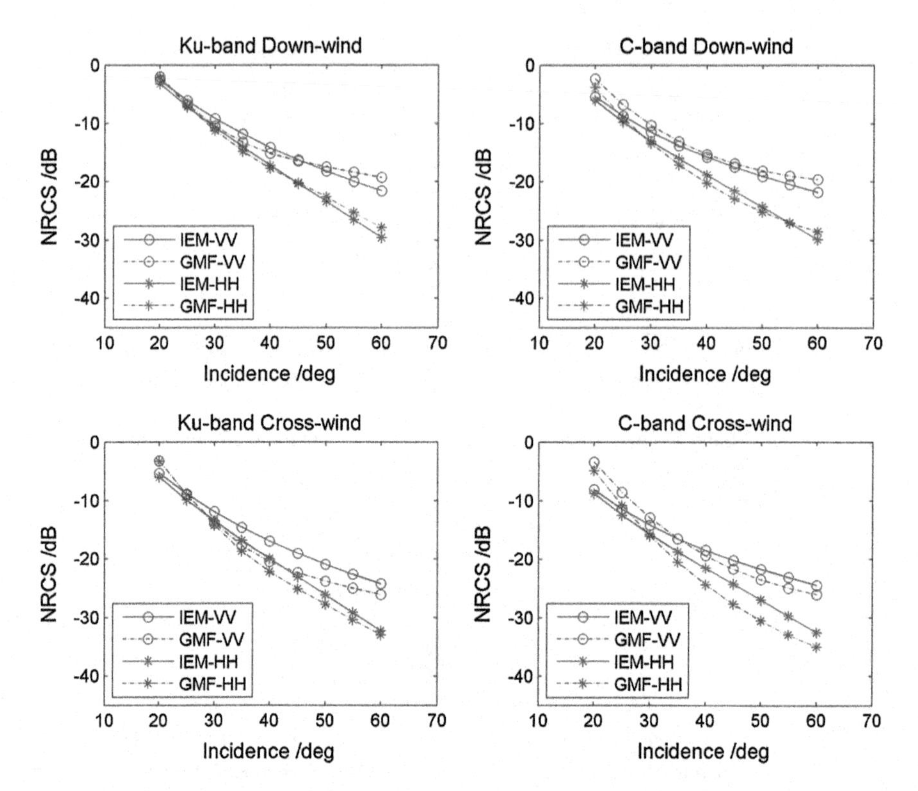

Fig. 13.7 Wind-induced sea surface gravity-wave backscattering calculated by our model versus that by the GMF empirical models at Ku-band and C-Band, and for down-wind and cross-wind circumstances, respectively; wind speed is 7 m/s; the empirical Fung-Lee sea surface spectrum is used; GMF at C-band is CMOD5; GMF at Ku-band is NSCAT-3 (KNMI)

Before proceeding to investigate the wind-rain combined effect, lets first validate the wind-only surface scattering model. In Fig. 13.7, the pure wind-driven surface scattering is compared with empirical Geophysical Model Functions (GMF), i.e. CMOD5 at C-band [40] and NSCAT-3 at Ku-band [41], where CMOD5 is extended to HH using the empirical VV/HH ratio by [42]. The IEM/Fung-Lee model matches well with GMF at both bands for the down-wind case. For cross-wind circumstances, the differences are observed as large as 5 dB, which might suggest that the Fung-Lee spectrum has a less accurate anisotropic factor.

Moore et al. (1979) conducted measurements at Ku-band of simulated rain over a wind-driven sea surface. Using the IEM with an incoherent wind-driven rough surface described by the Fung-Lee spectrum and a rain-induced rough surface described by the Log-Gaussian ring-wave spectrum, we are able to plot the results at the same settings as Fig. 13.1 of [5] (Fig. 13.8). It appears that the increment of backscattering due to rain-induced roughness is well captured by our model. For extremely high rain rates (120 mm/hr), our prediction is about ~2 dB higher. This might be attributed to

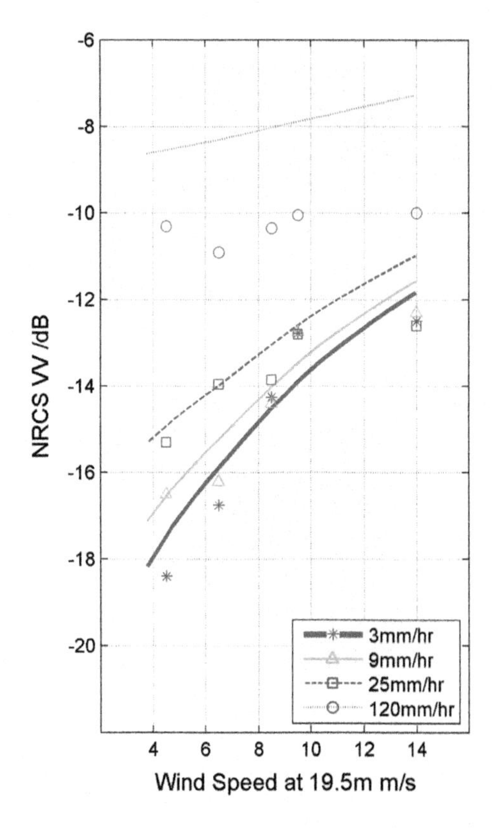

Fig. 13.8 Hybrid wind-rain induced sea surface backscattering calculated by our model at Ku-band compared to measurement data points from Fig. 1 of Moore et al. (1979) [5]

ignorance of the damping effect of strong rain applying to wind-driven roughness as well as the remaining nonlinear factors that are not taken into account.

13.4.2 Validation with Real SAR Data

To further evaluate the performance of the proposed scattering model, we present here a case study with ENVISAT ASAR data acquired offshore of New Orleans, Louisiana, USA at 16:02 UTC, June 19, 2010. The image used is a wide-swath mode image of 150 m resolution at VV polarization. The collocated wind data is obtained from the NOAA National Centers for Environmental Prediction (NCEP) Global Forecast System (GFS) weather forecast model product which has a spatial resolution of 0.5° and a temporal resolution of 1 hour. Both spatial and temporal interpolations are performed to get the collocated wind field.

The rain data is available through the NEXRAD Doppler radar network. The selected site is within the range of NEXRAD site 'KLIX' at New Orleans. The one-hour precipitation level-3 product at 4:02 PM of the same day is used.

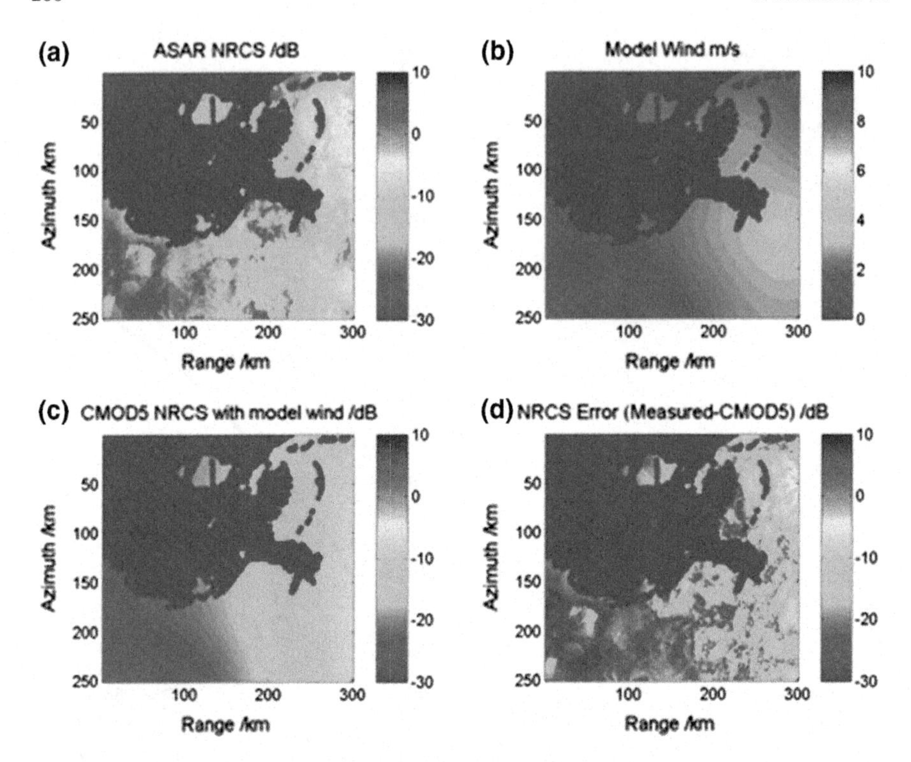

Fig. 13.9 **a** ENVISAT ASAR C-band image (radar looks from *left to right*); **b** Corresponding GFS wind field (wind directions are not shown here); **c** Simulated NRCS using CMOD5 and the GFS wind field; **d** error between CMOD5 simulation and the ASAR image

The ASAR image and wind data are plotted in Fig. 13.9. Shown in the same figure are the corresponding simulated NRCS using the CMOD5 geophysical model function [40] and the error. Looking at the NRCS error and the rain, we can visually identify some regions matched in space (see Fig. 13.10). However, due to the resolution error in the wind field and rain measurement (both spatial and temporal), the data is not sufficient for us to conduct a pixel by pixel comparison and analysis. Therefore, statistical analyses are conducted on one selected region where the pattern in the SAR image matches well with NEXRAD-observed rain cells. Note that large errors are observed in regions without rain which suggests error either in the model or in the input data.

Using the GFS wind data and NEXRAD rain data, three types of simulated NRCS data can be generated, namely, CMOD5 of wind-only scattering, IEM rough surface model with wind-only scattering, and the RTM scattering model with both wind and rain effects. These three simulated NRCS data fields are then compared with the SAR image data. The histograms and statistics are given in Fig. 13.11. Apparently, the IEM wind-only scattering model improves the mean bias. However, distinct negative tails on the left side can be observed in both CMOD5 and IEM wind-only results, which

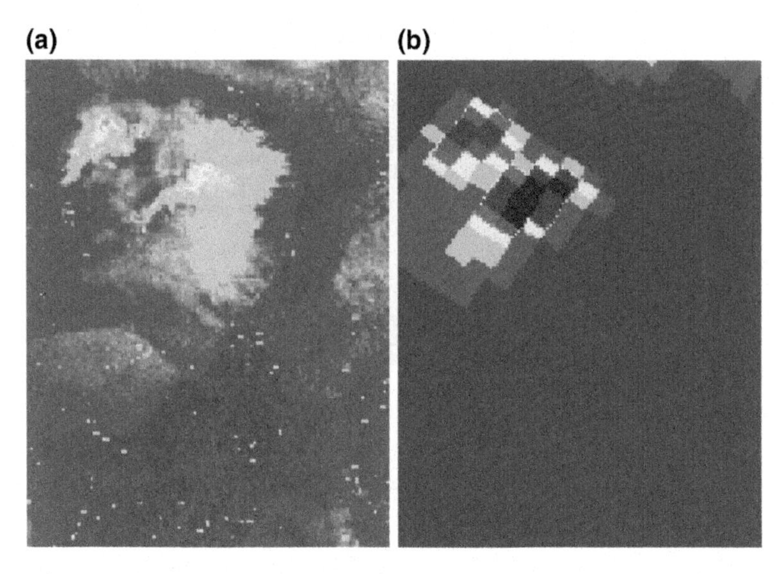

Fig. 13.10 **a** NRCS error of Fig. 13.9d with suspicious rain patches circled; **b** Corresponding rain rate data measured by NEXRAD Doppler radar; **c** Magnified selected region where the SAR pattern matches well with rain cells

indicates that wind-only models underestimate the scattering caused by rain in this case. By further incorporating rain effects in the overall wind-rain scattering model, the histogram is greatly balanced and the skewness is reduced. The net residual error is believed to be partially attributable to the mismatch in spatial resolution and the mismatch in observation times by SAR and NEXRAD, respectively.

Although a much preferred pixel-by-pixel quantitative comparison is not available due to the lack of high spatial/temporal resolution wind/rain measurement data, the overall statistics presented do, to some extent, evaluate the performance of our model. Additionally, a few representative range profiles are plotted in Fig. 13.12 where we can qualitatively evaluate how well the models match with measurements in space. Note that mismatch or shifting of rain cells is observed most likely as a result of the time difference of observation between SAR and NEXRAD. In Fig. 13.10c, we can notice the sharp circular 'gust front' [1, 20] associated with the rainfall downspread airflow. Such spatial patterns can perhaps be reproduced once the wind field alteration by rain cells is clearly understood and modeled.

To showcase the capability of the developed scattering model, we use regional WRF weather model to simulate the Hurricane Hermine occurred in North America at September 2016. The output from WRF is wind field at 10 m and rain rate data. As shown in Figs. 13.13, 13.14 and 13.15, wind and rain data are generated for September 2, 2016, UTC 00:00:00. A simulated SAR image can be calculated using the developed scattering model, which is shown in Fig. 13.16. A concurrent SAR image obtained by the ESA Sentinel-1A spaceborne SAR is also given in parallel for comparison in Fig. 13.17. The Sentinel-1A image is time stamped at September 1,

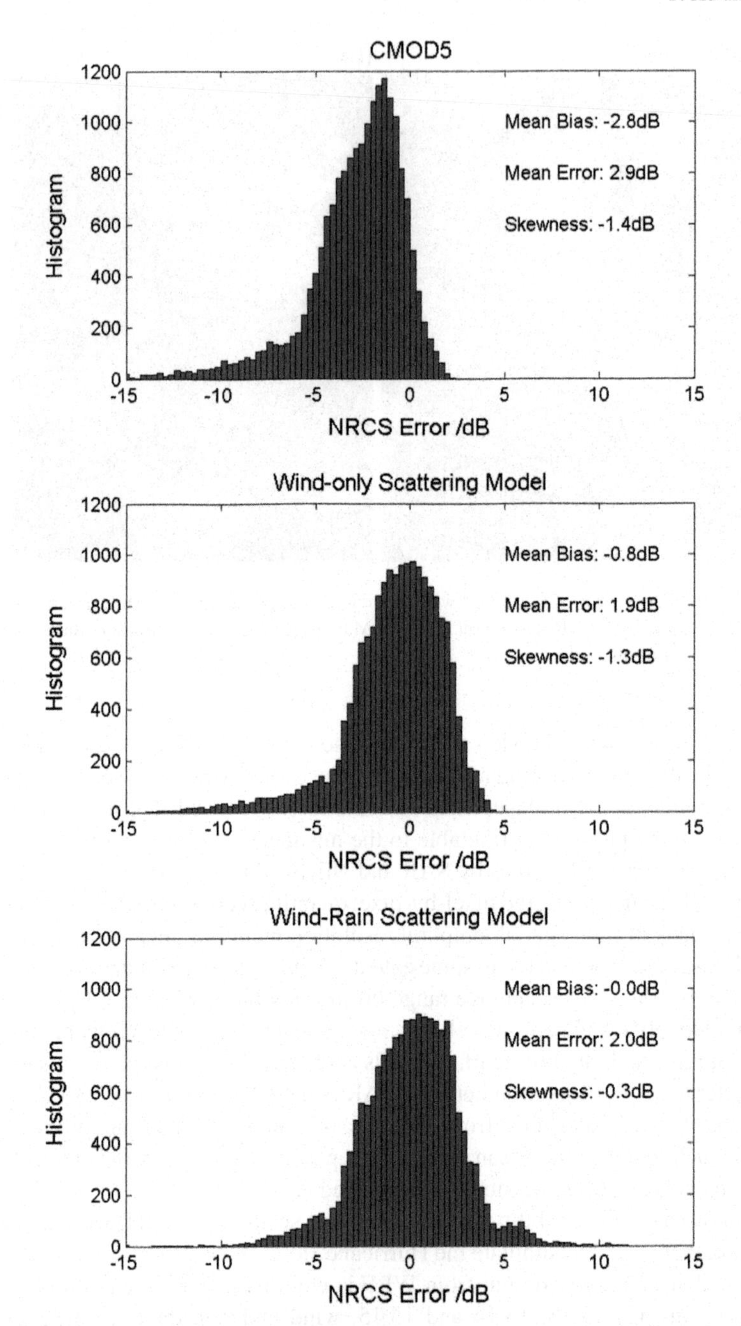

Fig. 13.11 Histograms and statistics of NRCS error of different models with respect to a SAR image of the selected region: CMOD5 (*top*), our model without rain data (*middle*), our model with rain data incorporated (*bottom*). The wind data is from GFS, while rain data is from NEXRAD

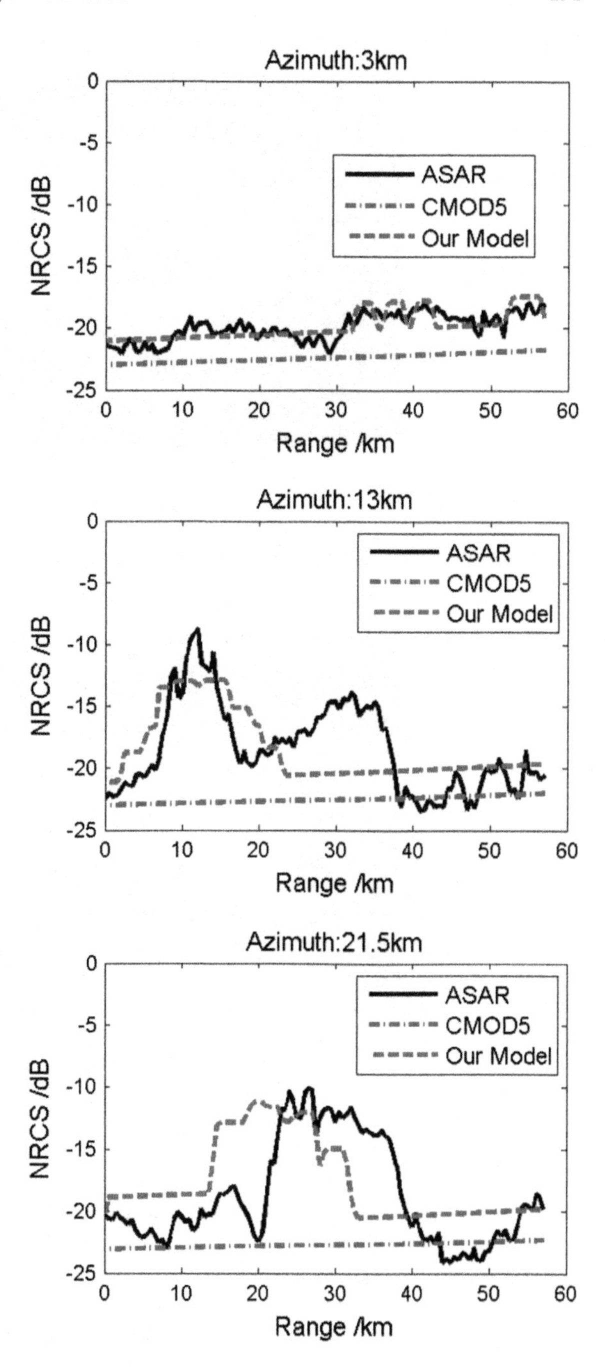

Fig. 13.12 Comparison of simulated and measured ASAR range profiles at selected azimuth positions

Fig. 13.13 WRF generated rain field of Hurricane Hermine

Fig. 13.14 WRF generated wind speed of hurricane Hermine

2016, 23: 45:17 which has only 15 min shift from the simulated image. Comparing the simulated SAR image of Hermine and the real observed one, it is found that the general level of NRCS near the eye region matches well with the real observation, i.e. in the range of [−13, −8] dB. Moreover, the general pattern of hurricane eye looks similar to what we observed from Sentinel-1A SAR image. Mismatches and shifts are likely due to time mismatch and WRF simulation error.

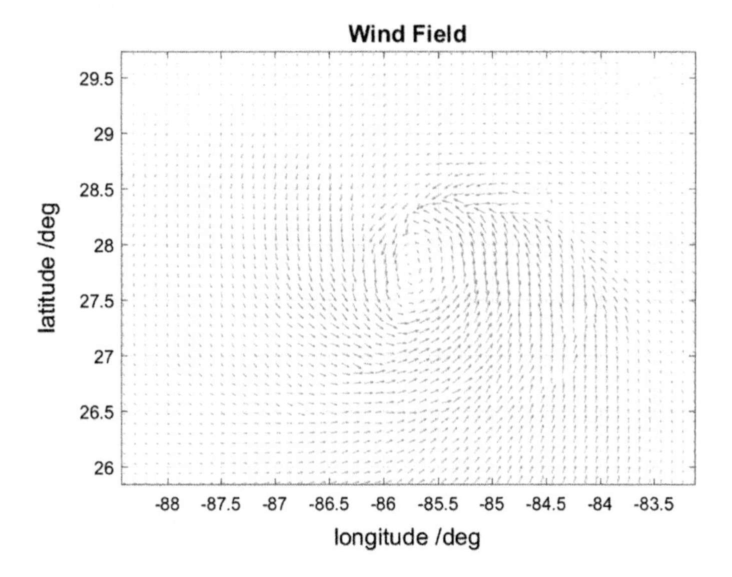

Fig. 13.15 WRF generated wind field of hurricane Hermine

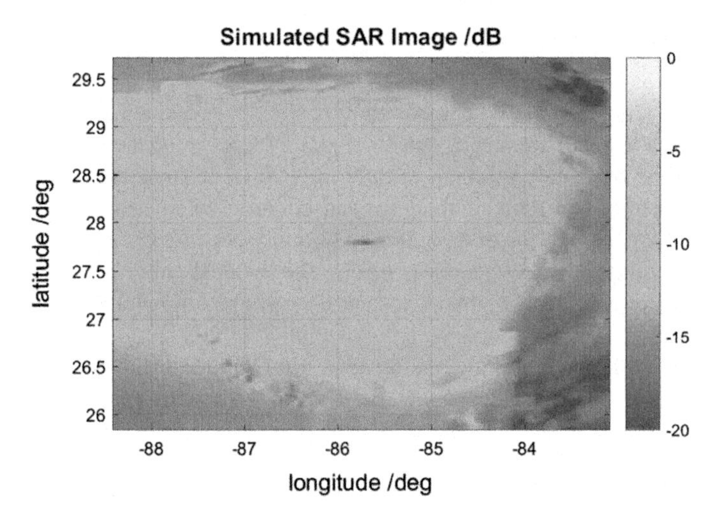

Fig. 13.16 Simulated C-band VV SAR image of hurricane Hermine using wind and rain data generated by WRF

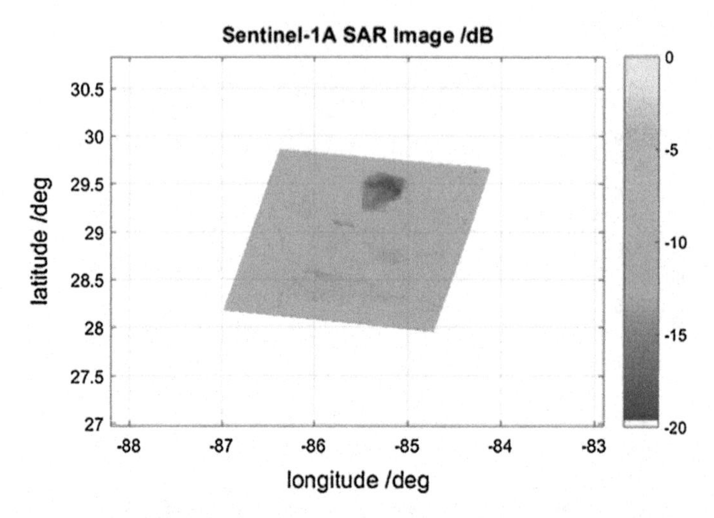

Fig. 13.17 Real SAR image of hurricane Hermine as observed by Sentinel-1A spaceborne C-band SAR (VV)

13.5 Numerical Analysis

Using the developed RTM model, we analyze the dependencies of SAR signal on rain rate, wind speed, frequency etc. In Fig. 13.18, we plot the dependencies of total scattering, wind-induced surface scattering (with rain attenuation), rain-induced surface scattering (with rain attenuation) and raindrop volumetric scattering as 3D curved surfaces on the plane of rain rate versus wind speed for C-, X- and Ku- bands, respectively. Only VV polarization is given. The incident angle is selected as 40°, while the uniform rainfall height is chosen as 4 km. The following observations can be made:

At C-band, the total scattering mainly consists of wind-induced surface scattering and rain-induced surface scattering, while raindrop volumetric scattering is negligible. A clear boundary can be found along which the two types of scattering compete for dominancy.

At X-band, raindrop volumetric scattering becomes stronger than rain-induced surface scattering for rain rates over ~6 mm/hr. Hence, raindrop volumetric scattering (RVS) replaces rain-induced surface scatterings (RSS) role to compete with wind-induced surface scattering (WSS) for dominancy. This is the case for Ku-band as well. Note that the boundary line between RVS dominancy and WSS dominancy appears to be linear, as opposed to the curved boundary between RSS dominancy and WSS dominancy.

In Fig. 13.19, we show the boundary lines which segment the wind/rain regions where different contributing factors dominate the backscattering. At C-band, wind-induced surface scattering quickly surpasses rain-related scattering for wind speeds >7 m/s. At X-band, wind-induced surface scattering and rain-induced surface

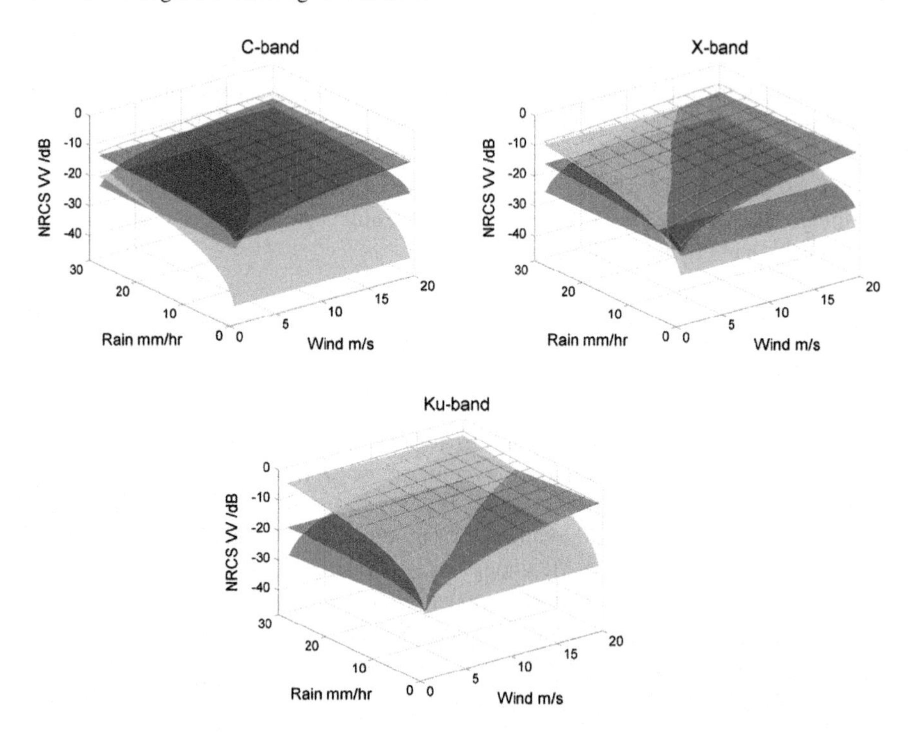

Fig. 13.18 Dependencies of different scattering contributions on wind speed (down-wind) and rain rate. (Legends: *Mesh surface* Total scattering; *Blue surface* Wind-induced surface scattering; *Red surface* Rain-induced surface scattering; *Green surface* Raindrop volumetric scattering)

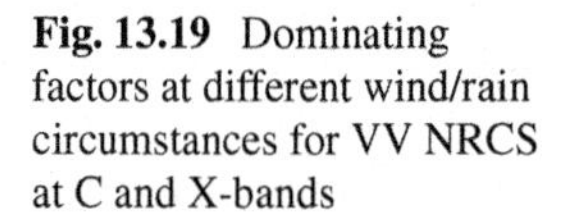

Fig. 13.19 Dominating factors at different wind/rain circumstances for VV NRCS at C and X-bands

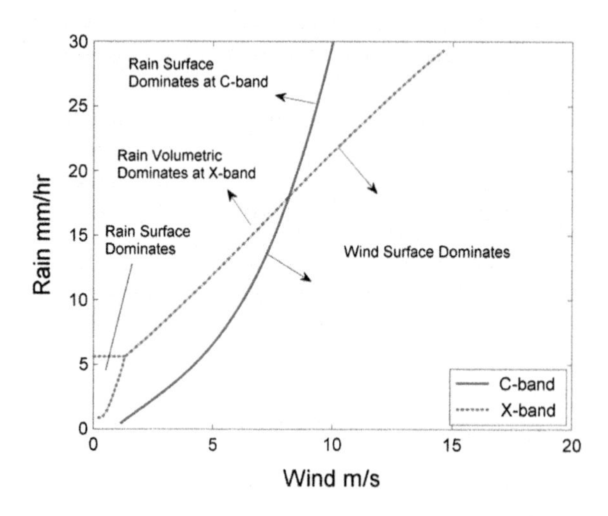

scattering increase in a linear fashion which indicates that both factors are important. Note that for the no-wind situation for X-band, raindrop volumetric scattering may dominate over surface scattering if rain rate is lower than 5 mm/hr. It should be pointed out that the analyses presented in this section have to be interpreted with caution as they do not reflect the two other rain-related factors, namely, the rain damping effect observable at L-band or small incident angles and the downspread airflow which alters the wind field.

13.6 Conclusions

A general physics-based scattering model for rainfall over the sea surface is established taking into account raindrop volumetric scattering and attenuation, rain-induced ring-wave scattering and wind-driven sea surface scattering. The model is validated against empirical models and measurements at C- and Ku-bands. A case study using an ENVISAT ASAR image with collocated wind and rain data shows that the proposed wind-rain scattering model is better than wind-only models. Using the newly developed model, numerical analyses reveal that rain-induced surface scattering plays an important role in radar backscattering at C-band, while raindrop volumetric scattering becomes dominant at higher frequency bands such as X- and Ku-bands.

This study provides the framework for incorporating different effects and factors into the modeling and lays the groundwork for future improvement in three aspects:

(a) incorporate a turbulence damping effect into the wind-driven spectrum, and validate it independently against lab/field measurements and L- and C-bands data;
(b) incorporate downspread airflow alteration to the wind field and validate it using the spatial features observed in SAR images;
(c) incorporate the geometric structure of different storms and compare them with that observed in high-resolution SAR images.

Last, but not least, to study these phenomena, Doppler sensing [43] of dynamic features of the ocean surface could potentially be used to decouple the various factors of rain-wind-wave interaction, which will be further explored.

References

1. Atlas, D. 1994. Footprints of storms on the sea: A view from spaceborne synthetic aperture radar. *Journal of Geophysical Research: Oceans* 99 (C4): 7961–7969.
2. Melsheimer, C., W. Alpers, and M. Gade. 1998. Investigation of multifrequency multipolarization radar signatures of rain cells derived from sir-c/x-sar data. *Journal of Geophysical Research* 103 (C9): 18867–18884.

3. Melsheimer, C., W. Alpers, and M. Gade. 2001. Simultaneous observations of rain cells over the ocean by the synthetic aperture radar aboard the ers satellites and by surfacebased weather radars. *Journal of Geophysical Research: Oceans* 106 (C3): 4665–4677.
4. Li, X., et al. 2013. Tropical cyclone morphology from spaceborne synthetic aperture radar. *Bulletin of the American Meteorological Society* 94 (2): 215–230.
5. Moore, R., et al. 1979. Preliminary study of rain effects on radar scattering from water surfaces. *IEEE Journal of Oceanic Engineering* 4 (1): 31–32.
6. Contreras, R.F., et al. 2003. Effects of rain on Ku-band backscatter from the ocean. *Journal of Geophysical Research: Oceans* 108 (C5).
7. Weissman, E., M.A. Bourassa, and J. Tongue. 2002. Effects of rain rate and wind magnitude on seawinds scatterometer wind speed errors. *Journal of Atmospheric and Oceanic Technology* 19: 738–746.
8. Contreras, R.F., and W.J. Plant. 2004. Ku band backscatter from the Cowlitz river: Bragg scattering with and without rain. *IEEE Transactions on Geoscience and Remote Sensing* 42 (7): 1444–1449.
9. Weissman, D.E., and M.A. Bourassa. 2008. Measurements of the effect of rain-induced sea surface roughness on the QuikSCAT scatterometer radar cross section. *IEEE Transactions on Geoscience and Remote Sensing* 46 (10): 2882–2894.
10. Stiles, B.W., and S.H. Yueh. 2002. Impact of rain on spaceborne ku-band wind scatterometer data. *IEEE Transactions on Geoscience and Remote Sensing* 40 (9): 1973–1983.
11. Draper, D.W., and D.G. Long. 2004. Evaluating the effect of rain on SeaWinds scatterometer measurements. *Journal of Geophysical Research* 109 (C12).
12. Nie, C., and D.G. Long. 2007. A C-band wind/rain backscatter model. *IEEE Transactions on Geoscience and Remote Sensing* 45 (3): 621–631.
13. Zhou, X., et al. 2012. Rain effect on C-band scatterometer wind measurement and its correction. *Acta Physica Sinica* 61 (14).
14. Nie, C., and D.G. Long. 2008. A C-band scatterometer simultaneous wind/rain retrieval method. *IEEE Transactions on Geoscience and Remote Sensing* 46 (11): 3618–3631.
15. Owen, M.P., and D.G. Long. 2011. Simultaneous wind and rain estimation for QuikSCAT at ultra-high resolution. *IEEE Transactions on Geoscience and Remote Sensing* 49 (6): 1865–1878.
16. Tournadre, J., and Y. Quilfen. 2003. Impact of rain cell on scatterometer data: 1. theory and modeling. *Journal of Geophysical Research: Oceans* 108 (C7).
17. Contreras, R.F., and W.J. Plant. 2006. Surface effect of rain on microwave backscatter from the ocean: Measurements and modeling. *Journal of Geophysical Research: Oceans* 111 (C8).
18. Wetzel, L.B. 1990. On the theory of electromagnetic scattering from a raindrop splash. *Radio science* 25 (6): 1183–1197.
19. Bliven, L.F., P.W. Sobieski, and C. Craeye. 1997. Rain generated ring-waves: measurements and modelling for remote sensing. *International Journal of Remote Sensing* 18 (11): 221–228.
20. Alpers, W., and C. Melsheimer. 2005. *Synthetic Aperture Radar Marine Users' Manual*, chapter Chapter 4 Rainfall. NOAA/NESDIS/STAR.
21. Craeye, C., P. Sobieski, and L. Bliven. 1999. Radar signature of the sea surface perturbed by rain. *Proceedings of IEEE 1999 International Geoscience and Remote Sensing Symposium, IGARSS'99*, 1: 1374–1389.
22. Lemaire, D., et al. 2002. Drop size effects on rain-generated ring-waves with a view to remote sensing applications. *International Journal of Remote Sensing* 23 (12): 2345–2357.
23. Capolino, F., et al. 1998. EM models for evaluating rain perturbation on the NRCS of the sea surface observed near nadir. *IEEE Proceedings-Radar, Sonar and Navigation* 145: 4.
24. Kudryavtsev, V.N., V.K. Makin, and B. Chapron. 1999. Coupled sea surface-atmosphere model: 2. spectrum of short wind waves. *Journal of Geophysical Research: Oceans* 104 (C4): 7625–7639.
25. Bringer, B.C., A. Mouche, and C.A. Guerin. 2014. Revisiting the short-wave spectrum of the sea surface in the light of the weighted curvature approximation. *IEEE Transactions on Geoscience and Remote Sensing* 52 (1): 679–689.

26. Portabella, M., et al. 2012. Rain effects on ASCAT-retrieved winds: toward an improved quality control. *IEEE Transactions on Geoscience and Remote Sensing* 50 (7): 2495–2506.
27. Xu, F., et al. 2015. A backscattering model of rainfall over rough sea surface for synthetic aperture radar. *IEEE Transactions on Geoscience and Remote Sensing* 53 (6): 3042–3054.
28. Fung, K., and K.K. Lee. 1982. A semi-empirical sea-spectrum model for scattering coefficient estimation. *IEEE Journal of Oceanic Engineering* OE–7 (4): 166–176.
29. Xu, F., and Y.Q. Jin. 2006. Imaging simulation of polarimetric SAR for a comprehensive terrain scene using the mapping and projection algorithm. *IEEE Transactions on Geoscience and Remote Sensing* 44 (11): 3219–3234.
30. Fung, A.K., and K. Chen. 2010. *Microwave scattering and emission models for users*. Boston: Artech House.
31. Jin, Y.Q., and F. Xu. 2013. *Polarimetric Scattering and SAR Information Retrieval*. New York: Wiley-IEEE.
32. Ishimaru, A., et al. 1982. Multiple scattering calculations of rain effects. *Radio Science* 17 (6): 1425–1433.
33. Czekala, H., and C. Simmer. 1998. Microwave radiative transfer with nonspherical precipitating hydrometeors. *Journal of Quantitative Spectroscopy and Radiative Transfer* 60 (3): 365–374.
34. Jin, Y.Q. 1993. *Electromagnetic Scattering Modeling for Quantitative Remote Sensing*. Singapore: World Scientific.
35. Chuang, C., and K.V. Beard. 1990. A numerical model for the equilibrium shape of electrified raindrop. *Journal of the Atmospheric Sciences* 47 (11): 1374–1389.
36. Chen, K.S., A.K. Fung, and D.A. Weissman. 1992. A backscattering model for ocean surface. *IEEE Transactions on Geoscience and Remote Sensing* 30: 4.
37. Apel, J.R. 1994. An improved model of the ocean surface wave vector spectrum and its effects on radar backscatter. *Journal of Geophysical Research: Oceans* 99 (C3): 16269–16291.
38. Elfouhaily, T., et al. 1997. A unified directional spectrum for long and short wind-driven waves. *Journal of Geophysical Research* 102 (C7): 811–817.
39. Petty, G.W. 2001. Physical and microwave radiative properties of precipitating clouds. part ii: A parametric 1d rain-cloud model for use in microwave radiative transfer simulations. *Journal of Applied Meteorology* 40 (12): 2115–2129.
40. Hersbach, H., A. Stoffelen, and S. de Haan. 2007. An Improved C-band scatterometer ocean geophysical model function: CMOD5. *Journal of Geophysical Research: Oceans* 112 (C3).
41. KNMI. Nscat-3 geophysical model function. 2013. http://www.knmi.nl/scatterometer/nscati$_$gmf/.
42. R. Thompson, T. M. Elfouhaily, and B. Chapron. 1998. Polarization ratio for microwave backscattering from the ocean surface at low to moderate incidence angles. *1998 IEEE International Geoscience and Remote Sensing Symposium Proceedings, 1998. IGARSS'98.*, 3:1671–1673.
43. Mouche, A., et al. 2012. On the use of doppler shift for sea surface wind retrieval from sar. *IEEE Transactions on Geoscience and Remote Sensing* 50 (7): 2901–2909.

Chapter 14
Synthetic Aperture Radar Observations of Extreme Hurricane Wind and Rain

Guosheng Zhang, Xiaofeng Li and William Perrie

Abstract Over the last decades, data from spaceborne Synthetic Aperture Radar (SAR) have been used in hurricane research. However, some issues remain: (1) many SAR images capture incomplete hurricane core structures; (2) the radar signal is attenuated by the heavy precipitation associated with hurricane; (3) wind directions retrievals are not available from the cross-polarized SAR measurements. When wind is at hurricane strength, the wind speed retrievals from co-polarized SAR may have errors because the backscatter signal may experience saturation and become double-valued. By comparison, wind direction retrievals from cross-polarization SAR are not possible until now. In this study, we develop a two-dimensional model, the Symmetric Hurricane Estimates for Wind (SHEW) model based on the mean wind profile in all radial directions, and combine it with the modified inflow angle model to detect hurricane morphology and estimate the wind vector field imaged by cross-polarization SAR. By fitting SHEW to the SAR derived hurricane wind speed, we find the initial closest elliptical-symmetrical wind speed field, hurricane center location, major and minor axes, the azimuthal (orientation) angle relative to the reference ellipse, and maximum wind speed. This set of hurricane morphology parameters, along with the speed of hurricane motion, are input to the inflow angle model modified with an ellipse-shaped eye, to derive the hurricane wind direction. A one-half modified Rankine vortex (OHMRV) model is proposed to describe the hurricane wind profile, particularly for those wind profiles with a wind speed maximum and an inflection point possibly associated with the degeneration of the inner wind maximum in the hurricane reintensification phase. The proposed method works well in area with significant radar attenuation by precipitation. Moreover, five possible mechanisms for

G. Zhang (✉) · W. Perrie
Fisheries and Oceans Canada, Bedford Institute of Oceanography,
Dartmouth, NS, Canada
e-mail: Guosheng.Zhang@dfo-mpo.gc.ca

W. Perrie
e-mail: William.Perrie@dfo-mpo.gc.ca

X. Li
GST, National Oceanic and Atmospheric Administration (NOAA)/NESDIS,
College Park, MD 20740, USA
e-mail: xiaofeng.li@noaa.gov

X. Li (ed.), *Hurricane Monitoring With Spaceborne Synthetic Aperture Radar*, Springer Natural Hazards, DOI 10.1007/978-981-10-2893-9_14

the rain effects on the spaceborne C-band SAR observations are investigated: (1) attenuation and (2) volume backscattering for the microwave transfer in atmosphere; as well as (3) diffraction on the sharp edges of rain products, and (4) rain-induced damping to the wind waves and (5) rain-generated ring waves on the ocean surface.

14.1 Hurricane Extreme Wind and Rain

Accurate analyses of hurricane (or typhoon) sea surface wind field, intensity and structure, are critical in enhancing readiness and mitigating risk for coastal communities worldwide. Previous theoretical and numerical studies have tried to understand why and how a hurricane forms, dynamically and thermodynamically, and how its eye interacts with the eyewall and the circulation in the outer core region [1–5]. However, determining the inner core and surface wind field structure of hurricanes remains a considerable operational challenge to the hurricane community [6], even when low-level aircraft reconnaissance data are available. This is partially because hurricane wind fields have high azimuthal variability and aircraft typically travel along radial legs relative to the eye, at roughly fixed azimuths [7–9].

Compared with optical satellite sensors, spaceborne synthetic aperture radar (SAR) has advantages in observing the two-dimensional sea surface wind field, with high resolution and large spatial coverage, in almost all-weather conditions. Since the first spaceborne SAR image became available in 1978 [10], hurricanes have been frequently observed by spaceborne SAR images. However, the number of SAR images covering the entire hurricane system has been limited until recently, when large numbers of hurricane images were acquired by the RADARSAT, ENVISAT, and Sentinel-1 SARs [11].

The characteristics of hurricane eyes (HEs) are strongly determined by the details of the life cycles and evolution of these tropical cyclones. Statistically, it has been shown that when hurricanes have higher intensity, their eyes tend to become more symmetric, and the area occupied by the hurricane eye, defined by the minimum wind field area, tends to be smaller [12]. Inner-core structure and intensity changes are also found to be correlated with eyewall replacement cycles [4]. Accurate determination of HEs, and accurate estimation of hurricane intensities and structures from SAR images, can contribute to better understanding of the dynamics of hurricane genesis, morphology and movement. Generally, NOAA aircraft fly through the hurricanes, providing measurements of the inner-core intensities and tangential wind profiles [8]. However, aircraft only can supply single point observations along the flight track. Thus, there might be some variance in the aircraft-measured tangential wind profiles, particularly for those hurricanes without symmetric structures. For example, we can assume that the true maximum wind speed exists somewhere on the major axis direction, for a non-symmetric hurricane with an elliptic eye. Therefore, if the aircraft flies along the minor axis of this hurricane, instead of the major axis, the maximum wind and the radius of maximum wind (RMW) observations are possibly not accurate.

With the two-dimensional hurricane surface wind estimation model, referred to as the Symmetric Hurricane Estimates for Wind (SHEW) model, and the developed inflow angle model, a complete hurricane core surface wind vector field from a SAR image will be estimated. The SHEW model is a continuation of a mean wind profile estimation method [13] which was proposed to analyze the wind speeds distributed along the radii.

It has been routinely assumed that hurricane eyes are circular, in studies of hurricanes with along-track observations made by the airborne Stepped Frequency Microwave Radiometer (SFMR) and the Global Positioning System (GPS) dropwindsondes [14]. This circular hurricane eye assumption was used to analyze the hurricane core dynamics, i.e., vortex Rossby wave dynamics [15], the eyewall replacement cycles [4], wind speed asymmetries [16] and hurricane pressure-wind model [9, 17, 18]. Based on the aircraft reconnaissance datasets, a set of continuous analytic functions was developed [15]. Recent research results indicate that most hurricane eye shapes are in the form of circles or ellipses [12], although there are a small number of hurricanes with different shapes of eyes from circles or ellipses [19]. In this study, we combine the assumed elliptical eye shape and the radial continuous analytic function to develop the SHEW model, which we suggest is close to the actual hurricane surface wind speed field.

In general, rain has two significant effects on the normalized radar-backscatter cross section (NRCS) of the SAR: (1) raindrops induce volumetric scattering and attenuation in the atmosphere and (2) rain alters the roughness of the ocean surface. The aim of this study is to understand each of these mechanisms under hurricane conditions. In the atmosphere, raindrops induce volumetric scattering and attenuation during microwave transfer. On the ocean surface, rain generates scatterers and induces damping. As a rain droplet impacts the water surface, it typically creates a crater with a crown then evolves into a vertical stalk. With the initial impact and the subsequent collapse of the crater, crown and stalk, concentric ring waves are created and radiate outward [20]. All of the splash products are potential scatterers of microwave radiation. On the other hand, rain has long been known to damp gravity waves. Tsimplis [21] provided a comprehensive description of possible damping mechanisms: change to nonlinear wave-wave interactions and the corresponding transfer of wave energy, alteration of the surface boundary condition, variations of kinematic viscosity associated with the different temperature and salinity of rain, and rain-induced turbulent damping. Additionally, it also attenuates the NRCS from the sea surface and the rain drops induce volume backscatter in the atmosphere. In this chapter, the hurricane rain effect on C-band SAR observations in terms of mechanisms related to the ocean surface part, as well as those related to the atmosphere are investigate. A composite model to include the rain effects in the atmosphere and also, the rain effects on the ocean surface is formulated.

The remainder of this chapter is organized as follows. Hurricane mean wind profile method is provided in Sect. 14.2. Then, we describe the SHEW model and revised inflow angle model in Sect. 14.3 and show the reconstructed idealized wind vector in Sect. 14.4, as well as possible mechanisms of rain effect on the SAR hurricane observation are provided in Sect. 14.5. And conclusions are given in Sect. 14.6.

14.2 Mean Wind Profile

14.2.1 Hurricane Eye Determination

For cross-polarized SAR images of hurricanes, low backscatters are associated with low winds in the hurricane eye region, while high backscatters are related to strong winds in the eyewall area. Thus, apparent NRCS gradient variations exist at the junction between the hurricane eye and the eyewall. These characteristics are very suitable for hurricane eye edge detection. Previous studies used HH-polarized RADARSAT-1 or VV-polarized ENVISAT ASAR images to determine hurricane eye characteristics [22, 23]. Cross-polarized (HV or VH) SAR observations have an advantage over those with HH- or VV-polarization because they do not become saturated under high wind conditions [24]. Here, we extract hurricane eye centers and their extents using an approach similar to that proposed in [23], with RADARSAT-2 wide swath SAR images as examples of Hurricanes Earl and Gustav acquired at VH polarization.

Before determining hurricane eye centers and extents, we extracted sub-scenes of observed hurricane eyes, as shown in Figs. 14.1a and 14.2a, which consist of 2000 pixels by 2000 pixels, covering a surface area of 100 km by 100 km. The darker region of the hurricane eyes were detected by using the first threshold (80% of the mean NRCS of the sub-scenes). The average of pixel longitudes and latitudes within the detected darker area are used as the initial hurricane eye center locations (Figs. 14.1b and 14.2b). Subsequently, we search for the Maximum Radiometric Gradient Points (MRGPs) along all radial directions from the initial center location for all integer angles ranging between 0° and 360°. The MRGPs are illustrated in Figs. 14.1c and 14.2c. The second threshold (mean value of these MRGPs) is finally used to estimate the extent of the eye area. Figures 14.1d and 14.2d show the locations of the hurricane eye centers, as estimated by computing the mean of all pixel locations in the eye regions. To assess the accuracy of hurricane eye determination, we compared the SAR-derived eye centers and diameters with those measured by SFMR. As shown in Table 14.1, they are in good agreement.

14.2.2 Hurricane Intensity and Structure Estimation

Tangential wind profiles observed by SFMR are occasionally affected by the hurricane non-axisymmetric structure characteristics, which are difficult to eliminate because the SFMR measures the hurricane surface winds only in the along-track direction. For example, if the hurricane eyes are elliptical in shape, SFMR measurements might be taken in the major axis direction, while the true maximum wind speed might exist along the minor axis direction. In this case, the estimated maximum wind and its radius from SMFR are possibly inaccurate. SAR can observe hurricanes in two dimensional directions, and thus we can obtain the wind profile in the each

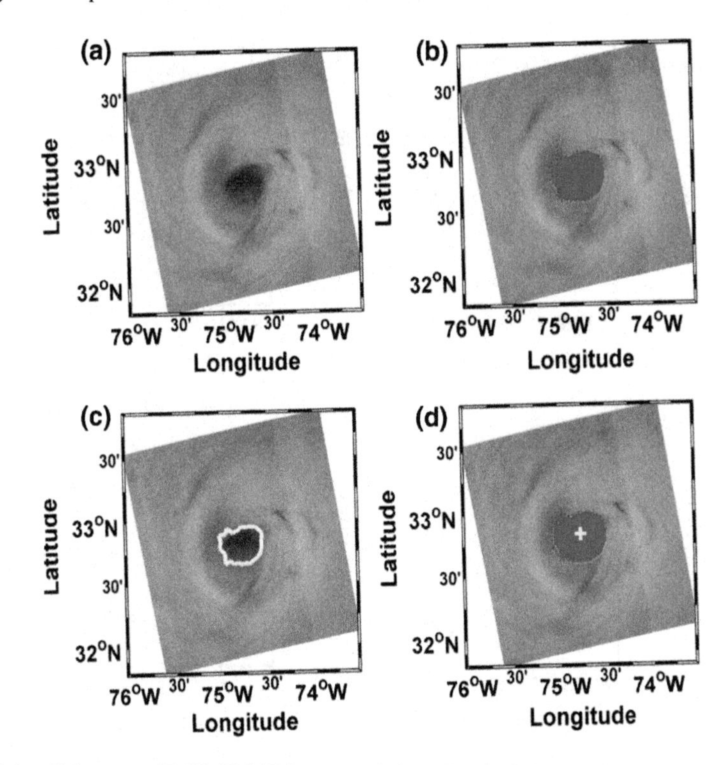

Fig. 14.1 **a** Sub-scene of RADARSAT-2 cross-polarized ScanSAR image with observed Hurricane Earl eye (22:59 UTC, September 2, 2010). **b** First guess region of hurricane eye (*blue area*). **c** The maximum radiometric gradient points (*white line*). **d** The extracted eye extent (*blue area*) and the center (*white plus sign*). RADARSAT-2 Data and Product MacDonald, Dettiler, and Associates Ltd., All Right Reserved

radial direction. Therefore, non-axisymmetric effects on estimates of intensity and structure can be removed by averaging wind profiles in all radial directions.

The hurricane tangential wind profiles are generally smoothed or approximated by continuous analytic functions with adjustable parameters that represent physically meaningful aspects of the profiles. Some typical continuous analytic functions, for example, the single modified Rankine vortex (SMRV) [4] and the double modified Rankine vortex (DMRV) [15] are used to characterize the hurricane wind profile when there is only one wind speed maximum, or two wind maxima. The hurricane eyewall replacement cycle (ERC) process includes three distinct phases [3]: (1) Intensification, (2) Weakening, and (3) Reintensification. The first phase is characterized by both inner and outer maxima intensifying, occurring at contracting radii, respectively. The second phase is bounded by the time-interval from the peak intensity of the inner eyewall until when the outer wind maximum surpasses the inner eyewall intensity. The first and second phases have similar features in that both include two wind maxima. Thus, the DMRV model is suitable to describe the wind profiles in these two phases. The basic characteristic of the third phase is that the inner wind

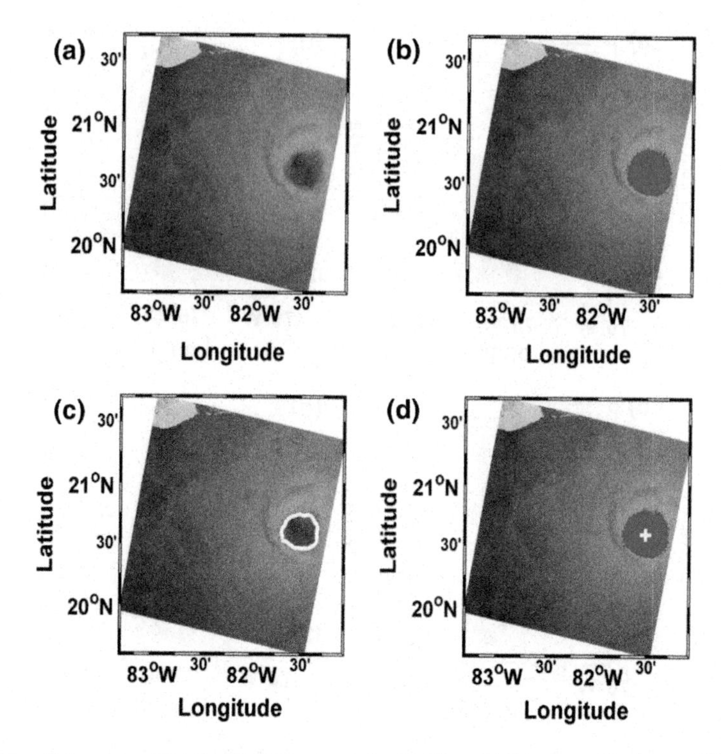

Fig. 14.2 **a** Sub-scene of RADARSAT-2 cross-polarized ScanSAR image with observed Hurricane Gustav eye (11:28 UTC, August 30, 2008). **b** First guess region of hurricane eye (*blue area*). **c** The maximum radiometric gradient points (*white line*). **d** The extracted eye extent (*blue are*) and the center (*white plus sign*). RADARSAT-2 Data and Product MacDonald, Dettiler, and Associates Ltd., All Right Reserved

Table 14.1 Comparisons between OHMRV model-derived eye center locations and diameters, intensity and RMW of Hurricane Earl and those measured by SFMR

	Eye center (degree)	Eye diameter (km)	Intensity (m/s)	RMW (km)
OHMRV	(32.841°N, 74.778°W)	113.4	32.1	56.7
SFMR	(32.790°N, 74.824°W)	115.9	35	61.2

maximum decays and eventually disappears in the eye. For this phase, the wind profile not only includes an outer wind maximum but also an inflection point. This inflection point is possibly generated by the degeneration of the inner wind maximum. In order to describe wind profile in this phase, we propose a one-half modified Rankine vortex (OHMRV) model,

$$V_r = \begin{cases} v_1\left(\frac{r}{r_1}\right), & (r \leq r_1) \\ (v_2 - v_1)\left(\frac{r-r_1}{r_2-r_1}\right) + v_1, & (r_1 < r \leq r_2) \\ v_1\left(\frac{r_2}{r}\right)^{\alpha_1}, & (r_2 < r \leq 150\,\text{km}) \end{cases} \tag{14.1}$$

where v_1 is the wind speed of the inflection point at radius r_1, v_2 is the maximum wind speed at radius r_2, and α_1 is the scaling parameter that adjusts the profile shape. The combination of r_1, r_2, v_1, v_2, and α_1 are selected in order to give the smallest root-mean-squared error between the parameterized fit and SAR-derived wind profiles in all radial directions. The OHMRV model is built on the SMRV model.

The SMRV model may be represented as

$$V_r = \begin{cases} v_1\left(\frac{r}{r_1}\right), & (r \leq r_1) \\ v_1\left(\frac{r_1}{r}\right)^{\alpha_1}, & (r_1 < r \leq 150\,\text{km}) \end{cases} \tag{14.2}$$

The function V_r is zero at the vortex center and increases to a maximum tangential wind value of v_1 at radius r_1. Thereafter, the function then decreases gradually away from the storm center according to the decay parameter α_1. As α_1 increases, the outer wind field quickly decays and the maximum intensity becomes more peaked in structure.

The double modified Rankine vortex (DMRV) is represented by the equation

$$V_r = \begin{cases} v_1\left(\frac{r}{r_1}\right), & (r \leq r_1) \\ v_1\left(\frac{r}{r_1}\right)^{\alpha_1}, & (r < r_1 \leq r_{moat}) \\ v_1\left(\frac{r_1}{r_{moat}}\right)^{\alpha_1} + \left[\frac{v_2 - v_1\left(\frac{r_1}{r_{moat}}\right)^{\alpha_1}}{r_2 - r_{moat}}\right](r - r_{moat}), & (r_{moat} < r \leq r_2) \\ v_2\left(\frac{r_2}{r}\right)^{\alpha_2}, & (r_2 < r \leq 150\,\text{km}) \end{cases} \tag{14.3}$$

Compared to the SMRV model, five new parameters (r_{moat}, v_{moat}, r_2, v_2 and α_2) are introduced in the DMRV model which allow the identification of an outer wind maximum. The portion of the fit in the SMRV model ceases at a specific pivot position between the two wind maxima. This position is labeled (r_{moat}, v_{moat}). The subscript 'moat' simply refers to any given position between the inner and outer wind maxima. Therefore, the position (r_{moat}, v_{moat}) is analogous to the vortex center in the SMRV model.

In this study, the procedure for estimating hurricane radial averaged intensity and structure parameters can be summarized as follows: (1) we estimate the wind speeds of Hurricanes Gustav and Earl with the cross-polarized SAR high wind speed retrieval model, (2) we obtain the mean wind profile by averaging wind profiles in all radial directions (using 1° intervals), (3) we use OHMRV or SMRV model to fit

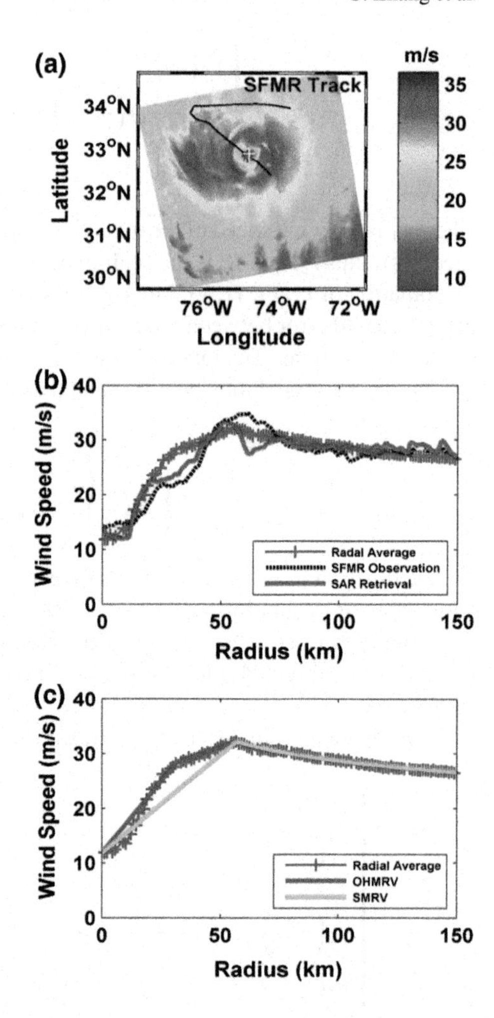

Fig. 14.3 **a** Wind speeds from VH-polarized SAR image of Hurricane Earl acquired at 22:59 UTC, September 2 2010, **b** SAR-derived mean wind profile (*blue asterisk line*), SFMR-measured wind profile at 23:27 UTC, September 2, 2010 (*black dash line*), and the SAR-retrieved wind profile in the SFMR track (*red solid line*). **c** Same as the *blue asterisk line* in Fig. 14.3b, the one-half modified Rankine vortex (OHMRV) model-fitted wind profile (*magenta solid line*), and the single modified Rankine vortex (SMRV) model-fitted wind profile (*green solid line*)

the mean wind profile, (4) we extract the maximum wind speed and its radius from the fitted wind profiles.

Figures 14.3a and 14.4a show the SAR-retrieved wind speeds with RADARSAT-2 cross-polarized SAR images of Hurricanes Earl and Gustav, respectively. The winds in the eye and eyewall regions are well retrieved because the cross-polarization ocean backscatters are not saturated under high wind conditions. Figures 14.3b and 14.4b illustrate the wind profiles from average of SAR-derived wind speeds in all radial directions (profile I), from the SFMR along-track measurements (profile II), and from SAR retrievals along the SFMR track (profile III).

It is notable that, whether for Hurricane Earl in Fig. 14.3b or for Hurricane Gustav in Fig. 14.4b, although the wind speed maximum of profile I is smaller than those in profiles II and III, it is more suitable to characterize non-axisymmetric hurricane wind profile than the other two. The reason for this result is that profile I averages

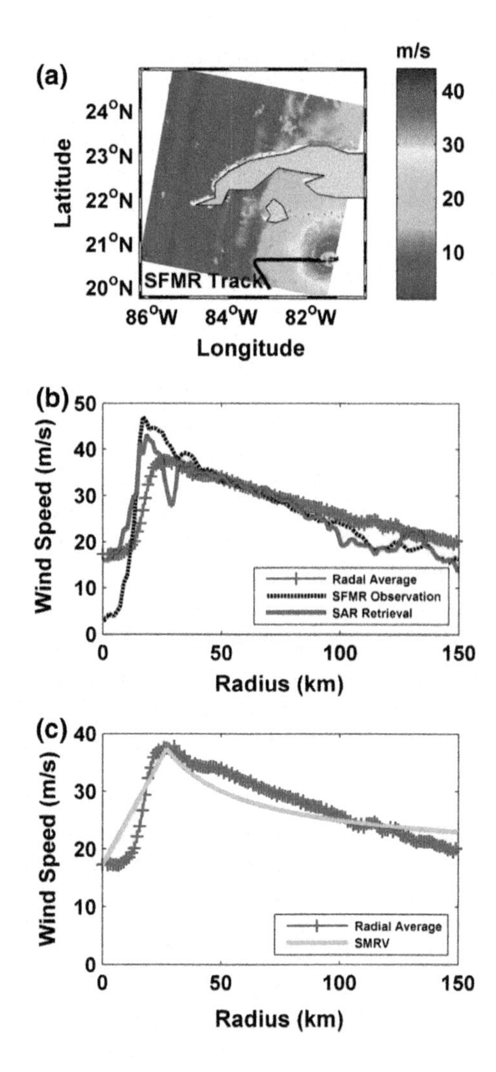

Fig. 14.4 **a** Wind speeds from SAR VH-polarized image of Hurricane Gustav acquired at 11:28 UTC, August 30 2 2008, **b** SAR-derived mean wind profile (*blue asterisk line*), SFMR-measured wind profile at 11:17 UTC, August 30, 2008 (*black dash line*), and the SAR-retrieved wind profile in the SFMR track (*red solid line*). **c** Same as the *blue asterisk line* in Fig. 14.4b, the single modified Rankine vortex (SMRV) model-fitted wind profile (*green solid line*)

the estimations in all radial directions, whereas the other two profiles represent local measurements along flight track. Although, profile III is close to profile II, obvious wind speed differences can be found. Intense precipitation, with high rain rates, can potentially account for part of the bias.

Figure 14.3b clearly shows that an inflection point and a wind maximum simultaneously exist in the wind profiles, from the SAR radial average and from the SFMR along track measurements. The inflection point in the wind profile is possibly associated with the inner wind maximum degeneration, during the reintensification phase in the ERC process. Previous studies showed that this phase could occur when the hurricanes tend to make landfall [3]. When Hurricane Earl was imaged by RADARSAT-2, it was approaching the southeast coast of North Carolina. By chance, the snapshot

Table 14.2 Comparisons between SMRV model-derived eye center locations and diameters, intensity and RMW of Hurricane Gustav and those measured by SFMR

	Eye center (degree)	Eye diameter (km)	Intensity (m/s)	RMW (km)
SMRV	(20.598°N, 81.508°W)	53.4	37.8	26.7
SFMR	(20.580°N, 81.508°W)	44.8	47	16.7

SAR image captured the reintensification phase, which corresponds to an outer wind maximum and a degenerated inner wind (inflection point) in the wind profile. Once the inner wind maximum can no longer be detected, the ERC is considered complete. Figure 14.3c shows model-fitted wind profiles from SMRV and OHMRV for Hurricane Earl, using the SAR-derived mean wind speeds in all radial directions. The SMRV model was developed in order to characterize the wind profile by only including one wind speed maximum. Thus, for this case, it is evident that the OHMRV model is more suitable to describe the hurricane wind profile in its third phase, than the SMRV model. When the reintensification phase is over, the SMRV model might be a good candidate model to describe the wind profile, because then the inner wind maximum can no longer be detected, the ERC is considered complete and the storm structure reverts to a single eyewall structure.

Figure 14.4c only shows the wind profile of Hurricane Gustav fitted by the SMRV model, since there is one wind maximum and no inflection point exists in the profile. The comparisons between model-fitted, and SFMR-measured, hurricane intensity and structure parameters are summarized in Tables 14.1 and 14.2. For Hurricane Earl, they match well; the maximum wind and its radius from OHMRV model and SFMR are 32.1 and 35.0 m/s, 56.7 and 61.2 km, respectively. However, for Hurricane Gustav, the differences in intensity and radius are 9.2 m/s and 10 km, respectively. It should be noted that the SAR-retrieved wind speed maximum along the SFMR track is 43.1 m/s, which is close to the SFMR-measured value (47.0 m/s). These larger discrepancies are possibly induced by four factors: (1) non-axisymmetric hurricane structure characteristic, (2) intensity and structure parameters that are derived from the SMRV model-fitted wind profile, with mean wind speed in all radial directions, (3) the effect of precipitation on the SAR wind retrievals, and (4) the acquisition time difference between the SAR observation and SFMR measurements.

Factors (1) and (2) account for the structural biases. For Hurricane Gustav, Fig. 14.3a clearly shows non-symmetric characteristics, the maximum wind speed occurs along the major axis direction, whereas SFMR measurements were carried out along the minor axis direction. Therefore, the maximum wind radius as suggested by SFMR data is smaller than that inferred from the SMRV model. Factors (3) and (4) are related to differences in estimated intensity (maximum wind speed). Since the SMRV-derived wind profile is based on the SAR-retrieved mean wind speeds in all radial directions, the average processing can reduce the estimated intensity.

Moreover, the observed rain rate was 13 mm/hr when the most egregious intensity difference exists, which also tends to dampen the radar backscatter and thus decrease the retrieved wind speed.

Besides Hurricanes Earl and Gustav, we also applied the OHMRV, SMRV and DMRV models to seven RADARSAT-2 cross-polarized SAR images of four additional Hurricanes (Bertha, Ike, Danielle and Igor) to calculate intensities and structure parameters. Results are summarized in Table 14.3. In this analysis, we used the following procedure: (1) for wind profiles with only one maximum, the fitted model is SMRV, (2) for wind profiles with two maxima, the fitted model is DMRV, and (3) for wind profiles with one maximum and an inflection point, the fitted model is OHMRV. It should be noted that no single model is able to describe all hurricane cases, whether OHMRV, SMRV, or even DMRV. Each of these three wind profile models is best associated with a specific family of hurricane wind profiles.

We also carried out numerical simulation experiments to analyze the errors in estimates to determine the location of the hurricane eyes. In this approach, we first assume that the error is $\pm$ 2 km in the estimate for the hurricane eye center. Subsequently, we construct a 5×5 window, including 25 grid points, centered on this estimate for the hurricane eye center, where the grid resolution is $0.01°$. These points are also considered as possible hurricane eye centers. The window core is the eye center estimated by the proposed methodology. We calculate the differences for the ISPs for all grid points. As summarized in Table 14.4, the largest errors in the maximum wind speeds and its radii are 2.11 m/s and 3.33 km, respectively. These errors do not significantly affect estimates for the mean wind profile and resulting ISPs.

14.3 Idealized Model

In this section, we continue the development of the SHEW model based on three assumptions: (1) the hurricane maximum winds occur at the elliptical-shaped eyewall locations, (2) the wind speed within a hurricane is a function of the radial distance to the center of a hurricane and the angle respected to the major axis of the ellipse, and (3) the maximum wind speeds on the eyewall are symmetric. The inflow angle model [25], which was originally based on a circular eye, was derived from wind vectors observed by over 1600 quality-controlled global positioning system (GPS) dropwindsondes. In this study, we expand the original one-dimensional SHEW model (mean wind profile) to the two-dimensional SHEW model and generalize the inflow angle model from a circular eye to an elliptical eye hurricane structure. We then combine the two models to derive the wind vectors for a given hurricane system.

Following the earlier methodology [13], the two-dimensional SHEW model is developed. This SHEW model is based on the modified Rankine vortex functions [15, 26] and an elliptical shape for the maximum wind speed contour around the eyewall. When the major axis is equal to the minor axis of the ellipse, it is a circle. Additionally, the inflow angle model is extended to simulate the surface wind direction by using the parameters of an elliptical eye estimated by SHEW model.

Table 14.3 Estimated hurricane eye locations, intensity and structure parameters (ISPs) from different tangential wind profile models (SMRV, DMRV and OHMRV), and RADARSAT-2 cross-polarized SAR imagery for six hurricanes (Bertha, Gustav, Ike, Danielle, Earl and Igor)

Hurricane name	Date (yy-mm-dd)	Time (UTC)	Eye center		Maximum wind and radius				Wind profile model
			Lat (°N) (degree)	Lon (°W) (degree)	r1 (km)	r2 (km)	v1 (m/s)	v2 (m/s)	
Bertha	2008/7/12	10:14:41	29.7149	62.5333	13	52	20.93	27.86	DMRV
Gustav	2008/8/30	11:28:04	20.5979	81.5075	30		37.8		SMRV
Ike	2008/9/10	23:54:57	24.6895	86.2809	10	65	26.25	23.81	DMRV
	2008/9/10	23:56:04	24.6849	86.277	10	63	26.07	26.65	DMRV
Danielle	2010/8/28	22:04:20	30.5976	59.6664	45		29.91		SMRV
Earl	2010/8/30	9:57:38	18.3794	62.531	29		30.72		SMRV
	2010/9/2	22:59:20	32.8409	74.7783	31	57	28.17	32.13	OHMRV
Igor	2010/9/12	21:22:40	17.595	46.6848	21		32.16		SMRV
	2010/9/14	9:19:42	18.0458	51.7908	22	34	35	33.87	DMRV
	2010/9/19	10:11:24	29.3127	65.2902	28	69	20.44	25.78	OHMRV

Table 14.4 Numerical experiment results of hurricane Earl eye center location determination errors. The detected hurricane eye center is (74.7783°W, 32.8409°N). The smallest errors for estimated wind speed maximum and its radius are 2.11 m/s and 3.33 km, respectively

Eye center	Lat = 32.82°N		Lat = 32.83°N		Lat = 32.84°N		Lat = 32.85°N		Lat = 32.86°N	
	r (km)	v(m/s)	r (km)	v(m/s)	r(km)	r(m/s)	r(km)	v(m/s)	r(km)	v(m/s)
Lon = 74.80°W	0	1.81	−1.11	1.6	−2.22	1.69	−1.11	1.51	−3.33	1.38
Lon = 74.79°W	0	1.82	−1.11	1.86	0	1.86	1.11	1.77	0	1.62
Lon = 74.78°W	1.11	1.88	1.11	1.87	0	1.98	1.11	1.84	0	1.79
Lon = 74.77°W	1.11	2.11	−1.11	−0.19	1.11	0.04	−1.11	0.44	−2.22	−0.03
Lon = 74.76°W	1.11	−0.41	0	−0.19	−1.11	−0.03	0	0.53	0	0.08

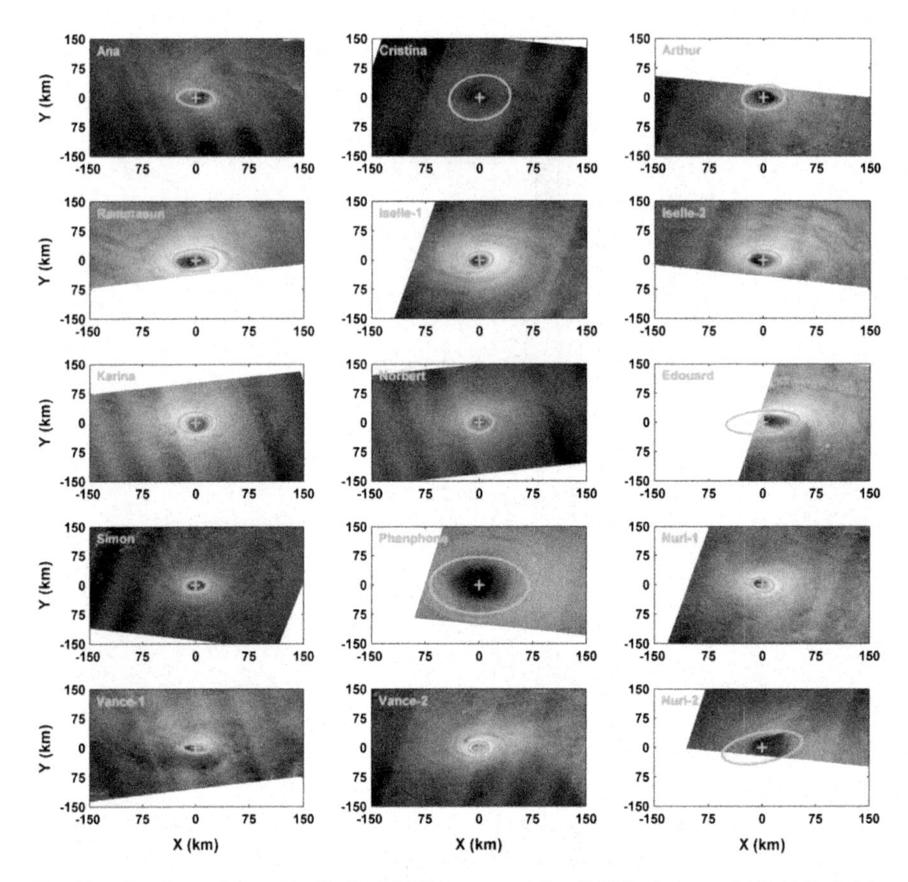

Fig. 14.5 Hurricanes imaged by RADARSAT-2 cross-pol ScanSAR in the year of 2014. The bright spots indicate land. RADARSAT-2 Data and Products ©MacDonald, Dettwiler and Associates Ltd. (2008–2009) - All Rights Reserved

14.3.1 SHEW Model

We applied the C-2PO wind speed retrieval model to the cross-polarization SAR images (Fig. 14.5), each of which may cover part of or whole of the hurricane core. The radial distributions of surface wind speeds within 150 km of the hurricane centers are displayed in Fig. 14.6, as well as the mean wind profiles.

All 15 SAR images have a relatively weak hurricane-vortex with maximum axisymmetric wind speeds on order of 25–35 m/s. The averaged radial wind profiles (in red) represent the axisymmetric wind structures, while the variance in radial wind profiles represents the azimuthal variations. The radii of the maximum wind speed (RMWs) in every 5° azimuth angle are displayed in Fig. 14.7, indicating that the shapes of most cases are close to ellipses.

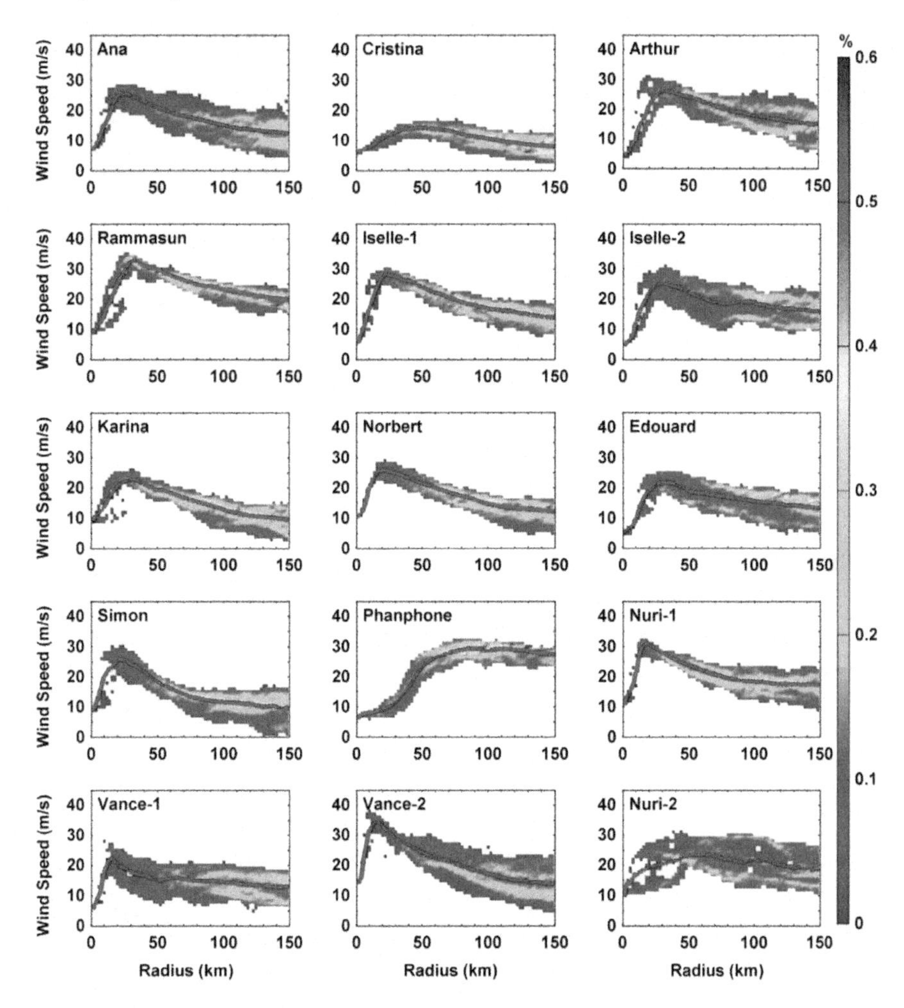

Fig. 14.6 SAR-retrieved wind speed distributions plotted as a function hurricane radius as well as the mean wind profiles (in *red*) for the 15 SAR images shown in Fig. 14.5

Therefore, if an elliptical-shaped eye is adapted to a continuous analytic function, a two dimensional analytic model may be developed to estimate the main structure of the hurricane eye shape (circle or ellipse). For the surface structure of a hurricane, the symmetry is normally referred to as rotational symmetry [12]. In contrast with previous studies, the symmetry in the SHEW model is noted as elliptical symmetry which is a reflectional, but not rotational symmetry. Therefore, the two dimensional RMWs are built in terms of a major and a minor axis of an ellipse as:

$$r_m(\theta) = a{\cdot}b/\sqrt{(b{\cdot}cos\theta)^2 + (a{\cdot}sin\theta)^2} \qquad (14.4)$$

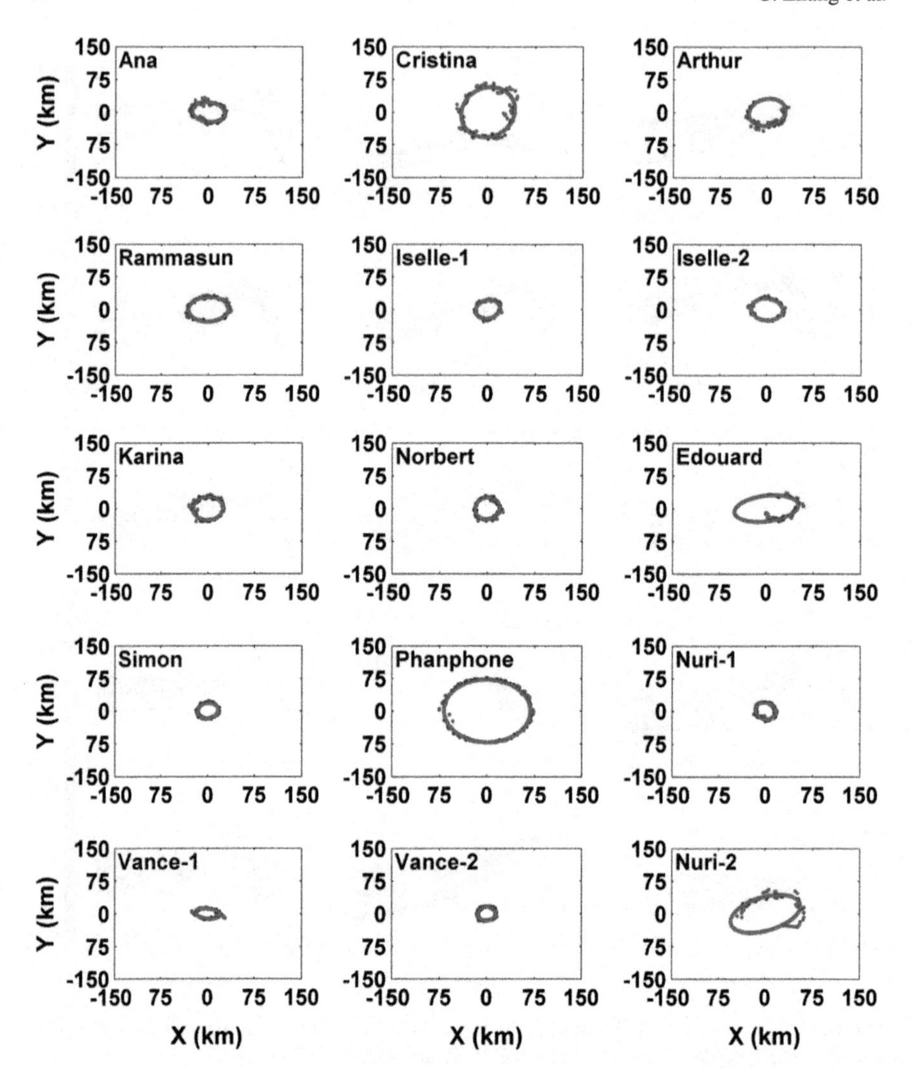

Fig. 14.7 The positions of the maximums in wind speed (*blue points*) derived by the C-band cross-polarization ocean model for SAR wind speed retrievals (C-2PO) and the reference ellipse of eyewall (in *red*) estimated by the SHEW model for the 14 SAR images shown in Fig. 14.5. The positions of maximum wind speeds were detected for every 5o azimuth angle where the wind speed maxima exist

where a is the major axis, b is the minor axis, both with units of km, and θ is the angle for each point with respect to the major axis (Fig. 14.8a). With this reference ellipse formulation, the one-dimensional continuous analytic functions are extended to two-dimensions. The surface wind field for an elliptical vortex is:

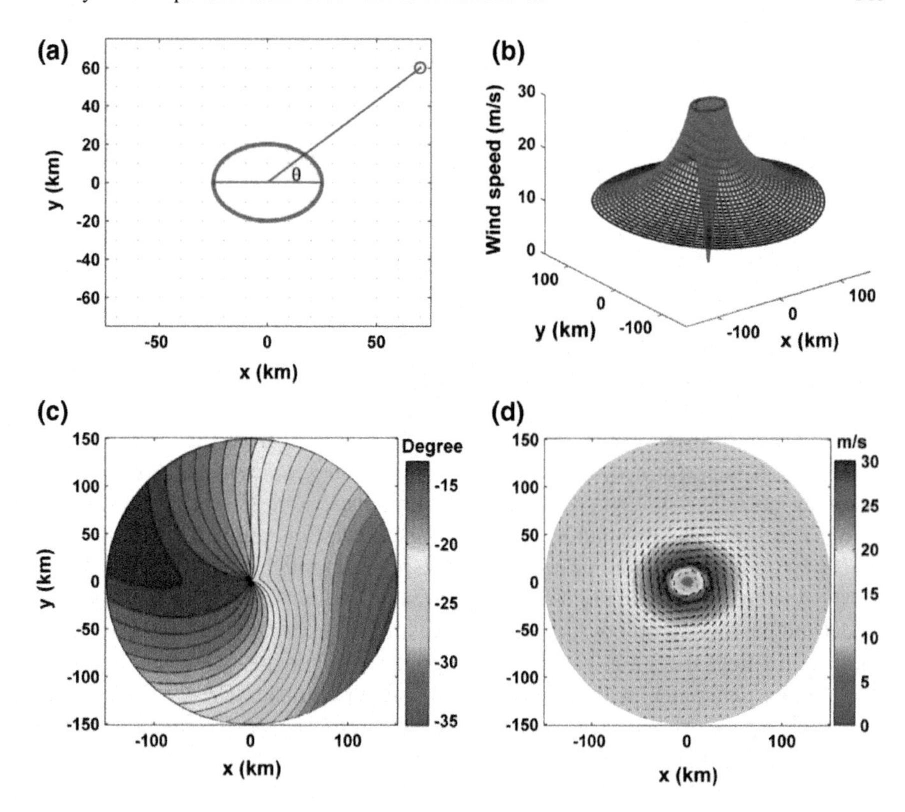

Fig. 14.8 **a** The angle for one grid with respect to the major axis, and **b** the reconstructed hurricane wind speed by the SHEW model, **c** inflow angle by the revised inflow angle model, and **d** wind vector field estimated by combination of these two models. The latter three cases have major axis of 25 km, minor axis of 20 km, the symmetric intensity of 30 m/s, and the hurricane moving speed of 2 m/s, toward north

$$V(r,\theta)=\begin{cases} V_{max}*\left[\frac{r}{r_m(\theta)}\right], & r\leq r_m(\theta) \\ V_{max}*\left[\frac{r_m(\theta)}{r}\right]^{\alpha}, & (r_m(\theta)<r\leq 150\,\text{km}) \end{cases} \tag{14.5}$$

where α is the decay parameter, r is the radial distance to the hurricane center (with unit of km), and V_{max} is the maximum wind speed which is assumed to occur at the elliptical eyewall. In Eq. (14.5), the wind speed for an elliptical symmetrical hurricane with one vortex can be reconstructed for given reference ellipse parameters of a and b, as well as the intensity parameters of V_{max} and α. For example, assuming a major axis of 25 km, a minor axis of 20 km, a maximum wind speed of 30 m/s, and a decay parameter of 0.5, the elliptical symmetric wind field constructed by the SHEW model is shown in Fig. 14.8b.

14.3.2 The Revised Inflow Angle Model

In the original inflow angle model [25], the radial distances are normalized by the axisymmetric RMW ($r^* = r/r_m$) assuming the eye and eyewall has a circle shape. For the elliptical-shaped eyewall, it is revised as:

$$r^*(\theta) = \frac{r}{r_m(\theta)} \tag{14.6}$$

By applying the hurricane motion speed and three morphology parameters (a, b and V_{max}) to the revised model, the inflow angles in a hurricane can be constructed. We vary the 6 morphology parameters in the SHEW model to calculate a wind field and then fit these values to the actual wind retrievals from the C-2PO model. A regression is performed to find the final morphology parameters. For example, assuming hurricane motion speed of 2 m/s, a major axis of 25 km, a minor axis of 20 km, and a maximum wind speed of 30 m/s, the inflow angle field is shown in Fig. 14.8c and wind vector constructed by the revised model is shown in Fig. 14.8d.

The flowchart shown in Fig. 14.9 displays the procedures needed to use the SHEW model and the revised Inflow Angle model to estimate the complete hurricane surface wind vector field from a C-band cross-polarized SAR image. When the hurricane morphology parameters with the elliptical eye are input to the SHEW model, a wind speed field can be estimated. At the same time, the inflow angle structure is estimated with a given hurricane motion speed. Thus, a wind vector field can be calculated with the wind speed from the SHEW model and the wind direction from the revised inflow angle model. By comparing with the wind speed retrieved from the VH-polarized SAR image based on the C-2PO algorithm, the closest surface wind vector field is estimated with the least squares methodology. Finally, the hurricane surface wind vector field is validated by aircraft measurements (SFMR and dropwindsonde). To simplify this process, an initialized wind field was firstly retrieved from the cross-polarized SAR image using the C-2PO algorithm. Secondly, morphology parameters are regressed by fitting the SHEW model to the initialized wind. Then the wind directions are simulated from the revised inflow angle model by providing the morphology parameters and the hurricane motion speed by the Best Track data (HURDAT2) from National Hurricane Center (http://www.nhc.noaa.gov/data/).

14.3.3 Validation

To validate the combined SHEW and inflow angle models, two SAR cases (Fig. 14.9) were matched with aircraft datasets for Hurricane Arthur (2014) and Hurricane Earl (2010). Here, we assume the structures of the hurricanes remain steady during the ± 3 h periods. This assumption was also used in previous studies that compared satellite surface wind retrievals to dropwindsonde data [13, 27, 28]. Thus, the storm-relative locations are detected by removing the physical radial locations of observa-

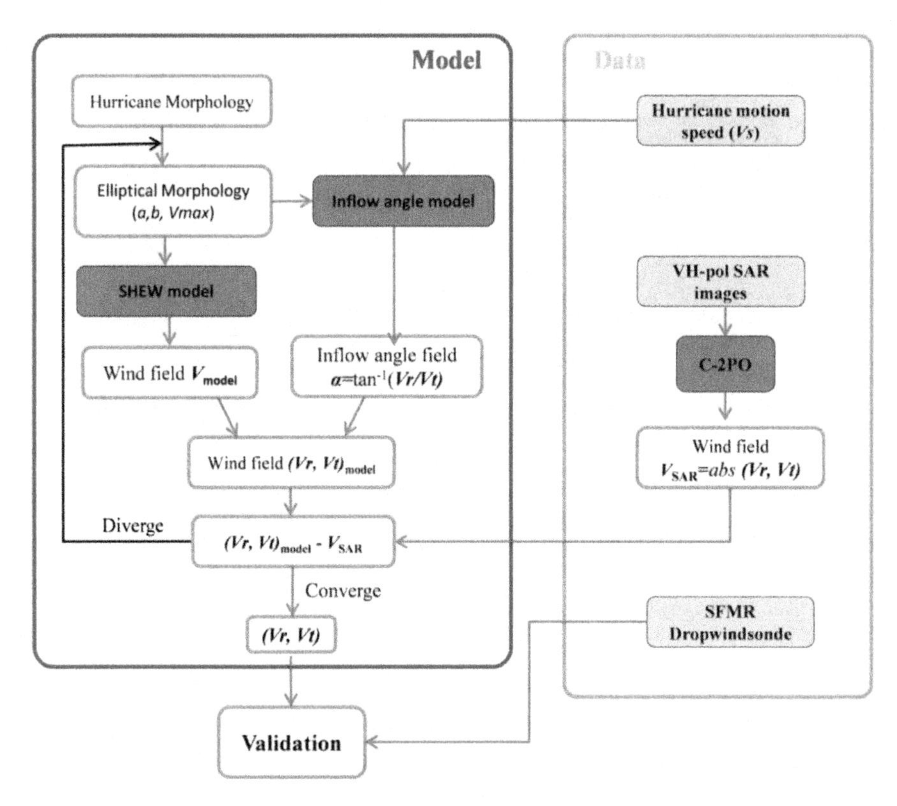

Fig. 14.9 A flowchart for the combination of the SHEW model and the revised inflow angle model to estimate a complete hurricane surface wind vector field. C-2PO is the C-Band cross-polarization SAR wind speed retrieval algorithm

tions from the hurricane center location, as calculated based on the linearly interpolated Best Track data. During ± 3 h, several wind profiles can be measured by SFMR. One complete radial profile observation, at the radius of 150 km, normally takes about 20 min. Therefore, the time to capture the closest profile for each case is restricted. The selected SFMR tracks and the storm-relative locations of the dropwindsonde during the ± 3 h of collocation time are shown in Fig. 14.10.

Two capabilities of the SHEW model are validated by the two hurricane cases: (1) estimated wind speeds from the SAR image covering only part of the hurricane core structure, and (2) estimated wind speeds from the SAR image under heavy precipitation conditions. The radial wind profiles observed by SFMR and estimated by the SHEW model with respect to the SFMR locations are shown in Fig. 14.10. The first capability, to estimate wind speeds from the SAR image covering only part of the hurricane core structure, is validated by Hurricane Arthur. The SAR image was acquired at 11:14 UTC (3 July 2014) and the matched SFMR data is measured during 11:04 to 11:27 UTC, 3 July 2014. As shown in Fig. 14.10a, more than half of the SFMR locations are outside the SAR image because it only captured part

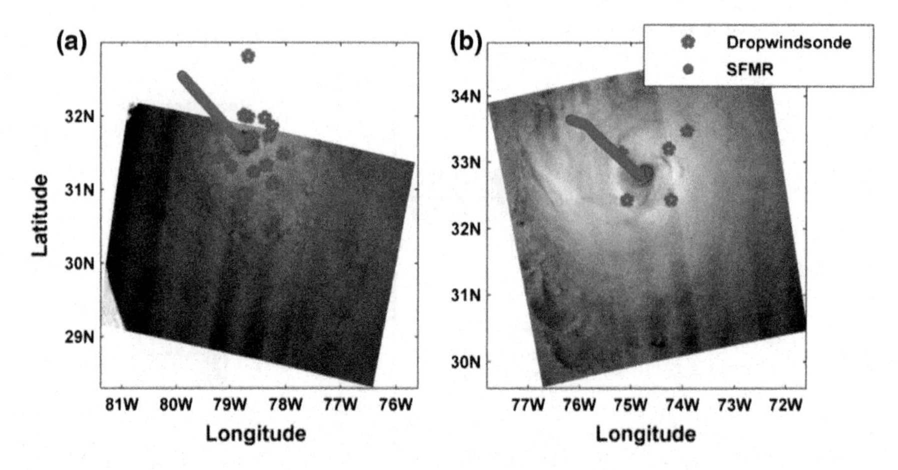

Fig. 14.10 RADARSAT-2 cross-polarized SAR images: **a** Hurricane Arthur (11:14 UTC, July 3, 2014), **b** Hurricane Earl (22:59 UTC, September 2, 2010); the positions of SFMR used here: **a** from 11:04 to 11:27 UTC (July 3, 2014), **b** from 22:59 to 23:19 UTC (September 2, 2010); and the relative positions of the dropwindsondes to the hurricane center during ±3 h of collocation time

of the hurricane core structure. As shown in Fig. 14.11a, the maximum wind speed estimated by the SHEW model (28.7 m/s) is close to that observed by the SFMR (27.6 m/s). The corresponding RMWs are 31 and 29 km, respectively, for the SHEW and SFMR results. The radial wind profile estimated by the SHEW model is found to be close to that observed by SFMR even when there is no SAR data. This shows the robustness of the SHEW model for 2D wind speed estimation.

The second capability of the SHEW model, to determine winds from SAR in the presence of heavy rain, is validated by Hurricane Earl. The SAR image is acquired at 22:59 UTC on September 2, 2010, and captured the complete hurricane core. However, wind speeds retrieved from the SAR image are underestimated due to heavy rainfall. As we learned in a previous study [29], heavy rain associated with a hurricane attenuates the radar signal of SAR. As seen in Fig. 14.11b, an obvious moat exists in the SAR wind profile coinciding with the heavy rain region, as derived by the C-2PO algorithm. Although the SHEW-estimated wind profile is also somewhat different from the SFMR-observed profile, it is much better than the C-2PO retrieved wind profile. Therefore, we draw the conclusion that the local effects due to the attenuation by rain can be reduced by adopting the SHEW model. This reduction in attenuation may result from the fact that the SHEW model is fitted to all azimuth angles, whereas the heavy rain band only exists at certain azimuth angles at the same radius (Fig. 14.12). Therefore, the wind profile (red line in Fig. 14.11b) estimated by the SHEW model, as fitted to all azimuth angles, is closer to the SFMR wind profile than that retrieved from SAR image directly.

Moreover, the dropwindsonde dataset is also used to validate the wind vector derived by the combined SHEW and inflow angle models. We calculated the wind vector (decomposed in zonal and meridional components) by using the wind speed

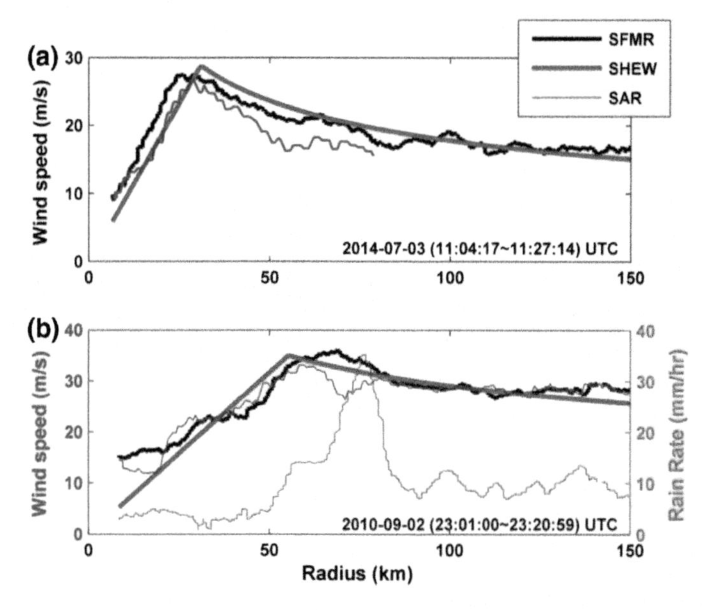

Fig. 14.11 Wind speed profiles measured by the SFMR (*black line*), estimated by the SHEW model (*red line*), and retrieved from C-band cross-polarized SAR image using the C-2PO algorithm (*blue line*) and the rain rate observed by SFMR (*green line*): **a** Hurricane Arthur (11:14 UTC, July 3, 2014), **b** Hurricane Earl (22:59 UTC, September 2, 2010)

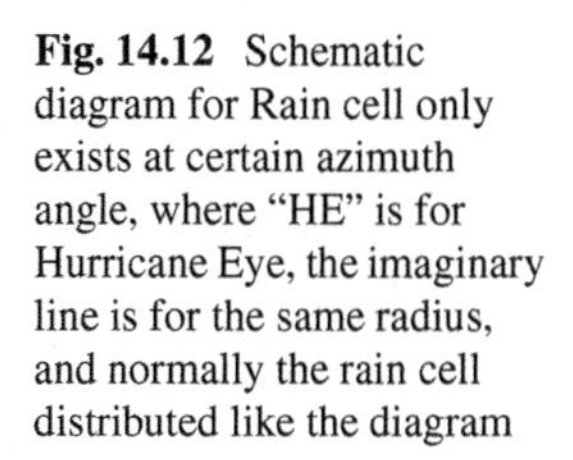
Fig. 14.12 Schematic diagram for Rain cell only exists at certain azimuth angle, where "HE" is for Hurricane Eye, the imaginary line is for the same radius, and normally the rain cell distributed like the diagram

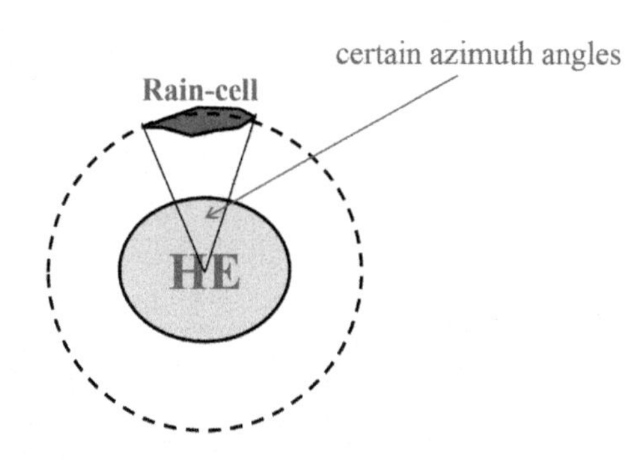

estimated by the SHEW model and wind direction estimated by revised inflow angle model. The wind vectors estimated by the two models and observed by dropwindsondes are shown in Fig. 14.13 for the two storms, which demonstrate good agreement between the model and observations. The statistics in terms of RMSE, bias and correlation coefficients suggest that the combined SHEW and inflow angle model excellently capture the observed wind vector distribution in both storms (Table 14.5).

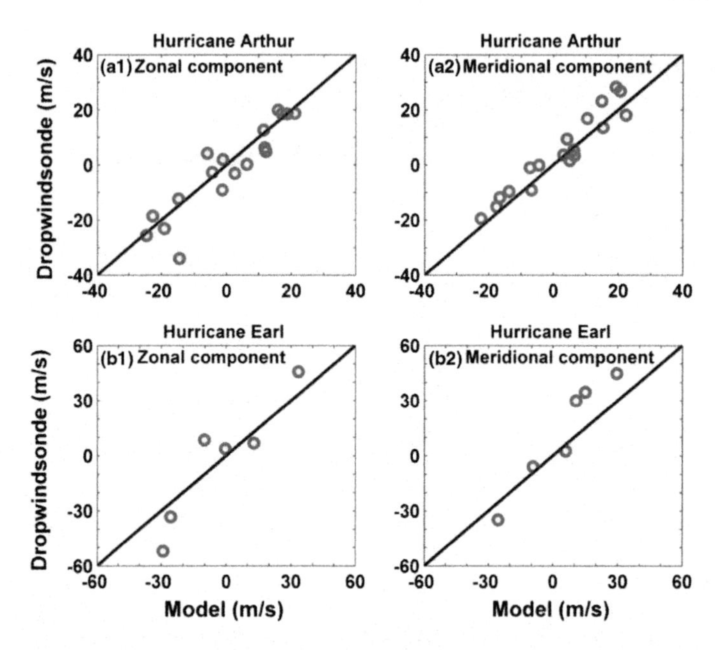

Fig. 14.13 Hurricane wind vector in terms of zonal and meridional components observed by the collocated dropwindsondes compared with that simulated by the combined SHEW and revised inflow angle models: **a** Hurricane Arthur (July 3, 2014), **b** Hurricane Earl (September 2, 2010)

Table 14.5 Statistics calculated by comparing the wind vector observed by Dropwindsondes and simulated by the two models

	Hurricane Arthur (2014)		Hurricane Earl (2010)	
	Zonal	Meridional	Zonal	Meridional
Number	18	18	6	6
Bias	1.73 m/s	−2.44 m/s	0.18 m/s	−7.43 m/s
RMSE	6.55 m/s	4.82 m/s	13.77 m/s	13.51 m/s
Correlation	91.85%	95.52%	93.21%	96.70%

14.4 Reconstructed Hurricane Winds

Following the flowchart in Fig. 14.9, the closest elliptical symmetrical wind speed fields for the 15 SAR images in the year of 2014 were detected (Fig. 14.14) by fitting SHEW model to the VH-polarized SAR image, as well as the 2D surface inflow angles for the 15 SAR images are estimated and shown in Fig. 14.15. The corresponding complete surface wind vector fields for the hurricanes acquired by the SAR images are also estimated (Fig. 14.16). The RMSEs and correlation coefficients between the elliptical symmetrical wind fields and C-2PO retrieved wind fields for the 15 cases are shown in Fig. 14.17.

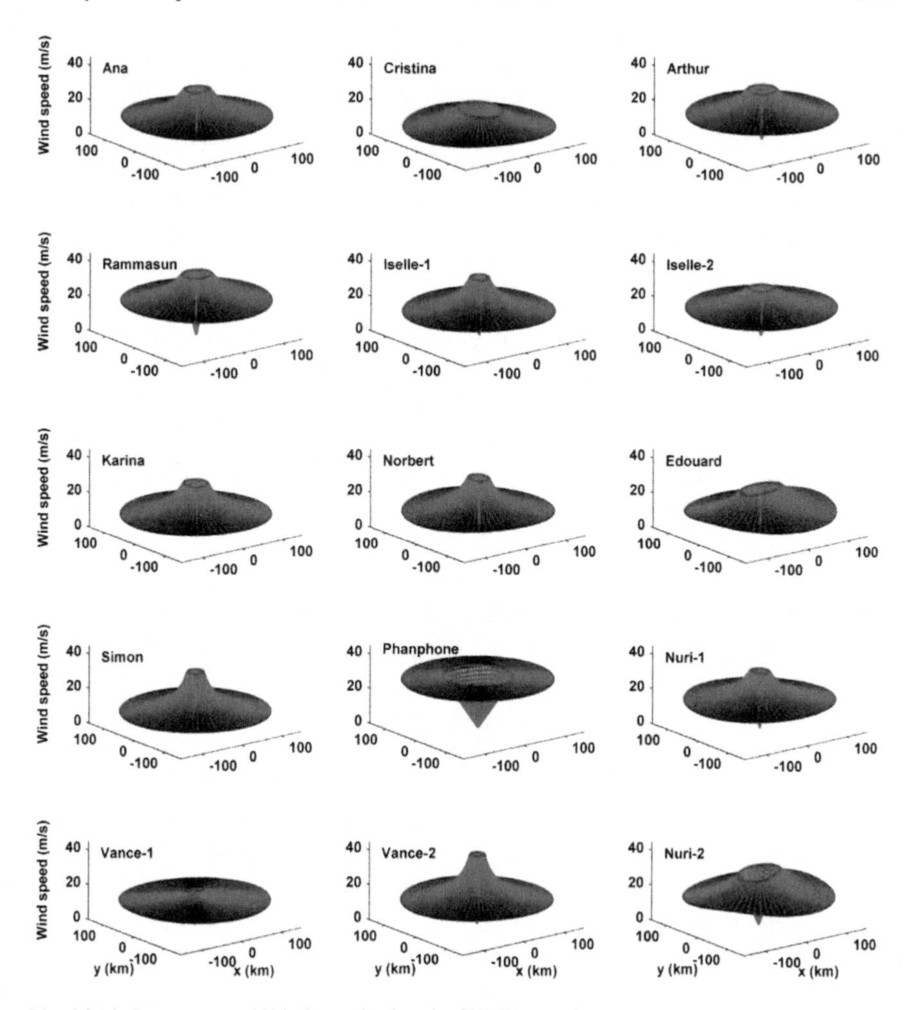

Fig. 14.14 Reconstructed Wind speed using the SHEW model for the 15 SAR images

The RMSEs are less than 4 m/s. The correlations are higher than 60%, except for the first SAR image for hurricane Vance (only 21.3%). The hurricane morphology and intensity parameters as derived for the 15 SAR images by SHEW model are shown in Table 14.6. Moreover, the hurricane elliptical morphology parameters of the closest wind speed field are determined. The reason for the low correlation of the first image for Hurricane Vance is notable and will be further studied in the future. The SHEW model is an elliptical symmetrical model. The less correlation between the SHEW simulations and SAR observations means a more asymmetrical hurricane structure. The first SAR image of Hurricane Vance provides us a chance to investigate the correlation between the asymmetry and intensity which may be helpful for the hurricane intensity change forecast.

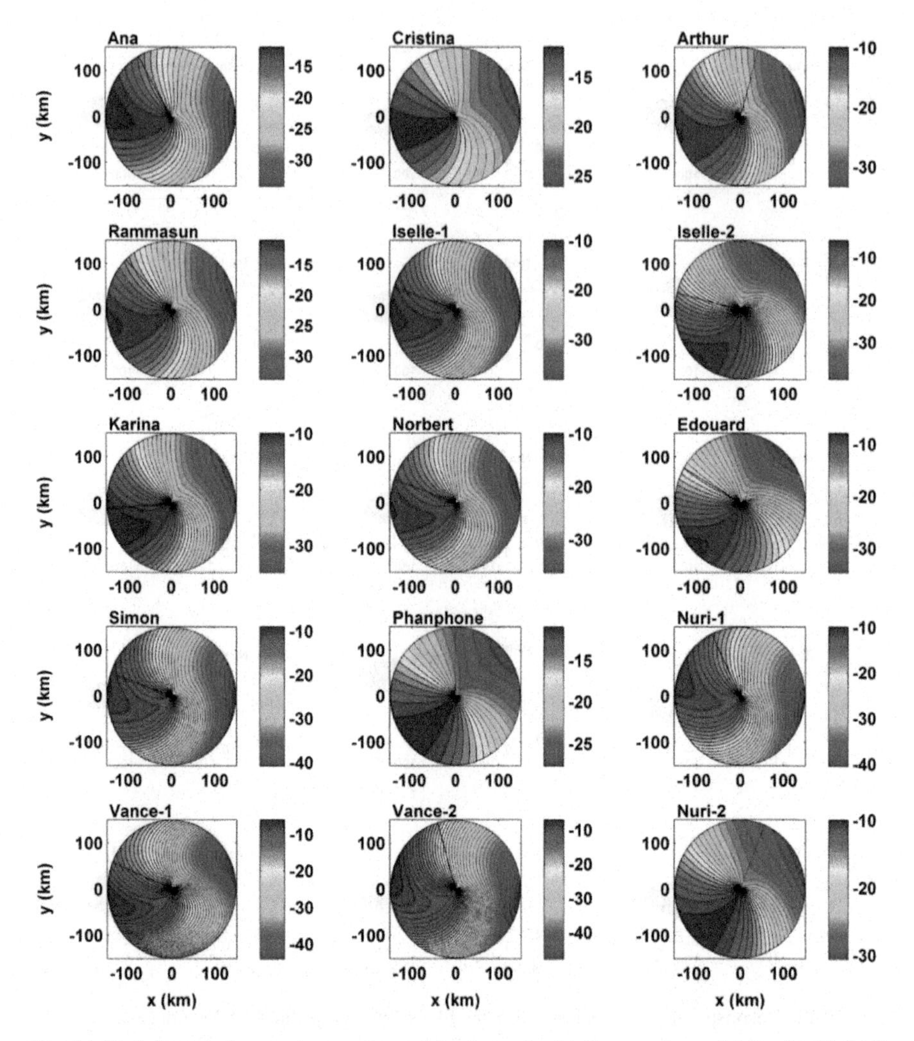

Fig. 14.15 Inflow angle structures estimated by the revised inflow angle model for the 15 SAR images shown in Fig. 14.5. The *colors* represent the inflow angle with the unit of degree

14.5 Rain Effects on the Hurricane SAR Images

In this chapter, five possible mechanisms (Fig. 14.18) for the rain effects on the spaceborne C-band SAR hurricane observations are considered: (1) attenuation and (2) volume backscattering for the microwave transfer in atmosphere; as well as (3) diffraction on the sharp edges of rain products and (4) rain-induced damping to the wind waves, and (5) rain-generated ring waves on the ocean surface.

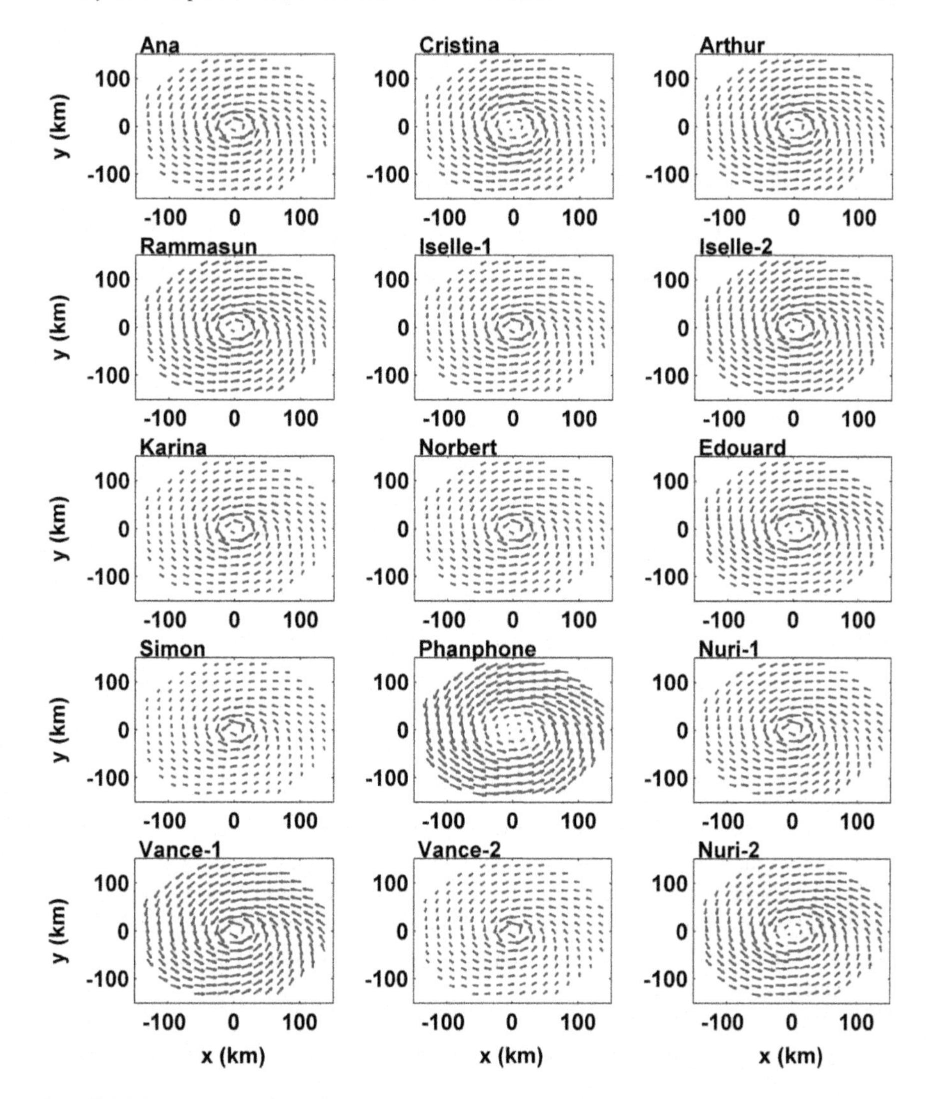

Fig. 14.16 Hurricane surface wind vector estimated by the combination of the SHEW and revised inflow angle models for the 15 SAR images shown in Fig. 14.5

14.5.1 In Atmosphere

When the microwave transfer in atmosphere, two possible mechanisms for the rain effects on the spaceborne C-band SAR hurricane observations are considered: (1) attenuation and (2) volume backscattering. Based on the atmosphere model, the rain effect in the atmosphere is described by:

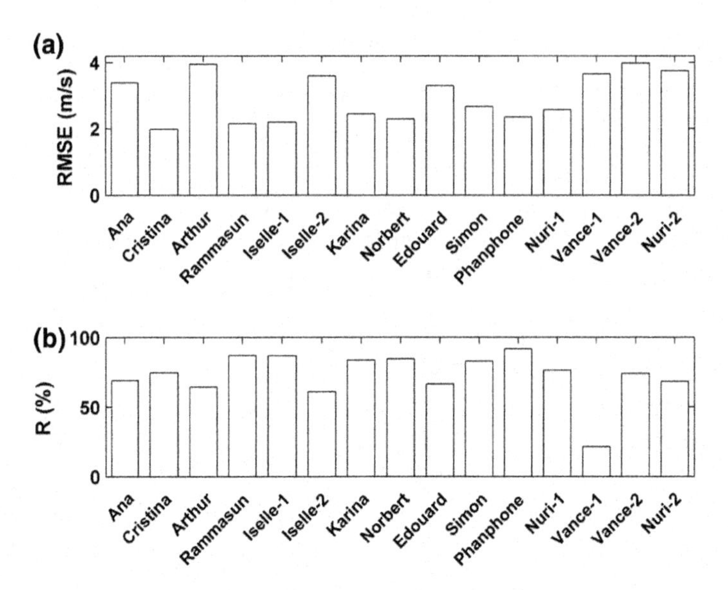

Fig. 14.17 Errors and correlation coefficients between the wind speed derived by the SHEW model and C-2PO SAR algorithm for the 14 SAR images shown in Fig. 14.5

$$\Delta\sigma_0(dB) = 10 \cdot lg(\sigma_{surf}\alpha_{atm} + \sigma_{atm}) - 10 \cdot lg(\sigma_{surf}) \tag{14.7}$$

Therefore, C-band NRCS affected by rain in the atmosphere are simulated and shown in Fig. 14.19 for the rain rates of 10, 30 and 50 mm/h respectively, with the assumption that rain rates were not changing along the radiative incidence track.

The VV-polarized hurricane measurements, which are mostly between −18 and −5 dB for winds above 5 m/s, are probably attenuated by the rain in the atmosphere; Fig. 14.19 suggests that the effects of attenuation and volume backscattering in the atmosphere are less than 1dB, even with a rain rate of 50 mm/h along the incidence track. When the NRCS is between −18 and −5 dB, the wind speed varies from 5 to 50 m/s. Moreover, the rain rate is not always the same along the entire radiative incidence track, as is often assumed as spiral rain band in a hurricane; therefore, the real effect of rain in atmosphere should be less than the results simulated (with assuming a maximum rain rate of 50 mm/h). Hence, it is concluded that the atmospheric part can be neglected for C-band VV polarization in the hurricane observations. When the sea surface backscattering NRCS is less than −15 dB, representing wind speeds of about 40 m/s for the cross-polarization, the NRCS is enhanced because of the volume scattering of raindrops.

Table 14.6 Hurricane morphology and intensity estimated by the SHEW model applied to 15 SAR images

Hurricane name	Date (mm-dd)	Time (UTC)	Hurricane center		Reference ellipse			Intensity	
			Latitude	Longitude	Major (km)	Minor (km)	Azimuth	u_m (m/s)	α
Ana	10–19	4:45	19.98°N	159.29°W	27.1	22.1	157°	28.6	0.45
Cristina	06–15	13:23	20.03°N	113.12°W	57.9	41.9	80°	17.1	0.71
Arthur	07-03	11:14	31.64°N	78.73°W	32.7	28.7	57°	28.8	0.41
Rammasun	07–17	10:28	17.36°N	114.52°E	32.3	28.3	7°	35	0.35
Iselle	08–03	14:35	15.53°N	132.57°W	22.5	18.5	64°	33.2	0.42
	08–07	15:59	18.63°N	150.99°W	26.7	24.7	143°	27.4	0.3
Karina	08–14	1:47	17.03°N	113.65°W	27.2	24.2	79°	27.7	0.56
Norbert	09–07	1:50	25.48°N	115.45°W	25.3	19.3	83°	30.9	0.48
Edouard	09–14	9:06	23.97°N	49.82°W	52.3	29.3	12°	25	0.51
Simon	10–03	13:15	18.38°N	109.47°W	18.4	16.4	60°	32.2	0.56
Phanphone	10–04	21:06	28.34°N	131.17°E	72.5	68.5	96°	29.8	0.08
Nuri	11–01	20:53	15.09°N	133.00°E	20	15.9	111°	33.1	0.33
	11–05	20:32	27.67°N	139.72°E	60.8	36.8	29°	29.9	0.47
Vance	11–02	1:12	10.25°N	104.87°W	20.6	11.6	169°	18.6	0.14
	11–03	13:12	15.00°N	110.65°W	16.2	14.2	79°	40.1	0.45

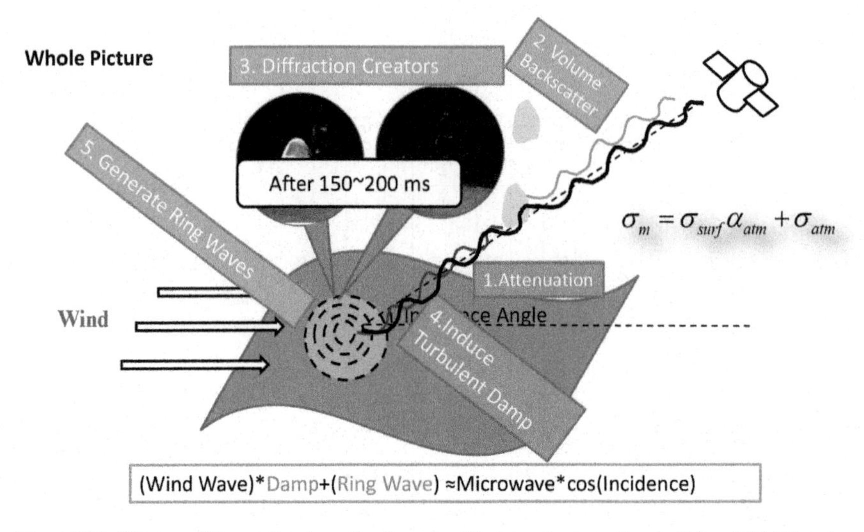

Fig. 14.18 Five possible mechanisms for the rain effects on the spaceborne C-band SAR hurricane observations

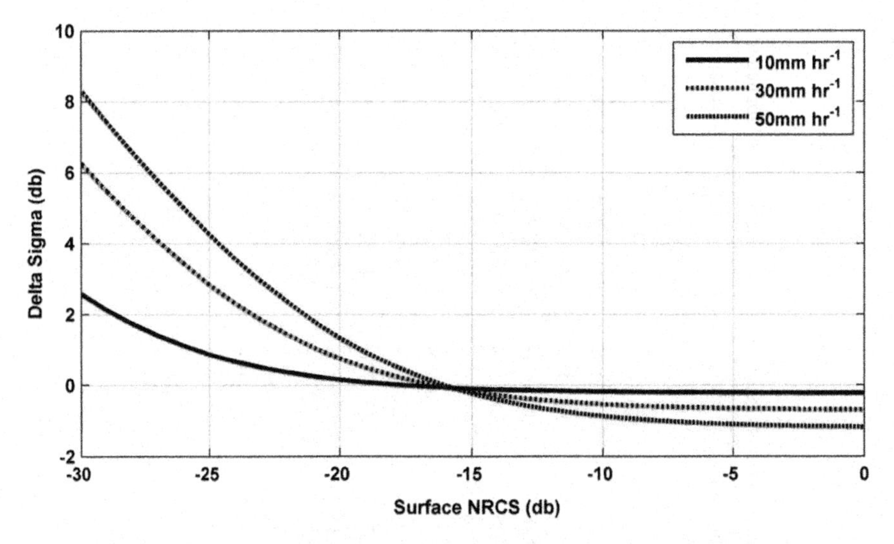

Fig. 14.19 Rain effects on attenuation and volume backscattering in the atmosphere as a function of the sea surface backscatter with three different rain rates

14.5.2 Rain Effects on the Ocean Surface

On the ocean surface, three possible mechanisms for the spaceborne C-band SAR hurricane observations are considered: (1) diffraction on the sharp edges of rain products, (2) rain-induced damping to the wind waves and (3) rain-generated ring waves on the ocean surface.

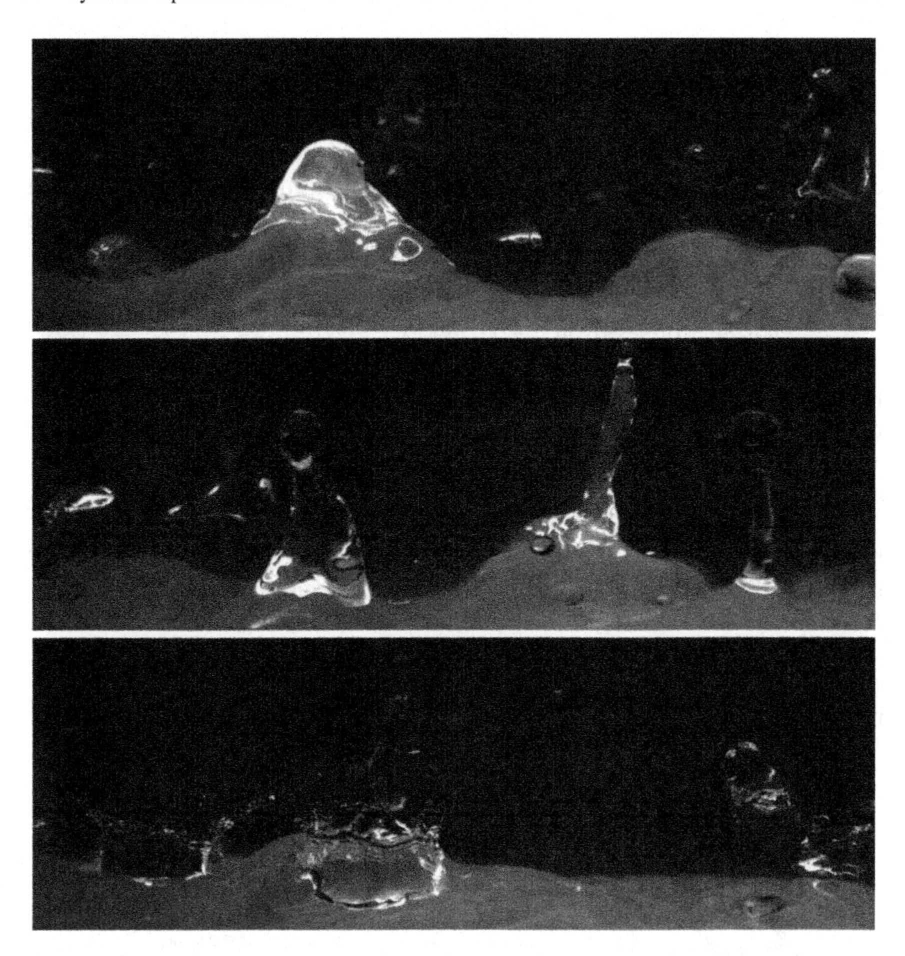

Fig. 14.20 Rain products: crater, crown and stalk [32]

14.5.2.1 Rain Products Diffraction

As long as we known, there are two sea surface backscattering mechanisms: Bragg resonance theory and Non-Bragg mechanism [30]. The non-Bragg mechanism mainly includes specular reflection and diffraction of radio waves on sharp wedges of wave crest breaks [31]. Here, if we assume the diffraction of radio waves on sharp wedges of wave crest breaks is similar as on sharp wedges of rain products (Fig. 14.20), the rain products diffraction can be analyzed by the difference between wind retrieval functions and the Bragg mechanisms.

For C-band VV-polarized SAR observations, the dependency of the NRCSs on the wind speed and the SAR geometries (e.g., incidence angle and azimuth angle with respect to wind direction) is generally expressed by geophysical model function (GMF), such as CMOD5 for neutral winds (CMOD5.N) [33, 34]:

$$\sigma_0 = A(U_{10}, \theta)[1 + b_1(U_{10}, \theta)\cos\varphi + b_2(U_{10}, \theta)\cos 2\varphi]^B \tag{14.8}$$

where σ_0 is the NRCS observed by SAR, U_{10} is the sea surface wind speed, φ is the azimuth angle which is the difference between the wind speed and radar moving direction, and θ is the incidence angle. For cross-polarization images (VH, HV), an empirical function denoted C-band Cross-Polarization Coupled-Parameters Ocean (C-3PO) model was developed for hurricane ScanSAR wind retrieval:

$$\sigma_0[dB] = [0.2983 \cdot U_{10} - 29.4708] \cdot \left[1 + 0.07 \times \frac{\theta - 34.5}{34.5}\right] \tag{14.9}$$

For the HH-polarization, no well-developed empirical function exists and the widely used approach is to develop a hybrid model function consisting of one of the GMFs and a polarization ratio [35, 36]. Therefore, HH-polarization SAR images are not discussed in this study.

Based on the assumption that the difference between the simulations results between the retrieval model and Composite Bragg theory (CB theory) model is caused by the non-Bragg mechanism, the model results are compared with the empirical functions, considering only the wind induced KHCC wave spectrum for the VV-polarization SAR images and the cross-polarization SAR images. Figures 14.21a (upwind) and 14.21b (crosswind) present the VV-polarized NRCSs simulated by radar scattering model (CB theory), in comparison with results from CMOD5.N.

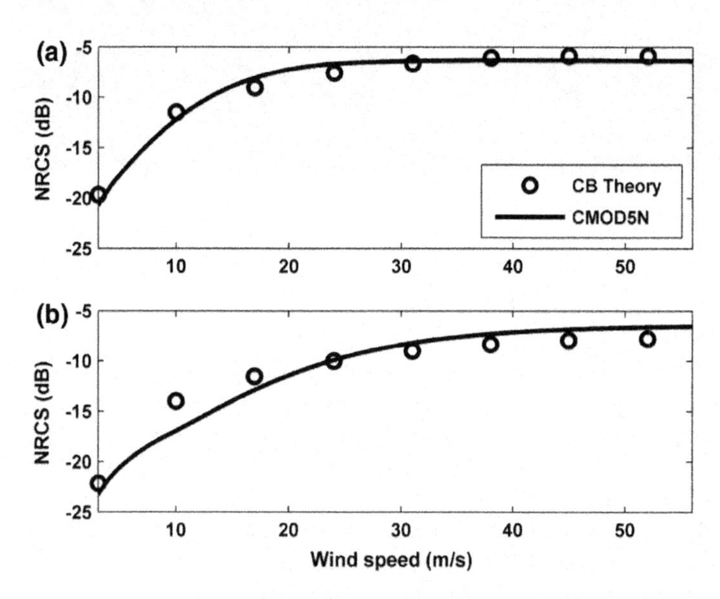

Fig. 14.21 NRCS as a function of the wind speed, comparing the composite model only, with KHCC wind wave spectrum (*black circles*) and the CMOD5N function. The incidence angle is 38° for C-band VV-polarization: **a** the relative wind direction is 0° (upwind) and **b** the relative wind direction is 90° (crosswind)

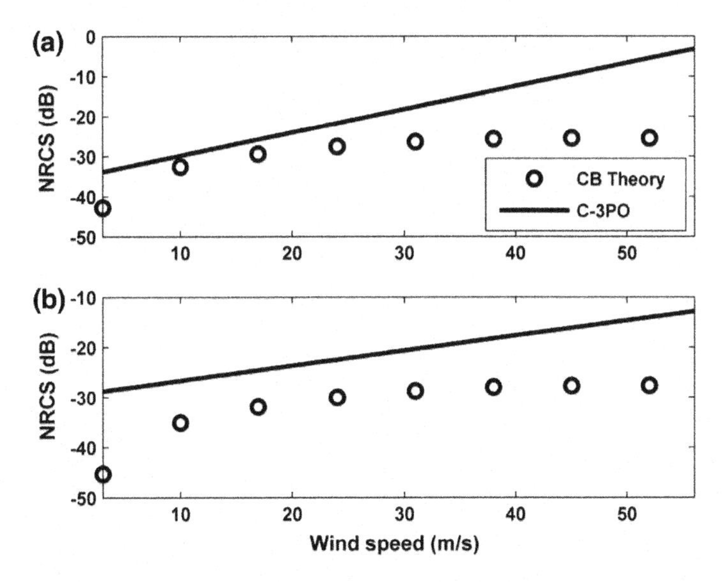

Fig. 14.22 NRCS as a function of the wind speed, comparing the composite model only, with KHCC wind wave spectrum (*black circles*) and the C-3PO function. The incidence angle is 38° for C-band VH-polarization: **a** the relative wind direction is 0° (upwind) and **b** the relative wind direction is 90° (crosswind)

One can see that the simulated NRCS results are close to those from CMOD5.N for both the upwind and crosswind cases, illustrating that this model can reliably simulate the NRCS, with respect to wind-generated wave spectra.

However, for cross-polarizations, the wind-induced NRCS modulations in both upwind (Fig. 14.22a) and crosswind (Fig. 14.22b) cases are much smaller than the results given by the C-3PO model. The reason is that the Bragg model is not suitable for wind-induced NRCS simulation for the cross pol imagery, and certainly does not include the rain's effect. The underestimated NRCSs values may be caused by non-Bragg scattering.

The non-Bragg scattering is not simulated in this study which is not related to polarizations. However, the VV-polarized NRCS values are much larger than those of cross-polarization images, with the same wind speed, which is evident, comparing Figs. 14.21 and 14.22. Hence, although the non-Bragg scattering impacts on the cross-polarization are significant, the impacts on the VV-polarization are slight. Just as for the crest breaks, we speculated that the diffraction of radio waves on the sharp wedges of craters, crowns and stalks produced by rain colliding with the sea surface can be neglected for C-band VV polarized SAR hurricane observations, but should be important for the cross-polarization images.

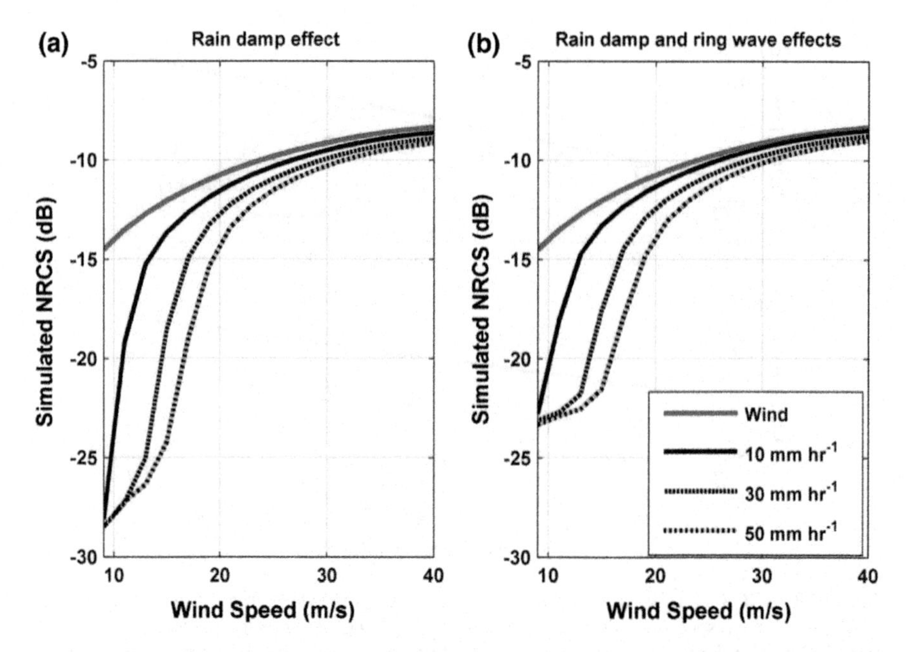

Fig. 14.23 Simulated NRCSs with different wave spectra: Only with wind wave spectrum (*red circles*), with the damped wind wave spectrum (*left panel*), with the damped wind wave and ring wave spectra (*right panel*). The incidence angle is 38°, and the relative wind direction is 0° (upwind) for C-band VV-polarization

14.5.2.2 Changes on the Wind-Induced Waves

We have found that the previous three mechanisms for the rain effects on the NRCS (1) attenuation and (2) volume backscattering for the microwave transfer in atmosphere; and (3) diffraction on the sharp edges of rain products can be neglected for C-band VV polarized SAR hurricane observations, but important for the cross-polarization. Then, the other two effects: (4) rain-induced damping to the wind waves and (5) rain-generated ring waves on the ocean are considered, based on a composite backscattering simulation model (will be described in Sect. 14.5.3).

The NRCSs with the rain damped wave spectrum, in comparison with the rain changed wave spectrum (that is, wave spectrum damped by rain and adding the ring waves generated by rain) for C-band VV-polarization are simulated shown in Fig. 14.23a for only damp effect, and Fig. 14.23b, for two effects on the wave spectrum. Therefore, even with the rain-induced ring waves, the NRCSs are also attenuated by the precipitation, for both C-band VV-polarization (Fig. 14.23) and VH-polarization (Fig. 14.24). However, the attenuation of the NRCSs values due to the rain damping effect decreases, with increasing wind speeds. Conversely, the total rain effect on the sea surface (wind wave damped by rain and the ring waves generated by rain) increases, with decreasing or during low wind speeds (<10 m/s),

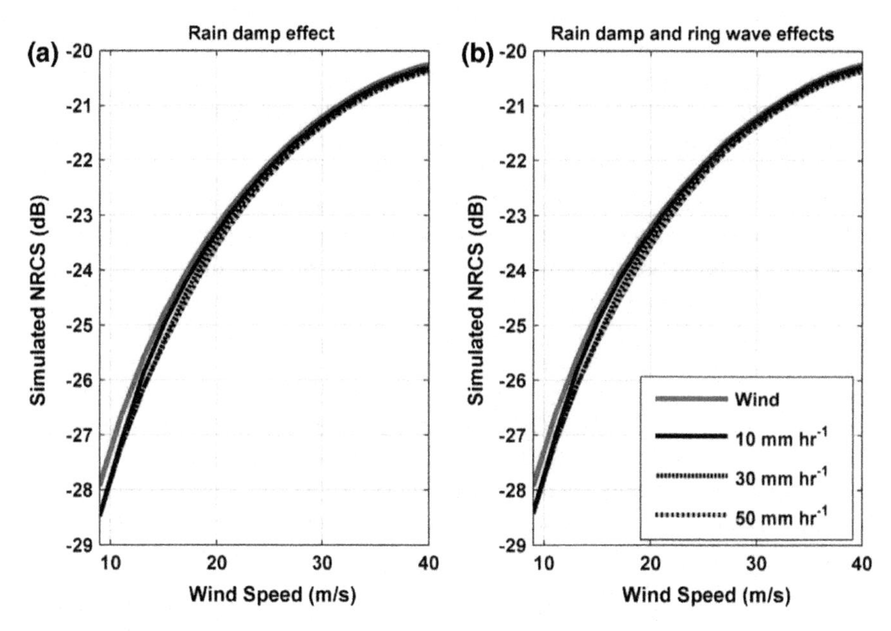

Fig. 14.24 Same as Fig. 14.23 but for C-band VH-polarization

but decreases for increasing, moderate and high wind speeds for VV-polarization (Fig. 14.23).

With the simulation results, the two impacts on the wind generated waves: (4) rain-induced damping to the wind waves and (5) rain-generated ring waves on the ocean are significant for the VV-polarization (Fig. 14.23), but slight for cross-polarization (Fig. 14.24).

14.5.3 The Radar Scattering Model

In the framework of Bragg resonance between the microwave and the sea surface wave and non-Bragg mechanism, we build a hybrid backscattering simulation model for a consistent explanation of C-band dual-polarization (VV and VH) ScanSAR images. For the Bragg resonance, the modeling approach used here consists of modifying the parameterization for wind wave spectra to include rain-induced turbulent damping and the enhancement of small gravity and gravity capillary waves due to the presence of ring waves. In total, there are five possible ways that rain can affect the NRCS: (1) rain droplets can attenuate the radar signal crossing the atmosphere; (2) rain droplets can cause volume scattering and thus increase the backscattering NRCS; (3) the radar waves can experience diffraction on rain products on the ocean surface including craters, crowns and stalks; and rain changes the ocean surface waves (4 damp and 5 ring wave) as generated and driven by wind. In this study, the former two ways are summarized as the atmosphere part, and the latter there ways, as the ocean surface part.

On the ocean surface, a NRCS model including the rain damping effect on the wind induced waves, and the rain enhanced ring wave spectra, was developed by Contreras and Plant [37]. They added the two rain effects to a semi-empirical wind wave spectrum model developed by Kudryavtsev et al. [30] (hereafter denoted the KHCC wave spectral model). However, the sea surface waves that satisfy Bragg resonance condition are tilted by the long waves, so the local incident angles and polarizations are altered [38–41]. This tilting effect is merely dealt with by averaging over the scales of the long waves in the Bragg scattering model adopted by Kudryavtsev et al. [30] and Contreras and Plant [37], instead of using a probability density function (PDF) methodology. As the geometric coefficients are zeros for the cross polarizations, the averaging process over the scales of long waves cannot simulate the cross-polarized NRCS caused by Bragg wave resonance. Recently, Kudryavtsev et al. [42] simulated the NRCS due to the Bragg resonance by adopting the PDF methodology. In the atmosphere, the raindrops attenuate the radar signal, and also intensify the NRCS through volumetric scattering. These two mechanisms were simply modeled by an empirical approach by Nie and Long [43] without rain effects on the ocean surface. Thereafter, a more sophisticated physics-based radiative transfer model was developed by Xu et al. [44], which is shown to give comparable results to those of the simple model of Nie and Long [43]. Although Xu et al. [44] simply modeled ring waves generated by rain with some success there is need for improvements; the mechanisms by which rain affects the ocean surface need further study. In Xu et al. [44], the rain-induced NRCS damping was not included and the ring waves contribution was linearly added as a modulation to the empirical GMF without considering the actual physical progresses. In this study, rains effects on the ocean surface including the rain-induced ring waves and the damping due to the wind waves were modeled by the Bragg scattering theory and the semi-empirical wave spectrum. Thus, we can analyze each NRCS modulation mechanism quantitatively.

In this study, we build a semi-empirical radar scattering model that takes into account the impacts of both sea surface wind and rain on the NRCS. The model considers the total NRCS as the summation of backscatters from both the ocean surface and the atmosphere. As shown in Fig. 14.25, there are five modules that are needed to account for these processes: (1) the composite sea surface Bragg and non-Bragg model given by Zhang et al. [29] and Plant [41] for its ability to simulate VV- and VH- polarized data; (2) the wind-driven KHCC gravity wave spectrum developed by Kudryavtsev et al. [30] without the rain effect; (3) the rain damping effect on the KHCC wave spectrum given by Contreras and Plant [37] and Tsimplis [21]; (4) the ring wave spectrum developed by Le Mhaut [45]; (5) an additive model developed by Nie and Long [43] for the atmospheric part. With this new composite model, the mechanisms of the rain effect on the NRCS are analyzed.

The composite Bragg model we formulated explains C-band NRCS measurements better and is less sensitive to the choice of the roughness spectral model than is the case for Ku-band measurements [46]. Although other wind-wave spectra are probably more commonly used [47], the KHCC spectrum has an advantage. The main reason

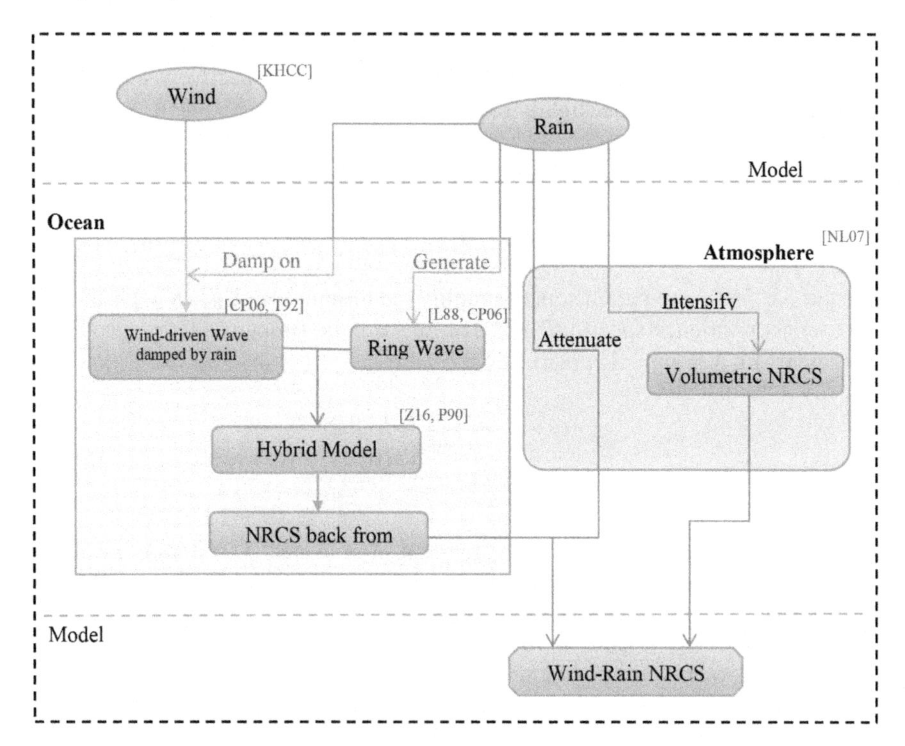

Fig. 14.25 Outline of the model. The *red* abbreviations mean the references: KHCC for Kudryavtsev et al. [30], CP06 for Contreras and Plant [37], T92 for Tsimplis [21], L88 for Le Mhaut [45], Z16 for Zhang et al. [29], P90 for Plant [41], and NL07 for Nie and Long [43]

for choosing KHCC is that it is developed from a balance of wind input, parasitic wave generation, and wave dissipation, and therefore is convenient to modify to include the rain damping effect on the wave dissipation.

14.5.3.1 The Surface Scattering Model

The Bragg resonance theory has been developed since the 1960s [38–41, 48, 49]. The backscatter cross section solutions for a slightly rough patch tilted by the longer waves are as follows:

$$\begin{aligned}
\sigma_{0HH}(\theta_i) &= 16\pi k^4\cos^4\theta_i\left|g_{HH}(\theta_i)\left(\frac{\alpha\cos\delta}{\alpha_i}\right)^2 + g_{VV}(\theta_i)\left(\frac{\sin\delta}{\alpha_i}\right)^2\right|^2 W(K_{Bx}, K_{By})\\
\sigma_{0VV}(\theta_i) &= 16\pi k^4\cos^4\theta_i\left|g_{VV}(\theta_i)\left(\frac{\alpha\cos\delta}{\alpha_i}\right)^2 + g_{HH}(\theta_i)\left(\frac{\sin\delta}{\alpha_i}\right)^2\right|^2 W(K_{Bx}, K_{By})\\
\sigma_{0VH}(\theta_i) = \sigma_{0HV}(\theta_i) &= 16\pi k^4\cos^4\theta_i\left(\frac{\alpha\sin\delta\cos\delta}{\alpha_i^2}\right)^2 |g_{VV}(\theta_i) - g_{HH}(\theta_i)|^2 W(K_{Bx}, K_{By})
\end{aligned} \tag{14.10}$$

where, k is the radar wave number, ψ and δ are the tilting angles of the long wave in the radar incident plane and perpendicular to this plane, respectively, and θ_i is the local incidence angle of the slightly rough patch:

$$\begin{aligned} \theta_i &= \arccos[\cos(\theta+\psi)\cos\delta] \\ \alpha_i &= \sin\theta_i \\ \alpha = \sin(\theta+\psi),\ \gamma &= \cos(\theta+\psi) \end{aligned} \tag{14.11}$$

where α_i, α, and γ are parameters to simplify the formula, $W(K_{Bx}, K_{By})$ is the 2D-wave number variance spectrum, g_{VV} and g_{HH} are the Bragg scattering geometric coefficients for VV and HH polarizations:

$$\begin{aligned} g_{VV}(\theta) &= \frac{(\varepsilon_r-1)[\varepsilon_r(1+\sin^2\theta)-\sin^2\theta]}{[\varepsilon_r\cos\theta+\sqrt{\varepsilon_r-\sin^2\theta}]^2} \\ g_{HH}(\theta) &= \frac{(\varepsilon_r-1)}{[\cos\theta+\sqrt{\varepsilon_r-\sin^2\theta}]^2} \end{aligned} \tag{14.12}$$

and where ε_r is the relative dielectric constant of seawater. Accounting for all the surface tilting, the backscatter cross section per unit area of the sea surface is:

$$\sigma_0(\theta) = int_{-\infty}^{\infty} int_{-\infty}^{\infty} \sigma_0(\theta_i) p(\tan\psi, \tan\delta) d(\tan\psi) d(\tan\delta) \tag{14.13}$$

where $p(\tan\psi, \tan\delta)$ is the joint probability density function (PDF) of the ocean surface slopes. We adopt the Gram–Charlier distribution [50] for the PDF function for p:

$$\begin{aligned} p(\tan\psi, \tan\delta) =& \frac{1}{2\pi s_c s_u} \exp\left(-\frac{\zeta^2+\eta^2}{2}\right) \\ &\cdot \begin{bmatrix} 1 - \frac{c_{21}(\zeta^2-1)}{2} - \frac{c_{03}(\eta^3-3\eta)}{6} + \frac{c_{40}(\zeta_4-6\zeta^2+3)}{24} \\ + \frac{c_{22}(\zeta^2-1)(\eta^2-1)}{4} + \frac{c_{04}(\eta^4-6\eta^2+3)}{24} \end{bmatrix} \end{aligned} \tag{14.14}$$

where $\eta = \tan\psi/s_u$ and $\zeta = \mathrm{rm}\delta/s_c$, are the normalized upwind and crosswind slope components respectively, and s_c^2 and s_u^2 are the crosswind and upwind Mean Square Slope (MSS) variables [49]. The skewness also given by Hwang et al. [49], as contained in the coefficients c_{21}, c_{40}, c_{22}, c_{04} and c_{03}, increases with wind speed from nearly zero at low wind to $c_{21} = 0.11$, $c_{40} = 0.4$, $c_{22} = 0.1$, $c_{04} = 0.2$, and $c_{03} = 0.42$ at 14 m/s; a linear approximation is used in the implementation such that:

$$\begin{aligned} c_{21} &= -0.11U_{10}/14,\ (U_{10} \le 14\ \text{m/s}) \\ c_{03} &= -0.42U_{10}/14,\ (U_{10} \le 14\ \text{m/s}) \end{aligned} \tag{14.15}$$

With the composite Bragg theory and the wave spectrum, the backscatter from the ocean surface can be simulated for different incident angles and polarizations.

14.5.3.2 Sea Surface Wave Model

The variance spectrum for the Bragg scattering model is related to the directional wave number spectrum $S(k, \varphi)$ by:

$$W(k, \varphi) = S(k, \varphi) + S(k, \varphi + \pi) \tag{14.16}$$

Here, we represent the wave spectrum with the saturation spectrum B (or the surface curvature spectrum), which is a function of the surface variance spectrum S:

$$B(\mathbf{k}) = k^4 S(\mathbf{k}) \tag{14.17}$$

where the rain-modulated sea surface wave spectrum B_s ranges from short gravity waves to capillary waves is:

$$B_s = B_{gc\_damped} + B_{cap} + B_{ring} \tag{14.18}$$

and where B_{gc_damped} is the gravity wave spectrum, damped by the precipitation, with a viscous dissipation rate for the rain-induced turbulence damping, B_{cap} is the spectrum in the gravity-capillary region, and B_{ring} is the spectrum for the rain-induced ring waves.

For the gravity and gravity-capillary region of the spectrum, the parameterization developed by Kudryavtsev et al. [30] is:

$$B_{eq}(k, \varphi) = \frac{\alpha}{2^{1/n}} \left[\beta_v(k, \varphi) + (\beta_v^2(k, \varphi) + 4I_{pc}(k, \varphi)/\alpha)^{1/2} \right]^{1/n} \tag{14.19}$$

where, k is the wavenumber, φ is the wind direction, n, α are two tuning parameters, I_{pc} is energy input due to generation of parasitic capillaries and is also a function of the effective growth rate $\beta_v(k, \varphi)$, and where effectively, $\beta_v(k, \varphi)$ is the difference between the wind growth rate $\beta(k, \varphi)$ and the viscous dissipation rate. In this model, the rain damping effect was added through the parameterization for the viscous dissipation rate. Therefore, the effective growth rate $\beta_v(k, \theta)$ may be described by:

$$\beta_v(k.\varphi) = \beta(k, \varphi) - \frac{4\nu_r k^2}{\omega} \tag{14.20}$$

where ν_r is the viscosity coefficient due to the rain damping, and we adopt the rain-induced damping model developed by Nystuen [51]:

$$\nu_r = \left(1 - e^{-2kd_{mix}}\right) \frac{\nu_e}{\nu} + e^{-2kd_{mix}} \tag{14.21}$$

where ν_e is the rain-induced eddy viscosity. Tsimplis [21] found ν_e to be about 3×10^{-5} m^2/s, when rain exists; not matter how high the rain rate is. Therefore, Contreras and Plant [37] described ν_e with a function of the rain rate as:

$$\nu_e = Cl\nu_0; \qquad \nu_0 = \left[\left(\frac{\int_0^\infty IF(D;RR)DdD}{\rho_w d_{mix}}\right)\right]^{\frac{1}{3}} \tag{14.22}$$

Here, l and v are differential length and velocity scales characteristic of the turbulence, $C = 0.2095$ is a constant coefficient, and IF is the energy flux into the eddy viscosity due to the rain impact:

$$IF(D;RR) = N(D;RR)\left(\frac{\pi}{12}\rho_w D^3 W^2(D)\right)W(D) \tag{14.23}$$

Here, N is the rain drop size distribution (DSD), W is the terminal drop velocity, and $D = l$ is the diameter of the rain droplet.

For the gravity wave region, the two tuning parameters (n and α) are constants. With constant tuning parameters, B_{gc_damped} was calculated by Eq. (14.19). However, for gravity-capillary waves, the tuning parameters are functions of wavenumber and then the B_{cap} is computed, also using Eq. (14.19). Details of the two tuning parameters and intervals for the gravity and gravity-capillary wavenumbers are the same as those of the KHCC spectrum [30].

The ring wave spectrum as developed by Le Mhaut [45] is also adopted in our model. Given a rain drop of radius R impinging upon water, the ring wave spectrum is:

$$B_{ring}(k, RR) = \begin{array}{l} \left\{\frac{64}{9}\frac{\omega^2}{g^2k^4}\int_0^\infty N(D)W^3(D)D^2\left[J_2\left(\frac{kD}{2}\right)\right]^2 dD\right\} \\ \cdot\left\{\frac{1}{8\nu_r k^2}(e^{-4\nu_r k^2(0.2)} - e^{-4\nu_r k^2\tau})\right\} \end{array} \tag{14.24}$$

where ω is the wave frequency, g is the gravitational acceleration, RR is the rain rate, D is the diameter of the raindrop, $N(D)$ is the drop size distribution [52], $W(D)$ is the terminal fall velocity of the drop, $J_2(kD/2)$ is the second order Bessel functions of the first kind, and the maximum limit of integration τ is set to 4. The ocean surface backscatter model including the simultaneous effects of wind and rain was developed using the sea surface spectrum in Eq. (14.19) and the composite Bragg theory.

Figure 14.26 shows the saturation spectra of wind waves at various wind speeds without the rain effect. Figure 14.27 compares the saturation spectra with the rain effect (solid lines), rain damping (dashed line) and the wind-wave only spectrum (dotted line) for a wind speed of 10 m/s with different rain rates. The rain effect on the waves contains wind-generated waves damped by rain and the rain-generated ring waves. Regarding the magnitude of the spectral values, although rain-affected wave spectra, are higher than the wave spectra only damped by rain (by adding the ring waves), the rain affected wave spectra are still lower than the wind-induced wave spectra. Hence, the composite rain effect on the ocean surface waves is attenuation of these waves.

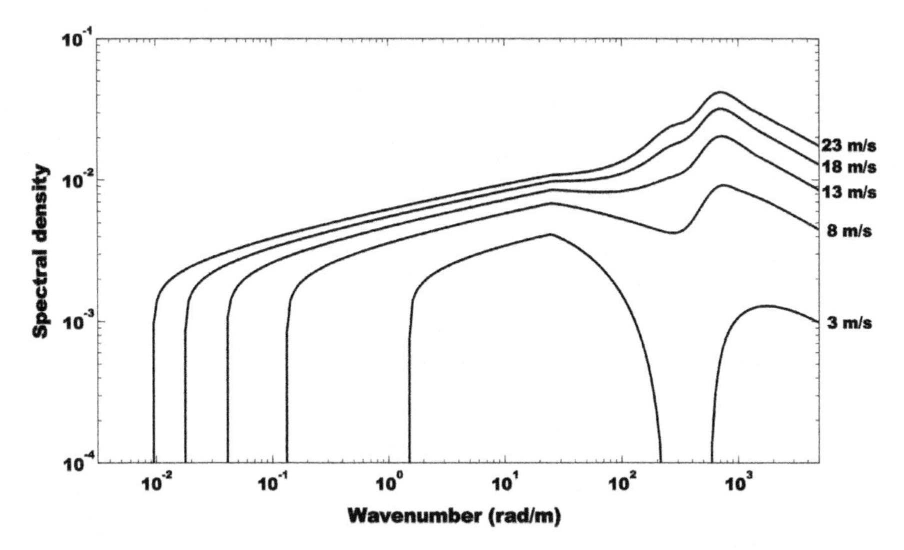

Fig. 14.26 Model saturation spectra of short wind waves at various wind speeds without the rain effect: 3–23 m/s from *lower to upper curves* with the resolution of 5 m/s

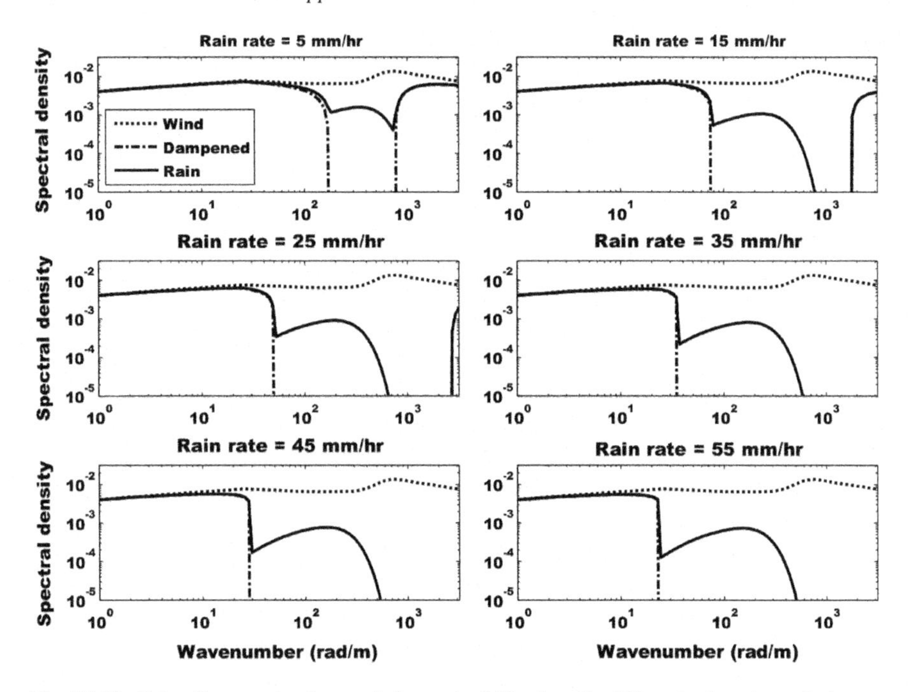

Fig. 14.27 Saturation spectra for a wind speed of 10 m/s with different rain rates: wind wave spectrum (*dotted*), damped wind wave spectrum (*dashed*), and damped wind wave spectrum with additive ring wave spectrum (*solid*)

14.5.3.3 Atmospheric Rain Effect on NRCS

In the atmosphere, raindrops attenuate the radar transmitting signal and also induce volume scattering. These two mechanisms compete by decreasing or increasing the NRCS, respectively. An additive model, including the rain's effect on the NRCS in the atmosphere, is given by Nie and Long [43]:

$$\sigma_m = \sigma_{surf}\alpha_{atm} + \sigma_{atm} \tag{14.25}$$

where, σ_m is the total NRCS received by radar, σ_{surf} is the backscatter from the ocean surface, α_{atm} is the rain-induced atmospheric attenuation, and σ_{atm} is the atmospheric volume backscatter.

14.5.3.4 Validation of the Composite Model

To validate the accuracy of this wind-rain model, datasets of C-band VV-polarized NRCSs, wind speeds and rain rates over two hurricanes were investigated. We have acquired two ENVISAT SAR images over Hurricane Gustav (03:56 UTC, September 01 2008) and Hurricane Ike (04:23 UTC, September 13 2008) from the European Space Agency (ESA). The SAR images are wide swath mode (WSM) with a medium resolution of 150 m and a swath width of 405 km at VV polarization. The ocean surface part of our model was based on Bragg resonance between the sea surface waves and the radiative waves; the sea surface waves were simulated by a wave spectrum, and therefore a particular scale of the ocean surface is needed to cover enough wave scales. Therefore, we calibrated the SAR image and then averaged the spatial resolution to 1 km (shown in Fig. 14.28) with the boxcar averaging method. Moreover, with the averaging, the SAR speckle noise is almost removed. The calibration process is also a methodology to transfer the SAR image pixels to latitude and longitude spatial location information, using the transfer function given by ESA.

At the two SAR imaging times, we also obtained wind speeds and rain rates, as measured by the stepped-frequency microwave radiometer (SFMR) on board the WP-3D research aircraft, which is employed for NOAA's operational surface wind and rain measurements [53]. SFMR provides along-track mapping of wind speeds at relatively high spatial (1.5 km) and temporal (1 Hz) resolutions. These winds are well validated by measurements from dropwindsondes, each equipped with GPS. The RMS error for SFMR measurements is less than ~4 m/s at the surface and less than ~5 m/s at 10-m height [54]. With six frequencies between 4.5 and 7.2 GHz used in the SFMR, the rain rate is also inferred with the wind speed. The rain rate error is less than ~5 mm/h, after algorithm correction and error analysis [55]. Along the flight track (blue dots in Fig. 14.28), longitude, latitude, sea surface wind speed, and rain rate are measured at the same time.

SFMR data selected here were collected during a flight that lasted one hour but two SAR image were captured almost instantaneously. During this one hour, the hurricane also moves. Therefore, our matchup process here was carried out to remove

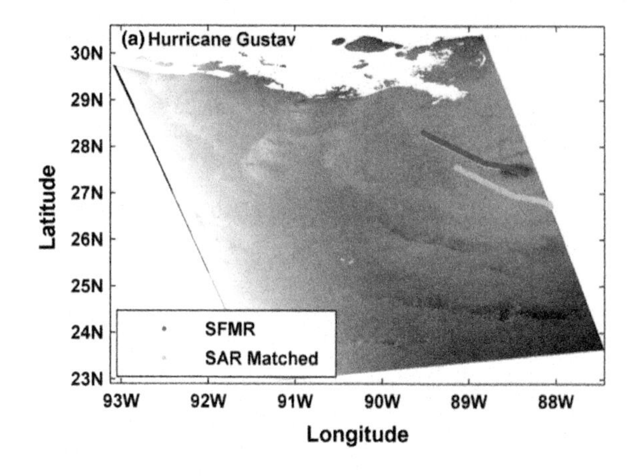

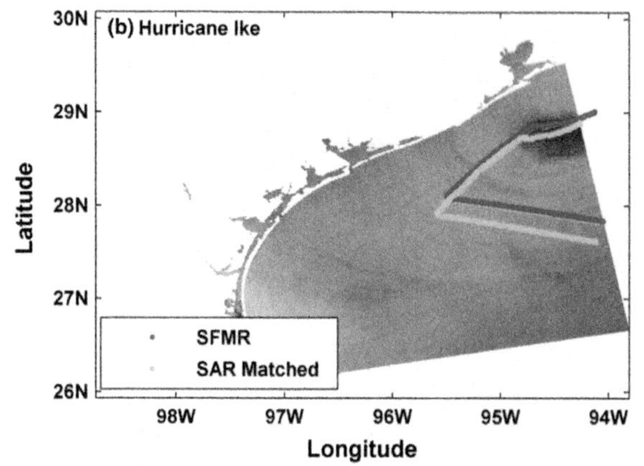

Fig. 14.28 ENVISAT SAR images for: **a** Hurricane Gustav (03:59 UTC, September 1, 2008) and **b** Hurricane Ike (04:23 UTC, September 13, 2008). The positions of SFMR adopted here were displayed in *blue points* as well as the SAR data matched up in *cyan points*

the effect of the relative movement between the aircraft carrying the SFMR and the hurricane. Moreover, we assume that the hurricane moves linearly along the Best Track (BT) during this one hour. For the hurricane moving continuously along a trajectory, we map the hurricane center observed by SAR to the BT position at each SFMR measurement time. The center locations of BT and of SAR hurricane images detected by Li et al. [12] were adopted, as shown in Table 14.7. To simplify the process, we assume that the hurricane moves linearly between two BT locations without rotation.

For each selected set of wind speeds and rain rates measured by SFMR, the mapping procedure consists of the following steps: (1) interpolate the BT center locations to the time of the SFMR observation; (2) move the SAR image to co-locate the SAR center and the interpolated BT center; (3) determine whether the SFMR dataset is within the mapped SAR image, or not; and (4) if yes, the dataset from the SFMR and the NRCS for the same location in the mapped SAR image are considered to be matched (cyan dots shown in Fig. 14.28). As shown in Fig. 14.29, there are 296

Table 14.7 Hurricane center information of SAR and related Best Track estimates

Name	Date	SAR			Best Track		
		Time	Latitude	Longitude	Time	Latitude	Longitude
Gustav	2008-09-01	3:59	27.0°N	88.4°W	0:00	26.9°N	87.7°W
					6:00	27.9°N	89.0°W
Ike	2008-09-13	4:23	28.0°N	94.6°W	0:00	28.3°N	94.0°W
					6:00	29.1°N	94.6°W

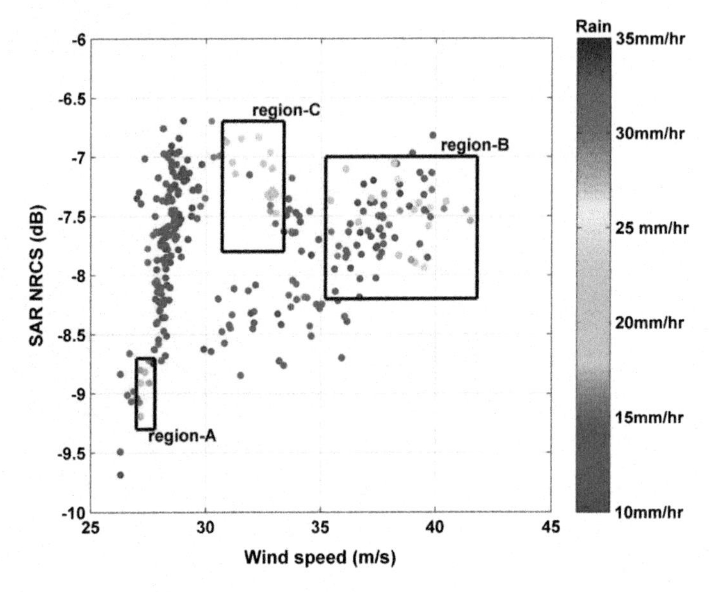

Fig. 14.29 The collected matchup datasets: wind speed and rain rate measured by SFMR, as well as NRCS observed by SAR

matchups of wind speed, rain rate and NRCS. The rain rates are between 10 and 35 mm/h, wind speeds are from 25 to 45 m/s and the NRCSs are between −10 and −5 dB.

To model the volume scattering and attenuation due to rain in the atmospheric column, three dimensional rain rates are needed, which are difficult to measure. Therefore, we modeled the NRCSs for the matched datasets using only the ocean surface part. The model results with: (a) wind only (Fig. 14.30a), and (b) with both wind and rain (Fig. 14.30b), are plotted against the matched dataset. For the wind-only model results, the comparison has a bias of 0.71 dB, a standard deviation of 0.60 dB and correlation coefficient of 80.9%. When the rain effect is taken into account, the simulation results improved; the bias is 0.20 dB with a standard deviation of 0.54 dB and correlation coefficient is 81.9%. In the comparisons shown in Fig. 14.30, near the SAR NRCSs at about −9.0 dB, the rain rate is about 20 mm/h (hereafter defined as region-A), and our composite model appears to simulate the attenuation of the

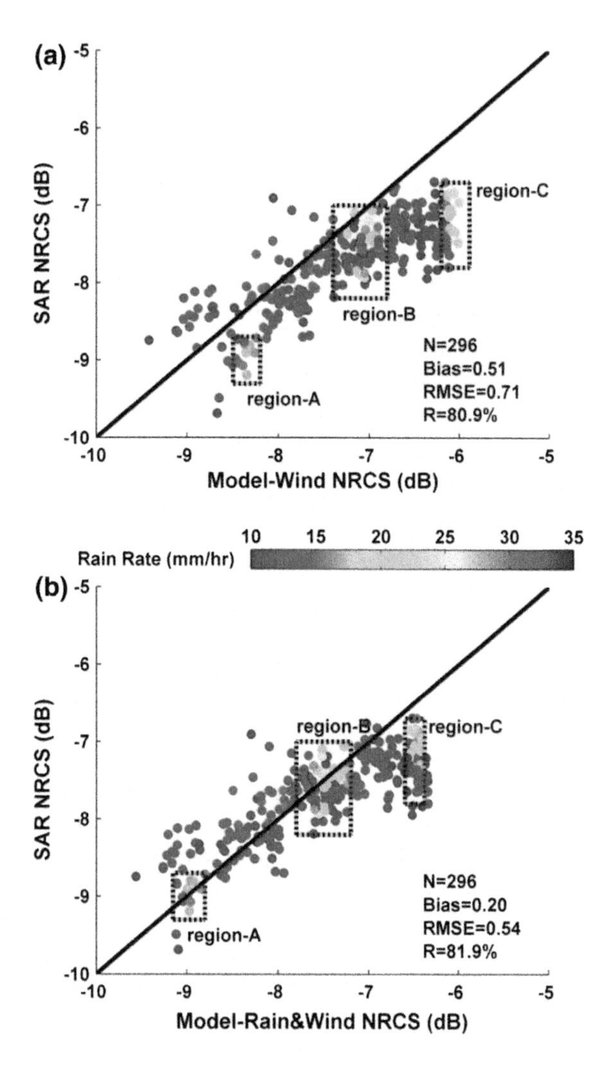

Fig. 14.30 Comparisons between simulated NRCS induced by the ocean surface part of the composite model and the ASAR observed NRCS: **a** simulation with only wind, and **b** simulation with wind and rain

NRCSs due to rain very well. When the NRCSs are about −7.7 dB, as observed by SAR, the effect of heavy rain is about 30 mm/h (hereafter defined as region-B), which also appears to be modeled appropriately. However, when the SAR NRCSs is −7.0 dB with rain rate of 20 mm/h (defined as Region-C), this case is not simulated very well, although the results presented in Fig. 14.30b are still better than those of Fig. 14.30a. In terms of winds, Fig. 14.29 implies that the wind speeds are about 27.5 m/s for region-A and from 35.0 to 42.0 m/s for region-B, and resulting SAR NRCSs are reasonably well simulated. The disagreement between simulated and measured NRCSs occurs near wind speeds of about 32.0 m/s. This demonstrates that the bias of region-C is not caused by the model saturation due to high wind speeds. A possible reason may be the asymmetry and rotational motion of the hurricane.

14.6 Conclusion

There are three possible problems when we process C-band cross-polarized SAR images: (1) many SAR images capture incomplete hurricane core structures; (2) the radar signal is attenuated by the heavy precipitation associated with hurricanes; (3) wind directions retrievals are not available from the cross-polarized SAR measurements. We note that the cross-polarization SAR signal appears to not saturate with increased of wind speeds, for the SAR data that has been collected thus far.

In this study, we have determined the locations of the eyes of hurricane centers on the ocean surface and the horizontal extents of their respective regions of maximum winds, using cross-polarization SAR measurements. To avoid possible effects of non-axisymmetric structure characteristics on estimates for hurricane ISPs, we estimate the average of wind profiles in all radial directions. Thus, we obtain mean wind profiles. To estimate an idealized complete wind vector field from a cross-polarized SAR image, a two-dimensional SHEW model is developed and the inflow angle model is revised, both based on the assumption of an elliptical shaped eye and eyewall. In the SHEW model, we assume an elliptical eyewall shape, where the maximum wind speeds exist, as a generalization of the one-dimensional wind profile in each radial direction.

We present a new model, the OHMRV model, to describe hurricane wind profiles that include both a wind maximum and an inflection point; the latter is associated with the degeneration of the inner wind maximum in the reintensification phase in the ERC process. The ISPs, such as maximum wind speed and RMW, can easily be calculated from the fitted wind profiles. In validating these results, we compared our estimates for HE centers, diameters, and ISPs with measurements from SFMR.

Our proposed OHMRV model complements the SMRV and DMRV models for hurricane wind profiles, previously described in the literature. The DMRV model is suitable to characterize the wind profile associated with the intensification and weakening phases in the ERC process. In these hurricane cases, there are both inner and outer wind maxima in these two phase profiles. By comparison, the OHMRV model is appropriate to describe wind profiles corresponding to the ERC reintensification phase. In these hurricanes, there are attenuated inner wind maxima (inflection point) and outer wind maxima. Once the inner wind maximum can no longer be detected, the ERC is complete, and the storm reverts to a single eyewall structure. In this case, the SMRV model can favorably describe the wind profile because there is only a single wind maximum.

When the SHEW model was applied to C-band RADARSAT-2 cross-polarized SAR images, the elliptical symmetrical wind speed field can be estimated and morphology parameters and the intensity parameters can be detected. To simulate the wind direction for cross-polarized SAR images, the inflow angle model is revised by adopting an ellipse-shaped eye. Thus, we replace the previously assumed circular-shaped eye. By providing the morphology parameters estimated by SHEW model and the hurricane motion vector from the Best Track data, the wind direction can be estimated by the revised inflow angle model. Combining the wind speed estimated by the SHEW model and wind direction derived by the revised inflow angle model, the idealized complete surface wind vector field of a hurricane is estimated from a given SAR image. For 15 SAR images of hurricanes observed in 2014, the closest elliptical symmetrical surface wind speed fields and surface wind vector fields were estimated. Thus, we determined six elliptical morphology parameters: hurricane center, reference ellipse parameters (major axis, minor axis, and azimuth angle), hurricane symmetric intensity and the decay parameter. Comparisons between the wind vectors based on our model and in situ observations show good agreement. Additionally, the SHEW model can address some aspects of the impact by rain. However, the complete impacts of rain on the ocean surface have still not been entirely addressed.

Five possible mechanisms (Fig. 14.18) for the rain effects on the spaceborne C-band SAR hurricane observations are considered: (1) attenuation and (2) volume backscattering for the microwave transfer in atmosphere; as well as (3) diffraction on the sharp edges of rain products and (4) rain-induced damping to the wind waves, and (5) rain-generated ring waves on the ocean surface. The four mechanisms except the third one caused by diffraction are simulated by a composite model. Comparisons between the observed NRCSs and the atmosphere part of our model imply that two possible effects in the atmosphere can be neglected for C-band VV-polarization SAR hurricane observations. However, the atmosphere part should be important at very low wind speed and the VV-polarized NRCS will be increased for the volume backscattering. Thus, we speculated that the diffraction of radio waves on craters, crowns and stalks, induced by rain on the ocean surface can be neglected for C-band VV-polarization, and that the NRCSs are essentially affected only by wind and can be simulated well for C-band VV-polarization without including the diffraction of the radar by the sharp wedges of wave breaking but not well for the cross-polarization (Fig. 14.23). However, the mechanism of diffraction on the sharp wedges of wave breaking and rain products needs to be further modeled to validate this speculation. Our composite model was validated with the matchup of observations of wind speed and rain rate from the SFMR data as well as the NRCSs and incidence angles of C-band VV-polarization SAR data over two hurricanes. Therefore, the most important mechanism for the rain effect on the C-band VV polarized SAR hurricane observations is through the influence of waves on the ocean surface.

It is of note that although our models capture the main features of a hurricane eye shape, there is still a good amount of unexplained variability which requires further study. The distribution of wind speeds estimated by the SHEW model has an idealized and elliptical-shaped structure. The wind speeds retrieved from the SAR image by using the C-2PO model are expected to represent the real wind speeds. Therefore, if the correlation between the two sets of wind speeds is high, the real hurricane structure is close to ideal. Conversely, when the hurricane structure is not ideal because of its high azimuthal variability like the first Vance image, results from the SHEW model are not as good as for more ideal hurricanes, in terms of following an idealized elliptical-shaped hurricane structure. Ongoing studies will continue the investigation of Hurricane Vance and the reasons that cause it to behave differently.

References

1. Smith, R. 1980. Tropical cyclone eye dynamics. *Journal of the Atmospheric Sciences* 37 (6): 1227–1232.
2. Shapiro, L.J., and H.E. Willoughby. 1982. The response of balanced hurricanes to local sources of heat and momentum. *Journal of the Atmospheric Sciences* 39 (2): 378–394.
3. Willoughby, H. 1990. Temporal changes of the primary circulation in tropical cyclones. *Journal of the Atmospheric Sciences* 47 (2): 242–264.
4. Sitkowski, M., J.P. Kossin, and C.M. Rozoff. 2011. Intensity and structure changes during hurricane eyewall replacement cycles. *Monthly Weather Review* 139 (12): 3829–3847.
5. Zhu, Z., and P. Zhu. 2015. Sensitivities of eyewall replacement cycle to model physics, vortex structure, and background winds in numerical simulations of tropical cyclones. *Journal of Geophysical Research: Atmospheres* 120 (2): 590–622.
6. Sanabia, E.R., B.S. Barrett, N.P. Celone, and Z.D. Cornelius. 2015. Satellite and aircraft observations of the eyewall replacement cycle in Typhoon Sinlaku (2008). *Monthly Weather Review* 143 (9): 3406–3420.
7. Kossin, J.P., and M.D. Eastin. 2001. Two distinct regimes in the kinematic and thermodynamic structure of the hurricane eye and eyewall. *Journal of the Atmospheric Sciences* 58 (9): 1079–1090.
8. Franklin, J.L., M.L. Black, and K. Valde. 2003. GPS dropwindsonde wind profiles in hurricanes and their operational implications. *Weather and Forecasting* 18 (1): 32–44.
9. Kossin, J.P. 2015. Hurricane wind-pressure relationship and eyewall replacement cycles. *Weather and Forecasting* 30 (1): 177–181.
10. Fu, L.L., and B. Holt. 1982. *Seasat views oceans and sea ice with synthetic aperture radar*, 81–120. Pasadena: JPL Publication.
11. Li, X. 2015. The first Sentinel-1 SAR image of a typhoon. *Acta Oceanologica Sinica* 34 (1): 1–2.
12. Li, X., J.A. Zhang, X. Yang, W.G. Pichel, M. DeMaria, D. Long, and Z. Li. 2013. Tropical cyclone morphology from spaceborne synthetic aperture radar. *Bulletin of the American Meteorological Society* 94 (2): 215–230.
13. Zhang, G., B. Zhang, W. Perrie, Q. Xu, and Y. He. 2014. A hurricane tangential wind profile estimation method for C-band cross-polarization SAR. *IEEE Transactions on Geoscience and Remote Sensing* 52 (11): 7186–7194.
14. Hock, T.F., and J.L. Franklin. 1999. The NCAR GPS dropwindsonde. *Bulletin of the American Meteorological Society* 80 (3): 407–420.
15. Mallen, K.J., M.T. Montgomery, and B. Wang. 2005. Reexamining the near-core radial structure of the tropical cyclone primary circulation: Implications for vortex resiliency. *Journal of the Atmospheric Sciences* 62 (2): 408–425.
16. Uhlhorn, E.W., B.W. Klotz, T. Vukicevic, P.D. Reasor, and R.F. Rogers. 2014. Observed hurricane wind speed asymmetries and relationships to motion and environmental shear. *Monthly Weather Review* 142 (3): 1290–1311.
17. Holland, G. 2008. A revised hurricane pressure-wind model. *Monthly Weather Review* 136 (9): 3432–3445.
18. Holland, G.J., J.I. Belanger, and A. Fritz. 2010. A revised model for radial profiles of hurricane winds. *Monthly Weather Review* 138 (12): 4393–4401.
19. Reasor, P.D., M.T. Montgomery, F.D. Marks Jr., and J.F. Gamache. 2000. Low-wavenumber structure and evolution of the hurricane inner core observed by airborne dual-doppler radar. *Monthly Weather Review* 128 (6): 1653–1680.
20. A. Worthington. A study of splashes: Including his 1894 lecture: The splash of a drop and allied phenomena. Macmillan, 1963. URL https://books.google.com/books?id=JxJRAAAAMAAJ.
21. Tsimplis, M. 1992. The effect of rain in calming the sea. *Journal of Physical Oceanography* 22 (4): 404–412.

22. Reppucci, A., S. Lehner, J. Schulz-Stellenfleth, and S. Brusch. 2010. Tropical cyclone intensity estimated from wide-swath SAR images. *IEEE Transactions on Geoscience and Remote Sensing* 48 (4): 1639–1649.
23. Du, Y., and P.W. Vachon. 2003. Characterization of hurricane eyes in RADARSAT-1 images with wavelet analysis. *Canadian Journal of Remote Sensing* 29 (4): 491–498.
24. Zhang, B., and W. Perrie. 2012. Cross-polarized synthetic aperture radar: A new potential measurement technique for hurricanes. *Bulletin of the American Meteorological Society* 93 (4): 531–541.
25. Zhang, J.A., and E.W. Uhlhorn. 2012. Hurricane sea surface inflow angle and an observation-based parametric model. *Monthly Weather Review* 140 (11): 3587–3605.
26. Wood, V.T., L.W. White, H.E. Willoughby, and D.P. Jorgensen. 2013. A new parametric tropical cyclone tangential wind profile model. *Monthly Weather Review* 141 (6): 1884–1909.
27. Stiles, B.W., R.E. Danielson, W.L. Poulsen, M.J. Brennan, S. Hristova-Veleva, T.P. Shen, and A.G. Fore. 2014. Optimized tropical cyclone winds from QuikSCAT: A neural network approach. *IEEE Transactions on Geoscience and Remote Sensing* 52 (11): 7418–7434.
28. Mai, M., B. Zhang, X. Li, P.A. Hwang, and J.A. Zhang. 2016. Application of AMSR-E and AMSR2 Low-Frequency Channel Brightness Temperature Data for Hurricane Wind Retrievals. *IEEE Transactions on Geoscience and Remote Sensing* 54 (8): 4501–4512.
29. Zhang, G., X. Li, W. Perrie, B. Zhang, and L. Wang. 2016. Rain effects on the hurricane observations over the ocean by C-band Synthetic Aperture Radar. *Journal of Geophysical Research: Oceans* 121 (1): 14–26.
30. Kudryavtsev, V., D. Hauser, G. Caudal, and B. Chapron. 2003. A semiempirical model of the normalized radar cross-section of the sea surface 1. Background model. *Journal of Geophysical Research: Oceans* 108 (C3): 8055.
31. Kalmykov, A., and V. Pustovoytenko. 1976. On polarization features of radio signals scattered from the sea surface at small grazing angles. *Journal of Geophysical Research* 81 (12): 1960–1964.
32. Liu, X., Q. Zheng, R. Liu, M.A. Sletten, and J.H. Duncan. 2017. A Model of Radar Backscatter of Rain-Generated Stalks on the Ocean Surface. *IEEE Transactions on Geoscience and Remote Sensing* 55 (2): 767–776.
33. Hersbach, H., A. Stoffelen, and S. De Haan. 2007. An improved C-band scatterometer ocean geophysical model function: CMOD5. *Journal of Geophysical Research: Oceans* 112 (C3).
34. Hersbach, H. 2010. Comparison of C-band scatterometer CMOD5.N equivalent neutral winds with ECMWF. *Journal of Atmospheric and Oceanic Technology* 27 (4): 721–736.
35. Zhang, B., W. Perrie, and Y. He. 2011. Wind speed retrieval from RADARSAT-2 quad-polarization images using a new polarization ratio model. *Journal of Geophysical Research: Oceans* 116 (C8).
36. Nunziata, F., M. Migliaccio, X. Li, and X. Ding. 2014. Coastline extraction using dual-polarimetric COSMO-SkyMed PingPong mode SAR data. *IEEE Geoscience and Remote Sensing Letters* 11 (1): 104–108.
37. Contreras, R.F., and W.J. Plant. 2006. Surface effect of rain on microwave backscatter from the ocean: Measurements and modeling. *Journal of Geophysical Research: Oceans* 111 (C8).
38. Valenzuela, G. 1968. Scattering of electromagnetic waves from a tilted slightly rough surface. *Radio Science* 3 (11): 1057–1066.
39. Valenzuela, G.R. 1978. Theories for the interaction of electromagnetic and oceanic wavesła review. *Boundary-Layer Meteorology* 13 (1–4): 61–85.
40. Bass, F., I. Fuks, A. Kalmykov, I. Ostrovsky, and A. Rosenberg. 1968. Very high frequency radiowave scattering by a disturbed sea surface Part II: Scattering from an actual sea surface. *IEEE Transactions on Antennas and Propagation* 16 (5): 560–568.
41. Plant, W.J. 1990. Bragg scattering of electromagnetic waves from the air/sea interface. In *Surface waves and fluxes*, 41–108. Springer.
42. Kudryavtsev, V., I. Kozlov, B. Chapron, and J. Johannessen. 2014. Quad-polarization SAR features of ocean currents. *Journal of Geophysical Research: Oceans* 119 (9): 6046–6065.

43. Nie, C., and D.G. Long. 2007. A C-band wind/rain backscatter model. *IEEE Transactions on Geoscience and Remote Sensing* 45 (3): 621–631.
44. Xu, F., X. Li, P. Wang, J. Yang, W.G. Pichel, and Y.Q. Jin. 2015. A backscattering model of rainfall over rough sea surface for synthetic aperture radar. *IEEE Transactions on Geoscience and Remote Sensing* 53 (6): 3042–3054.
45. Le Méhauté, B. 1988. Gravity-capillary rings generated by water drops. *Journal of Fluid Mechanics* 197: 415–427.
46. Hwang, P.A., and W.J. Plant. 2010. An analysis of the effects of swell and surface roughness spectra on microwave backscatter from the ocean. *Journal of Geophysical Research: Oceans* 115 (C4).
47. Elfouhaily, T., B. Chapron, K. Katsaros, and D. Vandemark. 1997. A unified directional spectrum for long and short wind-driven waves. *Journal of Geophysical Research: Oceans* 102 (C7): 15781–15796.
48. Wright, J. 1966. Backscattering from capillary waves with application to sea clutter. *IEEE Transactions on Antennas and Propagation* 14 (6): 749–754.
49. Hwang, P.A., B. Zhang, J.V. Toporkov, and W. Perrie. 2010. Comparison of composite Bragg theory and quad-polarization radar backscatter from RADARSAT-2: With applications to wave breaking and high wind retrieval. *Journal of Geophysical Research: Oceans* 115 (C8).
50. Cox, C., and W. Munk. 1954. Statistics of the sea surface derived from sun glitter. *Journal of Marine Research* 13 (2): 198–227.
51. Nystuen, J.A. 1990. A note on the attenuation of surface gravity waves by rainfall. *Journal of Geophysical Research: Oceans* 95 (C10): 18353–18355.
52. Marshall, J.S., and W.M.K. Palmer. 1948. The distribution of raindrops with size. *Journal of Meteorology* 5 (4): 165–166.
53. Uhlhorn, E.W., and P.G. Black. 2003. Verification of remotely sensed sea surface winds in hurricanes. *Journal of Atmospheric and Oceanic Technology* 20 (1): 99–116.
54. Uhlhorn, E.W., P.G. Black, J.L. Franklin, M. Goodberlet, J. Carswell, and A.S. Goldstein. 2007. Hurricane surface wind measurements from an operational stepped frequency microwave radiometer. *Monthly Weather Review* 135 (9): 3070–3085.
55. Jiang, H., P.G. Black, E.J. Zipser, F.D. Marks Jr., and E.W. Uhlhorn. 2006. Validation of rain-rate estimation in hurricanes from the stepped frequency microwave radiometer: Algorithm correction and error analysis. *Journal of the Atmospheric Sciences* 63 (1): 252–267.

Chapter 15
Detecting the Effects of Hurricanes on Oil Infrastructure (Damage and Oil Spills) Using Synthetic Aperture Radar (SAR) Imagery

Christopher R. Jackson and Oscar Garcia-Pineda

15.1 Introduction

In addition to their effects on coastal areas (flooding, erosion, property damage), tropical cyclones (tropical depressions, tropical storms and hurricanes) in the Gulf of Mexico can significantly impact offshore oil and gas infrastructure (platforms, rigs, pipelines). Past storms have sunk oil platforms and caused oil rigs to break free of their moorings and drift long distances, both of which represent a hazard to navigation. Damaged and sunken platforms and rigs have also resulted in oil spills which represent a significant environmental hazard. Rapid assessment of these effects after the passage of a storm event is important to help in prioritizing post hurricane response efforts and the mitigation of any possible environmental impacts. This chapter will examine the ways in which satellite synthetic aperture radar (SAR) imagery can be used to perform post hurricane assessments of missing oil platforms/rigs and detection and monitoring of oils spills.

15.1.1 Gulf of Mexico - Tropical Cyclone Events, Their Impact and Assessment Approaches

The oil and gas infrastructure located offshore within the United States Exclusive Economic Zone (EEZ) in the Gulf of Mexico includes more than 27000 miles of pipeline and more than 3000 facilities [1] which produce approximately 1.4 million

C.R. Jackson (✉)
Global Ocean Associates, Alexandria, VA, USA
e-mail: goa@internalwaveatlas.com

O. Garcia-Pineda
Water Mapping LLC, Tallahassee, FL 32306, USA
e-mail: oscar.oggp@gmail.com

X. Li (ed.), *Hurricane Monitoring With Spaceborne Synthetic Aperture Radar*, Springer Natural Hazards, DOI 10.1007/978-981-10-2893-9_15

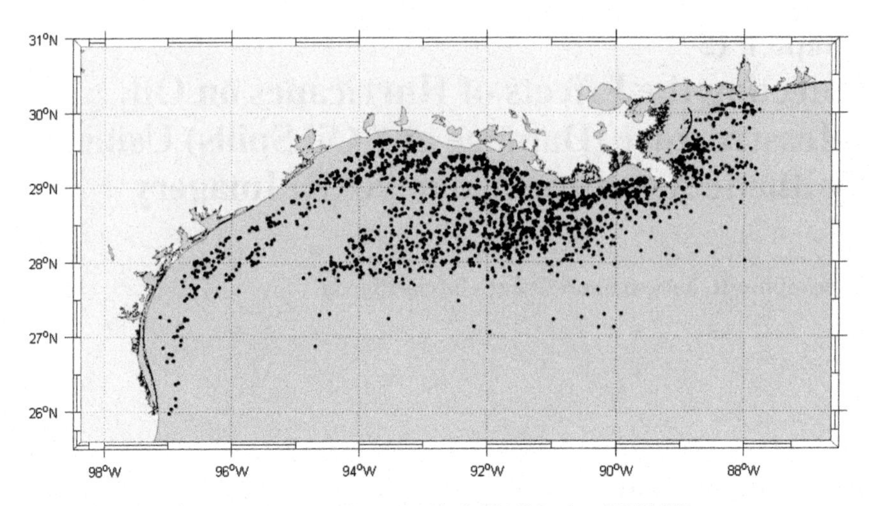

Fig. 15.1 Location of oil and gas platforms in the Gulf of Mexico (2009) [5]

barrels of oil per day [2]. The facilities span across the Gulf from 99°W to 88°W and extend to over 200 km offshore (Fig. 15.1). Nearly all tropical cyclones (tropical depressions, tropical storms and hurricanes) in the Gulf of Mexico can impact the oil and gas infrastructure (platforms, rigs, pipelines). These impacts can range from a temporary shutdown of production to the permanent loss (or sinking) of a platform or rig. A list of tropical cyclones affecting the Gulf infrastructure since 2002 has been compiled by the U.S. Bureau of Safety and Environmental Enforcement (BSEE) [3] and is presented in Table 15.1. The table shows most years have several storms that impact the infrastructure. Since 2009, the Gulf region has been fortunate in that it has only been subjected to 2 hurricanes, 6 tropical storms and 1 tropical depression. (A comprehensive examination of the history of hurricanes in the Gulf of Mexico can be found in Keim and Muller (2009) [4]).

BSEE reports also detail how these storms impact the offshore production and infrastructure. Table 15.2 contains the detailed impact from storms in 2005 (Major Hurricanes Katrina and Rita) and in 2008 (Category-2 Hurricanes Gustav and Ike). In these descriptions the term "destroyed" means that a facility's ability to produce oil or gas has been permanently ended, the platform's (or rig's) physical structure may (or may not) still be present on the surface of the Gulf. Extensive damage means that the facility will take 3–6 months to repair. While Gustav and Ike were only Category-2 storms when they crossed over the Gulf oil infrastructure, (compared to Rita (Category-4) and Katrina (Category-5)), they still resulted in considerable platform and rig destruction and damage.

Beyond the impact of the tropical cyclones on the oil infrastructure itself; damage to platforms, rigs and pipelines potentially has a more widespread impact on shipping and transportation and/or to the offshore environment as a whole. Sunken oil platforms and rigs set adrift represent hazards to navigation. Oil spills resulting from damage represent an environmental hazard with potentially severe effects. In

Table 15.1 Tropical cyclones which produced a significant impact on offshore oil and gas production (2002–2017) (https://www.bsee.gov/resources-tools/hurricane/hurricane-history)

Year	Tropical Event
2016	Tropical Storm Hermine
2015	No storms with significant impact on offshore oil and gas production
2014	
2013	
2012	Hurricane Isaac
	Tropical Storm Debby
2011	Tropical Storm Lee
	Tropical Depression 13
	Tropical Storm Don
2010	Tropical Storm Bonnie
	Hurricane Alex
2009	Tropical Storm Ida
2008	Hurricane Ike
	Hurricane Gustav
	Tropical Storm Edouard
	Hurricane Dolly
2007	Tropical Depression Ten
	Hurricane Dean
	Tropical Storm Erin
2006	Tropical Storm Ernesto
2005	Hurricane Wilma
	Hurricane Rita
	Hurricane Katrina
	Hurricane Emily
	Hurricane Dennis
	Hurricane Cindy
	Tropical Storm Arlene
2004	Hurricane Ivan
	Hurricane Frances
	Hurricane Charley
	Tropical Storm Bonnie
2003	Hurricane Erika
	Hurricane Claudette
	Tropical Storm Bill
2002	Hurricane Lili
	Tropical Storm Isidore

Table 15.2 Detailed impact (as reported by BSEE) of select storms oil and gas infrastructure

2008 Hurricane Gustav (Category-2) and Hurricane Ike (Category-2) [6, 7]
• 60 platforms destroyed, with 31 more suffering extensive damage
• 4 drilling rigs destroyed with one jack-up rig with extensive damage
2005 Hurricane Rita (Category-4) and Hurricane Katrina (Category-5) [8]
• 113 platforms destroyed with 52 more suffering extensive damage
• 19 rigs broke their moorings and were set adrift
• 3 platform rigs destroyed, 1 jack-up rig was sunk, 1 jack-up capsized, with 7 jack-ups, 2 semi-subs and 2 platform rigs suffering extensive damage
• 457 pipelines damaged

many cases these further impacts have been well documented. For example, on 15 September 2004, Hurricane Ivan sunk the Taylor Energy Oil platform which was located 12 miles south of the mouth of the Mississippi River. Oil slicks and sheens have been visible at the location ever since [9]. On 6 March 2009 the M/T SATILLA, a 900-plus foot oil tanker collided with the remains of the jack-up rig ENSCO 74, sixty-five miles offshore of Galveston Texas resulting in a 60-foot gash being opened up on the side of the vessel [10]. ENSCO 74 was ripped from its moorings 6 months earlier by Hurricane Ike and floated for approximately 92 miles before it sank to the bottom and came to rest at the location where it was struck by the M/T SATILLA. Fortunately no oil was spilled by the tanker.

A variety of state and federal agencies (e.g. U.S. Coast Guard, National Oceanic and Atmospheric Administration, BSEE) have the responsibility in the wake of tropical cyclones to perform off-shore damage assessment. Damage assessment is important to get in the hours to few days following a tropical event in order to ensure either safety of navigation and/or allow for the mitigation of any oil spills that might result from destroyed infrastructure. In the Gulf of Mexico, this can be a significant challenge to do in a timely manner given that a storm can cut a swath several hundred kilometers wide (with hurricane force winds and large significant wave height) and that the oil and gas infrastructure covers an expansive area within the Gulf of Mexico and includes more than 3000 offshore locations (Fig. 15.1).

One approach to doing the post event assessment, which can address the need to cover large areas and identify both oil spills and missing platforms/rigs quickly, is through the use of satellite remote sensing, in particular, synthetic aperture radar (SAR) imagery. SAR imagery can be acquired independent of both lighting and cloud cover conditions and provides quantitative measurements of both the ocean surface roughness and objects on the surface. There are currently 10 civilian SAR satellites (Table 15.3) on orbit all of which are capable of providing either wide-swath (>250 km) and/or medium to fine resolution (<25 m) imagery that could be used to identify the location of oil spills or missing platforms. While both oil and platform detection can be done at all viewing geometries, in general, lower incidence angle

Table 15.3 Current civilian SAR satellites on orbit

Satellite	Frequency band	Operating agency (Commercial source)
RADARSAT-2	C-Band	Canadian Space Agency (MDA Corporation)[a]
TerraSAR-X, Tandem-X	X-Band	German Aerospace Center (EADS Astrium)[a]
COSMO SkyMed-1,-2,-3,-4	X-Band	Agenzia Spaziale Italiana (e-GEOS)[a]
ALOS-2	L-Band	Japan Aerospace Exploration Agency
Sentinel-1A, Sentinel-1B	C-Band	European Space Agency

[a] Satellite operator accepts orders for commercial tasking

(closer to nadir) imagery works best to search for oil spills while higher incidence angle imagery is preferred for ocean surface platform/rig detection.

Given the potentially large areas and large number of platforms/rigs impacted by a storm event, the goal of using satellite SAR for post storm assessment would be to quickly: (1) identify platforms that are either missing or are substantially displaced from their known locations (rather than provide a detailed damage assessment of any particular location) and (2) identify the location and size of any resulting oil spills. These rapid identifications would allow subsequent resources to be properly allocated to the locations most heavily impacted for follow up assessment and response. The areas that would benefit most from the use of satellite SAR for post damage assessment are those that are farthest from shore.

15.2 SAR Platform Detection

15.2.1 Post Hurricane Platform/Rig Assessment Approach

The ability of SAR ability to detect ships on the ocean surface is well documented (Pichel et al. 2004 [11]; Vachon et al. 1997 [12]) and several of the satellite SAR imagery providers offer a "ship detection" product derived from their SAR imagery. Oil platforms and rigs are metal structures that reflect large amounts of radar energy and are therefore readily observed in SAR imagery. Figure 15.2 is an Envisat image acquired on 4 September 2008 (at 4:05 UTC) over the Gulf of Mexico in "Standard Mode" (100 km swath with 25 m resolution) centered near 91.6°W and 28.7°N. The image shows a collection of bright dots that stand out against a dark grey background. The bright dots represent the strong radar reflection (or backscatter) from objects on the surface of the ocean, in this case oil platforms or ships. The grey background represents the lower radar reflectivity of the ocean surface (with an approximately 3 m/s average surface wind speed).

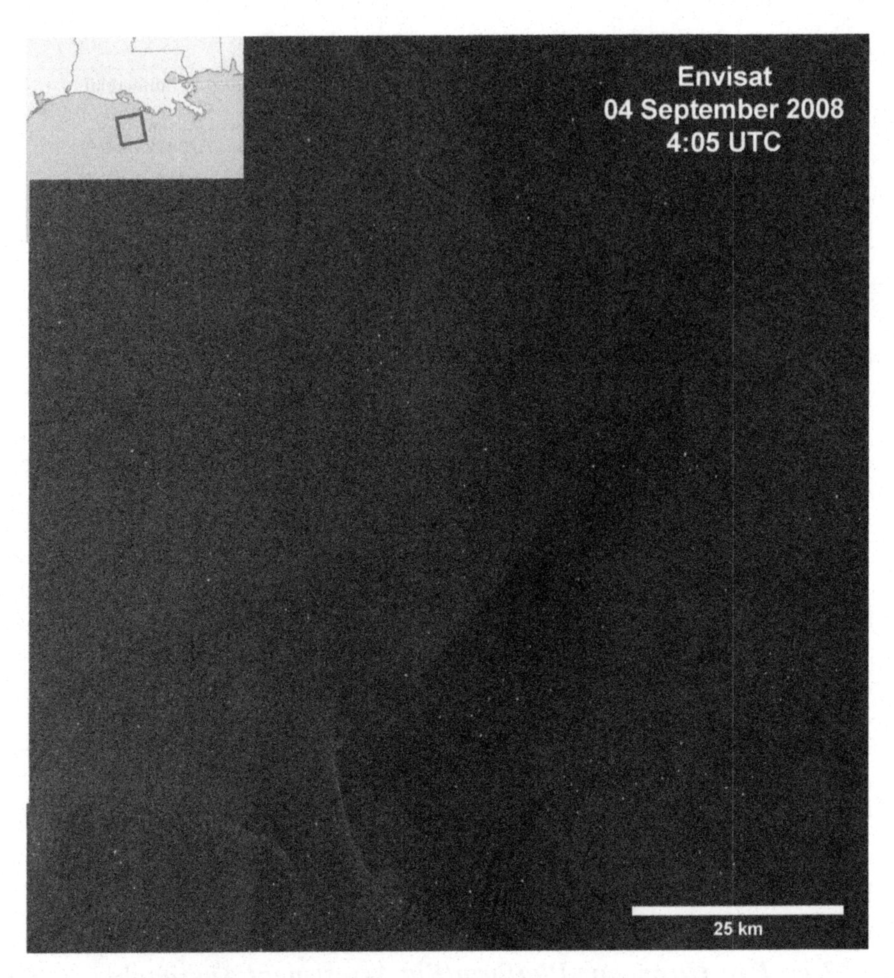

Fig. 15.2 Envisat standard mode image (25 m resolution) (centered near 91.6°W and 28.7°N) showing strongly reflecting surface objects (*bright dots*) (mostly oil platforms) in the Gulf of Mexico. Imaged area is approximately 100 km × 100 km

The ability to detect a particular object in a SAR image of the ocean surface is a function of the ocean surface roughness (the noise) and the normalized radar cross section of the structure itself (the signal). The full details of the theory underlying the detection of objects in SAR imagery is beyond the scope of this chapter and for the purposes of this discussion, it is assumed that oil platforms and rigs on the surface of the ocean (under low to moderate wind conditions (<10 m/s)) will be detectable in the imagery by some automated means. For post event assessment, the objective is to survey the largest possible area while maintaining the resolution requirements to ensure detection of the object of interest, either platforms or oil spills.

For detection of platforms, moderate (25 m) resolution works well, which allows for collection on 100 km (or greater) swath width. And while it may not be possible to detect every individual structure of every complex with 25 m resolution imagery, the largest platforms and rigs, whose destruction would cause the greatest impact, are expected to have a sufficiently high radar cross sections to be readily detected. Fine resolution (<3 m) SAR imagery has the potential to provide a more detailed structural assessment but requires an associated trade off in total imaged area and will not be examined here. Sub-meter satellite optical imagery might be better suited for detailed damage assessments. Oil spill detection in SAR imagery and its requirements is discussed in detail in the second part of this chapter.

The ability to detect oil platforms and rigs in post event SAR imagery is only the first part of the solution. There must be a way to determine if the detected object corresponds to a platform (or rig) as opposed to a ship. One scheme to do this involves performing a pre-season SAR imagery survey to identify all the locations in the Gulf with detectable structures. This would work well for platforms, which are large structures that remain at a fixed position, but fails for rigs which are mobile platforms and move regularly. This approach has other shortcomings as well. The first involves the need to collect a large amount of imagery over all the platform positions (Fig. 15.2) (a survey using imagery with a 100 km × 100 km footprint would require nearly 50 images to cover all the offshore oil platform locations). Second, it is impossible to know what detections in the survey image are from a ship versus a platform or rig, until a second pre-event image is collected. It would also be impossible to know post-event (with a single pre-event image) if the missing detection is a ship or a missing platform. Third, there is no way to characterize any inaccuracy in the image's earth location (since there are no control points on the ocean surface).

A better approach would be to use the known pre-event locations of platforms and rigs. The BSEE maintains a database of such position information for all the structures installed in the U.S. EEZ portion Gulf of Mexico (outside the 12 mile limit from the U.S. shore). A publically available version can be found at: https://www.data.bsee.gov/homepg/data_center/platform/platform.asp. In this database, the primary identification field is a "Complex identification number". Complexes are composed of one or more structures at a location in the Gulf of Mexico. Rigs which are single "units" are described by rig type. At moderate resolution (25 m) even multi-structure complexes appear as a collection of contiguous bright pixels (see Fig. 15.4). Combining the information in the BSEE database with the pre-season survey has the advantage of providing some idea of: (1) each complex's relative detectability and (2) quantify the earth location error in the SAR imagery by correlation or pattern matching all the detections in an image (or sub image) area to the database locations.

Figure 15.3 shows a comparison between BSEE database locations (green/black dots) covered by a portion of the Envisat image shown in Fig. 15.2. Forty-eight locations are noted and the BSEE locations are within approximately 200 m of the bright pixel locations. Overall the image locations are shifted towards the northeast relative the database positions, due to image geo-registration (projection) errors. There are two database locations without an associated bright pixel, both classified

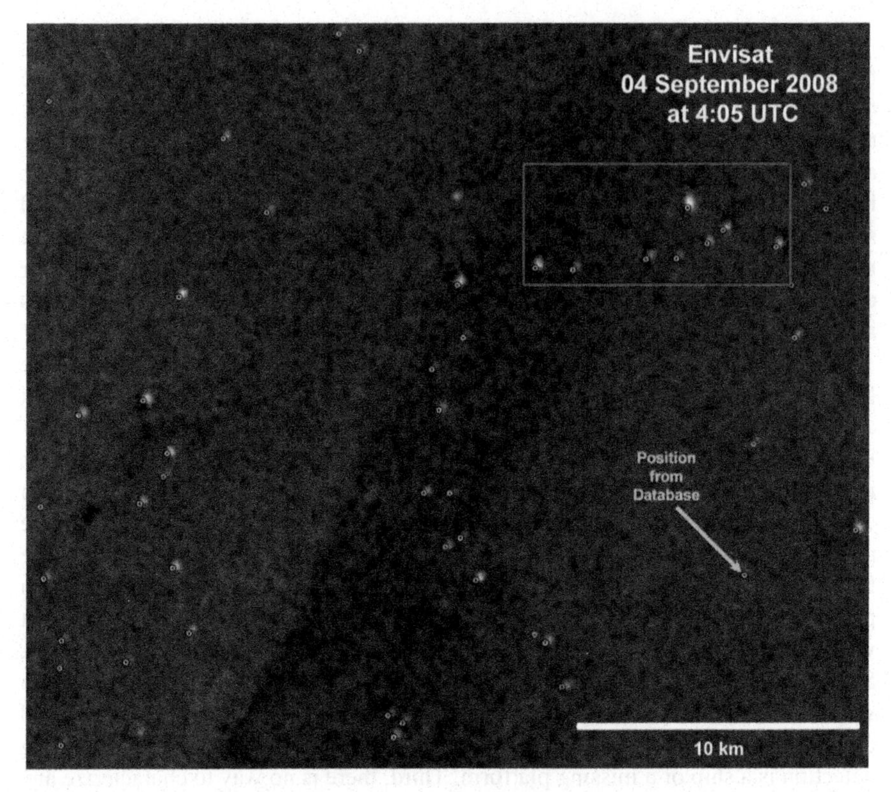

Fig. 15.3 Image with platform positions from the BSEE database (*green/black markers*). Forty-eight locations are noted. The offset between the location of the platform in the database and the pixel location is approximately 200 m. The *red box* indicates the area presented in Fig. 15.4

as non-major structures in the BSEE database, and two bright pixel groups without an associated database entry, assumed to be ships.

15.2.2 Concept of Operations - Post Hurricane Platform / Rig Assessment

A Concept of Operations for a post hurricane oil platform/rig assessment based on the use of satellite SAR imagery is outlined below. This assumes the ability to obtain satellite SAR imagery (with a suitable resolution and swath width) over the area of interest within a few days after a storm event, along with having a methodof reliably

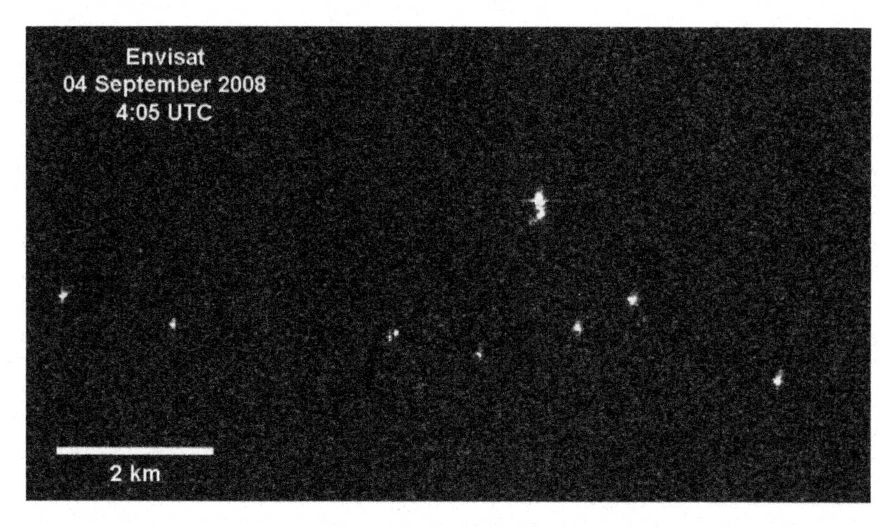

Fig. 15.4 Full resolution view of oil platforms in a 25 m resolution Envisat SAR image acquired on 4 September 2008. (This sub-scene from is outlined by the *red box* in Fig. 15.3). Most platforms appear as a contiguous collection of bright pixels

identifying and extracting the geo-location information of bright pixel objects from the ocean surface from that imagery. These operational steps include:

- Before the tropical event impacts an area containing the offshore infrastructure, use the available storm track and track prediction to plan for the first SAR acquisitions 12–24 h after the storm passes. Subsequent collection planning can be repeated with updated information on the storm track and then on actual areas impacted.
- Obtain the imagery and process it to extract the geo-location of bright pixel objects from the ocean surface in that imagery.
- Determine which platforms and rigs in the BSEE database fall within the footprint of the acquired imagery (with some buffer area to account for geolocation errors).
- Compare the platform detections to the database locations in order to compile a list of observed platforms, missing platforms and objects with no associated database entry. It is important to note objects with no associated database entry since rigs have been known to drift 10's of kms from their mooring positions after breaking free.
- Identify the location, and extent of any oil spills for the appropriate response agency.

15.2.3 Post Hurricane Assessment Example - September 2008

In the late summer of 2008, Hurricanes Gustav (31 August to 2 September) and Ike (11 to 13 September) passed over the oil infrastructure in the Gulf of Mexico. While both storms were only Category-2, with maximum sustained winds of around 100 mph during their passage, these storms destroyed 60 platforms and 1 rig (Table 15.2) [6]. Most of the platform damage was caused by Hurricane Ike due to the large areal extent of the storm and its associated high sea state.

Figure 15.5 shows the portion of the Gulf where the destroyed platforms were located and the extreme environmental conditions (wind and waves) produced by the storms to which the platforms and rigs were subjected. Figure 15.5a is a (1 min) maximum sustained surface wind produced by the H*Wind Tropical Cyclone Observing System [13]. The figure combines the H*Wind surface wind speed information from both Hurricane Gustav and Hurricane Ike, selecting the maximum speed value from the two storms at a particular location. The H*Wind fields geographically overlapped between longitudes 94°W to 90°W and latitude 25.5°N to 31.5°N. The white contour lines are the 43 m/s and the 33/ms (lowest Category-2) wind speeds. White and grey squares are platform locations from May 2009, taken from the BSEE database. More than 2000 platform positions are located within map area shown (white and grey boxes). The small solid black squares represent those platforms designated as destroyed by BSEE after Hurricane Ike.

Figure 15.5b shows significant wave height (H_s) information taken from the NOAA WAVEWATCH III [14] hind-cast at 0.5° sampling. The WW-III hind-cast product is output at 3 h intervals and Fig. 15.5b was produced by selecting the maximum significant wave height at each 0.5 deg cell (within the map area) between in the period 31 August 2008 through 13 September 2008. This captures both hurricanes Gustav and Ike. Here again, the small solid black squares represent those platforms designated as destroyed. The locations of the destroyed platforms are not well correlated with the location of the highest winds or the largest waves. The destroyed platforms were exposed to winds ranging between 32 and 49 m/s (or the Category-2 range of wind speeds) and maximum significant wave heights of between 3–11 m.

On 18 September 2008, 5 days after the passage of Hurricane Ike, the Japanese Space Agency's ALOS-1 spacecraft with its PALSAR instrument acquired 4 SAR scenes (at 25 m resolution) over 22 of the destroyed platform locations. The footprint of the ALOS image collections are shown by the large black squares in Fig. 15.5a. This data collection, with its temporal proximity to the passage of Hurricane Ike, created an opportunity to perform a SAR based post hurricane assessment on the oil infrastructure and verifying its performance against the post storm platform/rig assessment made by BSEE. This case is presented as a proof-of-concept demonstration since the ALOS data were obtained from the SAR archive at the Alaska Satellite Facility several months after Hurricane Ike and were not analyzed as part of any post hurricane response effort.

Applying the technique of using the pre-event known locations of the platforms from BSEE to the locations of bright returns observed from the SAR imagery, resulted in identifying 11 platforms as missing, from the 22 that were declared destroyed by BSEE. Strong SAR returns were found at the remaining 11 destroyed platform locations. In addition 3 of the missing platforms locations had visible oil spills as did 3 of the platform locations (2 destroyed) that had strong SAR signatures. Figure 15.6 shows a portion of the ALOS collection overlaid with the results of the platform assessment processing. Red circles in Fig. 15.6 represent the destroyed platform locations. Ten of the circles surround bright dots, meaning there is enough of a

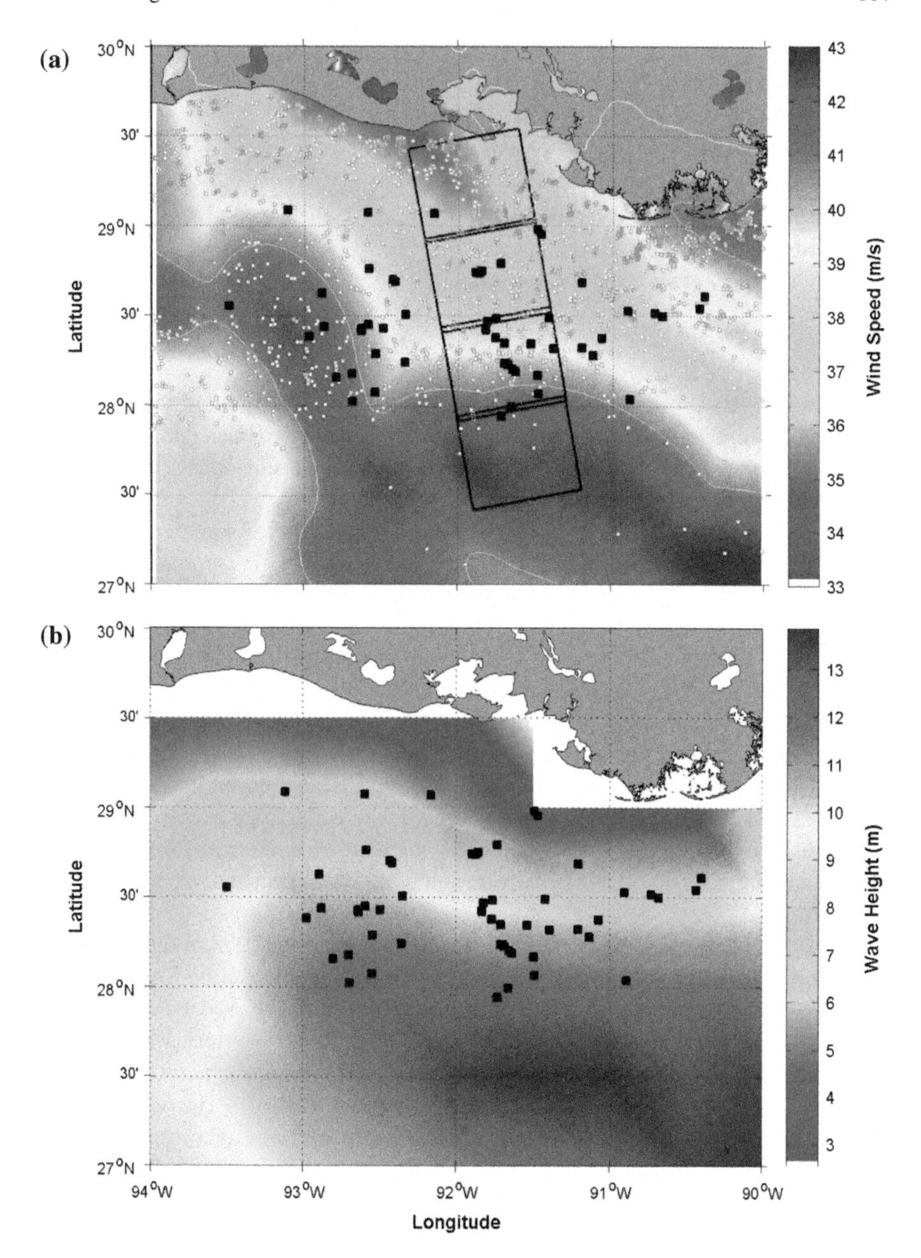

Fig. 15.5 **a** Combined H*Wind surface wind speed information from Hurricane Gustav and Hurricane Ike. Small *white* and *grey squares* are platform locations from the BSEE database. The small *solid black squares* represent those platforms designated as destroyed by Gustav/Ike in 2008. The *contour lines* delineate the 33 m/s and 43/ms wind speeds. **b** Significant wave height (H_s) information taken from the NOAA WAVEWATCH III [13] hind-cast. The values represent the maximum significant wave height between in the period 31 August 2008 through 13 September 2008

Table 15.4 Results of the examination of 4 ALOS SAR images in order to identify missing platform locations in the wake of Hurricane Ike

Platform locations covered by ALOS image footprint	532
Platform locations without an associated SAR detection	29
Destroyed platform locations within the SAR imagery	22
Destroyed platform locations without a SAR detection	11
Destroyed platforms with a SAR detection	11
Number of oil slicks observed in ALOS images	6

structure still present to produce a strong radar backscatter. The nine "M" markers indicate platform locations that do not have an associated SAR detection and would be flagged as missing by the procedure. Five oil slicks are also visible with 4 of them near destroyed platform locations.

The detection results for the full area covered by the ALOS imagery is presented in Table 15.4. The ALOS footprint from all 4 scenes contained approximately 523 platform locations. By using SAR imagery (detections) in combination with the known platform positions, the technique highlighted a total of 29 locations that would have needed further assessment if used in support a post tropical event response. These 29 locations represent only 5.45% of all the platform locations, thus achieving the goal of significantly reducing the number of platform locations that would require evaluation after the hurricane. The other 18 "missing" locations, (i.e. missing detections not associated with a destroyed platform (or false alarms)) were all within approximately 55 km of the shore. Four of them where classified as non-major fixed structures, the remaining 14 were caissons, which are concrete structures that serve as the foundation for an offshore platform and would be expected to have a low radar cross section.

The ALOS PALSAR data collection from 18 September 2008, 5 days after the passage of Hurricane Ike, provided a demonstration how of satellite SAR imagery, coupled with pre-event platform location information, could be used to rapidly determine the location of missing oil platforms. A total of 11 locations without SAR signatures were correctly identified with the proposed technique and these locations corresponded with the post hurricane assessment made by BSEE. The technique has the advantage of being able to assess large areas independent of lighting and weather conditions and has the potential to aid agencies and organizations in their disaster response.

15.3 Post-Hurricane Oil Detection with SAR

A major risk of the impact of hurricanes on oil infrastructure is the potential for oil spills. Oil discharged into the ocean threatens marine life forms at all levels of the

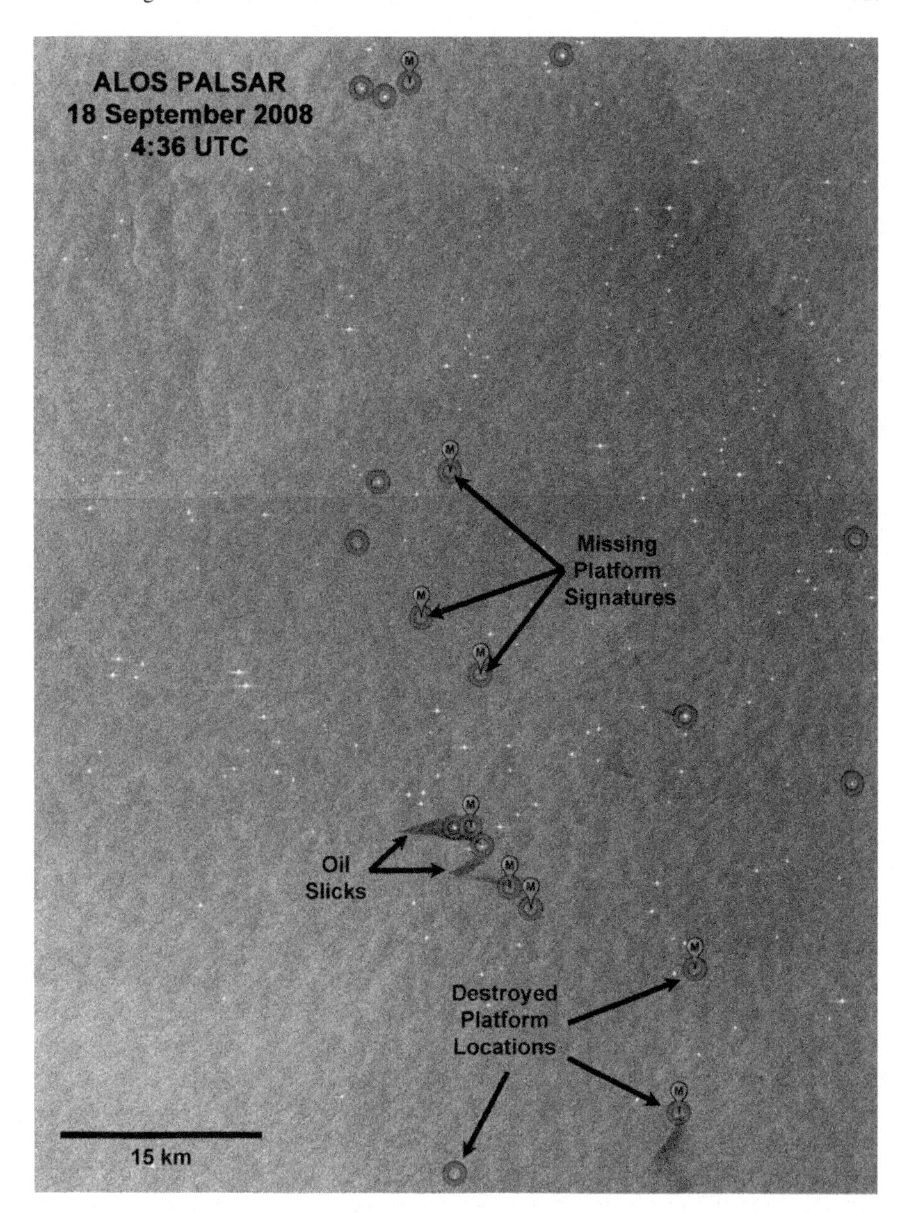

Fig. 15.6 A sub-scene of the ALOS collection from 18 September 2008. *Red circles* represent the destroyed platform locations. The "M" marker indicates platform locations that do not have an associated SAR detection. A total of 11 "destroyed" platform locations without SAR platform signatures were correctly identified with the proposed technique

ecosystem, from plankton and macro-algae to the largest marine mammals, fishes, and sea birds [15]. When the oil reaches the coastal zone it is destructive to shellfish beds, coral reefs, saltwater marshes, sea grass meadows, as well as other crucial marine and coastal habitats. The severe economic impact of oil spills result from a drastic reduction of tourism, closure of fisheries, and clean-up costs. In responding to a spill, accurate and rapid knowledge of spill magnitude, location and movement enables more effective and efficient clean up and therefore greatly reduced environmental impact and cost of response operations. This section will focus on describing the mechanism by which the oil on the water's surface is detected in SAR imagery and details the capabilities and limitations of SAR to monitor and characterize oil spills on the ocean surface.

An example of oil spills resulting from hurricane damage to oil infrastructure in the Gulf of Mexico with a lasting environmental impact occurred on 15 September 2004 when Hurricane Ivan knocked over the Taylor Energy Company platform rupturing several well heads [9]. As an illustration of the persistence of slicks near the Taylor location, Fig. 15.7a is a RADARSAT-2 quad-polarization (25 m resolution) image acquired more than 12 years after the hurricane (on 17 November 2016). The image contains a dark feature extending east/west for approximately 40 km (south from the Louisiana Peninsula) and this feature is the reduced backscatter signature of oil on the ocean surface. Since its sinking, many satellite SAR images have contained similar oil slick signatures near this location and Fig. 15.7b is a composite showing the outlines of these signatures from other images. The extent and orientation of the various slicks signatures are the result of the changing environmental conditions (wind speed, wind direction, ocean currents, and seepage rate).

15.3.1 SAR Detecting Oil at Sea

SAR is well known for its capability to detect oil spills, oil from natural seepage and maritime transit (vessels and other structures e.g. platforms) [16, 17] on the ocean surface. Interactions between the ocean surface and microwave radiation are affected by two main components: (1) the imaging geometry of the observed area and (2) the characteristics of the object, in particular its "roughness" at microwave wavelengths as well as its dielectric properties. At microwave wavelengths of a few centimeters, the dominant scattering from the sea surface is from wind driven capillary waves (ranging from 0.7 to 10 cm). These are waves that travel along the boundary of surface water, whose dynamics are dominated by the effects of the surface tension. When generated by light wind in open water, a gravity-capillary wave is influenced by both the effects of surface tension and gravity, as well as the fluid inertia [18].

Since these capillary waves and short gravity waves are of the same order wavelength as the SAR microwaves, they result in Bragg scattering, where the periodic spacing of the waves produces constructive interference resulting in enhanced radar reflection (or backscatter). Increasing wind speed produces increased roughness with a resulting increase in radar backscatter. The Bragg scattering effect has an angular

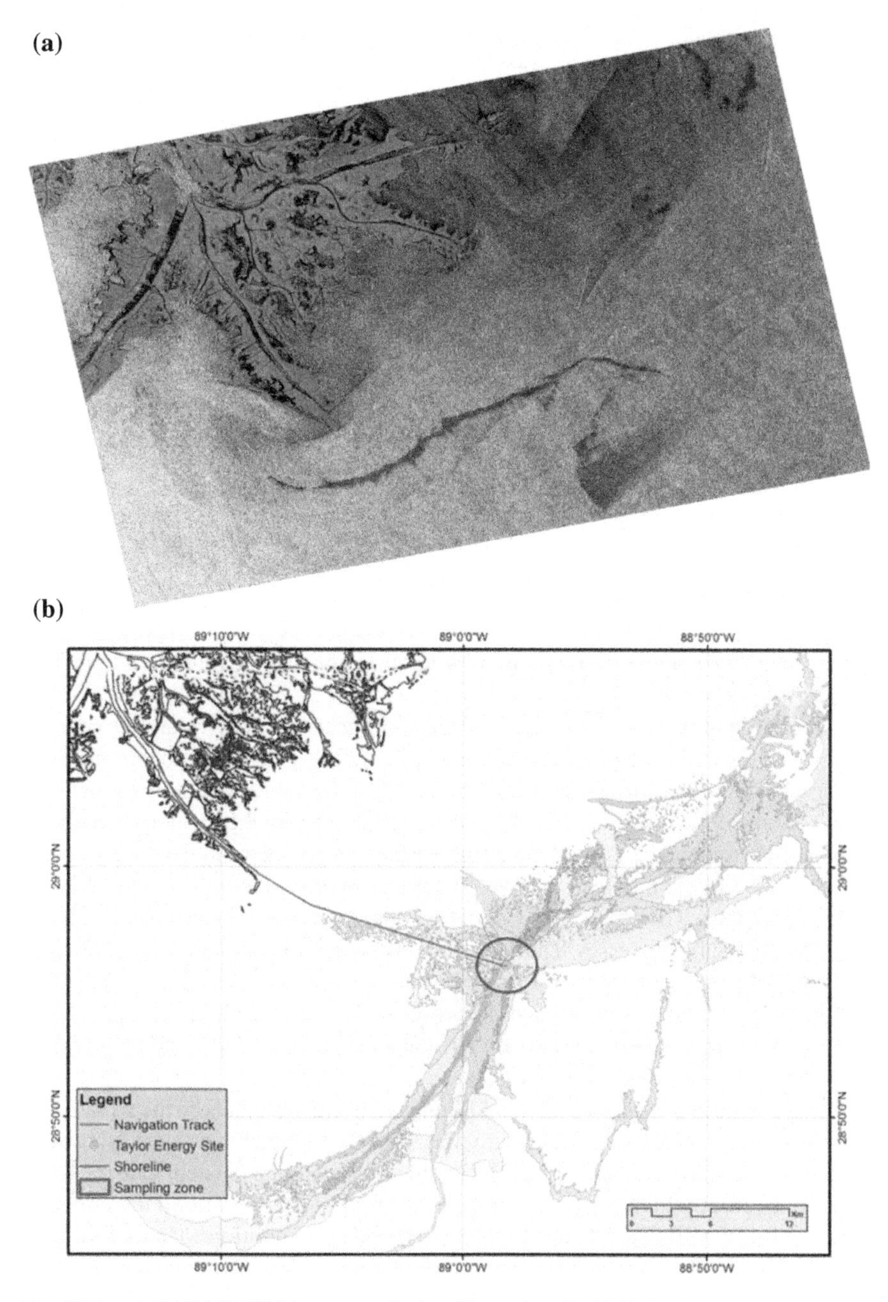

Fig. 15.7 **a** A RADARSAT-2 image acquired on November 17, 2016 showing an oil slick on the surface of the Gulf of Mexico near the Bird's Foot Delta. **b** A composite graphic containing the outlines of various slicks detected in satellite SAR imagery at this same location

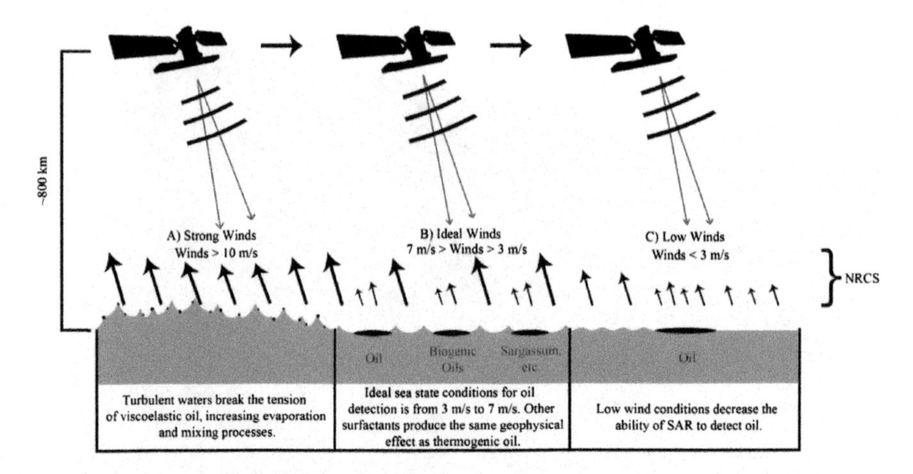

Fig. 15.8 Radar backscatter at different sea surface roughness. The wind conditions at the time of the data collection, constrain the NRCS values. Thin oil (sheen <400 nm) will persist for shorter period and will keep less tension on the water than thicker layers. The thicker the layer of oil, the more resistant to high winds [23]

dependence and is maximized when the wind direction and the radar look direction are aligned [19].

Modulation of these surface capillary and short gravity waves by various oceanographic and atmospheric processes produces the unique patterns visible in the SAR imagery of the ocean surface [20]. In the case of oil (or more generally surfactants) on the ocean surface, the radar backscatter is reduced when the capillary waves are reduced by viscoelastic properties of the surfactant producing a dark area on the image relative to the unaffected areas. The effect is illustrated in Fig. 15.8. Marine slicks are composed of two major types of hydrocarbons, mineral thermogenic oil and films from biogenic processes. Thermogenic oils come from multiple sources including: (1) discharges from storm-water urban run-off, (2) spills from ships, (3) drilling platforms, (4) pipelines, and (5) natural hydrocarbon seeps. Mineral oils spread into thin layers through gravity and surface tension and evaporate and weather over time [21].

The ability to detect surface oil in SAR images strongly depends on the wind speed at the sea surface. Much of the literature refers to a threshold of wind speeds between 2–12 m/s to be able to identify oil [22], but the ideal conditions to delineate oil slicks reduces that threshold to between 3 and 7 m/s [16]. Detectability of oil under strong wind conditions will depend on the oils density and concentrations. Under low wind conditions, the slicks may not be distinguishable from the surrounding ocean since wind speeds on the order of 2–4 m/s are required to generate sufficient small-scale capillary waves needed to produce a radar backscatter above the system noise floor of most operational instruments.

Radar has been used to study the physical basis of the interaction of waves with both mineral and biogenic slicks [18]. In addition to theoretical studies, many studies have examined differentiating slicks from ambiguous ocean returns (false positives) with SAR, which generally requires repeat imaging, wind information, and knowledge of the source composition [23]. Other SAR studies have also sought detection algorithms that use image classification tools, neural network methods for classification, feature vectors, wavelet transforms, and Geographic Information Systems techniques [19–25].

The ability to use SAR imagery to locate oil slicks after a tropical event is just the first step in the response. The quantitative nature of the SAR signal allows for application of oil spill segmentation algorithms (e.g. textural classifier neural network algorithm (TCNNA) [23]) to produce consistent results during an emergency response for spill detection. After detection, the next step in this process is to identify the presence and location of thick patches of oil emulsions (within the extent of the oil slick) knowledge of which is often a priority in oil spill response.

15.3.2 Weathering of Floating Oil and Effects on SAR Detection

When crude oil reaches the surface of the ocean, it reacts dynamically with biological, physical, and chemical processes [21]. Collectively the result is known as "weathering" of the oil [24]. The weathering processes include the following: spreading, evaporation, dissolution, dispersion, photochemical oxidation, emulsification, sinking, biodegradation, adsorption to suspended matter, and deposition on to the seafloor [26]. Among these, the three dominant processes that change the physical characteristics with respect to SAR detection are spreading, evaporation, and emulsification.

After spreading, the light volatile components of the crude oil start evaporating. The rate of evaporation depends upon the vapor pressure and the volatile components present in the crude oil. Previous studies have shown that oil thickness can significantly impact the rate of oil evaporation [25]. Further, the remaining concentrated portion of crude oil mixes with the surface seawater under the influence of wind and wave action to form an emulsion [27]. The material formed can be either water in oil (w/o) or oil in water (o/w) emulsion. Water in oil emulsions may be extremely stable because the water droplets (1–10 um diameter range) are held in a semi-rigid structure by asphaltenes, waxes, and resins or similar components [26]. The inclusion of water in the crude oil emulsion is an important step in the weathering process because it increases the volume and viscosity of the emulsion and it changes its dielectric constant. These emulsions can also have serious impacts on coastal activities by creating hazardous conditions and nuisance on the shore [28], so detection of emulsions patches is often a priority in oil spill response. Emulsion patches can be found using fully polarimetric (also known as 'quadpol') collection modes [29]. This imagery, in addition to the regular single polarization amplitude or intensity, allows for the phase among the different polarizations (HH, VV, HV, or VH) to discern oil and oil-emulsions mixtures [30]. Concepts previously described in the literature [30] where later used [23] to process imagery in search of oil emulsions.

For an oil-in-water (o/w) emulsion, water is the continuous phase and, in seawater, the bulk electrical conductivity of the o/w emulsion is large (~10 S/m), compared to pure oil, which is an electric insulator. Absorption and Bragg scattering from the SAR electromagnetic pulse responds to variations in electrical conductivity. Our hypothesis is that due to an increase in the electrical conductivity and volumetric scattering, as well as altered surface texture, thick patches of floating emulsions will scatter SAR energy more effectively than a thin layer of pure oil that normally damps capillary waves [21].

Less total energy will be reflected back to the SAR antenna from a lower conductive surface (i.e. pure crude oil). Therefore, the Bragg normalization model needs to include the incidence angle, wave tilt, and wind strength. Once the normalization and enhancement of the Bragg scattering inside the oil slick is achieved, various threshold values could be determined from trial and error to assign relative oil thickness classes within the oil slick. The assumption is that increasingly higher pixel values in the new Oil Emulsion Detection Algorithm (OEDA) [21] correspond to increasingly thicker emulsions. In order to understand under which conditions each of the different SAR wavelengths (X, C, and L) are able to detect the oil emulsion, Bragg scattering is normalized on the SAR imagery based on the following hypothesis: if thick oil is mixed with water within a thin layer (a few millimeters for a radar wavelength) below the surface, the oil will reduce the effective dielectric constant

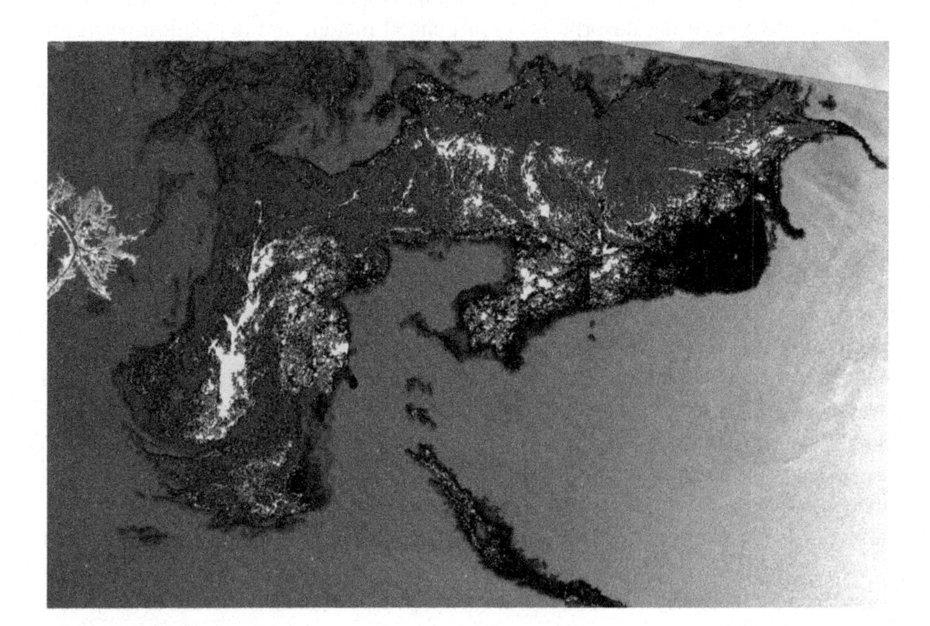

Fig. 15.9 Classification of oil thickness inside the oil spill extent using an ALOS L-SAR image acquired on May 24, 2010. The *red* outline (TCNNA) shows the extent of the surface oil presence and the *yellow* areas represent the thick oil emulsions (OEDA)

of the ocean surface because the dielectric constant of oil is much lower than that of sea water. Following this assumption, the brightest pixels are thresholded in the new normalized and stretched OEDA raster to classify features of higher backscatter (thickest oil emulsion). Figure 15.9 shows a sample OEDA model output where yellow represents the anomalies inside the overall oil slick that may correspond to thicker layers.

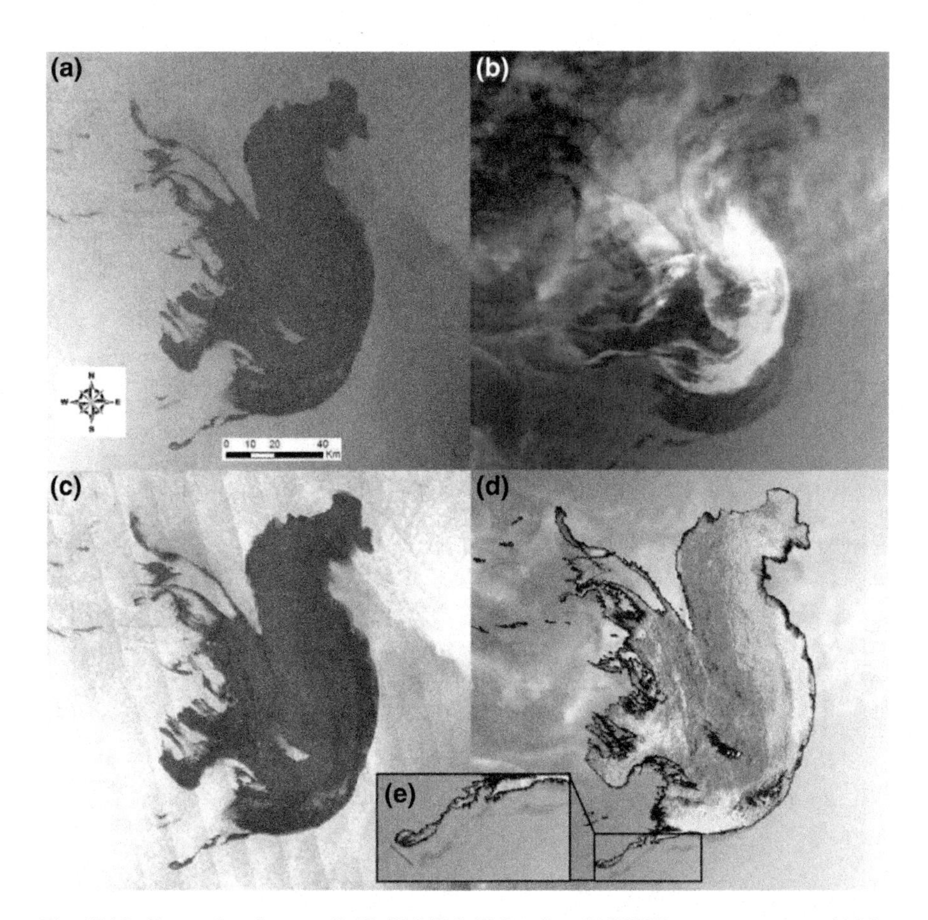

Fig. 15.10 Comparison between RADARSAT-2 (C-Band) and MODIS measurements on 10 May 2010. **a** SAR image (23:53 UTC) showing the surface oil extent; **b** MODIS image (16:50 UTC) showing the various oil contrast under thin clouds. The image was generated as a *Red-Green-Blue (RGB)* composite. The bright features are thought to be from emulsified thick oil, while the *black* features are thought to result from surface oil films; **c** The same SAR image normalized to incident angle; (**d**) OEDA classification output. A 6 km displacement was observed between the two snapshots (*blue line* in **e**)

15.4 Comparison with Optical Remote Sensing Imagery

The OEDA results from the SAR imagery were compared with optical remote sensing imagery under different wind conditions. An example of an OEDA model output for a SAR image collected by RADARSAT-2 (C-Band) on 10 May 2010 (23:53 UTC) is shown in Fig. 15.10a. A MODIS optical image was collected on the same day at 16:35 UTC, approximately 7 h before the SAR snapshot (Fig. 15.10b). The normalization of the SAR data by OEDA (Fig. 15.10c) was then used to threshold oil thickness

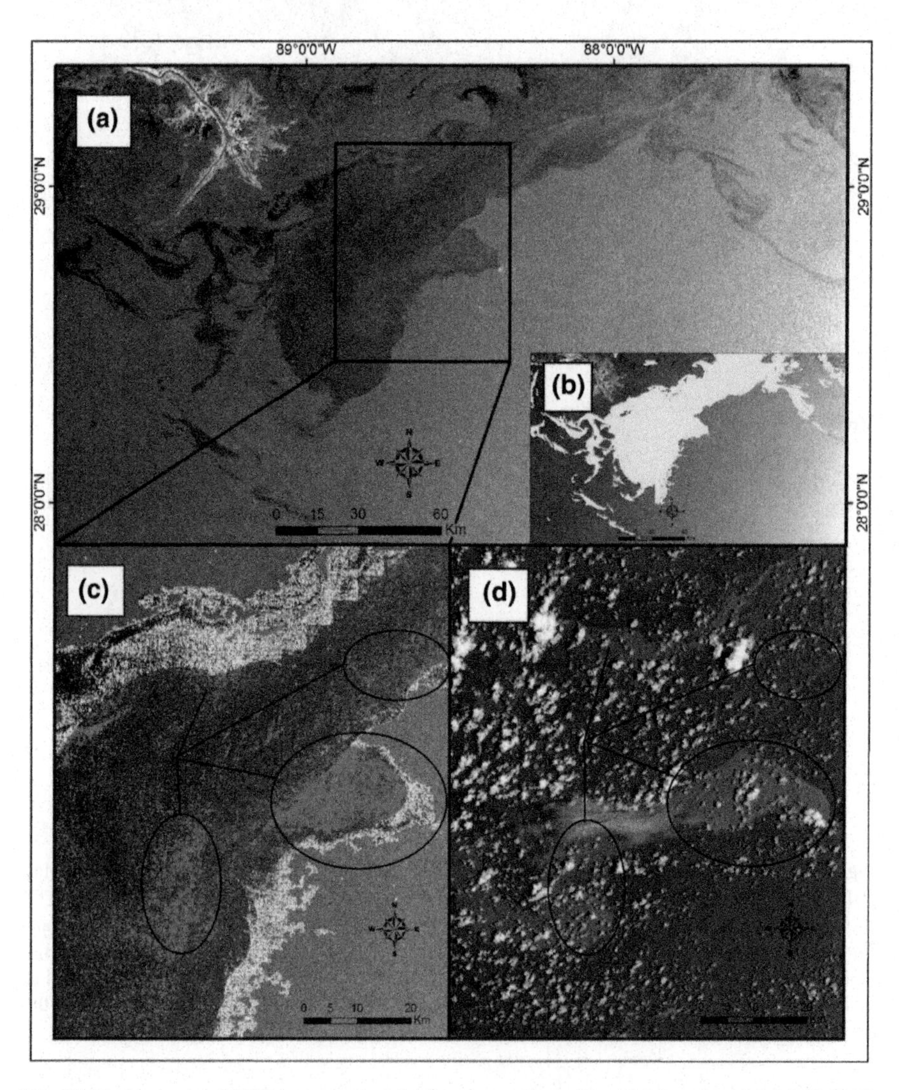

Fig. 15.11 Example of OEDA output for a SAR image collected by ALOS on June 26, 16:25 UTC. The signatures detected by the two methods show a general agreement in terms of location, size, and scale of the thick patches of oil emulsions

classes as shown on Fig. 15.10d. Comparing the OEDA output with the optical image, a relative agreement is found in the directionality and size of the signatures detected by the two different sensors which are thought to be thick oil emulsions. The time gap between the two acquisitions also affects the relative position of the drifting thick oil detected. Satellite imagery within the optical range like MODIS will be limited to detect presence/absence of oil and oil-emulsions under specific viewing angles also dependent on the illumination source. In this case shown below, although there were some cloud obstructions, the optical sensor MODIS was able to capture the stronger reflectance created by the thick oil due to the sunglint.

Figure 15.11 shows another example of the OEDA output and its comparison with Landsat observations. The L-Band ALOS PALSAR image was acquired on 26 June (16:25 UTC). While the oil extent showing in Fig. 15.11a was delineated by the Textural Classifier Neural Network Algorithm (TCNNA) [23] (Fig. 15.11b), the normalization from the OEDA model resulted in the classified oil image in Fig. 15.11c where two oil categories were used: Sheen (yellow outline) and thick oil emulsion (red pixels). Compared with the Landsat RGB image taken 8 min before the SAR snapshot (16:17 UTC), the SAR classification results in Fig. 15.11c showed qualitative agreement in the thick oil patterns (outlined in black circles) revealed in the Landsat imagery.

While oil presence/absence on the oil surface is relatively easy to delineate in remote sensing imagery, estimating oil thickness is much more difficult. One approach is based on measuring thermal energy by IR sensors. Oil absorbs solar radiation differently from the ambient ocean and re-emits a portion of this energy as longwave radiation in the infrared spectral band. Early studies report that IR sensors observe thick oil as hot spots, intermediate thick oil as cool, while thin oil is not detectable [31]. Some preliminary results were obtained by using airborne measurements [32] covering partially the extent of the slick along flight paths. The approach presented here demonstrates the potentials of using SAR imagery for qualitative classifications through examining variations of the Bragg scattering due to changes in the conductivity and surface roughness of oil emulsions. It is important to point out that under specific viewing and illumination conditions, either IR, visible, or hyperspectral sensors will be able to detect more information about the oil-emulsions constituency (e.g. thickness) than microwave sensors. SAR and OEDA needs to be seen as a complement for those temporal-space situations where no other sensors are available or capable of detecting oil (at night, under cloud cover, or off-sunglint conditions).

The most likely cause of the structural characteristics delineated by the OEDA algorithm are due to differences in emulsion formations and floating volumes; i.e. thicknesses. While this cannot be definitive verified over the entire areas of the images analyzed, it is possible to ground truth the OEDA anomalies where focused aerial dispersant missions and burning operations targeted thick oil layers based on direct responder observations (Fig. 15.12). In addition to this indirect validation, a qualitative agreement was found in the oil distribution patterns observed between the SAR OEDA outputs, the optical imagery from MODIS and Landsat, and airborne thermal imagery. Differences on the size and distribution of classified thick oil patches

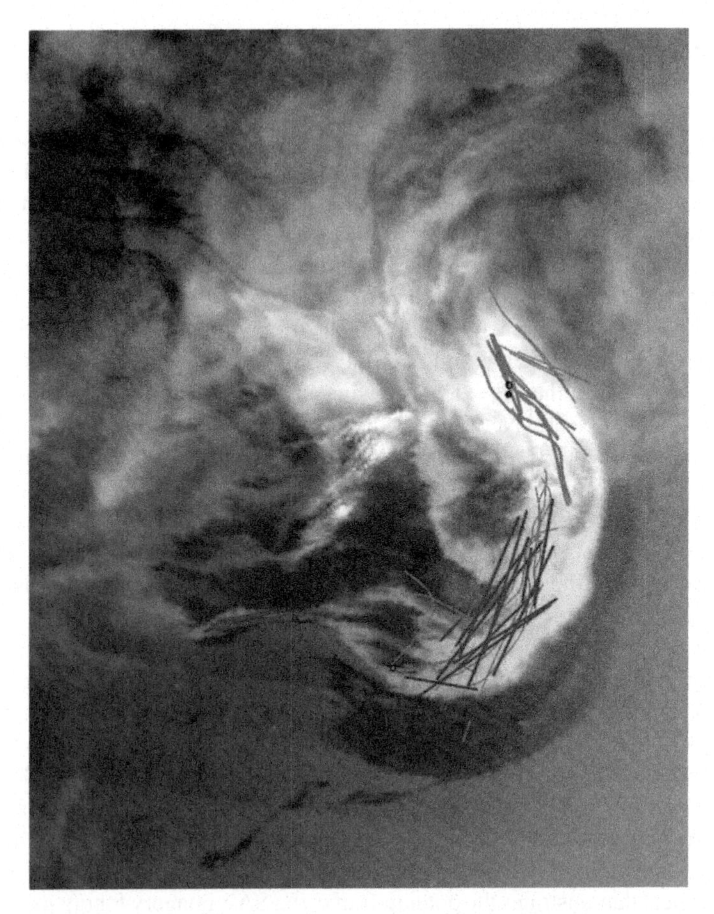

Fig. 15.12 Overlaying the dispersant flight lines and the burning operations shows that the OEDA identifies possible oil thickness anomalies in the identical areas targeted by the responders

could be due to the time gap between image acquisition, the presence of un-emulsified thick oil (undetectable by SAR), and other geophysical constrains for each sensor (i.e. sun glint, wind).

The opportunity to compare optical or thermal imagery concurrent in time with OEDA outputs is unique. This offered a unique capability to explore the OEDA algorithm performance throughout different incidence angles and wind conditions. As expected, when two images are collected by different sensors at nearly the same time, more agreement on the thick oil emulsions signatures can be observed. A qualitative agreement was found in the oil distribution patterns observed between the SAR OEDA outputs and the optical imagery from MODIS and Landsat. Differences on the size and distribution of classified thick oil patches could be due to the time gap between image acquisition, the presence of un-emulsified thick oil (undetectable by SAR), and other geophysical constrains for each sensor (i.e. sun glint, wind).

The Deep Water Horizon oil spill provided an opportunity to compare SAR/OEDA outputs with other remote sensing techniques such as optical remote sensing through MODIS and Landsat observations. Although the results are preliminary, OEDA shows the potential to identify signatures of thick patches of floating oil for feature tracking and modeling efforts. In addition, the OEDA may offer valuable applications for many of the oil spill response operations. For example, the capability to classify thick patches of oil emulsion can result in more effective targeting for skimming, or burning operations. These operations typically require the knowledge of both oil presence/quantity and the oil's trajectory in the ocean. While the former may be achieved through remote sensing and algorithm development, the latter must incorporate oil spill models that are specifically designed to model oil trajectories. The classified maps provide important initial condition and validation data for these models.

Typical oil spill models, in their simplest implementations, simulate the movement of oil on the ocean surface by tracking the movement of a large number of particles that travel at a velocity that is some combination of the ocean surface velocity and possibly surface wind. Particles can be initially placed in the domain based on maps of surface oil slicks, and additional particles can be added to the surface to simulate leaking oil. The simulated oil particles can also be removed based on certain characteristics such as age, or any other modeled processes such as biodegradation, burning, or mechanical removal. Although oil is a fluid, its treatment as discrete particles implies that each particle is representative of a certain quantity of oil. This requires knowledge (or often gross estimates) of the amount of oil reaching the surface from a leaking vessel or subsurface blowout, or the thickness of the oil covering a given area. For example, if it is assumed that the mean thickness of an oil slick is one micron, and each particle represents 1 m^3 of oil on the surface, then each particle would represent an oil slick of area of 1 km^2. However, the thickness of an oil slick is definitely not uniform factor, and the thickness of a certain quantity of oil may change over time through spreading, emulsification, or other processes. Thus, it can be useful to track the thickness of a quantity of oil represented by a discrete particle.

Not only is the thickness of an oil slick important for interpreting simulated discrete particles as a surface oil slick, but the thickness of the oil may also have implications for the dynamics of the oil spill model. As oil spill models become more sophisticated involving coupled ocean-atmospheric models with surface flux algorithms that take into account surface oil, knowledge of, and ability to simulate, thickness may improve the simulations. Oil drift models that are able to simulate oil thickness need to be initialized with oil thickness distribution inside the oil spill and this is why OEDA outputs can be very useful. This information incorporated into initial fields will improve numerical predictions.

15.5 Oil Spill Conclusion

During/after the passage of hurricanes, using SAR as tool for oil spill detection is a well stablished task. The utilization of oil spill segmentation algorithms like TCNNA adds consistency on the operation during an emergency response. The value of adding an extra step on this process to detect not only the oil extent but the presence of thick patches of oil emulsions (within the extent of the oil slick) is crucial not only for contingency/recovery operations but also for damage assessment of the natural resources. This is why the Oil Emulsion Detection Algorithm (OEDA) as an operation tool is important. The operating premise for OEDA development is that there are radar backscatter differences due to the reflectivity of floating oil that result from both the roughness of oil layer and to the dielectric constant of the oil layer. Compared with oil sheen, oil water emulsions have greater surface roughness and higher electrical conductivity due to the salt water in the emulsion. Through trial and error, we found an empirical relationship between the SAR incidence angle and the SAR sensitivity to the enhanced radar return from oil emulsions. It is important to point out that each beam mode in each satellite produces its own image configuration (pixel resolution, incident angle range) and therefore different equalization polynomials are used for different beam modes. Further versions of the OEDA model need to incorporate wind information to produce a more accurate normalization and thresholding of the oil thickness classification. Nevertheless, the limited examples shown here, although preliminary in nature, demonstrate the potential of using SAR for classifying thick oil emulsions in the ocean environment.

References

1. BSEE. Gulf of Mexico - Bureau of Safety and Environmental Enforcement. https://www.bsee.gov/stats-facts/ocs-regions/gulf-of-mexico.
2. EIA. Federal Offshore–Gulf of Mexico Field Production of Crude Oil. https://www.eia.gov/dnav/pet/hist/LeafHandler.ashx?n=PET&s=MCRFP3FM2&f=M.
3. BSEE. Hurricane History - Bureau of Safety and Environmental Enforcement. https://www.bsee.gov/resources-tools/hurricane/hurricane-history.
4. Keim, B.D., and R.A. Muller. 2009. *Hurricanes of the Gulf of Mexico*. Baton Rouge: LSU Press.
5. BSEE. Platform/Rig Information - Bureau of Safety and Environmental Enforcement. https://www.data.bsee.gov/Main/Platform.aspx.
6. BSEE. MMS Completes Assessment of Destroyed and Damaged Facilities from Hurricanes Gustav and Ike - News Release US DOI MMS Office of Public Affairs, 26 November 2008. https://www.bsee.gov/sites/bsee.gov/files/news/hurricanes/081126a.pdf, 2008a.
7. BSEE. Minerals Management Service Releases Details of Drilling Rigs Destroyed from Hurricane Ike - News Release US DOI MMS Office of Public Affairs 30 September 2008. https://www.bsee.gov/sites/bsee.gov/files/news/hurricanes/080930.pdf, 2008b.
8. BSEE. Impact Assessment of Offshore Facilities from Hurricanes Katrina and Rita. https://www.bsee.gov/sites/bsee.gov/files/press-release/news-item/060119.pdf, 2006.
9. NOLA. Taylor Energy oil platform, destroyed in 2004 during Hurricane Ivan, is still leaking in Gulf. http://www.nola.com/environment/index.ssf/2013/07/taylor_energy_oil_platform_des_1.html.
10. NOAA. Tanker Vessel SKS SATILLA. https://incidentnews.noaa.gov/incident/7989.

11. Pichel, W.G., P. Clemente-Colón, C. Wackerman, and K.S. Friedman. 2004. Ship and wake detection. In *Synthetic aperture radar marine users manual*, pp. 277–303.
12. Vachon, P., J. Campbell, C. Bjerkelund, F. Dobson, and M. Rey. 1997. Ship detection by the RADARSAT SAR: Validation of detection model predictions. *Canadian Journal of Remote Sensing* 23 (1): 48–59.
13. Powell, M.D., S.H. Houston, L.R. Amat, and N. Morisseau-Leroy. 1998. The HRD real-time hurricane wind analysis system. *Journal of Wind Engineering and Industrial Aerodynamics* 77: 53–64.
14. Tolman, H.L., and D. Chalikov. 1996. Source terms in a third-generation wind wave model. *Journal of Physical Oceanography* 26 (11): 2497–2518.
15. MacDonald, I. 2010. Deepwater disaster: How the oil spill estimates got it wrong. *Significance* 7 (4): 149–154.
16. Caruso, M.J., M. Migliaccio, J.T. Hargrove, O. Garcia-Pineda, and H.C. Graber. 2013. Oil spills and slicks imaged by synthetic aperture radar. *Oceanography* 26 (2): 112–123.
17. Alpers, W., and H. A. Espedal. 2004. Oils and surfactants. *Synthetic aperture radar marine users manual*, pp. 263–275.
18. Garcia-Pineda, O., I. MacDonald, M. Silva, W. Shedd, S.D. Asl, and B. Schumaker. 2016. Transience and persistence of natural hydrocarbon seepage in Mississippi Canyon, Gulf of Mexico. *Deep Sea Research Part II: Topical Studies in Oceanography* 129: 119–129.
19. Frank, M.M., and B. Robert. 2004. Wind speed and direction. *Synthetic Aperture Radar Marine User's Manual* 305–320.
20. Jackson, C.R., and J.R. Apel. 2004. Synthetic aperture radar: Marine user's manual. US Department of Commerce, National Oceanic and Atmospheric Administration, National Environmental Satellite, Data, and Information Serve, Office of Research and Applications.
21. Garcia-Pineda, O., I. MacDonald, C. Hu, J. Svejkovsky, M. Hess, D. Dukhovskoy, and S.L. Morey. 2013a. Detection of floating oil anomalies from the Deepwater Horizon oil spill with synthetic aperture radar. *Oceanography* 26 (2): 124–137.
22. Espedal, H. 1999. Satellite SAR oil spill detection using wind history information. *International Journal of Remote Sensing* 20 (1): 49–65.
23. Garcia-Pineda, O., I.R. MacDonald, X. Li, C.R. Jackson, and W.G. Pichel. 2013b. Oil spill mapping and measurement in the Gulf of Mexico with Textural Classifier Neural Network Algorithm (TCNNA). *IEEE Journal of Selected Topics in Applied Earth Observations and Remote Sensing* 6 (6): 2517–2525.
24. Fingas, M.F. 2004. Preface. *Journal of Hazardous Materials* 1 (107): 1.
25. Brekke, C., and A.H. Solberg. 2005. Oil spill detection by satellite remote sensing. *Remote Sensing of Environment* 95 (1): 1–13.
26. Fingas, M., and B. Fieldhouse. 2012. Studies on water-in-oil products from crude oils and petroleum products. *Marine Pollution Bulletin* 64 (2): 272–283.
27. Thibodeaux, L.J., K.T. Valsaraj, V.T. John, K.D. Papadopoulos, L.R. Pratt, and N.S. Pesika. 2011. Marine oil fate: knowledge gaps, basic research, and development needs; a perspective based on the Deepwater Horizon spill. *Environmental Engineering Science* 28 (2): 87–93.
28. Atlas, R.M., and T.C. Hazen. 2011. Oil biodegradation and bioremediation: A tale of the two worst spills in US history. *Environmental Science and Technology* 45 (16): 6709–6715.
29. Zhang, B., X. Li, W. Perrie, and O. Garcia-Pineda. 2017. Compact polarimetric synthetic aperture radar for marine oil platform and slick detection. *IEEE Transactions on Geoscience and Remote Sensing* 1–17. doi:10.1109/TGRS.2016.2623809.
30. Minchew. B. 2012. Determining the mixing of oil and sea water using polarimetric synthetic aperture radar. *Geophysical Research Letters* 39 (16).
31. Fingas, M.F., and C.E. Brown. 1997. Review of oil spill remote sensing. *Spill Science and Technology Bulletin* 4 (4): 199–208.
32. Svejkovsky, J., W. Lehr, J. Muskat, G. Graettinger, and J. Mullin. 2012. Operational utilization of aerial multispectral remote sensing during oil spill response. *Photogrammetric Engineering and Remote Sensing* 78 (10): 1089–1102.

Chapter 16
Tropical Cyclone Eye Morphology and Extratropical-Cyclone-Forced Mountain Lee Waves on SAR Imagery

Qing Xu, Xiaofeng Li, Shaowu Bao and Guosheng Zhang

Abstract This chapter introduces an objective method for determining the center of the tropical cyclone (TC) from spaceborne synthetic aperture radar (SAR) data based on the structures of the well-defined TC eyes in the SAR images. A series of Radarsat-1 SAR images are used, which capture the TCs over the world ocean basins during the years from 2001 to 2007. Also, a case study of the atmospheric gravity waves over the Kuril Islands observed in a Sentinel-1A SAR image during the passage of an extratropical cyclone will be presented together with the use of the state-of-the-art atmospheric numerical model. The objective is to obtain a more complete understanding of the generation mechanism and the dynamics governing the gravity waves.

16.1 Introduction

The characteristics of tropical cyclone (TC) eyes are strongly correlated with the cyclone's behaviors [1–4]. Particularly, precise TC center locating is critical for operational track and intensity forecasting, and visualization [5]. Since 1978, when the first spaceborne microwave synthetic aperture radar (SAR) was deployed onboard National Aeronautics and Space Administration (NASA) Seasat satellite, TCs have been observed on many follow-on SAR images. Different from visible and infrared sensors, the high-resolution (a few to tens of meters) SAR can see through most

Q. Xu
College of Oceanography, Hohai University, Nanjing, China

X. Li (✉)
GST, National Oceanic and Atmospheric Administration (NOAA)/NESDIS, College Park, MD, USA
e-mail: xiaofeng.li@noaa.gov

S. Bao
School of Coastal and Marine Systems Science, Coastal Carolina University, Conway, SC, USA

G. Zhang
Fisheries and Oceans Canada, Bedford Institute of Oceanography, Dartmouth, NS, Canada

X. Li (ed.), *Hurricane Monitoring With Spaceborne Synthetic Aperture Radar*, Springer Natural Hazards, DOI 10.1007/978-981-10-2893-9_16

clouds and thus observe the sea surface signatures of a TC. Extracting quantitative information of TCs (e.g., center, shape and area of the TC eye) from SAR images has been a focus of several studies over the past years [1, 6–11].

The TCs, or the extratropical cyclones (ETC), once move over the mountains or isolated islands, might be a generation source of the atmospheric gravity waves (AGWs) [12]. The AGWs play an important role in establishing atmospheric circulation and forming the vertical structure of wind, temperature and moisture fields [13]. In addition to affecting atmospheric circulation and structure, AGWs could also lead to aviation hazards [14, 15]. AGWs induce the spatial variations of the ocean surface wind field, and thus, modulate the short-scale sea surface roughness. These rough-and-smooth sea surface patterns associated with the AGWs are imaged by SAR as alternate bright-and-dark features [16]. SAR images provide valuable information of the AGWs in all weather and day-night conditions with high spatial resolution. Various types of AGWs, which often appear as well organized groups of waves with wavelength ranging from a few to several tens of kilometers, have been observed by SAR onboard satellites such as European Remote Sensing Satellite-1 and -2 (ERS-1/2), RADARSAT-1/2, and Environmental Satellite (ENVISAT) (e.g., [8, 13, 17–22]). Among them, the lee waves occurring in the downstream of the obstacle are most frequently reported.

In this chapter, an objective method for TC center locating from a series of Radarsat-1 SAR images is first described in Sect. 16.2. Then a case study of extratropical-cyclone-induced AGWs is presented in Sect. 16.3 based on SAR observations and numerical model results, followed by a summary in Sect. 16.4.

16.2 Tropical Cyclone Eye Detection from SAR

16.2.1 Data

A. SAR data

The C-band RADARSAT-1 ScanSAR wide mode images of tropical cyclones have been acquired over the world ocean basins through the Canadian Space Agency's Hurricane Watch program [23]. These images provide a high-resolution synoptic view of these intense storms and new insights into their morphology (e.g., [8]). Thirty-six of these SAR images, which capture the complete structure of the TC eyes over the basins of Atlantic Ocean (13 images), Eastern Pacific Ocean (7 images) and Western Pacific Ocean (16 images), during the period from 2001 to 2007, were selected for estimating the TC center locations. Sub-scenes of these SAR images are shown in Fig. 16.1. One can see that the well-defined TC eyes appear as dark oval or circular regions in the SAR images. Detailed information of the images is listed in Table 16.1.

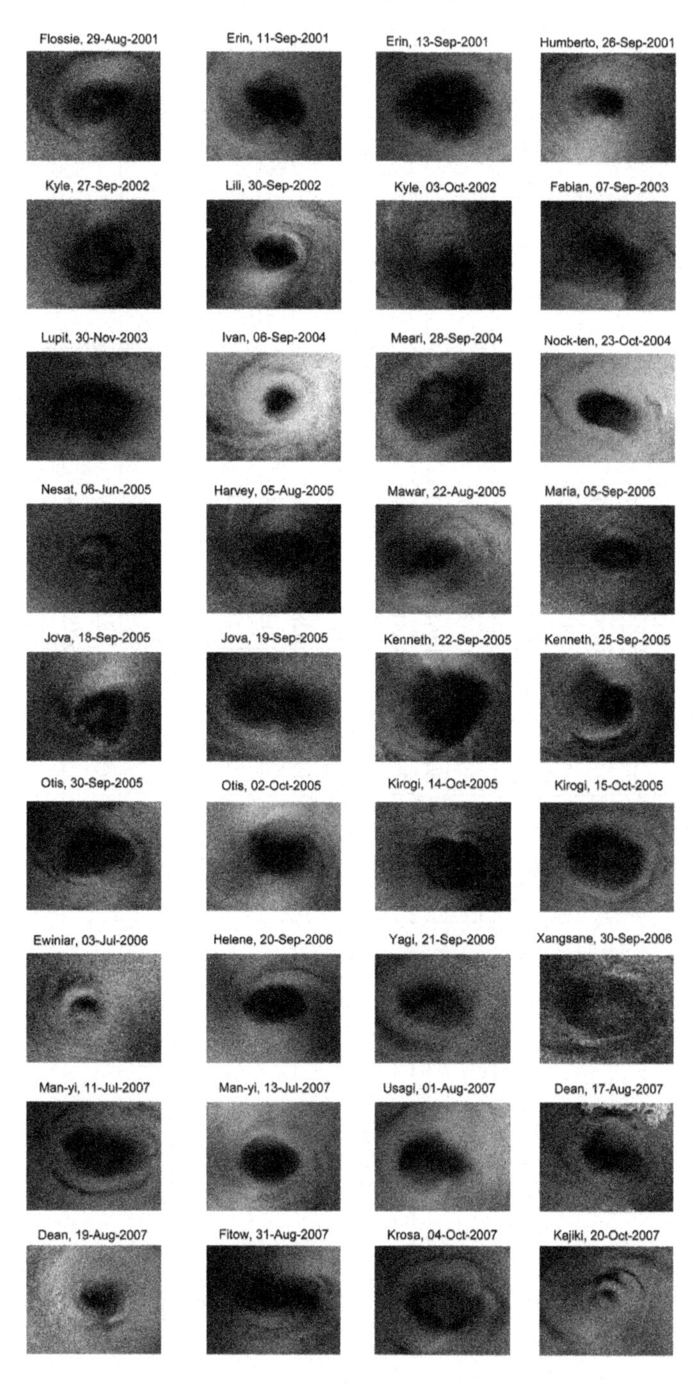

Fig. 16.1 Sub-scenes of 36 RADARSAT-1 ScanSAR images containing complete well-defined TC eyes, as summarized in Table 16.1 (© Canadian Space Agency/Agence spatiale canadienne)

Table 16.1 Information of RADARSAT-1 SAR images and estimated TC parameters

Tropical cyclone		Image acquired		Eye center location		Eye region	Reference ellipse parameter			
Number	Name	Date (Y/M/D)	Time (UTC)	Long. (°)	Lat.N (°)	A (km^2)	a (km)	b (km)	R	θ (°)
1	Flossie	2001/08/29	01:43	115.26W	19.77	435.0	32.4	17.1	1.9	186.2
2	Erin	2001/09/11	22:19	64.48W	37.75	2328.9	67.2	44.1	1.5	149.2
3	Erin	2001/09/13	10:03	61.03W	38.54	5079.9	89.1	72.6	1.2	186.8
4	Humberto	2001/09/26	21:42	56.07W	42.03	283.8	22.9	15.7	1.5	142.6
5	Kyle	2002/09/27	22:03	61.20W	26.65	1191.2	45.2	33.6	1.3	221.8
6	Lili	2002/09/30	11:07	79.40W	19.60	302.9	24.0	16.1	1.5	167.4
7	Kyle	2002/10/03	22:28	67.89W	29.42	1173.0	47.3	31.6	1.5	103.7
8	Fabian	2003/09/07	09:44	54.67W	39.91	1262.6	60.4	26.6	2.3	126.0
9	Lupit	2003/11/30	20:43	137.53E	26.93	4631.9	89.6	65.8	1.4	190.3
10	Ivan	2004/09/06	09:06	51.49W	10.92	76.2	12.1	8.0	1.5	248.9
11	Meari	2004/09/28	09:26	127.42E	29.53	3225.7	80.5	51.0	1.6	199.8
12	Nock-ten	2004/10/23	21:18	126.59E	19.67	628.1	35.1	22.8	1.5	155.4
13	Nesat	2005/06/06	09:03	133.67E	21.65	213.2	23.0	11.8	1.9	169.5
14	Harvey	2005/08/05	21:42	57.15W	32.45	1647.9	60.9	34.4	1.8	166.7
15	Mawar	2005/08/22	20:38	138.02E	25.12	1171.1	51.1	29.2	1.7	170.9
16	Maria	2005/09/05	21:37	56.75W	32.52	232.6	21.4	13.9	1.5	185.4
17	Jova	2005/09/18	03:25	139.26W	13.16	434.6	26.0	21.3	1.5	211.2
18	Jova	2005/09/19	15:22	142.56W	15.27	681.6	39.9	21.7	1.8	168.7
19	Kenneth	2005/09/22	03:08	134.11W	15.00	6806.3	104.3	83.1	1.3	228.4
20	Kenneth	2005/09/25	03:21	139.27W	16.37	150.7	15.3	12.5	1.2	95.4
21	Otis	2005/09/30	13:19	110.66W	20.89	1489.5	50.8	37.4	1.4	204.2
22	Otis	2005/10/02	01:38	111.86W	22.00	751.1	36.9	25.9	1.4	192.1

(continued)

Table 16.1 (continued)

Tropical cyclone		Image acquired		Eye center location		Eye region	Reference ellipse parameter			
Number	Name	Date (Y/M/D)	Time (UTC)	Long. (°)	Lat.N (°)	A (km^2)	a (km)	b (km)	R	θ (°)
23	Kirogi	2005/10/14	09:12	131.37E	22.90	996.4	40.5	31.3	1.3	142.2
24	Kirogi	2005/10/15	21:04	132.73E	24.16	2374.8	63.3	47.8	1.3	143.2
25	Ewiniar	2006/07/03	20:53	133.17E	15.00	18.3	6.5	3.6	1.8	202.2
26	Helene	2006/09/20	21:52	56.98W	26.28	800.7	43.1	23.7	1.8	177.1
27	Yagi	2006/09/21	20:17	144.23E	22.93	642.9	34.2	24.0	1.4	159.8
28	Xangsane	2006/09/30	22:38	108.99E	16.00	271.5	22.5	15.4	1.5	139.3
29	Man-yi	2007/07/11	21:13	129.54E	20.54	2852.4	74.5	48.7	1.5	148.1
30	Man-yi	2007/07/13	09:33	127.66E	28.15	1490.6	50.8	37.3	1.4	172.7
31	Usagi	2007/08/01	20:57	133.15E	29.77	1230.9	46.7	33.5	1.4	151.2
32	Dean	2007/08/17	09:50	60.99W	14.30	207.1	21.6	12.2	1.8	129.5
33	Dean	2007/08/19	23:17	77.65W	17.40	246.5	20.2	15.5	1.3	169.8
34	Fitow	2007/08/31	19:42	152.81E	27.87	3760.9	94.9	50.4	1.9	153.1
35	Krosa	2007/10/04	21:33	125.38E	20.35	1589.7	51.8	39.0	1.3	141.5
36	Kajiki	2007/10/20	20:25	142.79E	26.85	45.8	13.3	4.3	3.0	201.3

Note all variables as defined in the text. A: the area of the TC eye; a, b, and R: the major axis, minor axis, and aspect ratio of the reference ellipse, respectively; θ orientation of major axis, i.e., the angle between major axis and North (clockwise)

B. Best Track data

To assess the accuracy of the TC center estimation method, the SAR-derived TC center locations were compared with that from the Best Track (BT) datasets. The BT data used for validation come from Shanghai Typhoon Institute/China Meteorological Administration (STI/CMA) and National Hurricane Center/National Oceanic and Atmospheric Administration (NHC/NOAA), which contain the 6-hourly center locations for tropical storms and hurricanes over the Western Pacific Ocean, and the Atlantic Ocean and Eastern Pacific Ocean, respectively. A linear interpolation approach was used to obtain the best track center locations at SAR imaging time. This approach assumes that (a) the movement of the cyclone follows a straight line between two 6-hourly locations and (b) the speed of motion between any two points is constant for a given 6-h period.

16.2.2 TC Eye Detection Method

Two characteristics of the well-defined TC eye observed in SAR images are the darker appearance of the eye and the discontinuous of wind speed at the edge [1]. Based on these characteristics, an automatic method is proposed to estimate the center of the TCs. The procedure consists of four steps:

(1) Use a filter (here we use 3 × 3 Lee filter) to reduce the noise in the SAR subimage which covers the complete TC eye region.
(2) Detect the darker region of the TC eye roughly by using a threshold of the radiometric value, i.e., the normalized radar cross section (NRCS) (e.g., 90% of the mean radiometric value of the whole scene). The geometric center of all the pixels within the TC eye is described as the initial TC center location (x_0, y_0) (see Fig. 16.2), where x_0 and y_0 denote the longitude and latitude, respectively.
(3) In the polar coordinate centered at (x_0, y_0), calculate the radial radiometric gradient at each pixel in the SAR image. The region surrounded by the pixels with the maximum gradient along each radial direction is the new region of the TC eye (see Fig. 16.2), and the geometric center of all the pixels within the eye is the new center location (x_1, y_1).
(4) Iterate step 3 until a convergent condition is satisfied, e.g., the distance between (x_i, y_i) and (x_{i-1}, y_{i-1}) is smaller than a pre-defined threshold. Then the final TC eye and center location (x_c, y_c) are determined.

16.2.3 Results and Analysis

Using the above method, the TC eye region and center location were estimated from 36 RADARSAT-1 SAR images. Figure 16.3a–f are some examples of the results. Once the spatial extent of the TC is determined, its area (*A*) can be calculated as

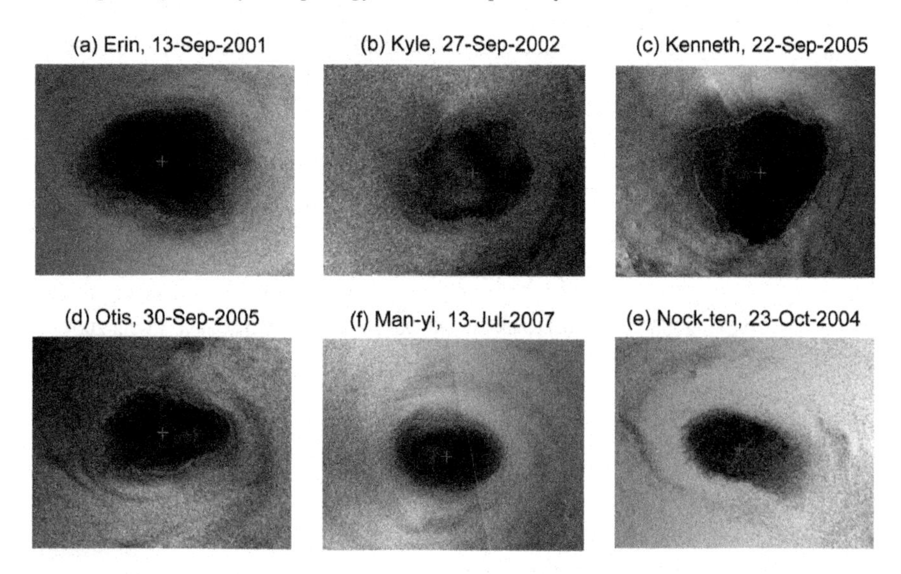

Fig. 16.2 Initial TC center location (x_0, y_0) (*red plus*) and pixels with maximum radiometric gradient (*blue pluses*) along each radial direction, overlaid with filtered sub-images of the tropical cyclones: Hurricanes Erin (Sep. 13, 2001) (**a**) and Kyle (Oct. 3, 2002) (**b**) over the Atlantic Ocean; Hurricanes Kenneth (Sep. 22, 2005) (**c**) and Otis (Sep. 30, 2005) (**d**) over the eastern Pacific Ocean; typhoons Nock-ten (Oct. 23, 2004) (**e**) and Man-yi (Jul. 13, 2007) (**f**) over the western Pacific Ocean

the number of pixels within the spatial extent multiplied by the pixel size of the SAR image. The area, together with the SAR detected center location of all TCs are listed in Table 16.1. Although irregular, the shape of most TCs is close to elliptical. By using the largest distance (L_d) between two points on the detected edge of the TC eye, the major axis (a) of the reference ellipse can be estimated as $a = L_d/2$. Set $A = \pi ab$, then the minor axis b may be calculated. Since the center point of the reference ellipse coincides with that of the TC, the aspect ratio of the reference ellipse ($R = a/b$) can be used as a rough descriptor of the shape of the TC. All these parameters are listed in Table 16.1.

To assess the performance of the automatic method for TC center estimation from SAR images, the distance between the SAR-detected TC center and the temporally interpolated center from the BT datasets was calculated. The distance between the TC center locations, D, is defined as [24]

$$D = r_0 \cdot cos^{-1}\begin{Bmatrix} sin(\varphi_1) \cdot sin(\varphi_2) \\ cos(\varphi_1) \cdot cos(\varphi_2) \cdot cos(\lambda_1 - \lambda_2) \end{Bmatrix} \tag{16.1}$$

where D is measured using the distance between two geographic points (λ_1, φ_1) and (λ_2, φ_2) on the Earth surface. The λ and φ are the longitude and latitude of the TC center, respectively; r_0 is the radius of the Earth (6.4×10^3 km).

The comparison result is shown in Fig. 16.4a. Figure 16.4b is the histogram of the distances. One can see the SAR-derived TC centers are very close to BT centers, with

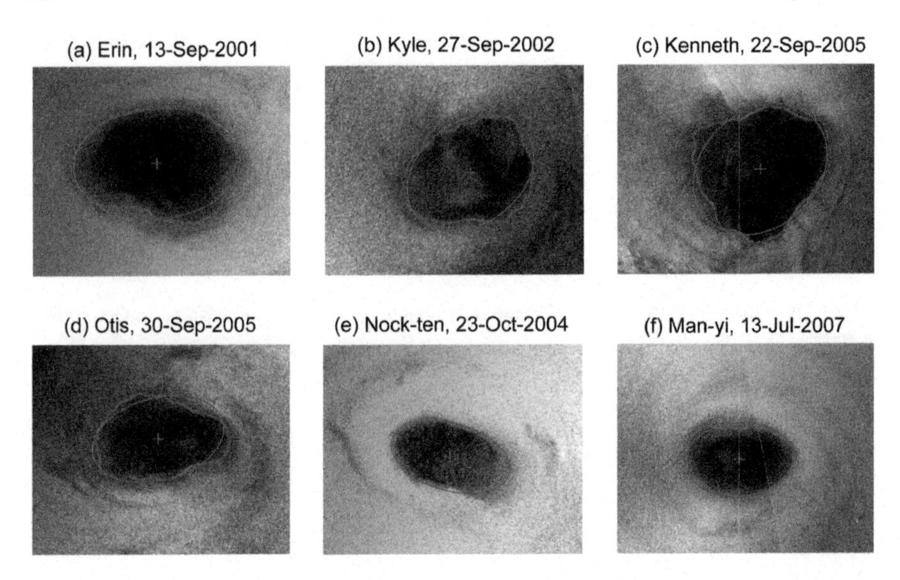

Fig. 16.3 The edge (*blue pluses*) and center location (*red plus*) of the TC eye estimated from SAR images for tropical cyclones in Fig. 16.2. The *red ellipse* is the matched reference ellipse with the same area as the TC eye

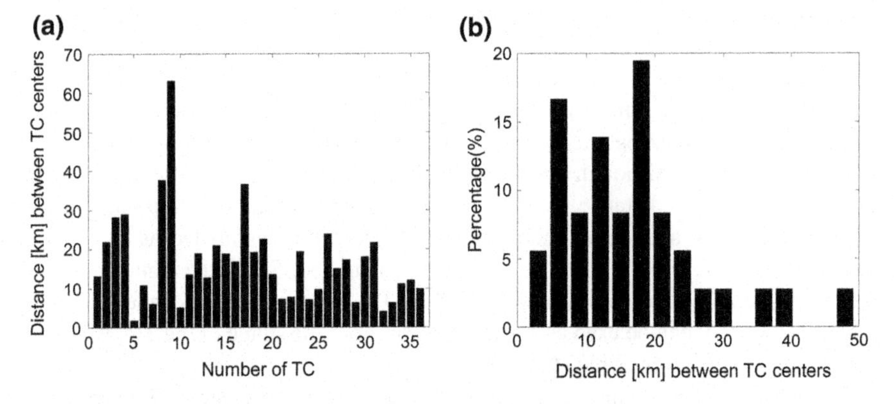

Fig. 16.4 **a** Distance (km) between the TC center determined from SAR image and BT center. **b** Histogram of the distance

over 90% of the distances less than 30 km. This implies that the proposed method performs well in determining the center of the well-defined TCs observed by SAR.

The TC centers estimated by the proposed method were also compared with that manually extracted from the SAR images by [8]. The result in Fig. 16.5 demonstrates that the automatic method can effectively reduce the subjective errors in the procedure of TC center determination from SAR images.

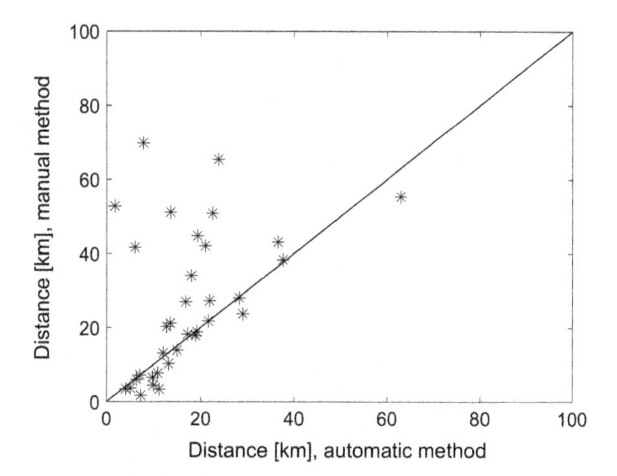

Fig. 16.5 Comparison of the distance (km) between the BT center and the SAR-derived center determined by the automatic method (X-axis) and manual method (Y-axis), respectively

16.2.4 Conclusions

Based on the characteristics of the well-defined TC eyes observed in SAR images, an automatic TC eye detection method was proposed. The center locations extracted from a series of RADARSAT-1 SAR images are very close to the temporally interpolated BT center locations. Most of the SAR and BT TC center distances are within 30 km. The method has also been proved to be most effective in reducing the subject errors generated with the manual method.

16.3 Extratropical-Cyclone-Forced Mountain Lee Waves

During the passage of an ETC over Kuril Islands (see Fig. 16.6) on June 1, 2015, several groups of atmospheric gravity waves (AGWs) were observed on a Sentinel-1A SAR image (see Fig. 16.7) on the lee side of the mountains located at the islands. The Kuril Islands consist of a chain of isolated islands, separating the Sea of Okhotsk from the North Pacific Ocean. The heights of the mountains on these islands vary from 800 to 1300 m (898, 812 and 1171 m for Ketoy, Russhua and Matua Islands, respectively) above the sea level. Studies show that the Kuril Islands are characterized by high cyclonic activity and extremely variable weather [25]. Although AGWs have been frequently observed by satellite images in this area, they have not been scientifically reported in the literature.

Satellite observations can provide the spatial scales of the AGWs and allow the estimation of wave parameters such as the wavelength and amplitude (e.g., [21, 22]). However, the vertical pattern and lifelong evolution of the AGWs cannot be revealed from the limited satellite snapshots of the phenomenon. With the rapid advance in atmospheric model development in recent years, the community atmospheric

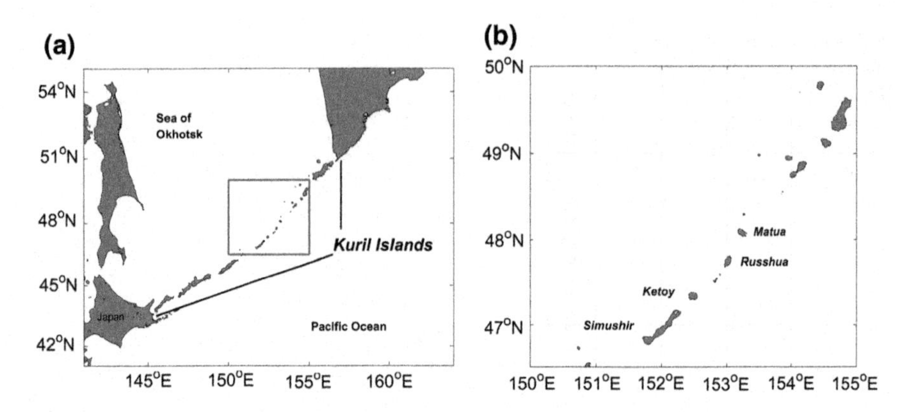

Fig. 16.6 **a** Map showing the Kuril Islands, which separate the Sea of Okhotsk from the North Pacific Ocean. The *red rectangular* denotes the study area. **b** Distribution of the islands in the study area

mesoscale numerical models have been applied to simulate SAR-observed AGWs and improved our understanding of these waves (e.g., [19, 20]). In this section, the generation mechanism and characteristics of the AGWs observed by SAR will be studied together with the use of the Weather Research and Forecasting (WRF) model.

16.3.1 Satellite Observations

Figure 16.7 is a VV-polarized Sentinel-1A SAR image acquired at 07:39 UTC on June 1, 2015. The Sentinel-1A was launched in April 2014 and the onboard C-band SAR provides medium (25 × 100 m) to high (5 × 5 m) spatial resolution images with four operational modes at single polarization (HH or VV) and dual polarization (HH+HV or VV+VH). The SAR image is centered at 152.8°E, 42.2°N, and covers parts of the Kuril Islands. The isolated islands are shown as brighter areas in the SAR image. A distinguished feature in the SAR image is several groups of alternating bright-dark patterns to the southeast of the isolated islands, extending to about 150 km offshore. Are these features sea surface signatures of AGWs or oceanic internal waves? Let us look at some quasi-simultaneously acquired satellite visible images. Figure 16.8a, b are two MODIS images at 500 m spatial resolution acquired about 5 (02:35 UTC) and 7 (00:50 UTC) hours before the SAR image acquisition time, respectively. A true-color image from the Enhanced Thematic Mapper Plus (ETM+) onboard Landsat-7 satellite is also overlaid on the MODIS image in Fig. 16.8b. The visible image has a spatial resolution of 30 m and was acquired at 00:35 UTC on June 1, 2015, i.e., about 15 min before the MODIS image (Fig. 16.8b) acquisition time. From Fig. 16.8, we can find similar wave-like cloud structures associated with the AGWs. Furthermore, the wave patterns are in the lee side of the islands (see Figs. 16.9 and 16.14a), which is

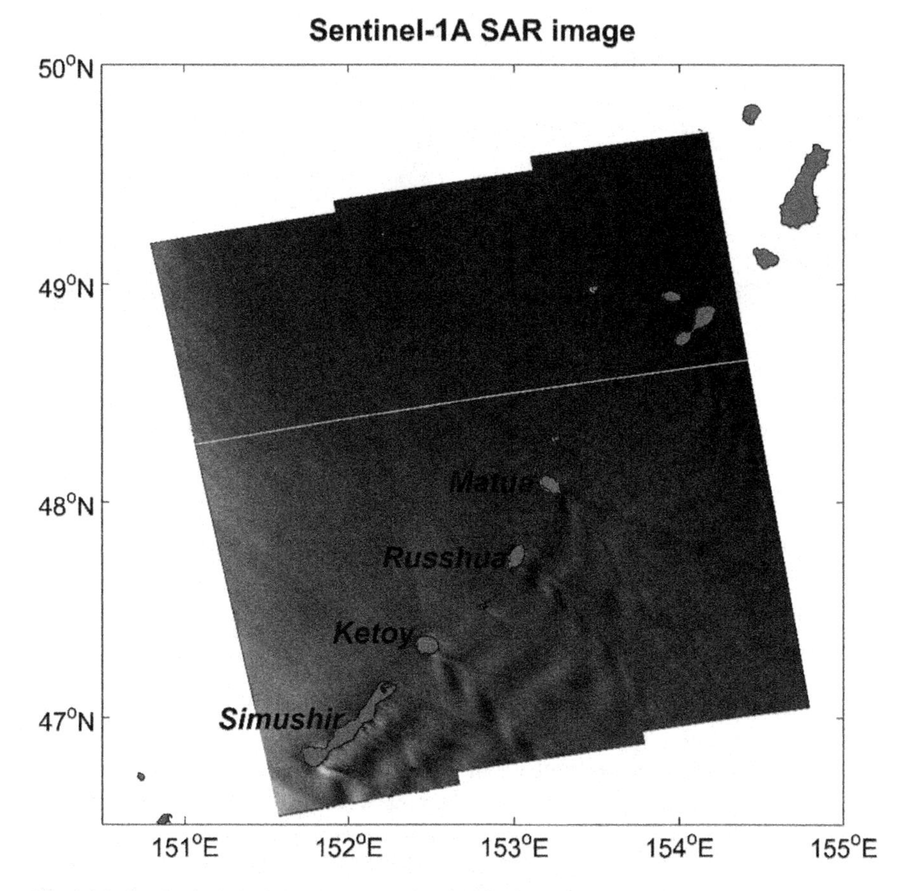

Fig. 16.7 Sentinel-1A SAR image acquired at 07:39:19 UTC on June 1, 2015

one of the criteria in favor of AGWs [26]. These evidences suggest that the wave-like features in the SAR imagery are sea surface imprints of AGWs.

The SAR image not only reveals the AGW patterns but also provides direct measurements of sea surface wind speeds and the variations associated with the AGWs (e.g., [27–30]). Figure 16.9 is the SAR-derived sea surface wind speed (at 10-m height) using the C-band geophysical model function (GMF) CMOD4 [31] with wind direction from the NOAA Global Forecast System (GFS) model ($\sim 0.5°$ spatial resolution) [32]. The wind directions agree well with that from the higher-resolution WRF model (see Fig. 16.10a). The horizontal sea surface wind associated with the transverse AGWs oscillates between 10 and 20 m/s in the trough and peak areas, indicating they are strong waves. The winds in the diverging areas offshore of Ketoy, Russhua and Matua Islands are slightly higher. An interesting phenomenon is that the wind oscillation associated with the right arm of the diverging wave is stronger than that with the left arm in this case. What causes the asymmetry in the diverging

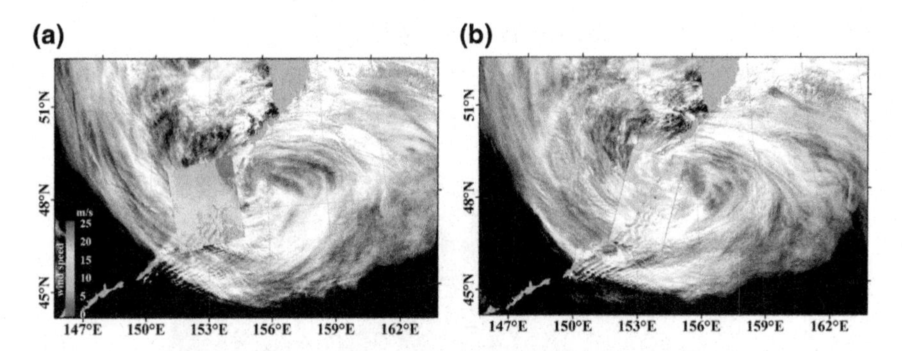

Fig. 16.8 **a** Overlay of the SAR wind image (see Fig. 16.9) with the MODIS true-color image (bands 1, 4 and 3) acquired about 5 h prior to the Sentinel-1A pass. **b** Overlay of the Landsat ETM+ true-color image (bands 4, 3 and 2) acquired at 00:35 UTC on June 1, 2015 with a MODIS image acquired 15 min later

wave? This will be discussed later in Sect. 16.3.4 based on the numerical simulation results.

On satellite images and SAR wind images, one can both see the coexistence of the transverse AGW parallel to the coastline of the elongated Simushir Island and the diverging waves on the lee side of the mountains with the wave crest oriented outward from the center of the wake. The wavelengths of the diverging waves in the SAR image are from 20 km (over Matua Island) to 25 km (over Ketoy Island) and that of the transverse wave is about 30 km. The diverging angle between the two arms of the diverging wave, i.e., the wedge angle, ranges from 30° (over Ketoy Island) to 45° (over Matua Island), indicating that the longer wave has a smaller wedge angle.

Another prominent atmospheric feature at the SAR imaging time is the existence of a cyclonic system shown on MODIS images, i.e., the extratropical cyclone (ETC), which was moving north-eastward offshore of the Kuril Islands. The high wind fields associated with the ETC are clearly shown in the SAR-derived sea surface wind image. If we compare the SAR image and ETM+ image acquired 7 h later, we find an apparent change of the orientation of the diverging waves, indicating a possible change of atmospheric circulation due to the evolution of the ETC. We will describe in detail the cyclonic process later in Sect. 16.3.3 and discuss the reason causing the change of the wave parameters.

16.3.2 WRF Model Simulation of the AGWs

To understand the dynamic mechanisms that induce the AGWs on the lee side of the Kuril Islands, the WRF model [33] is implemented to simulate the low-level atmospheric circulation in the study area. We use a two-way interactive, triply nested grid technique. The configuration consists of a 9-km outer coarse domain, a 3-km

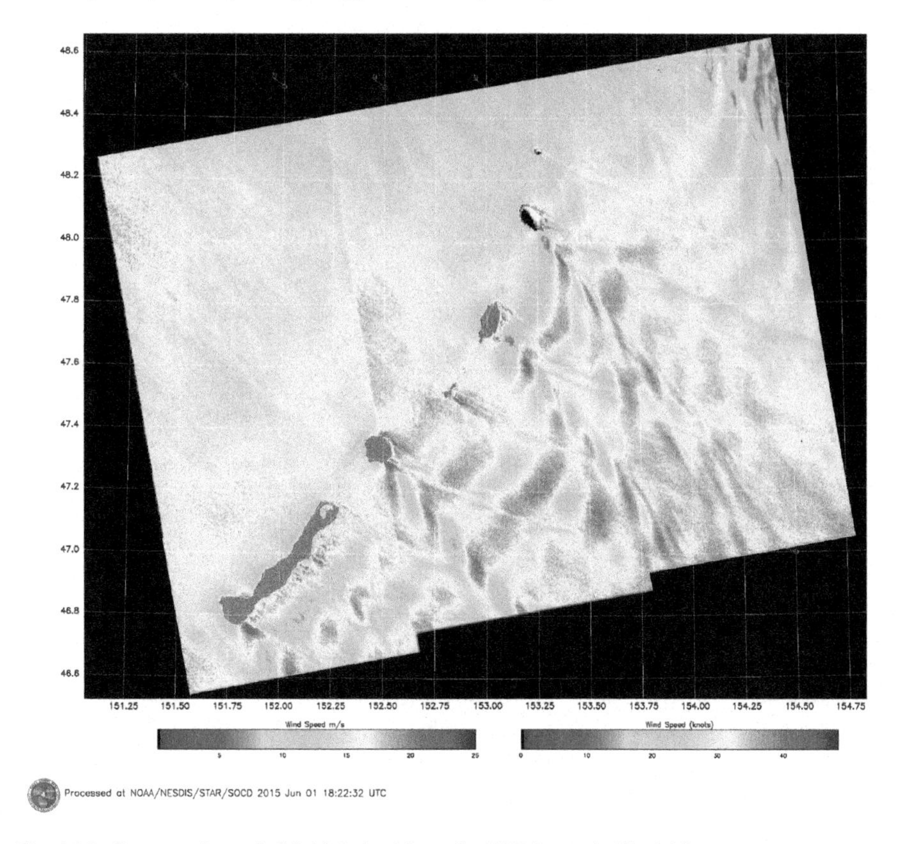

Fig. 16.9 Ocean surface wind field derived from the SAR image in Fig. 16.7

medium domain and an inner high-resolution 1-km domain. The physical parameterizations of the model are the same as those contained in previous studies (see [19, 20, 34]). Refer to [12] for more details of the model configuration. The model begins at 00:00 UTC on 31 May 2015 and then integrates continuously for 3 days.

The WRF simulated sea surface wind field in the SAR coverage area about 20 min after the SAR imaging time is shown in Fig. 16.10a. The transverse and diverging wave patterns are clearly shown on the wind map and resemble the SAR observations. The histograms of the WRF-simulated and SAR wind speeds (see Fig. 16.10b) are similar in shape, but there is a mean bias of about 2.5 m/s. The systematic discrepancy is understandable because both wind retrieval schemes are complicated and the actual wind field changed a little due to the movement of the ETC in 20 min.

Figure 16.11 shows the WRF simulation of the horizontal wind speed, vertical wind speed, potential temperature, and planetary boundary layer (PBL) height in the lowest-level domain (1 km) at 08:00 UTC on June 1. The coexistence of transverse and diverging waves is also shown. Compared with the other three variables, the vertical wind displays a much clearer wavelike motion. The AGW induced perturbations

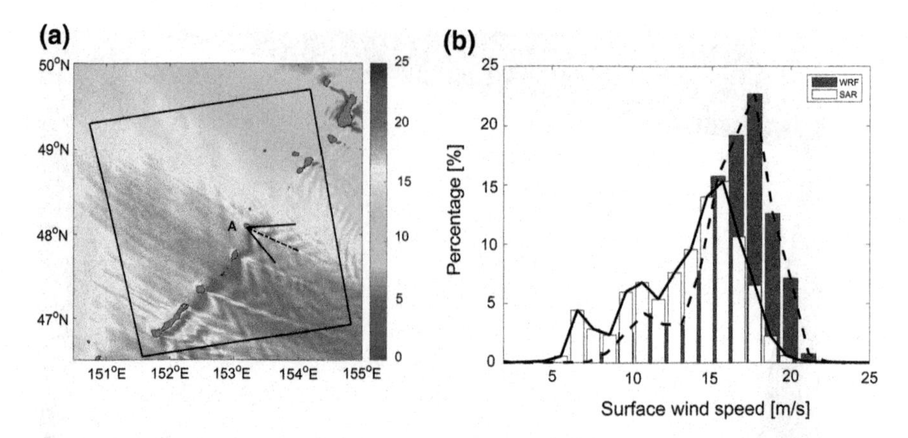

Fig. 16.10 **a** WRF simulation of the sea surface wind speed (m/s) in the SAR imaging area (the *black rectangular*) at 08:00 UTC on June 1, 2015. The wind speeds along the two arms of the diverging waves in the lee of the Matua Island (cross section A) and the Ketoy Island (cross section B) are extracted and plotted in Fig. 16.15. **b** Histograms of the WRF-simulated and SAR-derived sea surface wind speeds

are strongest near the mountains, and gradually decay away from the mountains. The vertical wind perturbations along the right arm of the diverging waves are larger, similar to that observed by SAR. The wedge angle of the diverging wave over the Ketoy Island is apparently smaller than that over the Matua Island. The depth of the PBL, which is related to the local buoyant flux and static stability as well as convective movement, closely matches the vertical wind field. The AGW patterns on the potential temperature map are also very distinct, implying strong temperature variation associated with the AGW motion. In general, the WRF simulation captures the variation of the low-level atmospheric circulation associated with the gravity waves observed by SAR well.

16.3.3 Generation Mechanism of the AGWs

Both MODIS images (Fig. 16.8) and daily surface pressure map from the NCEP (Fig. 16.12) show that an ETC is present over the south of the Kuril Islands on May 31, the same day that the AGWs were generated (see Fig. 16.13). The ETC impinged on the entire Kuril Islands the next day when the AGWs were observed by SAR and visible satellite sensors. It moved to the east on June 2 and the AGWs dissipated accordingly.

The WRF model also captured the cyclonic process. Figure 16.13 shows the generation, development and dissipation of the AGWs with the evolution of the ETC. At 06:00 UTC on May 31, the diverging waves are first visible on the lee side of the mountains at Ketoy, Russhua and Matua Islands, with the horizontal wind speed

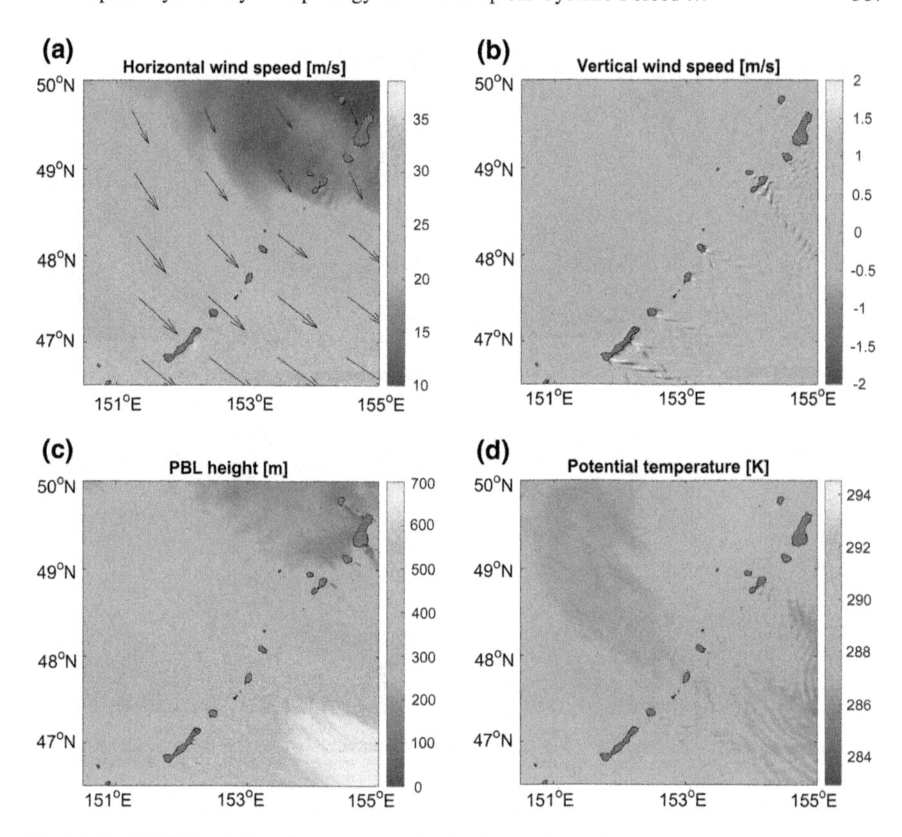

Fig. 16.11 WRF simulation of the atmospheric circulation at 1 km in the study area at 08:00 UTC on June 1, 2015. **a** Horizontal wind field. **b** Vertical wind speed. **c** PBL height. **d** Potential temperature

higher than 15 m/s. The disturbance of the left arm is much stronger than that of the right arm. Both the transverse and diverging waves are distinguishable in the lee of the islands with an opposite orientation one day later (i.e., 06:00 UTC on June 1) when the synoptic wind changes its direction from the southeast to the northwest and becomes much stronger due to the movement and deepening of the cyclone. At this time, the fluctuation of the right arm of the diverging wave is larger. The AGWs are still clearly seen when the northwesterly wind decreases at 18:00 UTC on June 1. They then become significantly weakened further 6 h later and begin to dissipate at 06:00 UTC on 2 June while the surface wind becomes much smaller (less than 5 m/s). The life span of the AGWs over the Kuril Islands is about two days, consistent with that of the ETC.

Both satellite observations and the WRF simulation reveal that the AGW patterns react to the synoptic movement of the ETC and the location is always on the lee side of the mean wind field. The reorientation of the waves follows the change of the wind direction. The wavelength, amplitude and the wedge angle of the diverging wave change with the atmospheric condition as well.

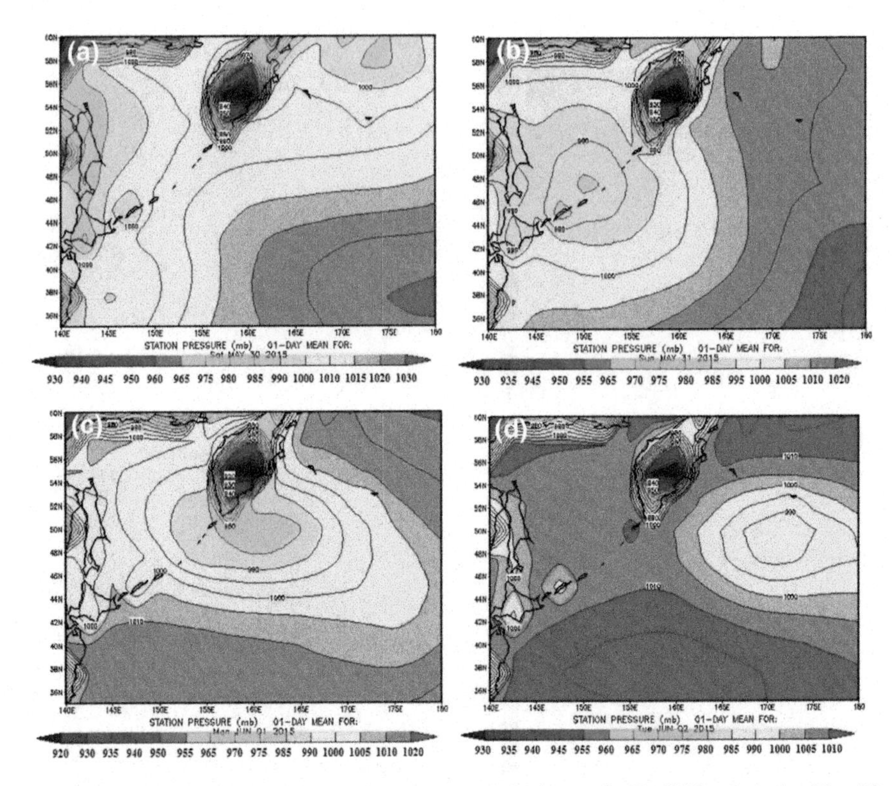

Fig. 16.12 NCEP analysis of daily surface pressure (mbar) over the Kuril Islands during May 30 to June 2, 2015 (https://www.esrl.noaa.gov/psd/data/histdata/)

For the AGWs coexisting with cyclones, there are two possible generation mechanisms: cyclone forcing and orography forcing. Cyclone-associated airflow over the islands or the mountains could induce lee waves. The cyclone itself might also lead to the generation of the gravity waves due to the associated cumulus heating. These waves are called cyclone or convectively generated gravity waves (CGWs). The CGWs in the lower stratosphere or upper troposphere are believed to be generated through different mechanisms including deep heating, obstacle effects due to turrets in the presence of strong wind shear (like lee waves but the obstacle is nonstationary), and mechanical oscillatory effect [35–37]. Hence a question here is, for the gravity waves observed by the satellites in this study, are they lee waves or CGWs? To answer this question, we carried out another numerical experiment in which the Russhua Island was removed. Simulation results show that there is no gravity wave over the Russhua Island in the absence of the orography (see Fig. 16.14). Therefore, the AGWs are apparently triggered by the interaction of the ETC-associated airflow with the island. They are lee waves that are aligned perpendicular to the wind direction and locked on the lee side of the island. The ETC provides favorable

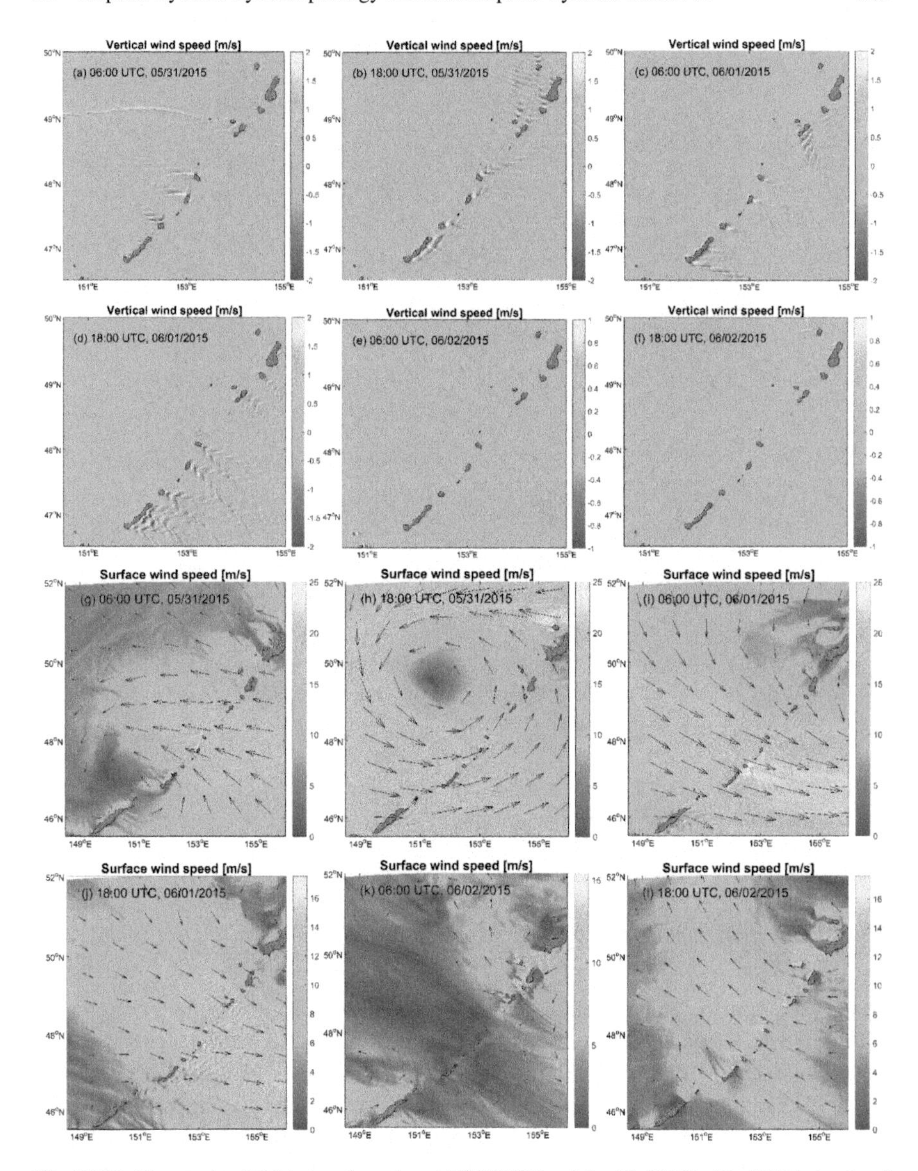

Fig. 16.13 Time series (12-h interval starting at 06:00 UTC on May 31, 2015) of the WRF-simulated vertical wind speed (a–f) in the lower troposphere (1 km above sea level) and sea surface wind field (g–l)

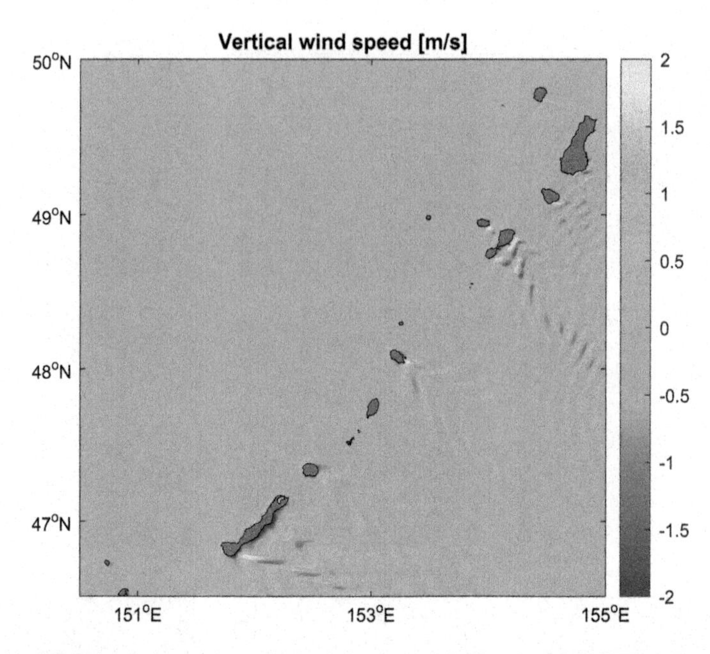

Fig. 16.14 WRF simulation of the vertical wind speed at 1 km level at 08:00 UTC on June 1, 2015. The Russhua Island centered at (153.0°E, 47.75°N) was removed in this numerical experiment

atmospheric conditions for the generation of the lee waves. This is different from CGWs that move in step with the cyclones.

16.3.4 Characteristics of the AGWs

As mentioned in the above section, a prominent feature of the diverging wave visible in both the SAR image and the WRF simulation results is the asymmetry of the two arms. This is shown more clearly in Fig. 16.15. One can see that the fluctuation of the AGW modulated wind speed along the right arm is much stronger. In addition, the wavelength of the diverging wave offshore of the Matua Island with a higher mountain height is smaller than that offshore of the Ketoy Island. Similar features are seen from the vertical structure of the wave patterns in Figs. 16.16 and 16.17. The right arm shows a larger perturbation at all levels at the SAR imaging time.

The distributions of the vertical velocity field also show that the wave motion near the Ketoy Island can reach above 8–9 km over the sea surface, whereas that near the Matua Island propagates up to a height of about 6–7 km. This implies that the wave with shorter horizontal wavelengths is trapped at a slightly lower level, which is consistent with the findings of [38]. The potential temperature contours in Fig. 16.16 show a similar but weaker pattern.

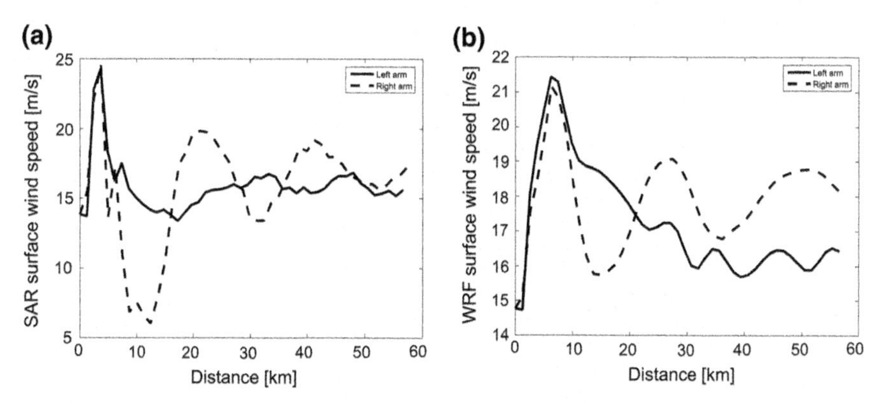

Fig. 16.15 The variation of SAR-derived (**a**) and the WRF-simulated (**b**) surface wind speeds (m/s) at 08: 00 UTC on June 1, 2015 along the two arms of diverging wave A in Fig. 16.10a

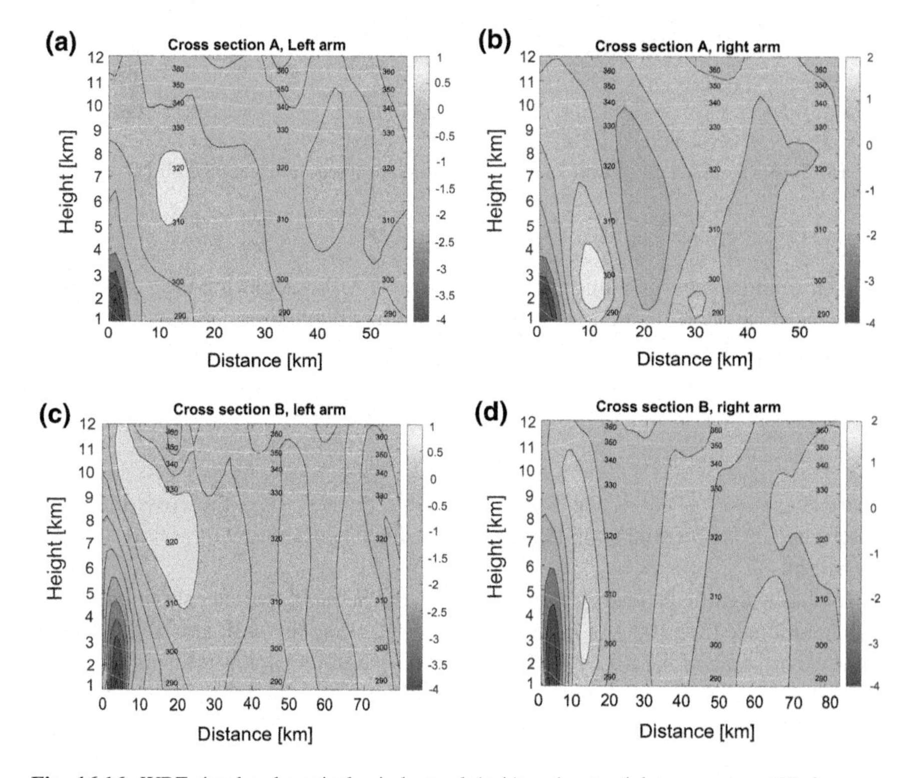

Fig. 16.16 WRF-simulated vertical wind speed (m/s) and potential temperature (K) for cross sections A and B at 08:00 UTC on June 1, 2015

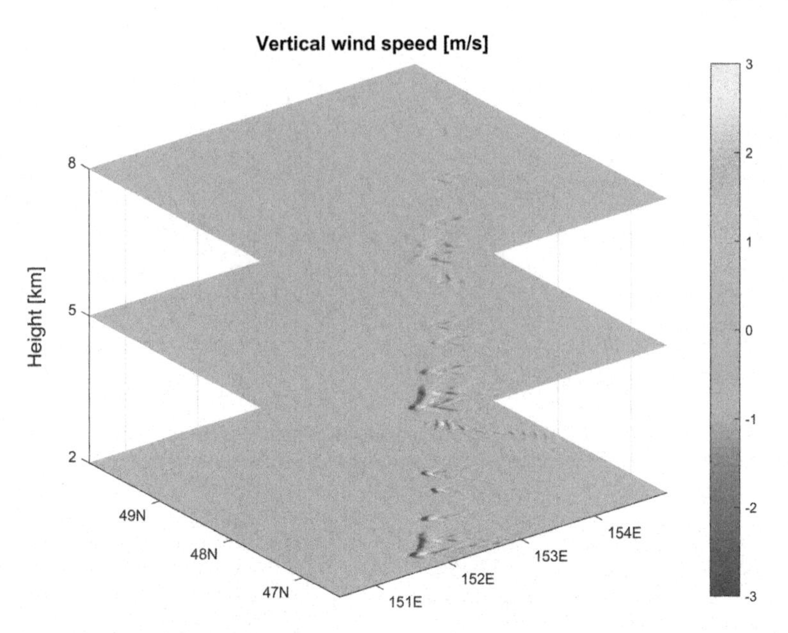

Fig. 16.17 WRF-simulated vertical wind speed (m/s) at different levels at 08:00 UTC on June 1, 2015

To investigate the dynamic mechanism that is responsible for the radiation of wave energy away from the low levels, we calculate the Scorer parameter l^2, which is related with the wind speed, its vertical shear and the atmospheric stability [39],

$$l^2(z) = \frac{N^2(z)}{U^2} - \frac{1}{U}\frac{\partial^2 U}{\partial z^2} \tag{16.2}$$

where $N(z) = (\frac{g}{\theta}\frac{d\theta}{dz})^{1/2}$ is the Brunt-Vaisala Frequency, U is the cross-mountain wind speed, θ is potential temperature, g is the local acceleration of gravity, and z is the height.

To obtain significant energy at the low level, the Scorer parameter should decrease with height (see [39, 40]). As shown in Fig. 16.18, along the right arm of the diverging wave offshore of the Ketoy Island, the Scorer parameter at lower levels decreases gradually with height, corresponding to a condition favorable for the upward propagation of the gravity waves. The AGW is trapped within a waveguide and does not propagate upward when the Scorer parameter becomes sufficiently small (close to zero values) [41] at a height of about 8 km. There is no motion in the region where the Scorer parameter increases rapidly with height at higher levels. For the lee waves near Matua Island, the vertical profile of the Scorer parameter demonstrates that the waves are trapped at a relatively lower level, which confirms our finding from the WRF simulation results.

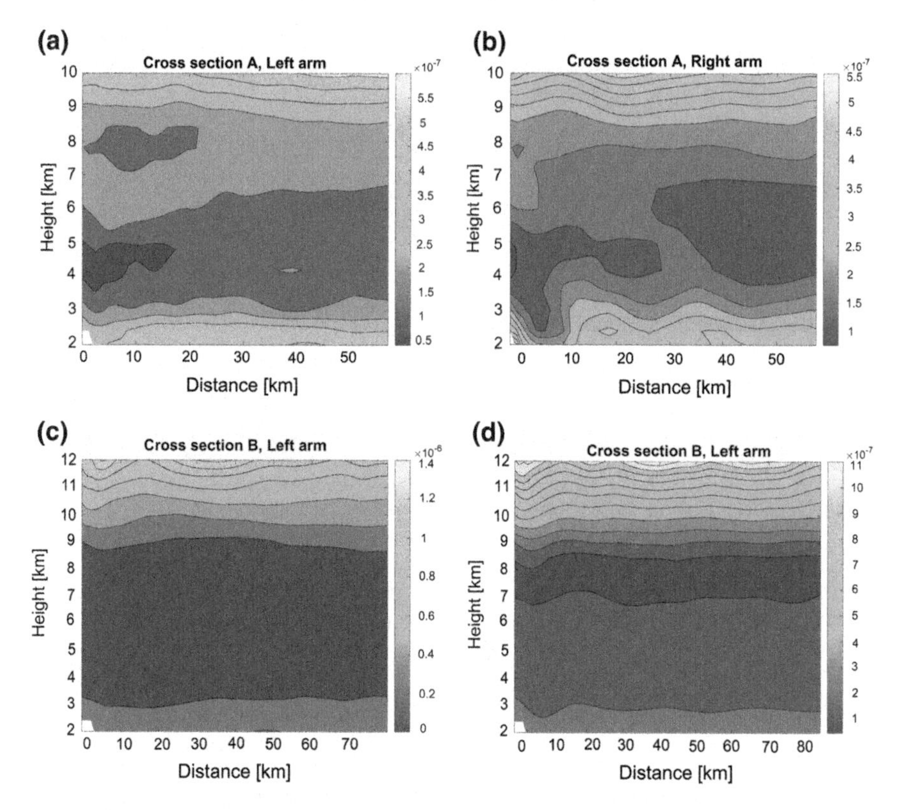

Fig. 16.18 Scorer parameter calculated from the WRF model simulation for cross sections A and B at 08:00 UTC on June 1, 2015

As for the asymmetry in the amplitude of SAR-observed mountain diverging waves, it might be associated with several factors, e.g., the shape of the mountain, the SAR imaging geometry, the earth rotation, or the atmospheric conditions of environmental wind and stability. In this case, the height of mountain is in general symmetric about the axis perpendicular to the wind direction (not shown) at the SAR imaging time. Except for the SAR observation, the numerical simulation also shows the asymmetry of the diverging waves. Furthermore, the feature of the asymmetry changes at different times, i.e., the wave motion of the right arm is not always stronger than that of the left arm during the entire life span of the gravity waves. All these facts suggest that the diverging wave characteristics are closely related with the atmospheric conditions. As revealed in Fig. 16.18, the asymmetry in the horizontal scale of the wave-like distribution of the Scorer parameter may lead to different modulations of the two arms of the diverging wave.

When strong airflows pass over a mountain, the pattern and amplitude of the AGWs depend upon another important parameter, the Froude number F_r, which is a measure of the stratification [18]. F_r is a dimensionless number defined as [42],

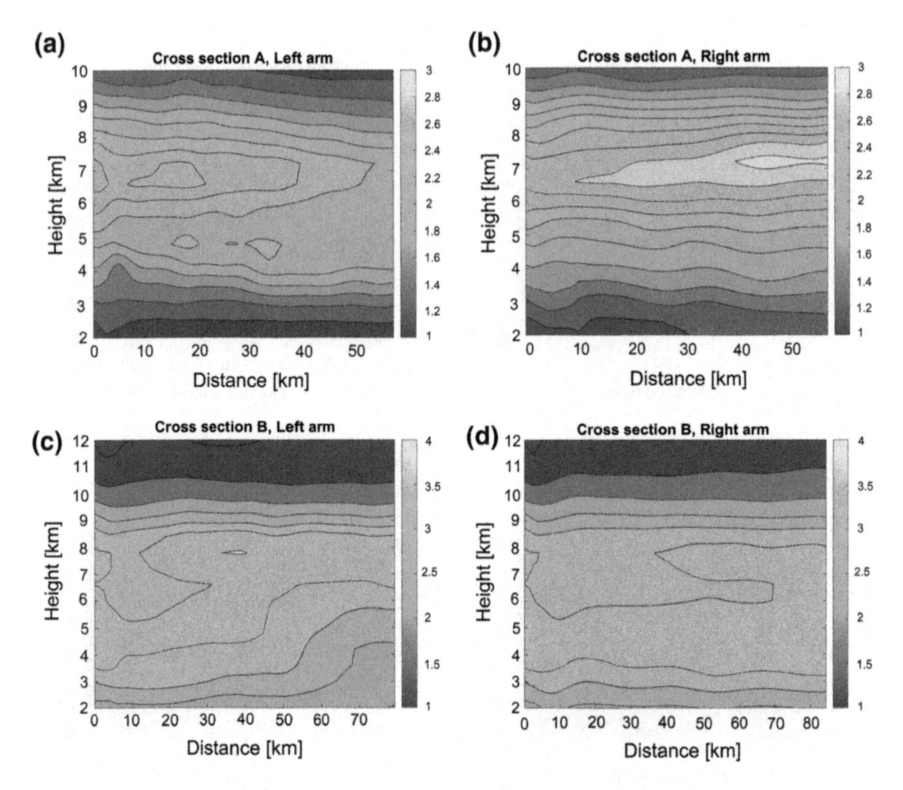

Fig. 16.19 Froude number calculated from the WRF model simulation for cross sections A and B at 08:00 UTC on June 1, 2015

$$F_r = U/\sqrt{Nh} \tag{16.3}$$

where h is the characteristic height of the mountain.

Figure 16.19 shows the Froude number along the cross sections A and B based on the WRF simulation results. In general, smaller amplitude along the left arm of the diverging wave is associated with a relatively lower Froude number. One can also see that the higher Froude number partly due to the lower value of mountain height leads to a longer wavelength or the narrower wedge angle of the diverging wave offshore of the Ketoy Island. This is consistent with the findings from the theoretical analysis [13] and the rotating tank experiments [43]. Similar to the AGWs, a recent study [44] based on the observation of airborne images has also shown that the wave angle formed by ship wakes seems to decrease as the Froude number increases, and the amplitude of the diverging waves increases instead.

16.3.5 Conclusions

A high-resolution SAR image captures several groups of AGWs on the lee side of the mountains located at the Kuril Islands during the passage of an ETC. By employing the community WRF numerical model, the generation mechanism and characteristics of the AGWs were investigated. The WRF simulation results reveal that the AGWs are mountain lee waves forced by the air flow within an ETC that transected past the mountains. The time span of the lee waves is about two days, consistent with that of the ETC over the region. Due to the movement of the ETC, the orientation of the lee waves follows the change of the wind direction instantaneously, and rotates 360° as the ETC passed by the region.

A distinguishing feature of the diverging wave is its asymmetry in the wave motions of the two arms, which depends on the atmospheric conditions. In this case, the right arm has a larger perturbation at the SAR imaging time. However, this is not always the case during the entire life span of the waves. Numerical simulation shows that the AGWs are not trapped within the low-level waveguide usually capped by the inversion layer at 1 km height, but rather propagate upward to 8–9 km above the sea level. The waves with shorter wavelength are trapped at a slightly lower level. As shown in the vertical profile of the Scorer parameter and Froude numbers, these phenomena are mainly caused by the variation in the wind speed and atmospheric stratification.

16.4 Summary

A variety of satellite SAR images of TCs have been acquired over the past few years, which show the imprint of these storms on the ocean surface roughness with high resolution at sub-kilometer scales. An objective method based on the radiometric characteristics of the well-defined TC eyes observed in SAR images has been developed for estimating the parameters that quantitatively and objectively describe TC eyes. The center location and the eye wall derived from SAR allow study of the cyclones properties both inside and outside the TC eye. Furthermore, characterization of TC eyes using the objective method can promote the use of SAR images for the study of TC morphology and dynamics.

This chapter also demonstrates the advantage of the synergy between high-resolution SAR observations and the numerical model, in AGW studies. Employing both tools, the 3-D structure and temporal evolution of the AGWs triggered by the interaction of the air flow within an ETC with the orography have been investigated in detail. This could lead to an improved understanding of the generation mechanism and characteristics of the AGWs.

References

1. Du, Y., and P.W. Vachon. 2003. Characterization of hurricane eyes in RADARSAT-1 images with wavelets analysis. *Canadian Journal of Remote Sensing* 29 (4): 491–498.
2. Friedman, K., and X. Li. 2000. Storm patterns over the ocean with wide swath SAR. *Johns Hopkins University Applied Physics Lab Technical Digest* 21 (1): 80–85.
3. Li, X. 2015. The first Sentinel-1 SAR image of a typhoon. *Acta Oceanologica Sinica* 34 (1): 1–2.
4. Li, X., W. Pichel, M. He, S. Wu, K. Friedman, P. Clemente-Colon, and C. Zhao. 2002. Observation of hurricane-generated ocean swell refraction at the Gulf Stream North Wall with the RADARSAT-1 synthetic aperture radar. *IEEE Transactions on Geoscience and Remote Sensing* 40 (10): 2131–2142.
5. Wimmers, A.J., and C.S. Velden. 2010. Objectively determining the rotational center of tropical cyclones in passive microwave satellite imagery. *Journal of Applied Meteorology and Climatology* 49 (9): 2013–2034.
6. Cheng, Y.H., S.J. Huang, A.K. Liu, C.R. Ho, and N.J. Kuo. 2012. Observation of typhoon eyes on the sea surface using multi-sensors. *Remote Sensing of Environment* 123: 434–442.
7. Jin, S., S. Wang, and X. Li. 2014. Typhoon eye extraction with an automatic SAR image segmentation method. *International Journal of Remote Sensing* 35: 3978–3993.
8. Li, X., J. Zhang, X. Yang, W. Pichel, M. deMaria, D. Long, and Z. Li. 2013. Tropical cyclone morphology from spaceborne synthetic aperture radar. *Bulletin of the American Meteorological Society* 94 (2): 215–230.
9. Xu, Q., G. Zhang, Y. Cheng, and L. Ju. 2006. Satellite SAR detection of Hurricane Helene. In *The Twenty-third International Offshore and Polar Engineering Conference, 2013*, 865–868. International Society of Offshore and Polar Engineers.
10. Zheng, G., J. Yang, A.K. Liu, X. Li, W.G. Pichel, and S. He. 2016. Ku band backscatter from the Cowlitz river: Bragg scattering with and without rain. *IEEE Transactions on Geoscience and Remote Sensing* 54 (2): 1000–1012.
11. Lee, I., A. Shamsoddini, X. Li, and J.C. Trinder. 2016. Extracting hurricane eye morphology from spaceborne SAR images using morphological analysis. *ISPRS Journal of Photogrammetry and Remote Sensing* 7: 115–125.
12. Xu, Q., X. Li, S. Bao, and L.J. Pietrafes. 2016. SAR observation and numerical simulation of mountain lee waves forced by an extratropical cyclone. *IEEE Transactions on Geoscience and Remote Sensing* 54 (12): 7157–7165.
13. Chunchuzov, I., P.W. Vachon, and X. Li. 2000. Analysis and modeling of atmospheric gravity waves observed in RADARSAT SAR images. *Remote Sensing of Environment* 74 (3): 343–361.
14. de Villiers, M.P., and J. Heerden. 2001. Clear air turbulence over South Africa. *Meteorological Applications* 8 (1): 119–126.
15. Ralph, F.M., P.J. Neiman, and D. Levinson. 1997. Lidar observations of a breaking mountain wave associated with extreme turbulence. *Geophysical Research Letters* 24 (6): 663–666.
16. Valenzuela, R.G. 1978. Theories for the interaction of electromagnetic and ocean waves - A review. *Boundary-Layer Meteorology* 13 (1): 61–65.
17. Gan, X., et al. 2008. Coastally trapped atmospheric gravity waves on SAR, AVHRR and MODIS images. *International Journal of Remote Sensing* 29 (6): 1621–1634.
18. Li, X., C. Dong, P. Clemente-Colon, W.G. Pichel, and K.S. Friedman. 2004. Synthetic aperture radar observation of the sea surface imprints of upstream atmospheric solitons generated by flow impeded by an island. *Journal of Geophysical Research: Oceans* 109 (2).
19. Li, X., W. Zheng, X. Yang, Z. Li, and W.G. Pichel. 2011. Sea surface imprints of coastal mountain lee waves imaged by synthetic aperture radar. *Journal of Geophysical Research: Oceans* 116 (2).
20. Li, X., et al. 2013. Coexistence of atmospheric gravity waves and boundary layer rolls observed by SAR. *Journal of the Atmospheric Sciences* 70 (11): 3448–3459.

21. Vachon, P.W., O.M. Johannessen, and J.A. Johannessen. 1994. An ERS 1 synthetic aperture radar image of atmospheric Lee waves. *Journal of Geophysical Research: Oceans* 99 (1115): 22483–22490.
22. Zheng, Q., et al. 1998. Coastal lee waves on ERS-1 SAR images. *Journal of Geophysical Research: Oceans* 103 (415): 7979–7993.
23. Iris, S., and G. Burger. 2004. RADARSAT-1: Canadian space agency hurricane watch program. *2004 IEEE International Geoscience and Remote Sensing Symposium, 2004*, vol. 4, 2742–2745, IGARSS'04 Proceedings.
24. Song, J.J., Y. Wang, and L. Wu. 2010. Trend discrepancies among three best track data sets of western North Pacific tropical cyclones. *Journal of Geophysical Research: Atmospheres* 115 (D12): 12128.
25. Rozhnoi, A., M. Solovieva, B. Levin, M. Hayakawa, and V. Fedun. 2014. Meteorological effects in the lower ionosphere as based on VLF/LF signal observations. *Natural Hazards and Earth System Sciences* 14: 2671–2679.
26. Alpers, W., and W. Huang. 2011. On the discrimination of radar signatures of atmospheric gravity waves and oceanic internal waves on synthetic aperture radar images of the sea surface. *IEEE Transactions on Geoscience and Remote Sensing* 49 (3): 1114–1126.
27. Lin, H., Q. Xu, and Q. Zheng. 2008. An overview on SAR measurements of sea surface wind. *Progress in Natural Science* 18 (8): 913–919.
28. Xu, Q., et al. 2010. Assessment of an analytical model for sea surface wind speed retrieval from spaceborne SAR. *International Journal of Remote Sensing* 31 (4): 993–1008.
29. Yang, X., et al. 2010. Comparison of ocean surface winds retrieved from QuikSCAT scatterometer and Radarsat-1 SAR in offshore waters of the U.S. West Coast. *IEEE Geoscience and Remote Sensing Letters* 8 (1): 163–167.
30. Yang, X., X. Li, W.G. Pichel, and Z. Li. 2011. Comparison of ocean surface winds from ENVISAT ASAR, MetOp ASCAT scatterometer, buoy measurements, and NOGAPS model. *IEEE Transactions on Geoscience and Remote Sensing* 49 (12): 4743–4750.
31. Stoffelen, A., and D. Anderson. 1997. Scatterometer data interpretation: estimation and validation of the transfer function CMOD4. *Journal of Geophysical Research: Oceans* 102 (315): 5767–5780.
32. Monaldo, F.M., C.R. Jackson, X. Li, and W.G. Pichel. 2015. Preliminary evaluation of Sentinel-1A wind speed retrievals. *IEEE Journal of Selected Topics in Applied Earth Observations and Remote Sensing* 9 (6): 2638–2642.
33. Skamarock, W.C., et al. 2008. A description of the advanced research WRF version 3. *NCAR Technical Note NCAR/TN-4751STR, Boulder, Colorado, USA*, 1–113.
34. Li, X., W. Zheng, W.G. Pichel, C.Z. Zou, and P. Clemente-Colon. 2007. Coastal katabatic winds imaged by SAR. *Geophysical Research Letters* 34 (3).
35. Kim, S.H., H.Y. Chun, and W. Jang. 2014. Horizontal divergence of typhoon-generated gravity waves in the upper troposphere and lower stratosphere (UTLS) and its influence on typhoon evolution. *Atmospheric Chemistry and Physics* 14: 3175–3182.
36. Kuester, M.A., M.J. Alexander, and E.A. Ray. 2008. A model study of gravity waves over Hurricane Humberto (2001). *Journal of the Atmospheric Sciences* 65: 3231–3246.
37. Fritts, D.C., and M.J. Alexander. 2003. Gravity wave dynamics and effects in the middle atmosphere. *Reviews of Geophysics* 41 (1).
38. Doyle, J.D., and Q. Jiang. 2006. Observations and numerical simulations of mountain waves in the presence of directional wind shear. *Quarterly Journal of the Royal Meteorological Society* 132 (619): 1877–1905.
39. Scorer, R.S. 1954. Theory of airflow over mountains: Part III: Airstream characteristics. *Quarterly Journal of the Royal Meteorological Society* 480 (345): 417–428.
40. Crook, N.A. 1988. Trapping of low-level internal gravity waves. *Journal of the Atmospheric Sciences* 45 (10): 1533–1541.
41. da Silva, J.C.B., and J.M. Magalhes. 2009. Satellite observations of large atmospheric gravity waves in the Mozambique Channel. *International Journal of Remote Sensing* 30 (5): 1161–1182.

42. Smolarkiewicz, P.K., and R. Rotunno. 1989. Low Froude number flow past three-dimensional obstacles. Part I: Baroclinically generated lee vortices. *Journal of the Atmospheric Sciences* 46 (8): 1154–1164.
43. Gao, S., and H. Chen. 2000. The studies of lee waved over a big topography through the rotating tank experiments (in Chinese). *Advances in Atmospheric Sciences* 58 (6): 653–665.
44. Rabaud, M., and F. Moisy. 2014. Narrow ship wakes and wave drag for planning hulls. *Ocean Engineering* 90: 34–38.